SAN JOAQUIN DELTA COLLEGE LIBRARY
Y0-AGU-871

The Gasohol Handbook

V. Daniel Hunt
Director, The Energy Institute

Industrial Press Inc.
200 Madison Avenue
New York, N.Y. 10157

Notice

This book was prepared as an account of work sponsored by the Industrial Press. Neither the Industrial Press nor The Energy Institute, nor any of their employees, nor any of their contractors, subcontractors, consultants, or their employees, make any warranty, expressed or implied, or assumes any legal liability or responsibility for the accuracy, completeness or usefulness of any information, apparatus, product or process disclosed, or represents that its use would not infringe on privately owned rights.

The views, opinions, and conclusions in this book are those of the author and do not necessarily represent those of the United States Government or the United States Department of Energy or Department of Agriculture.

Public domain information and those documents abstracted or used in full, edited or otherwise used, are noted in the acknowledgments or on specific pages or illustrations.

Library of Congress Cataloging in Publication Data

Hunt, V. Daniel.
The gasohol handbook.

Bibliography: p.
Includes index.
1. Gasohol—Handbooks, manuals, etc. I. Title.
TP358.H86 662'.669 81-6543
ISBN 0-8311-1137-2 AACR2

FIRST PRINTING
THE GASOHOL HANDBOOK

Table of Contents

	PREFACE	*vii*
	ACKNOWLEDGMENTS	*ix*
1	INTRODUCTION TO GASOHOL	1
2	DECISION TO PRODUCE GASOHOL	50
3	BASIC PRODUCTION PROCESSES	83
4	FUNDAMENTALS OF FEEDSTOCKS	139
5	MARKETS FOR ETHANOL AND COPRODUCTS	183
6	ECONOMICS OF ETHANOL PRODUCTION	222
7	OVERVIEW OF PROCESS TECHNOLOGY	245
8	ETHANOL-PLANT DESIGNS	273
9	ENVIRONMENTAL AND SAFETY IMPACTS	380
10	UTILIZATION OF GASOHOL	406
11	BUSINESS PLANNING	456
12	PREPARATION OF FEASIBILITY ANALYSIS	478
13	INTERNATIONAL DEVELOPMENT	497
A	RESOURCE PEOPLE AND ORGANIZATIONS	513
B	REFERENCE SOURCES	526
C	BIBLIOGRAPHY	533
D	GLOSSARY AND ACRONYMS	539
E	TECHNICAL REFERENCE DATA	556
	INDEX	570

Preface

During the past several years, I have been associated with the national movement to develop, demonstrate, and document the real potential of alcohol fuels for transportation applications.

My association with the Solar Energy Research Institute, Department of Energy, National Alcohol Fuel Commission, and Department of Agriculture has encouraged me to produce in a comprehensive volume the technical, economic, production, environmental, and utilization information for gasohol.

In recent years, fermentation ethanol produced from agricultural feedstocks, in particular grains such as corn, has been demonstrated to be one of our most promising near-term options for producing synthetic fuels. For all practical purposes, gasohol, which is a blend of 10 percent anhydrous agricultural-derived fermentation ethanol and 90 percent unleaded gasoline, is an opportunity to reduce our dependence on foreign oil in the near-term.

V. Daniel Hunt
Fairfax Station, Virginia

Acknowledgments

The information in *The Gasohol Handbook* has been obtained from a wide variety of authorities who are specialists in their respective fields.

I wish to thank Mr. Paul Notari and his staff at the Solar Energy Research Institute for the provision of data, reports, photographs, and illustrations, which are used extensively in this volume.

Also, I thank the contributors whose cooperation in providing source material was indispensable in preparing this handbook. In particular, those consultants and individuals who participated in the preparation of *Fuel from Farms—A Guide to Small Scale Ethanol Production, A Guide to Commercial Scale Ethanol Production and Financing,* and *Small-Scale Fuel Alcohol Production,* all of which this book is based upon.

We acknowledge the significant contributions to this effort by Dr. Harlan Watson, Subcommittee on Energy, Nuclear Proliferation and Federal Services; Mr. Steven J. Winston, Energy Incorporated; Mr. Milton David, Development Planning and Research Associates; Mr. Samuel Eakin, Energy Research Group; Dr. Jean-Francois Henry, TRW Energy Engineering; Mrs. Ann Heywood, RPA Inc.; Mr. Don Fink, U.S. Department of Agriculture; Mr. Ted D. Tarr, Vulcan Cincinnati; and Dr. Raphael Katzen, Raphael Katzen Associates International Inc.

A book of this magnitude is dependent upon excellent staff, and I have been fortunate. Judith A. Anderson has effectively served as coordinator and technical editor of this effort. Mr. B.A. "Dusty" Rhodes and Allen Higgs and their staff at Guild, Inc. have done an excellent job in preparing the graphics. Special thanks are extended to Anne Potter for the composition and typing of the manuscript.

Credits for the photographs and illustrations are indicated where appropriate throughout this book. The Department of Energy and the Solar Energy Research Institute are credited with public domain illustrations from *Fuel from Farms—A Guide to Small Scale Ethanol Production* and *A Guide to Commercial Scale Ethanol Production and Financing.* We thank the U.S. Department of Agriculture for the preparation of *Small-Scale Fuel Alcohol Production* which we have utilized where possible.

Chapter 1

Introduction to Gasohol

In recent years, alcohol produced from agricultural feedstocks, in particular grains such as corn, has been demonstrated to be our most promising near-term option for producing synthetic fuels. There are other forms of alcohol fuels such as methanol; however, this handbook stresses the role of gasohol. For all practical purposes, gasohol,[1] a blend of 10 percent anhydrous (water-free) fermentation ethanol and 90 percent unleaded gasoline (Fig. 1-1) delivers comparable engine performance as 100 percent unleaded gasoline, with the added benefit of superior antiknock properties. Gasohol has an antiknock index (pump octane) rating two to three numbers higher than that of unleaded gasoline. The production of fermentation ethanol from agricultural crops or wastes is supported strongly by the farming community, and the initial response of the general public to the introduction of gasohol as an alternative automotive fuel has been favorable. The federal government has recognized the potential of fermentation ethanol as a fuel additive or substitute and, in November, 1978, reduced the excise tax on gasohol by four cents per gallon. Primarily as a result of this encouragement,

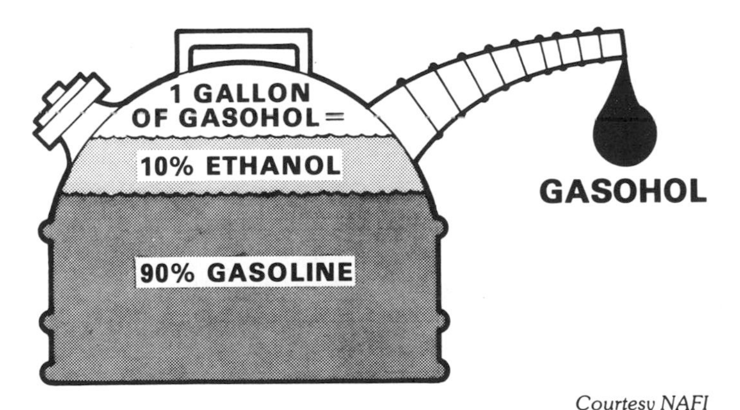

Courtesy NAFI

Figure 1-1. Gasohol is a mixture of gasoline and ethanol.

[1]The name "gasohol" is a trademark of the Nebraska Agricultural Products Industrial Utilization Committee (APIUC).

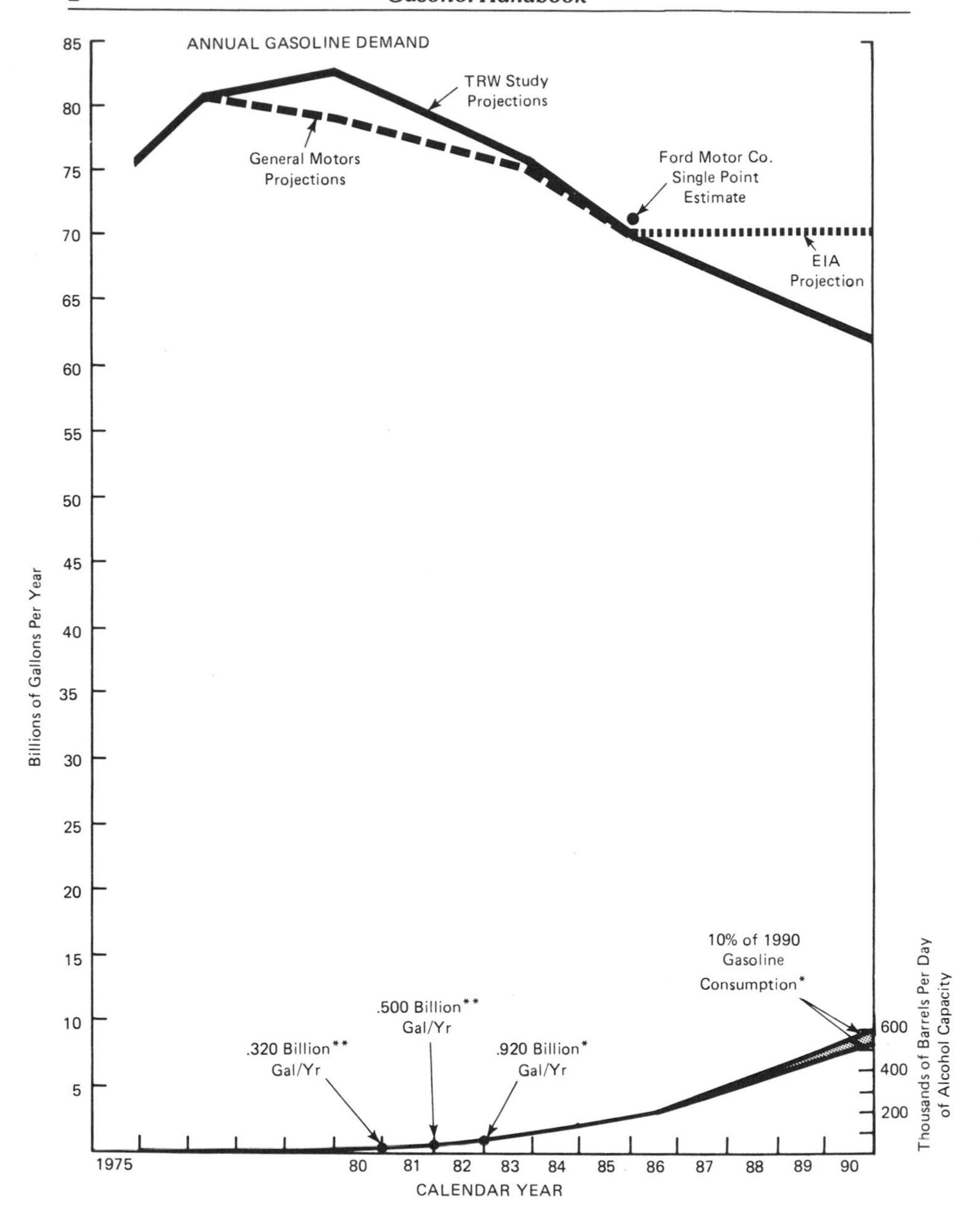

*Public Law 96-294 Goals
**President's Goals from Jan '80 Address

Courtesy NAFI

Figure 1-2. U.S. goals for alcohol-fuel production from biomass.

fermentation ethanol capacity increased from a few million gallons to about 80 million gallons annually by the end of 1979.

Goals for increased production of fermentation ethanol for blending with gasoline were established by President Carter; legislation that encourages the production of fermentation ethanol for fuel use has been adopted. Figure 1-2 shows the projected contribution of gasohol to the automotive gasoline market and indicates the magnitude of the President's objectives for ethanol-fuel production. Meeting the 1982 goal will increase the potential fermentation ethanol-fuel market about 10 times over the end of 1979s production.

Perspective

United States dependence on imported oil is illustrated in Fig. 1-3. The United States can no longer afford such a degree of dependence on oil reserves outside its control to meet its energy needs. Such dependence is hazardous to the U.S. economy and security. Therefore, during the past several years, there has been increased emphasis on domestic alternatives and pressure to develop synthetic fuels such as fermentation ethanol from biomass resources.

Statistics show a drop in crude oil imports over the past four years, perhaps implying that the United States is responding to the President's calls to conserve energy. However, increased development of U.S. reserves (particularly in Alaska), increases in the price of gasoline, and public response to gasohol play a role in bringing about this decrease.

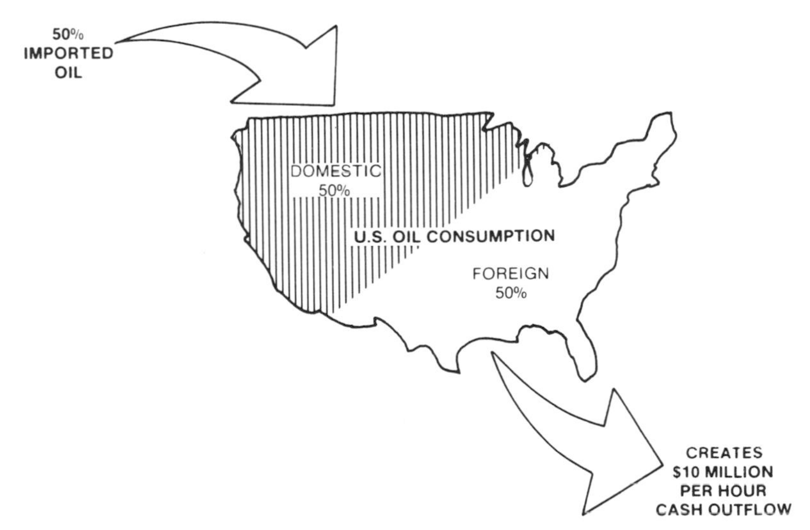

Source: *A Guide to Commercial Scale Ethanol Production and Financing*, SERI, Denver, 1980.

Figure 1-3. Impact of imported oil consumption.

Although little gasohol was marketed in the mid-1970s, by the end of 1979 about 80 million gallons of fermentation ethanol, which could be blended with unleaded gasoline, were produced and sold across the United States.

In the current political and economic climates, fermentation ethanol used as gasohol has become increasingly attractive. The Department of Energy (DOE) and the U.S. Department of Agriculture (USDA) have supported increased production of fermentation ethanol in small-scale facilities. DOE has also expanded its role by encouraging fermentation ethanol production in commercial-scale facilities. Small-scale is currently defined as under 15 million gallons of fermentation ethanol per year; this category includes the farmer producing 10 to 25 gallons per day strictly for his or her own use. Commercial-scale plants may range in size from 15 million to 100 million gallons annually. Figure 1-4 shows a breakdown of average fermentation ethanol production costs for the larger scale plants.

In recent years, the increase in gasoline price and the desire for energy independence have made alcohol attractive as a transportation fuel, even though this is not a new idea. Henry Ford was an early supporter of using "home-grown" fuels, and his Model-T could be adjusted to burn pure alcohol or gasoline. During World War II, many countries relied on alcohol as a fuel source. Since 1975, Brazil has had a huge program in operation to produce a national change from the use of petroleum to the use of alcohol fuels. Brazil is now using 20 percent alcohol blends on a regular basis, and large fleets of government- or industry-owned cars run on 100 percent alcohol. Figure 1-5 shows the trends in ethanol production in Brazil. It is apparent that, since the announcement of the Brazilian National Alcohol Program in November, 1975, the total production of ethanol, as well as anhydrous ethanol for blending, has increased dramatically. This experience shows that a dedicated effort to reduce oil imports by developing national renewable resources can produce significant results in a short period of time.

If the need arises, the United States could develop an intensive ethanol program similar to Brazil's program. A significant reduction in gasoline use could be achieved by changing to gasohol. The 10 percent alcohol/90 percent unleaded gasoline blend requires no engine modifications and has the potential to reduce oil imports.

The raw materials for commercial-scale ethanol production are readily available in various forms; the major feedstocks are corn, wheat, grain sorghum, barley, potatoes, sugar beets, sweet sorghum, sugarcane, and fodder beets. Since it is possible to use a wide variety of feedstocks for ethanol production, the plant location is closely linked to the locally available type of feedstock. For example, a plant in Iowa would use corn as feedstock, while one in Texas might use grain sorghum, and a plant in Idaho could consider potatoes or sugar beets.

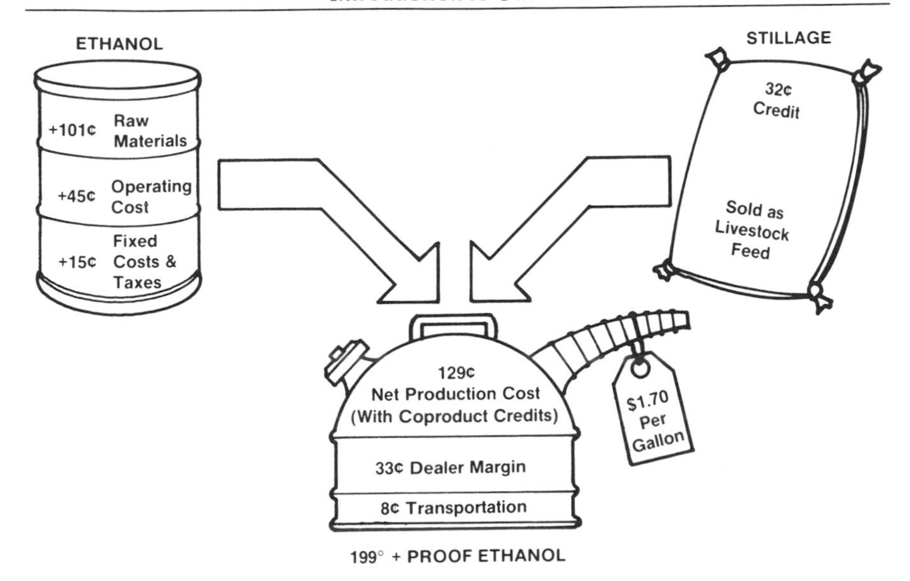

Source: U.S. DOE 1st Annual Report to Congress (Revised).

Figure 1-4. Breakout of production costs for fermentation ethanol.

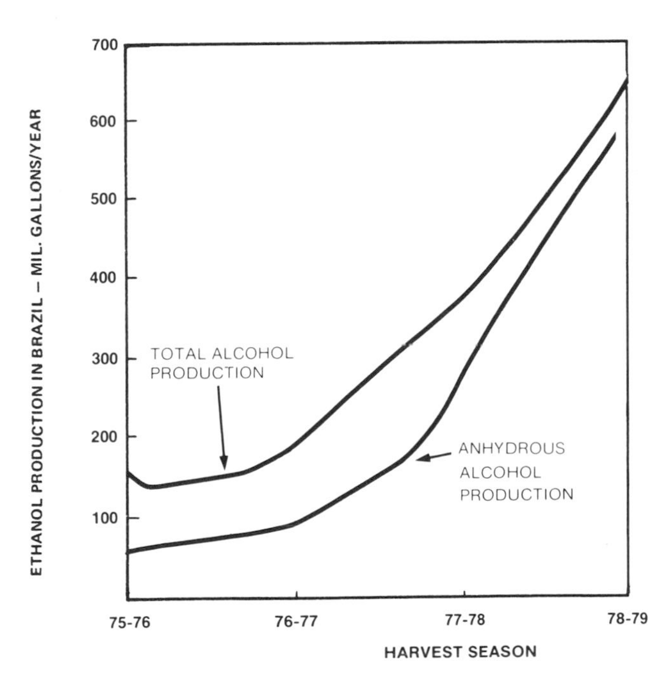

Source: *A Guide to Commercial Scale Ethanol Production and Financing*, SERI, Denver, 1980.

Figure 1-5. Ethanol production in Brazil.

To meet the President's goal of 500 million gallons of ethanol for 1981, production of fermentation ethanol must be increased. Table 1-1 shows the number of plants of various sizes required to attain annual production goals ranging from 200 to 2000 million gallons. It is apparent from the figures used to develop the chart that a significant fraction of ethanol fuel is expected to be produced in commercial-sized plants (the medium- and large-sized plants mentioned in the chart).

The primary market for fermentation ethanol mixed as gasohol is as a transportation fuel. Fermentation ethanol used in a gasohol blend has successfully penetrated the transportation fuel market and established itself as the first usable synfuel. The marketing effort for gasohol has resulted in the establishment of a national network of gasohol retail stations as shown in Fig. 1-6. Annual production capacity for biomass-derived ethanol for use in a gasohol-type blend (Fig. 1-7) has expanded to 320 million gallons per year (equal to 3.2 billion gallons of gasohol). The long-term viability of fuel-grade ethanol production is under active review by many firms in the U.S. private sector. Several firms, including beverage-grade alcohol producers, members of the food processing industry, and some major oil companies such as Texaco, have made long-term commitments to fuel-grade ethanol production. The contribution of biomass-derived alcohol as a means of satisfying near-term demand for automotive fuel, and therefore reducing petroleum demand, is generally accepted.

Through the end of August, 1980, the demand for gasoline continued at a staggering 80 billion-gallon-per-year rate as shown in Fig. 1-2. Of the total petroleum use, about one-half is imported, representing a minimum projected cash drain of about 78 billion dollars for a 12-month period. It can

Table 1-1. Number of Plants Required to Meet Goal for Alcohol Production

PLANT SIZE	CY 1980		CY 1982		CY 1985	
	TOTAL NO. PLANTS	PRODUCTION CAPACITY, MILLION GALS/YEAR	TOTAL NO. PLANTS	PRODUCTION CAPACITY, MILLION GALS/YEAR	TOTAL NO. PLANTS	PRODUCTION CAPACITY, MILLION GALS/YEAR
10,000 – 15 Million Gallons/Year	96	175	200	375	300	600
> 15 – 50 Million Gallons/Year	6	150	13	450	23	900
> 50 Million Gallons/Year	1	100	2	200	5	500
Cumulative Total Installed Capacity (Projected in Million Gallons Per Year)		425		1,025		2,000
National Goals Million Gallons/Year		300*		1,000**		2,000*

*President's Goals Announced 11 January 1980.

**P.L. 96-294.

Source: *A Guide to Commercial Scale Ethanol Production and Financing,* SERI, Denver, 1980.

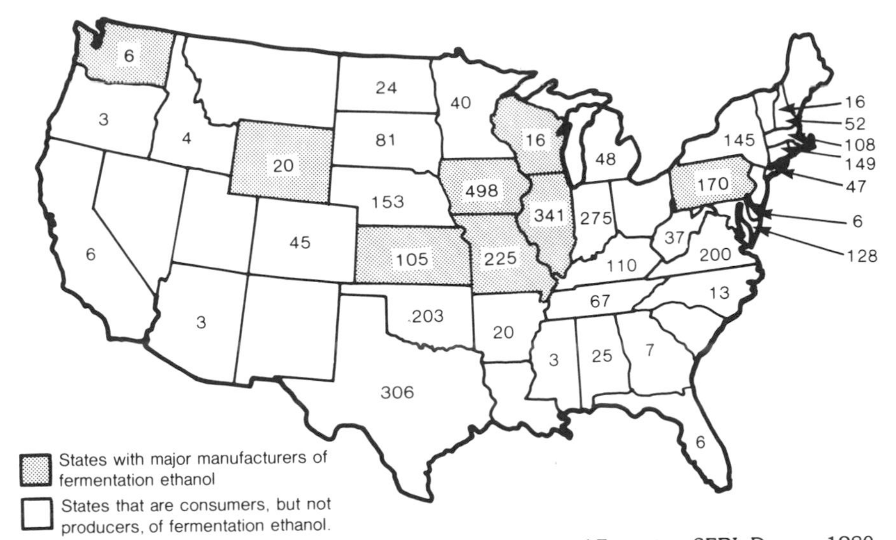

Source: *A Guide to Commercial Scale Ethanol Production and Financing*, SERI, Denver, 1980.

Figure 1-6. National network of retail gasohol stations.

Courtesy: DOE Photo by Jack Schneider

Figure 1-7. Gasohol being sold at co-op station in Wheaton, Maryland.

be expected that gasoline demand will continue to grow during the next 20-year period and beyond. The projected demand for gasoline is shown in Fig. 1-2. Two major trends dictate the shape of the curve. Between 1980 and 1990, there is a slight decline in total consumption, resulting from increased fuel economy of automobiles originally dictated by Federal Corporate Average Fuel Economy (CAFE) mileage standards. If expected sharp real-dollar price increases in gasoline occur, demand can be expected to be reduced further. Beyond 1990, the continuing steady growth of the automobile fleet will be the predominant effect, and total consumption will again increase. During the next 20-year period, annual gasoline consumption will fluctuate around an average of about 70 billion gallons per year. Therefore, during the next 20-year period, there will continue to be a domestic shortfall of about 35 billion gallons per year of gasoline produced from domestic reserves. As a point of reference, it would take over 700 grain ethanol plants of 50-million-gallon-per-year capacity to replace this shortfall, or about 200 commercial-scale plants to produce an amount of ethanol sufficient to produce a 10/90 gasohol blend for all fuel sales. A very sizable, stable market exists for transportation synfuels such as gasohol.

The initial success of fermentation ethanol in penetrating the transportation fuel market is based on several key factors:

- The existence of industries related to fuel-grade ethanol (e.g., potable alcohol producers) and grain processors (e.g., corn starch producers) which have an existing business base.
- The technical feasibility of commercial-scale fuel-grade alcohol processes. Fuel-grade alcohol technology is a modification of potable alcohol processes wherein the overall process has been reoptimized to produce anhydrous ethanol with higher alcohol yields and lower energy consumption.
- A relatively short plant construction time of about 2 years.
- The ready compatibility of ethanol with gasoline and the current automobile fleet. Ethanol has already been used in the past as a blending agent for gasoline to improve octane.
- The competitive price of fermentation ethanol relative to petroleum-derived ethanol sold as a chemical feedstock.
- A combination of political support, government subsidy, and consumer acceptance and preference.
- The short-term shortage of gasoline which exists in times of international conflicts.

The History of Alcohol Fuels[2,3]

The use of alcohol in motor vehicles is not a new technology; an exhaustive international bibliography prepared on the subject has entries which date before 1920. Alcohol fuels have been used in both wartime and peacetime in our country, and are currently attracting renewed attention as petroleum prices increase and supplies remain uncertain.

Because it was clean and odorless, alcohol fuel replaced whale oil in lamps in the mid-1800s. By the end of the century, alcohol was being considered as a prime automobile fuel. The first modern internal combustion engine, the Otto cycle of 1876, ran on alcohol as well as gasoline.

Henry Ford was an early champion of alcohol fuels. He thought that the agricultural feedstocks used in alcohol production would help make farmers energy producers. In the 1880s, he designed one of his earliest automobiles—the quadri-cycle—to burn alcohol. His subsequent Model T was built with an adjustable carburetor that could easily be modified to use pure alcohol in the engine. Despite the intense price competition from gasoline, alcohol fuels were used to power American cars well into the 1920s and 1930s.

Falling farm crop prices after the 1929 stock market crash kept interest in alcohol alive because alcohol production for industry offered an alternative market for farm crops. Hiram Walker, in 1934, sold a gasoline-alcohol mixture called Alcoline. By 1938, the Atchison Agrol plant in Kansas was producing 18 million gallons of ethanol for distribution to about 2000 independent service stations in the Midwest. Dealers sold the ethanol/gasoline mixture under the trade name Agrol. The Agrol venture ultimately proved unprofitable and was abandoned at the time the United States was emerging from the Depression.

In December, 1938, the U.S. Department of Agriculture issued a noteworthy document identified as Miscellaneous Publication No. 327, titled *Motor Fuels From Farm Products*. The publication contained information on a variety of crops, including cereal grains, sugarcane, beets, tubers, and fruits. The data presented for the various materials included crop yields, fermentable starch (sugar) content (average and high), and probable commercial yields of 199-proof fermentation ethanol.[4]

Even wider use was being made of alcohol fuels in other countries. Worldwide, over 40 nations were blending alcohol (usually ethanol) into their fuel base. In Czechoslovakia, Hungary, France, Germany, Sweden, Italy,

[2]Material reproduced and edited from "A History of Alcohol Fuels" in the January 1980 issue of DOE's *The Energy Consumer*.

[3]National Alcohol Fuels Commission, *Farm and Cooperative Alcohol Plant Study*, Raphael Katzen Associates International, 1980.

[4]National Alcohol Fuels Commission, *ibid*.

Poland, Austria, Brazil, and Chile, the use of 10-25 percent alcohol blends was mandatory. Prior to World War II, more than 4 million European automobiles ran on alcohol fuels. In Australia, International Harvester proudly advertised motor trucks powered with engines especially designed to burn alcohol. Simultaneously, Chrysler was shipping to New Zealand Detroit-made cars that had a modified manifold and different carburetor jets to permit the use of pure alcohol.

The onset of World War II prompted an even greater interest in alcohol-fuel sources. Throughout much of Europe, alcohol provided a primary fuel source to power motor vehicles. Hitler converted Germany's aircraft and much of the other war machinery to alcohol fuels after his East European refineries were destroyed. In the United States, though, it was the need for synthetic rubber that brought about a surge in the domestic industrial alcohol market. With the aid of two refugee scientists, a major synthetic rubber industry was created and three new grain alcohol plants were constructed to support the effort.

Furthermore, a Government decree ordered the nation's whiskey distilleries to modify their plants to produce industrial alcohol for the war effort. Torpedoes and submarines were fueled by both grain and wood alcohol. Alcohol was also used as an additive to jet fighter fuels to prevent carburetors from icing. And in China, the U.S. Army purchased alcohol for use in its jeeps, generators, and Land Rovers. In total, during the war years, the U.S. alcohol industry managed a sixfold increase in production. By 1944, the United States was producing almost 600 million gallons of alcohol, not primarily for fuel, but rather as a solvent and as a chemical intermediate.

Federal officials quickly lost interest in alcohol-fuel production: Many of the wartime distilleries were dismantled, while others were converted back to beverage alcohol plants. By 1949, less than 10 percent of U.S. industrial alcohol was made from grain; the rest was distilled with a new and cheaper process using natural gas.

After World War II, interest in the use of agricultural crops to produce liquid fuels decreased. Fuels from petroleum and natural gas, and synthetic ethanol from these sources, became available in large quantities at low cost and eliminated economic incentives for the production of liquid fuels from crops. The United State's use of petroleum products grew without any apparent limit. Simultaneously, the farming community enjoyed bountiful harvests, leading to accumulation of large grain reserves for which there was no ready market.

Throughout the 1950s and 1960s, few automotive uses were made of alcohol other than for racing cars—alcohol has powered race cars run at the Indianapolis 500.

Then, in the early 1970s, rising gasoline prices coupled with the continuing search for new markets for agricultural products sparked a renewed interest in ethanol production for fuel. In 1971, American farmers, in an effort to create markets for these accumulating grain reserves, put forth the idea that excess grains (corn and wheat) might be used in the production of liquid-fuel supplements. The movement was initiated in Nebraska, but economics associated with the production of liquid fuels from agricultural products were still unfavorable. This version of the "ethyl alcohol from agricultural crops" scenario became known as the "Gasohol" movement. In Nebraska in 1973, legislation went into effect to reduce that state's gasoline tax by 3 cents per gallon for fuels that contained a minimum of 10 percent agricultural ethanol of at least 190 proof. In 1975, Nebraska's new Agricultural Products Industrial Utilization Committee was established to coordinate alcohol-fuel commercialization in the state; it conducted a two-week marketing experiment, selling over 90,000 gallons of gasohol. The committee also initiated a 3-year, 2-million-mile test program to compare the performance and properties of unleaded gasoline and gasohol.

The "OPEC Oil Crisis" of 1973 made the American people aware of the extent of their dependence on foreign oil imports. The decision by the OPEC countries to escalate the price of oil placed a growing burden on the American economy, and the United States began to look for means of relieving our foreign energy dependence. Farm groups revived their interest in the possibility of using excess crop production to produce fuel-grade alcohol.

Other states, among them Virginia, Maine, and Indiana, began studies or demonstration projects to assess the feasibility of using alcohol in automobiles. In the U.S. Congress, Senators Carl Curtis and Robert Dole successfully cosponsored an amendment to a 1977 farm bill that directed the USDA to develop four pilot alcohol-production plants.

The USDA reluctantly accepted the mandate, arguing that alcohol-fuel production was not economically viable. The DOE took a similarly dim view of alcohol fuels, claiming that they were not price-competitive with gasoline and that it required more energy to produce ethanol than it could yield. President Carter's 102-page National Energy Plan in 1977 did not mention ethanol fuels.

Taking a more optimistic view, the Department of Commerce decided in 1977 to experiment with alcohol fuels. A grant was awarded to construct an alcohol-production plant that could salvage diseased crops produced that year. This project was the first of a rapidly growing number of experimental alcohol-fuel projects. The number of service stations selling the 10 percent alcohol/90 percent unleaded gasoline blends now called "gasohol" multiplied rapidly.

In October, 1978, the Congress approved the National Energy Act, which included a provision exempting gasoline containing at least 10 percent alcohol produced from agricultural products or wastes from the 4 cent per gallon federal gasoline excise tax. By early 1979, over 1000 service stations nationwide were selling gasohol. In Iowa alone, gasohol sales had jumped from 600,000 to 5.6 million gallons in just 5 months.

The growing grass roots support for gasohol finally prompted the White House to modify its position. In July, 1979, the DOE released "The Report of the Alcohol Fuels Policy Review," which observed that "development of alternative sources of liquid fuels will be necessary. Alcohol fuels represent sources that can substitute for petroleum products now and increasingly through the 1980s drawing on abundant supplies of renewable materials and coal."

While much of the fuel-alcohol production will be accomplished in large-scale agriindustrial facilities, there are incentives which encourage individual farmers and small co-ops to consider installing their own small-scale fuel-alcohol plants. The idea that fuel-alcohol plants can be built and operated simply and cheaply, regardless of size, is generally accepted opinion. Many systems and equipment units for producing alcohol from crops on a small scale have recently appeared in the marketplace.

Until recently, the relatively inexpensive price of gasoline has made alcohol fuel uneconomical to produce. Now, however, with the increased price of gasoline and the new tax incentives, which reduce the cost difference between gasoline and alcohol fuels, substantial public and private activity has centered on "gasohol."

Basics of Ethanol Production

Ethanol is the product of three processes—cooking, fermentation, and distillation—as shown in Fig. 1-8. Cooking converts the carbohydrates in feedstocks to fermentation sugars; fermentation converts simple sugars by biological action to ethanol and carbon dioxide; and distillation concentrates the dilute ethanol and separates it from the water and nonfermentable materials.

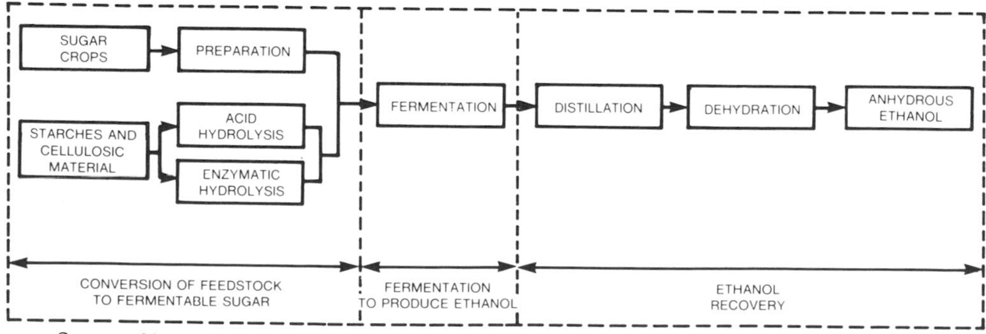

Source: National Alcohol Fuel Commission

Figure 1-8. Ethanol is the product of cooking, fermentation, and distillation.

The amount of ethanol that can be produced depends on a number of variables. The ethanol producer should give the most attention to preparing the right form of sugar through grinding and cooking, preparing a mash that contains the proper concentrations of sugar and yeast nutrients, and adjusting the controlling fermentation conditions such as temperature and pH (acidity/alkalinity) values to those preferred by the yeast.

CONVERSION OF CARBOHYDRATES TO SIMPLE SUGAR

The preferred form of sugar for fermentation is glucose, the building block for carbohydrates and cellulose. Because green plants produce both carbohydrates and cellulose, the entire range of plant life is a potential resource for ethanol production. It is easier to break down carbohydrates into glucose than it is to break down cellulose. Thus, in order of preference (after plants that directly produce sugar such as sugarcane) are plants that produce large quantities of carbohydrates (corn), followed by those that primarily yield cellulose. In practical terms, this means that cereal grains, potatoes, and other starchy tubers are preferred with present technology over cornstalks, wood, or pulp as ethanol feedstock.

The amount of ethanol that can be produced from any feedstock depends not only on the carbohydrate content of the feedstock, but also on the efficiency of fermentation and distillation. Although there is no reason to presume that large-scale systems are inherently superior to the small-scale systems, many of the small-scale systems currently in operation are not capable of consistently producing ethanol yields as high as large commercial-scale plants. The better the small-scale producer understands the principles and objectives of each separate operation to produce ethanol, the more likely that a high yield will result. However, even a perfect understanding will not overcome the limitations of poor equipment.

The carbohydrates are converted to simple sugars by preparing a "mash" from potable water and crushed grain or other feedstock as shown in Fig. 1-9. The pH of the mash is then adjusted according to the particular enzyme being used. Liquefying enzymes are added to the mash, which is agitated and heated until it is liquefied. A saccharifying enzyme is then added, and the mash is held at a constant temperature until the conversion of the carbohydrate to glucose is complete.

Although all potential ethanol agricultural feedstocks will break down to fermentable sugars owing to natural bacterial action, none of them can be directly placed into a fermenter whole with any real expectation of efficient ethanol production. Grain must be ground to remove the husk and break the starch granules into small pieces; potatoes must be mashed; and the sugar must be extracted from sugar beets or sugarcane.

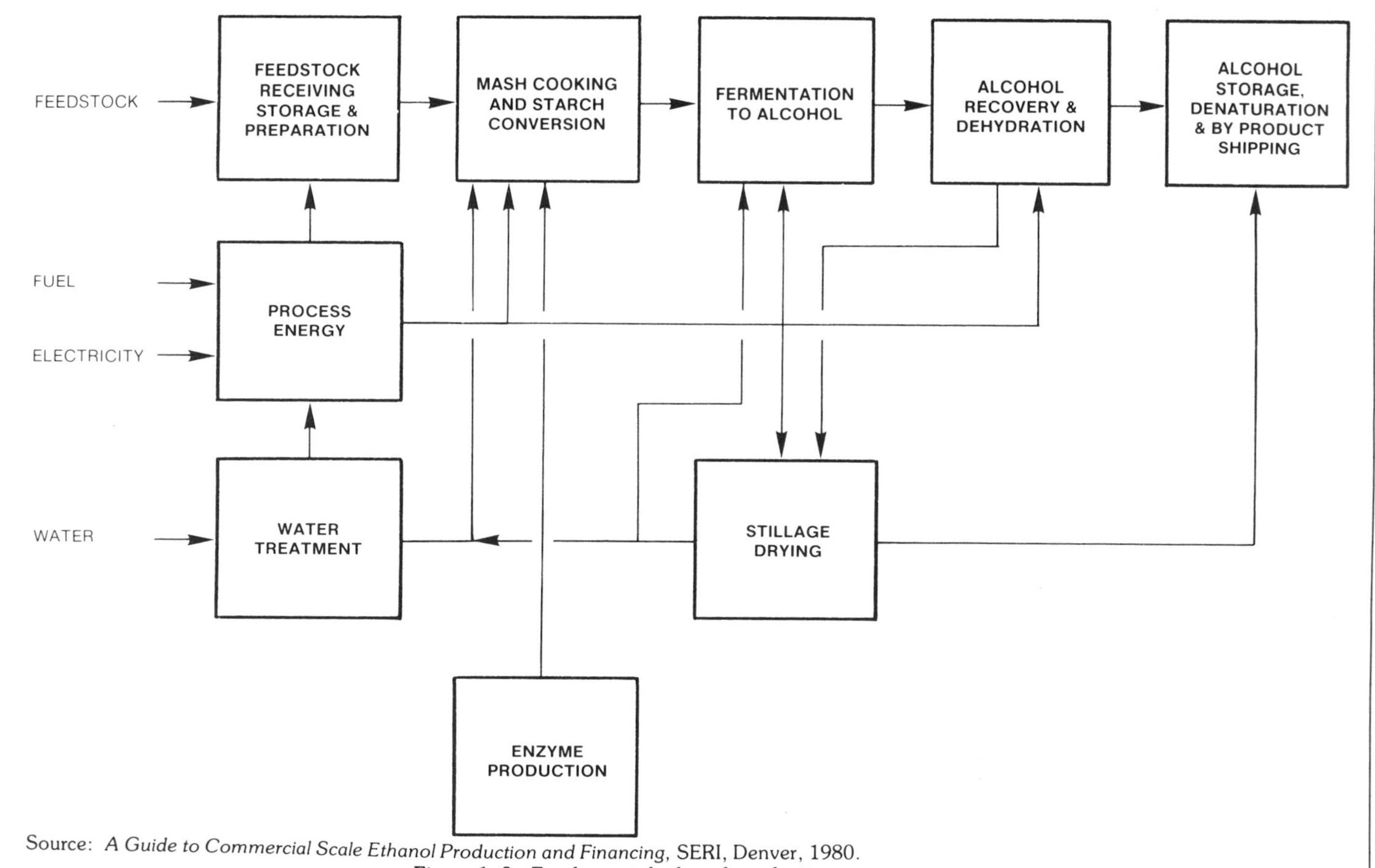

Source: *A Guide to Commercial Scale Ethanol Production and Financing*, SERI, Denver, 1980.

Figure 1-9. Fundamental ethanol-production processess.

Feedmills will suffice for small-scale grain grinding. Although these do not produce a highly consistent range of particle sizes, they are available on farms, and the impact on fermentation efficiency is not irreconcilable.

The mashing of potatoes can be accomplished in a hammer mill or ball mill. Sugar beets and sugarcane must have the sugar extracted from the tuber and the stalk, respectively, which can be done by slicing and leaching or by squeezing through roller presses.

FERMENTATION

The conversion of simple sugars to ethanol through fermentation (Fig. 1-10) is a complex process caused by enzymes produced by specific varieties of yeast.

Fermentation goes through four characteristic phases: (1) the lag phase while the yeast cells become acclimated to their new environment; (2) the exponential growth phase in which the yeast cells propagate most rapidly; (3) a stationary phase; and (4) a death phase in which the alcohol concentration is high and the available sugar for yeast metabolism is low. If proper conditions are maintained throughout all phases, relatively complete fermentation can be achieved in 48 hours. The beer that results is generally about 6-12 percent ethanol by volume.

DISTILLATION

A mixture that is 6-12 percent ethanol and 88-94 percent water will not burn. The ethanol must be concentrated. This is usually accomplished by distillation shown in Fig. 1-11, which is a series of successive evaporation and condensation steps that concentrate the more volatile compound (ethanol) in

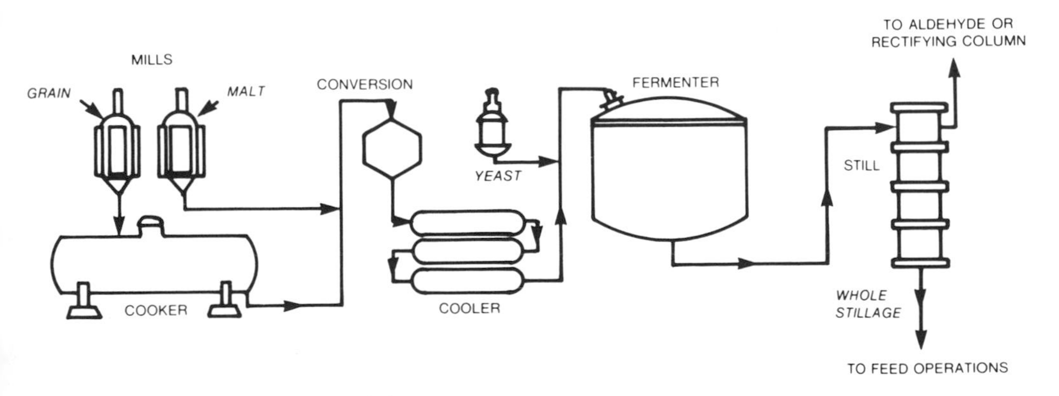

Source: National Alcohol Fuel Commission

Figure 1-10. Basic cooking-fermentation process.

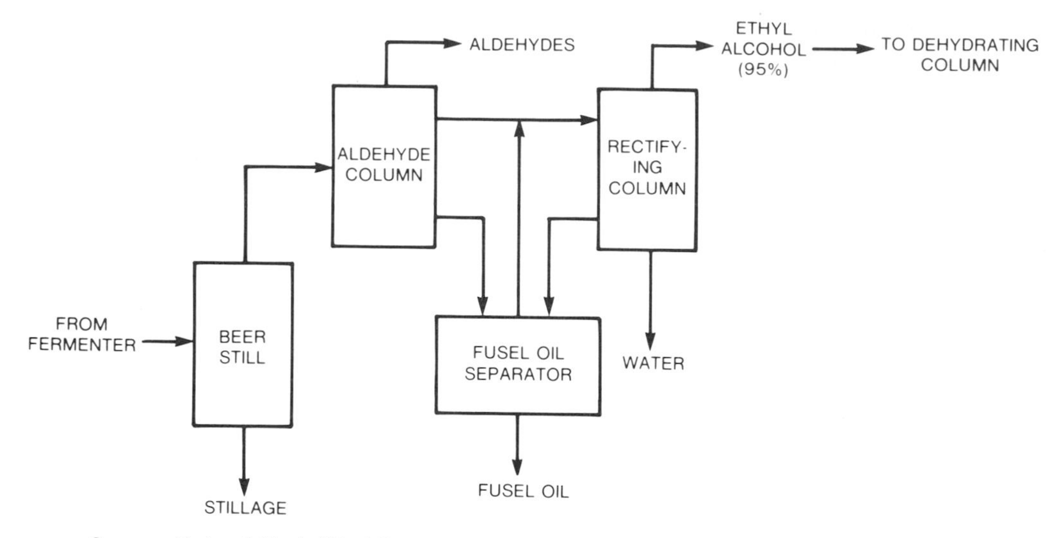

Source: National Alcohol Fuel Commission

Figure 1-11. Basic distillation process.

one stream and the less volatile compound (water) in the other. In effect a distillation column is a series of boiling pots and condensers.

It is not possible to produce 200-proof ethanol in a conventional atmospheric distillation column of the sort found in currently available small-scale stills. Efficient small-scale stills should be able to produce 175- to 180-proof ethanol. By using a community-based "topping cycle" facility, small-scale plants can less expensively obtain 200-proof ethanol. Large-scale plants provide 200-proof ethanol.

STILLAGE RECOVERY AND USE

Not all of the material contained in the feedstocks can be converted to fermentable sugar. The amount that ultimately gets processed (Fig. 1-12) through the fermenter depends on the specific feedstock and the system of operation selected.

The nonfermentable materials can be removed prior to fermentation by wet milling (a process that is economically feasible for the large-scale producer) or by screening the mash after cooking.

The most common method for recovering stillage solids is removal from the bottom of the still. The stillage contains about 90 percent water and 10 percent suspended solids as well as soluble proteins and fatty acids. The solids can be separated from most of the liquid by passing the entire quantity

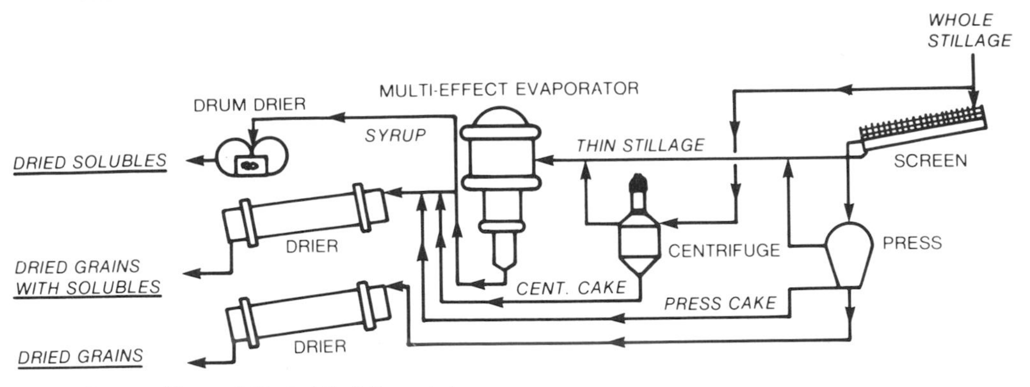

Source: National Alcohol Fuel Commission

Figure 1-12. Basic stillage recovery.

generated through filters or over slant screens (hydro-sieves). The wet solids can be pressed to remove additional absorbed water. After this the solids feel dry to the touch, but still retain up to 60 percent moisture.

The liquid, called thin stillage, can be applied to fields as a soil amendment, concentrated by evaporation and then dried in a drum dryer with the solids, or disposed of as a waste effluent. It is generally impractical for small-scale ethanol plants to dry thin stillage because of the large energy requirement to evaporate the water. Consequently, it is common to spray thin stillage directly on fields to build soil tilth.

Because the protein is not consumed during the fermentation process, the solids have a greatly increased protein content. Furthermore, the fermentation process renders the protein more digestible, for most animals, than it would be in raw grain. Hence, animals fed with some portion of stillage solids in the feed ration gain weight more rapidly than those fed straight grain. Dairy cattle produce proportionately more butterfat.

Stillage solids can be fed directly to animals without drying. However, residual sugars in the stillage can cause spoilage that renders the stillage inedible within a day or two in warm climates.

The stillage solids from grain can contain up to 30 percent protein. About 7 pounds of dry solids are produced for every gallon of ethanol produced.

The feed value of stillage varies from animal to animal depending on principal nutritional needs. Ruminants (cattle and sheep) are able to derive the most benefit from stillage because they can obtain their energy requirements (carbohydrates) from low protein cellulosic materials such as straw. Fowl, which have rapid metabolisms and relatively large energy requirements, generally cannot consume any appreciable proportion of stillage without suffering nutritional deficiencies.

CAUTIONS

A few words of caution are appropriate for the individual unfamiliar with the process of ethanol production. Although the process is well known and has been used on a commercial scale, ethanol production through fermentation is a complex process employing biochemical reactions as well as sophisticated engineering concepts. Consistent and reliable operation can be achieved, provided the prudent operator follows precautions required by biological processes and operates the facility according to established engineering practice, which includes cleanliness, close control of operating conditions such as temperature, and careful storage of product. Plant operating personnel must have the background and experience to deal with the problems related to the common, but complex, unit processes.

Ethanol production involves hazardous materials, and plants must be designed and operated to minimize risks. The prospective entrepreneur must be cognizant of these aspects of ethanol production and assure that the vendors retained have the experience and knowledge required to advise him in the design and operation of the proposed plant.

The Federal Gasohol Plan

The Federal Gasohol Plan is contained in the Joint U.S. Department of Agriculture and Department of Energy Plan for Biomass Energy Production and Use. It was prepared in accordance with the Energy Security Act, Public Law 96-294, Title II, Subtitle A, Section 211(a). "General Biomass Energy Development, Biomass Energy Development Plans." This joint plan presented the objectives, organization, strategy, and implementation plan to achieve the December 31, 1982, alcohol production and use goal established in the Act. Also included is a discussion of major variables and emerging problems that point to the need for accelerated research and development.

GOALS

The Energy Security Act sets a specific goal of 60,000 barrels per day (bbl/day) alcohol production by December 31, 1982. This level is equal to production of 0.920 billion gallons per year (gal/yr) of alcohol, or can be equated to 90,000 bbl/day of oil equivalent. The Act also includes a goal of achieving a level of alcohol production within the United States equal to at least 10 percent of the level of 1990 gasoline consumption. The goals established by the Act, together with the President's announced goals, are presented in Fig. 1-13. Private industry has made significant contributions toward meeting these goals even prior to the enactment of the Energy

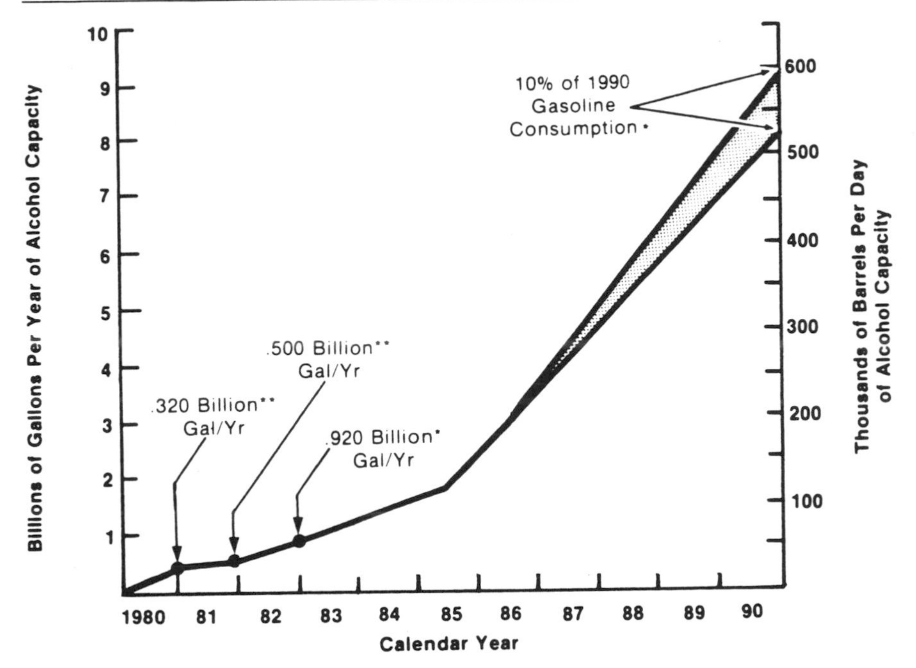

Figure 1-13. National alcohol-fuel-production goals.

Security Act. Nonetheless, federal financial incentives are definitely needed to meet the national goal for 1982.

BACKGROUND

The biomass alcohol industry has grown over the last year from an 80-million-gallon-per-year capacity to about 320 million gallons per year. The fact that this industry is rapidly emerging at present is the result of three major factors. First, recent innovation in grain-ethanol-production processes has converted long existing alcohol processes and gain-processing technology into higher-yield, more-energy-efficient processes for producing transportation fuels. Anhydrous ethanol produced by these fuel-grade-alcohol processes can be readily blended in gasoline, with a resulting improved octane. Second, the improved energy efficiency and yield of the fuel-grade-alcohol production process, combined with federal and state subsidies, allows reasonable return on investment. Third, prior to enactment of the Energy Security Act, a limited number of companies entered the fermentation ethanol industry or expanded their production without benefit of loan guarantees provided by the Act.

To date, federal and state legislation and initiatives have provided substantial subsidies and incentives for biomass-derived alcohol production. The first major incentives were provided by the Energy Tax Act of 1978 (Public Law 95-618) and the Crude Oil Windfall Profit Tax Act of 1980 (Public Law 96-223). The exemption of the 4-cent-a-gallon excise tax on motor fuels provided by the Energy Tax Act of 1978 and the additional 10 percent renewable energy investment tax credit provided by the Crude Oil Windfall Profit Tax Act substantially improved the economic viability of fuel-grade ethanol production.

However, it has been the enactment of the Energy Security Act which has permitted the federal government to assume a portion of the investment risk in fermentation ethanol energy projects through loan guarantees. This risk would have continued to prevent substantial investment in this new industry.[5]

Figure 1-14 shows that private sector alcohol production unaided by federal loan guarantees (the dark areas) will be significant, but will fall short of achieving the December 31, 1982, goal. This production, together with the additional production gained from plants financed by federal loan guarantees (the light shaded areas) awarded by DOE and USDA, will achieve the 1982 production goal. Plant design and construction lead times dictate that the loan guarantees negotiated in 1981 will result in initial production in early 1983.

Uncertainty in several areas of alcohol production and utilization may prevent alcohol from achieving its full potential as a biomass energy alternative. These aspects include economic viability of alcohol-fuel plants without incentives; unknown new investment risk for the financial community; variable feedstock availability and price impact; regulatory disincentives; lack of public awareness of the advantages of alcohol-fuel use; current marginal price competitiveness of gasohol relative to unleaded gasoline; and undeveloped markets for coproducts. Strategies to deal with these concerns are addressed in the plan.

In summary, both USDA and DOE are confident that their plan will achieve the 1982 production goal. However, some concern exists because several events have occurred since the goals were set, such as reduced domestic demand for gasoline, decreased grain feedstock production due to poor weather, and rising interest rates, all of which have a direct impact on alcohol production and use.

ORGANIZATIONAL RELATIONSHIPS

The Energy Security Act specifies that the Secretary of Agriculture has responsibility for determining national, regional, and local agricultural policy impacts of biomass energy efforts on agricultural supply, production, and use.

[5]The Reagan Administration is planning significant reductions in loan guarantee programs.

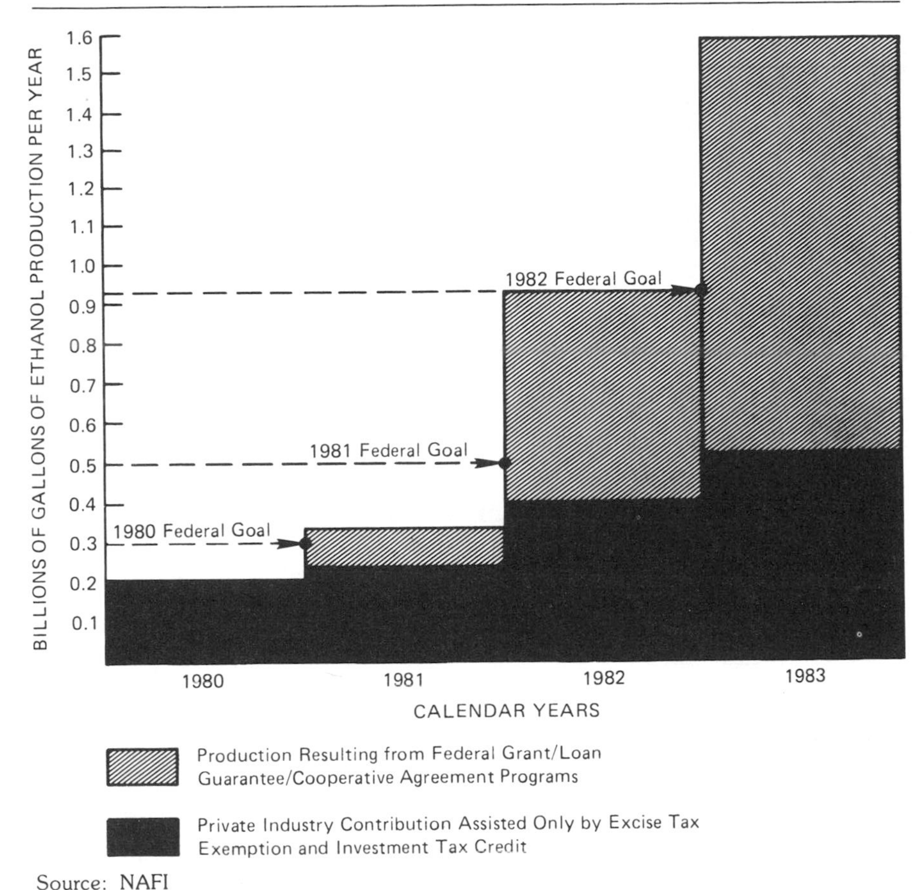

Source: NAFI

Figure 1-14. Expected biomass-derived alcohol production capacity.

The Secretary of Energy has responsibility for national energy policy impacts of biomass energy efforts and for assuring the technical feasibility of these efforts.

The Secretaries of Agriculture and Energy have initiated the management structure and designated the individuals to carry out the production and use program. The lead persons for USDA are the Director of Economics, Policy Analysis and Budget, who is responsible for USDA policy in the Biomass Fuels Production and Use Plans, and the Assistant Secretary for Rural Development, who is in charge of implementing these plans. The lead person for DOE is the Director of the Office of Alcohol Fuels who is responsible for all biomass-derived alcohol production and use efforts.

Figure 1-15 summarizes the USDA and DOE responsibilities and interrelated actions as specified in the Energy Security Act.

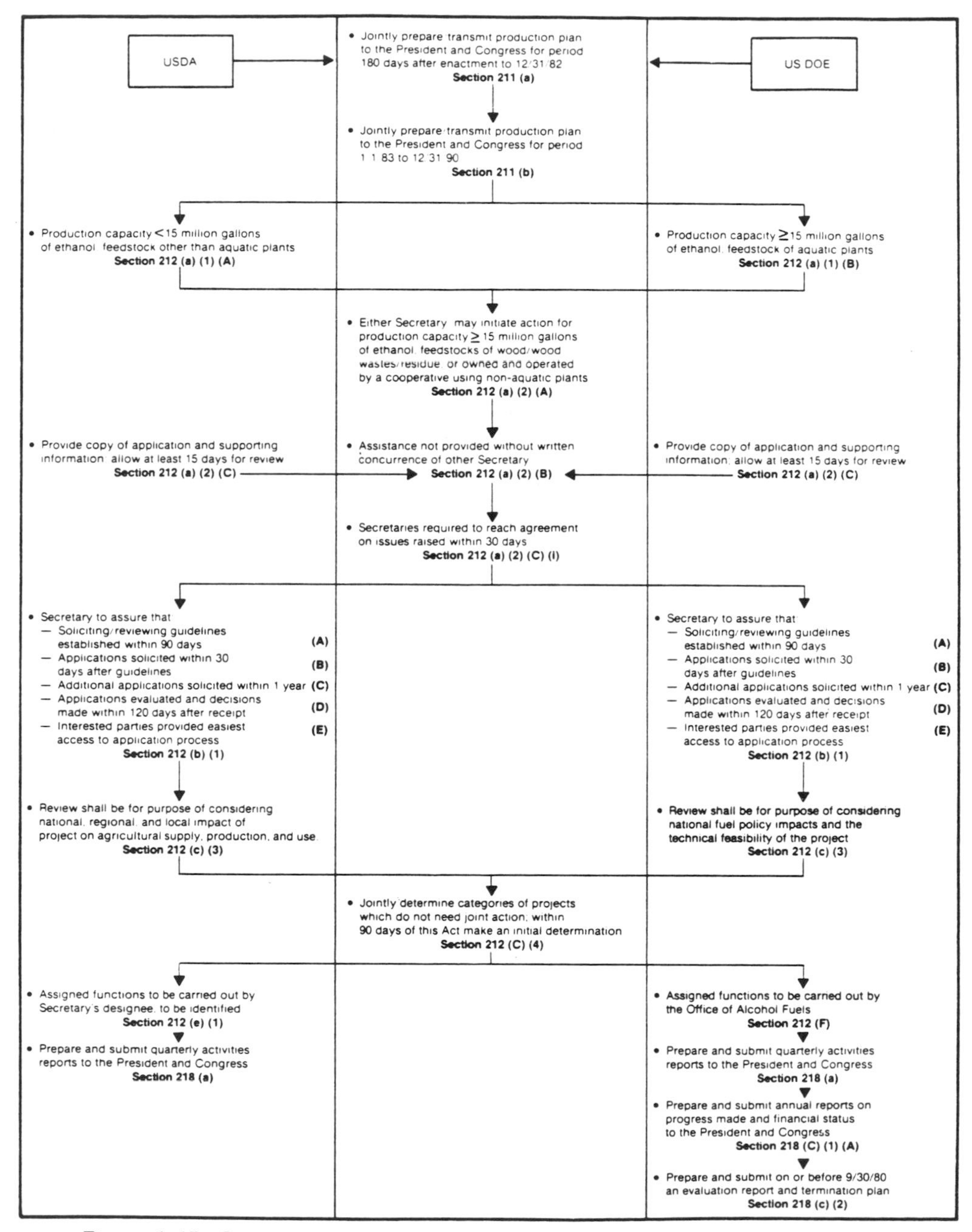

Figure 1-15. Organizational responsibilities and interactions specified in the Energy Security Act.

STRATEGY

Tables 1-2 and 1-3 summarize the federal objectives, strategy elements, actions, and applicable legislation initiatives to achieve the 1982 biomass energy production and utilization goals.

A detailed description of the implementation actions for the federal loan guarantee program is summarized in this section.

PROGRAM FOR ACHIEVING GOALS

The main thrust of the alcohol fuels portion of the Energy Security Act is to provide federal financial incentive programs to stimulate private sector development of fermentation ethanol production facilities.

Table 1-2. Summary of Near-Term Federal Strategy To Stimulate the Production of Alcohol Fuels

OBJECTIVE	FEDERAL STRATEGY ELEMENTS	FEDERAL ACTIONS	APPLICABLE FEDERAL LEGISLATION OR INITIATIVE
Stimulate Production of Alcohol Fuels	Improve Marginal Economic Viability	Provide an incentive for construction of alcohol production plants by providing an additional 10% energy investment tax credit (for a total investment tax credit of 20%).	Provided by Energy Tax Act of 1978. P.L. 95-618, and the Crude Oil Windfall Profit Tax Act of 1980, P.L. 96-223.
	Reduce Investment Risk Assumed by Private Sector	Make available funds to private industry primarily through the use of loan guarantees, to assist in financing of new alcohol fuel production plants.	The Energy Security Act, P.L. 96-294 authorizes up to $1.2 billion in funds to be administered by the Department of Energy and the Department of Agriculture.
	Reduce Investment Uncertainty	Provide funds to private industry for performing feasibility studies and cooperative agreements for alcohol production plants.	Funds for feasibility studies and cooperative agreements were provided by two Public Laws: Title II of the Department of Interior and Related Agencies Appropriations for Fiscal Years 1980, P.L. 96-126; and Department of Energy, Alternative Fuel Production Supplemental Appropriations, P.L. 96-304.
	Remove regulatory disincentives	Provide a cash transfer (entitlement) worth about 5 cents per gallon of alcohol to the national producer.	*Crude Oil Entitlements Program (43 FR 2429, May 18, 1978)
	Ensure adequate supply of feedstocks	Legislation provides authority to establish grain reserve to ensure adequate supply of grain feedstock materials for alcohol production.	The authority is provided in Public Law 95-279, Wheat, Feed Grains, and Upland Cotton, Emergency Assistance. It has not been utilized by USDA since a single reserve is maintained to provide grain stocks for feed, food, and export purposes.
	Address environmental issues related to alcohol production	Perform an Environmental Impact Assessment (EIA) of a 60,000 barrel per day production level.	The EIA is being performed in order to obtain compliance with the National Environmental Policy Act (NEPA).

*The Crude Oil Entitlements are being phased out.
This authority will expire at the end of September 1981.

Table 1-3. Summary of Near-Term Federal Strategy for the Utilization of Alcohol Fuels

OBJECTIVES	FEDERAL STRATEGY ELEMENT	FEDERAL ACTION	APPLICABLE FEDERAL LEGISLATION OR INIATIVE
Stimulate Utilization of Alcohol Fuels	Subsidize the marketing of alcohol fuel such as gasohol	Provide subsidy to the marketing of gasohol-type blends to make price competitive with gasoline.	Exemption of the Federal 4-cent-a-gallon Federal Excise tax on motor fuels which amounts to 40 cents per gallon of alcohol blended with gasoline. (Energy Tax Act of 1978 - P.L. 95-618 and Crude Oil Windfall Profit Tax Act of 1980 - P.L. 96-223).
	Increase public awareness and knowledge	Perform public dissemination of information and other consumer oriented programs to promote use of alcohol.	The Department of Energy, Office of Alcohol Fuels sponsors the National Alcohol Fuels Information Center located at the Solar Energy Research Center in Golden, Colorado. The Department of Agriculture will use existing organizations including the Agricultural Cooperative Extension Service.
	Mandate alcohol use	Require that motor vehicles which are owned or operated by Federal agencies and are capable of operating on gasohol, shall use gasohol where available at reasonable prices and in reasonable quantities.	Use of gasohol is mandated by Section 271 of the Energy Security Act, P.L. 96-294. The President will issue an Executive Order requiring the use of gasohol in Federal vehicles.
	Address technical, environmental, and safety issues.	Resolve technical issues related to utilization of alcohol fuels	The Department of Energy, Office of Alcohol Fuels is sponsoring technical programs at Bartlesville Energy Technology Center in Oklahoma to evaluate the use of gasohol-type blends, and other programs.

The alcohol production goals can be met if

1. DOE/USDA approval time is expedited,
2. Plants can become operational within 24 months,
3. Interest rates are reduced in the near term.

Expected contributions from recent federal legislation and existing authority are shown in Table 1-4.

The following sections discuss the available federal funding under the Energy Security Act and the USDA and DOE actions to use those funds to help meet the 1982 production goal.

Available Funds

Beginning October 1, 1980, the Energy Security Act authorized a two-year appropriation of $600 million to USDA and $600 million to DOE. The Act stipulates that these amounts represent the combined value of insured loans and loan guarantees and the contract value of possible price guarantees and purchase agreements. Neither the statute nor the appropriation statutes specifically restrict the allocation of funds among the various types of financial assistance. To date, both USDA and DOE have concluded that the benefits of loan guarantees and insured loans outweigh the potential advantages of price guarantees and purchase agreements. Accordingly, it was jointly decided not

Table 1-4. Summary of Alcohol Production Contributions From All Sources

CONTRIBUTIONS TO PRODUCTION GOALS (APPROPRIATED FEDERAL FUNDS, IN MILLIONS)	CALENDAR YEAR CUMULATIVE INSTALLED ANNUAL PRODUCTION CAPACITY (BY YEAR'S END) MILLIONS OF GALLONS					
	1980	1981	1982	1983	1985	1990
Production Resulting from Energy Security Act. Public Law 96-294 Loan Guarantees						
– DOE Programs ($740)[1]	0	0	50	350	(?)[4]	(?)[4]
– USDA Programs ($515)[2]	0	30	176	283	(?)[4]	(?)[4]
Production Resulting Directly from Federal Grant/ Loan Guarantee Program other than Public Law 96-294						
– DOE Programs[3]						
P.L. 96-126 ($56)	0	21	81	81	(?)[4]	(?)[4]
P.L. 96-304 ($72)	0	0	91	91	(?)[4]	(?)[4]
– USDA Programs Business & Industry ($364)[2]	0	50	143	263	(?)[4]	(?)[4]
Production by Private Industry Only. Assisted by Federal/State Excise Tax Waver and Investment Tax Credit	160	250	380	510		
TOTALS	160	351	921	1,578		

Source: NAFI

[1]Represents level of loan guarantee authority to be obligated to achieve 1983 production level. Leveraging authority will allow an additional 400 million for production which will be on-line in 1984.

[2]Includes appropriations for both alcohol and non-alcohol projects.

[3]Fifty-five feasibility studies totaling over one billion gallons per year of capacity are not included in estimates.

[4]Budget to be determined.

to implement price guarantees and purchase agreements at this time, but to concentrate instead on the insured loans and loan guarantee authorities. Therefore, Congress has appropriated $525 million each to USDA and DOE to be used for loan guarantees and insured loans.

Consistent with the provisions of the Energy Security Act, USDA will direct a total of $525 million to small- and intermediate-scale plants, less administrative costs of $10 million. Small-scale plants are those of less than 1.0 million gal/yr capacity and intermediate-scale plants are defined as being in the range of greater than 1.0 and less than 15 million gal/yr. DOE has allocated $525 million appropriated by Congress (less administrative costs of $10 million) to large-scale biomass energy plants, i.e., those plants with an anticipated production capacity equal to or greater than 15 million gal/yr.

If funds appropriated for Title II are not used before October 1, 1982, the statute allows for their carry-over into future years until expended, provided they are committed by September 30, 1984. The President signed Public

Law 96-514 on December 12, 1980, which provides that funds appropriated for implementation of the Energy Security Act may be leveraged for loan guarantees on a 3 to 1 basis. In essence, USDA and DOE can now guarantee three times as much in loans for biomass energy development with the same level of appropriations. This added capability will further promote increased production investment with perhaps proportionately more emphasis on nonalcohol biomass projects.

Division of Funds Within USDA

USDA was not allocated a specific level of loan guarantee funds between alcohol- and nonalcohol-fuel biomass projects. Since the application procedure for different types of biomass energy projects is essentially the same, each project will be evaluated on criteria established in the Final Rule (*Federal Register*, October 30, 1980), including the encouragement of those projects using alternative feedstock materials.

The statute provides that up to one-third of the USDA funding may be reserved for small-scale biomass energy projects. Such a division would allocate $175 million of the current appropriation to small-scale projects and $350 million to intermediate-scale projects, less administrative costs.

An analysis of inquiries to the Farm Home Administration (FmHA) concerning loans for biomass projects under previously existing authorities showed that only about 9.5 percent of the activity was for loans or guarantees for small-scale alcohol-fuel projects. Given this level of activity, it is expected that the great majority of applications under Title II will be for proposed projects of 1 to 15 million gallons annual capacity. It appears unlikely that FmHA would use the entire $175 million for small-scale projects before exhausting the remaining two-thirds ($350 million) for intermediate- and large-scale projects. Therefore, it has been tentatively decided to allocate approximately $150 million for small-scale projects, $365 million for intermediate-size projects, and $10 million for administrative costs. So little is known about the possible demand that the split of funds between small- and intermediate-scale projects is necessarily arbitrary and may need to be revised later. It may also be that the amount reserved for administrative expenses is more than needed, in which case the excess will revert to use as loan guarantees.

Division of Funds Within DOE

To date, given leveraging authority provided for in Public Law 96-514, $157.5 million will be available to fund biomass energy projects, and $1,140 million will be available to fund alcohol-fuel projects. Of this amount, at least $400 million will be available to fund projects under a new solicitation in early 1981, which would result in on-line production in 1984.

Loan Application Procedures

Title II of the Energy Security Act allows a choice of procedures for soliciting and processing loan applications, provided that whatever procedure or combination of procedures is selected will result in the processing of applications "expeditiously." It also provides that projects should be categorized by size and that simple application procedures be adopted for small-scale producers.

USDA Small- and Intermediate-Scale Projects

FmHA is receiving and processing applications on a modified continuous basis. The agency will officially receive applications on a series of specified dates, the first of which was on November 17, 1980. The second receipt date was December 17, 1980, with applications officially received every 60 days thereafter, until funds are expended. A decision will be made on each application within 120 days of its official receipt date.

Applicants are encouraged to participate in preapplication discussions and conferences with FmHA prior to the submission of an application; the specific application receipt dates are intended to provide a period for such discussions and conferences to occur in a timely manner.

DOE Large-Scale Projects

The DOE is using a batch review of proposals, which requires several steps. Initially, notification of available funds is made public through solicitation. Following the close of the solicitation period, there is an evaluation cycle. A team of evaluators reviews and ranks the technological, environmental, business, marketing, economic, and financial sections of the proposals. The initial selections are made following the application of program policy factors. Following the selection procedure, further discussions may be held with the selectees. Final selection depends on departmental judgment made consistent with the program regulations. Final selection is signified by conditional commitment after which negotiations begin. After completion of negotiations, guarantees will be approved for the winners and construction may begin.

The evaluation cycle generally takes 3 to 6 months depending on the number of applicants; the negotiation process takes approximately 3 to 6 months; and design/construction lead times for large-scale plants are estimated to take from 24 to 36 months.

Schedules

The optimistic schedules for USDA and DOE loan guarantee programs under Title II are presented in Fig. 1-16. The figure shows the different

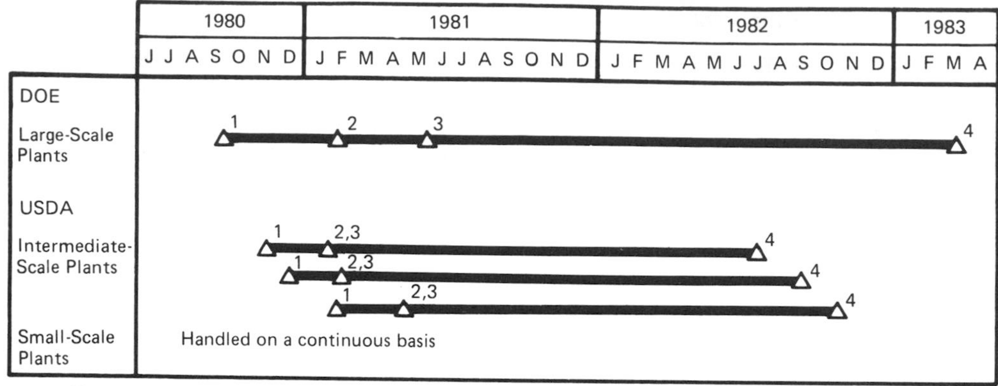

Key 1 Solicitation closed
2 Initial selections announced
3 Negotiation complete
4 Construction complete, production on-line

Source: NAFI

Figure 1-16. Schedules for loan guarantee awards by USDA and DOE under Title II of the Energy Security Act.

phases of the programs for the large- and intermediate-scale projects. Negotiations of intermediate-scale projects of the USDA take place prior to the award announcement. Small-scale projects are handled on a continuous basis and are therefore not delineated on the figure.

Eligibility

Priority will be given by both USDA and DOE to biomass energy projects using a primary fuel other than petroleum or natural gas in the production of fuel. These primary fuels include geothermal and solar energy, waste heat, coal, wood, wood wastes, crop residue, and waste. Priority will also be given to projects using new technologies that use different biomass feedstock, or produce new forms of biomass energy, or produce biomass fuel using improved or new technologies. This priority does not, however, necessarily exclude from financial assistance a project not using an alternative primary fuel or not applying a new technology.

The British thermal unit (Btu) content of motor fuels used in the project must not be greater than the Btu content of the biomass fuel produced. This applies only to the biomass energy process plant and excludes motor fuels used in the production and transportation of feedstocks. Any displacement of motor fuel or other petroleum products which occurs after the biomass fuel is produced may also be considered.

For a project to receive financial assistance, the USDA must find that necessary feedstocks are available, and that it is reasonable to expect they will continue to be available in the future. A project which extracts the protein content of the feedstock for use as food or feed for readily available markets is preferred by USDA.

Financial Terms

DOE and USDA may guarantee, against loss of principal and interest, loans made to provide funds for the construction of alcohol-fuel projects. USDA will approve loan guarantees for permanent financing after construction is complete. The following provisions apply to all loan guarantees made by both departments for alcohol-fuel and biomass energy projects:

- A guaranteed loan may be for up to 90 percent of the project's estimated construction costs and the amount guaranteed cannot exceed 90 percent of the loan.
- An insured loan cannot exceed 90 percent of the total estimated construction of the project. (Only USDA is providing insured loans.)
- In the event that total estimated construction costs exceed the originally estimated costs, the department may guarantee an additional loan for an amount up to 60 percent of the difference between the currently estimated costs and the total costs originally estimated.
- The borrower must establish that, without the department's guarantee, the lender is unwilling to extend credit at reasonable rates and terms for construction of the project. The need for the guarantee will be decided by departmental judgment.
- The lender must bear a reasonable degree of risk in the financing of the project. This is to ensure (1) that the lender fully participates in the financing, (2) that the lender will fully evaluate and scrutinize the loan for viability, and (3) that the lender will fully service the loan during the life of the loan.
- In the event that the department determines that the borrower is unable to meet payments but is not in default, then the department may elect to pay to the lender the amount of principal and interest the borrower is obligated to pay. However, the borrower must first agree to reimburse the department on terms and conditions the department deems necessary to protect the financial interests of the United States.

To obtain more information on current gasohol planning, contact the respective organizations in the USDA and DOE.

RESEARCH AND DEVELOPMENT

During the past several years, attention has been placed on producing fermentation ethanol. This area has had grass roots support, and both small-scale and large-scale plants have rapidly become operational. There are limits to fermentation ethanol production, however; some believe that 2 to 10 billion gallons per year is the limit. In Fig. 1-17, we show that fermentation ethanol is the only near-term choice to produce alcohol. However, other

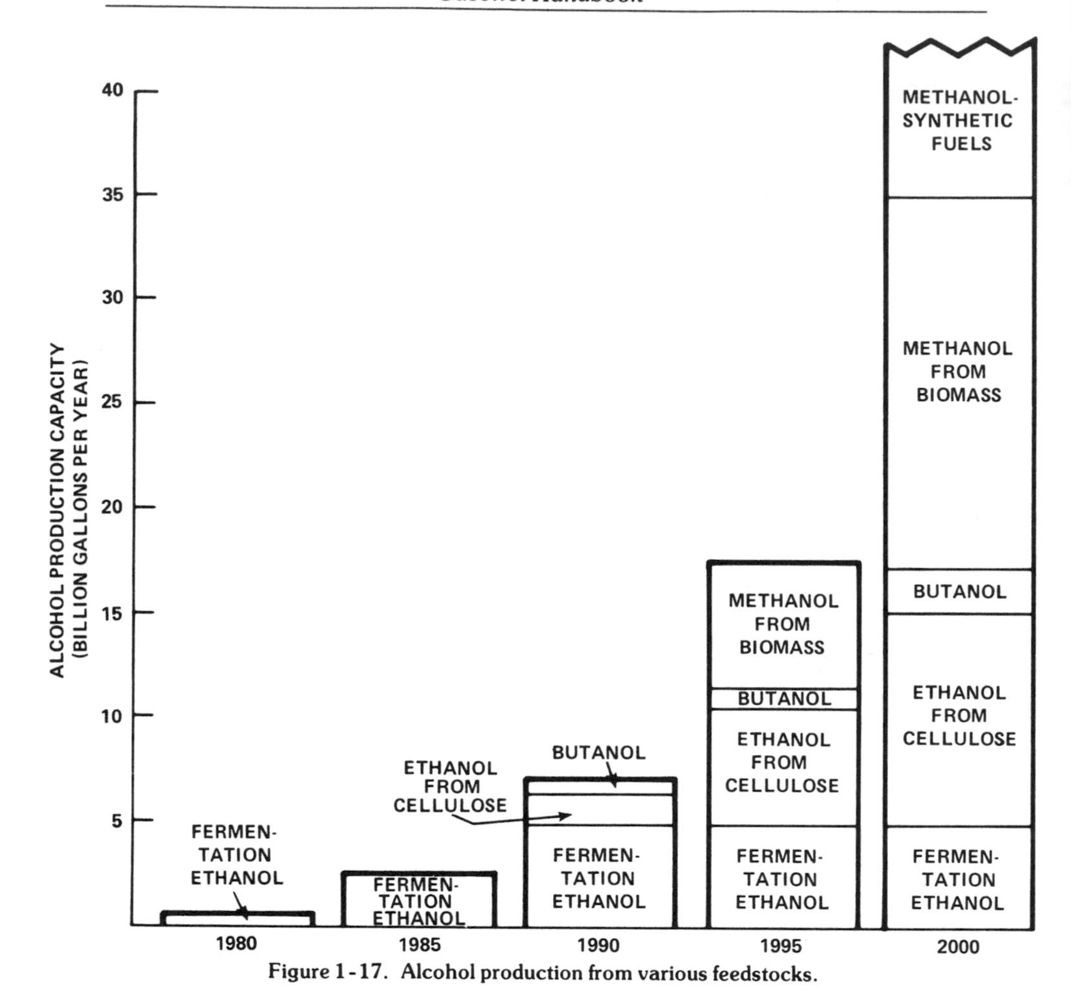

Figure 1-17. Alcohol production from various feedstocks.

alcohol synthetic fuels have considerable potential beginning in the 1990s. In order to develop these process technologies, a concerted research and development program stressing cellulose conversion, feedstock improvement, stillage marketing, and methanol environmental and process developments must be funded now.

Significant Issues

Gasohol initially encountered considerable resistance from the petroleum and automobile interests. However, they now seem to accept the fact that gasohol (1) is viable, (2) performs satisfactorily, and (3) can make a real contribution to octane improvement. However, several key issues need to be resolved:

- Vehicle performance with gasohol
- Food vs fuel
- Availability of lignocellulose technology
- Energy balance

These four issues are discussed briefly in the following sections.

VEHICLE PERFORMANCE WITH GASOHOL

The differing performance of ethanol and ethanol-gasoline blends in conventional spark ignition engines compared to straight gasoline can be attributed to their differences in chemical and physical properties. While gasoline is a mixture of about 4-12 carbon atom hydrocarbons, ethanol is a single compound with uniquely and narrowly defined properties. Differences in engine performance and system compatibility between gasoline, ethanol, and ethanol blends can be attributed predominantly to flash point, boiling point, octane quality, heat of vaporization, heating value, stoichiometric air/fuel ratio required for combustion, and water solubility. As a greater percentage of ethanol is added to straight gasoline, the deviation of characteristics is approximately proportional to the percentage of ethanol.

The octane-boosting properties of ethanol in gasoline were particularly attractive at a time when higher octane lead-free gasolines were in short supply and other octane enhancers such as MMT (methylcyclopentadienyl manganese tricarbonyl) and lead are under restrictions. For example, a 3 percent ethanol addition increases the octane by 2 to 3 points, depending on the composition of the gasoline. Ethanol has been permitted by the Environmental Protection Agency for use as a gasoline additive under Section 211(f) of the Clean Air Act. Two other chemicals, TBA (tertiary butyl alcohol) and MTBE (methyl tertiary butyl ether), also have been permitted as octane enhancers. However, both are currently made largely from petroleum.

Performance of Ethanol-Gasoline Blends in Spark Ignition Engines

A number of gasohol fleet tests have been run to evaluate performance of the ethanol-gasoline blend under normal driving conditions. In addition, a number of investigators have compiled technical information or performed laboratory and performance tests to evaluate various proportional blends of gasoline and ethanol. The gasohol fleet tests so far have not been performed under the degree of rigorous controls and testing procedures that would ensure complete accuracy. However, none of the fleet testing which has been performed indicates any major problems with the use of gasohol in normal automotive use. The DOE has received the results of extensive tests conducted by the state governments of Illinois, Nebraska, and Iowa and also by the American Automobile Association. These tests indicate that the great majority of unmodified vehicles tested ran as well or better with gasohol than with fuel previously used. Specifically

- The American Automobile Association test showed that 88 percent of the vehicles ran as well or better on gasohol, which suggests that 12 percent of the vehicles showed negative results and would require some minor modification for a comparable or improved performance on gasohol.
- Illinois state government officials indicate that their test, which has used approximately 1800 state vehicles including state police cars, has yielded positive results in each category tested.
- The Iowa Development Commission (a state agency) conducted a 90-day gasohol marketing (customer opinion) test from June 15, 1978, to September 15, 1978, in which 232,608 gallons of gasohol were sold at five stations (also offering unleaded regular) across the state. Results show: (a) 67 percent of users reported improved performance, (b) three of every five users were repeat customers, (c) 90 percent of users would purchase gasohol if it were available at most stations, and (d) gasohol outsold unleaded regular 3.9 to 1.
- The state of Nebraska ran state vehicles 2 million miles on gasohol from December, 1974, to October, 1977, in different altitudes and under varying weather conditions. Their published results on performance state that drivers reported no starting problems, vapor lock, or driveability problems. Examination of the engines during and after the program revealed no unusual engine wear or carbon buildup.

Gasohol is being used in automobiles without modification. With normal precautions to maintain a moisture-free blend during refining/mixing, during transportation, and at the service station, the tendency of the ethanol-gasoline mixture to undergo phase separation can be minimized. The presence of alcohol in the gasoline increases the water tolerance of the gasoline. Although, only a maximum of a few parts per million of water will mix freely with gasoline, a gasohol mixture will tolerate nearly 0.25 percent water (depending on temperature) before phase separation takes place. Because most automobiles have some small percentage of water in their gas tanks, the use of anhydrous ethanol will minimize phase-separation problems.

Certain minor driveability problems can occur when using ethanol-gasoline blends in engines which are set up for optimum performance with gasoline. With gasohol, under certain conditions, driveability problems may include the following:

- Greater tendency of stalling and stumbling during engine warm-up
- Increased tendency toward hesitation during acceleration for late model cars

- Increased vapor-lock tendency in hot climates or at high altitude

Leaning of the air/fuel ratio in a gasoline engine is known to impair driveability. Driveability problems resulting from a lean air/fuel mixture can include increased engine stalling and stumbling during engine warm-up, increased tendency to hesitate during acceleration, and increased tendency for engine surge during constant speed driving. The use of ethanol in automotive fuel leans the air/fuel mixture because the ethanol molecule contains oxygen and, therefore, increases the oxygen content in the engine combustion chamber. In late model vehicles which are adjusted to run lean in order to minimize pollution, the further leaning induced by use of blends could cause driveability problems. The same tendency is not typically present with older vehicles which have a richer initial air/fuel ratio.

A quantitative measure of warm-up driveability indicates that the use of a 10 percent by volume ethanol blend can be expected to reduce warm-up driveability of standard (i.e., nonadjusted and nonmodified) cars with carburetors. Furthermore, at 22 percent ethanol, warm-up driveability was degraded significantly in three cars which were tested recently. In the test, the three cars were tested by a procedure which measures driveability with a total weighted demerit (TWD) value. (Higher values of TWD indicate poorer engine driveability.) Carburetor modifications performed on these cars to achieve an air/fuel ratio for the blend which is equivalent to straight gasoline will result in normal driveability.

For engines which are adjusted or modified to operate at equivalent air/fuel ratios, the fuel economy differences between gasohol and gasoline are apparently negligible. However, the leaning effect of ethanol may result in fuel economy or exhaust emission effects ranging from significant increases to significant decreases, depending on the adjustment of the engine. An analysis of state and private test data received by DOE indicates that use of a 90/10 blend of unleaded regular gasoline and ethanol (compared to 100 percent unleaded regular) resulted in similar miles per gallon or a small increase in mileage by a majority of the vehicles tested. A substantial mileage increase or decrease was found with very few vehicles. Differing results in mileage are attributable to the differing age, size, condition and adjustment, weather conditions, and quality of ethanol and gasoline.

Effect of the Use of Ethanol Gasoline Blends on Engine Emissions

On December 16, 1979, EPA approved the use of gasohol under Section 211(f)(3) of the Clean Air Act of 1977 and found that there was no significant environmental risk associated with the continued use of gasohol. Furthermore, new emissions control systems, such as the "threeway catalyst with exhaust oxygen sensors for carburetion feedback for air-fuel control," have been shown to be equally effective using either gasoline or gasohol.

EPA and DOE have conducted a cooperative gasohol testing program to obtain and evaluate environmental impact data. On the basis of these tests, EPA concluded that the addition of 10 percent ethanol to gasoline:

- Slightly decreases hydrocarbon emissions
- Significantly decreases carbon monoxide emissions
- Slightly increases nitrogen oxides emissions
- Increases evaporative hydrocarbon emissions

The results to date have generally been favorable with respect to the use of gasohol in automobiles. However, in a recent technical memorandum, the Office of Technology Assessment of the Congress of the United States stated that the "mixture of observed emissions reductions and increases, and the lack of extensive and controlled emissions testing, does not justify a strong value judgment about the environmental effect of gasohol used in the general automotive population (although the majority of analysts have concluded that the net effect is unlikely to be significant)."

Compatibility of Ethanol/Gasoline Blends with Automotive Fuel/Engine Systems

Experience with gasohol has indicated that the solvent properties of ethanol loosen corrosion and dirt from the walls of fuel tanks and fuel lines of automobiles. This makes it advisable to flush and dry all storage tanks used with ethanol-gasoline blends. Vehicle tanks, particularly with older vehicles, should be flushed with ethanol or gasohol, and the fuel filter may require replacement after the first or second tankful. The use of neat ethanol or ethanol blends may potentially cause minor problems with corrosion of metal fuel system materials, particularly aluminum, copper, iron, lead, and zinc. In addition, clear polymid, used in fuel systems for such items as fuel filter housings, has been reported to degrade in service with ethanol blends.

FOOD vs FUEL

There has been considerable discussion and argument concerning the competitive use of grains in food and fuel markets. These arguments and discussions pertain to a conflict which should not exist. A brief examination of the animal feed-human food cycle, and of the part which grains play in the cycle, will clarify the issue.

Most of the grain produced for domestic use or export, particularly corn, is used (Fig. 1-18) for animal feed, primarily to develop production of protein as the animal grows; this applies to feeding cattle, hogs, or poultry. In such feeding, the protein is of far more value than the starch. This can be verified readily by comparing the price of corn with the prices of such high-protein feeds as soybean meal, fish meal, meat scraps, and yeast. The relative prices of these feedstocks are somewhat proportional to their protein content.

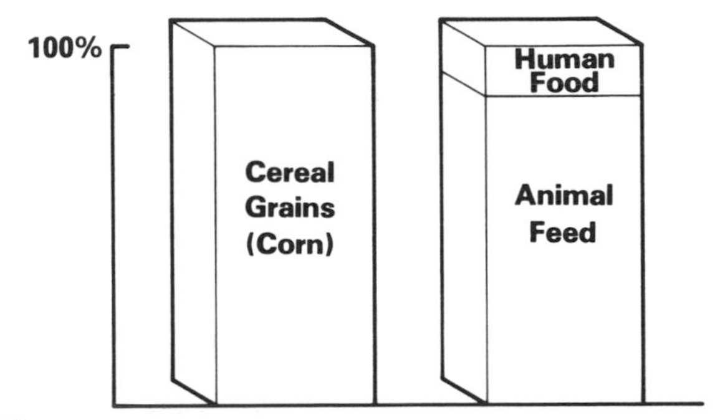

Source: NAFI

Figure 1-18. Most cereal grains (corn) are used for animal feed.

Distiller's dried grains (DDG) and distiller's dried grains with solubles (DDGS) are in the same category as alternate protein feedstocks. To show that there is no conflict can best be conceived by visualizing (Fig. 1-19) part of the United States's grain crop passing first through fuel-grade ethanol plants for removal of most of the starch, yielding a more-concentrated protein feedstock in the form of DDG and DDGS. When ready for use as feed, the DDG and DDGS can be supplemented with local forage or silage high in

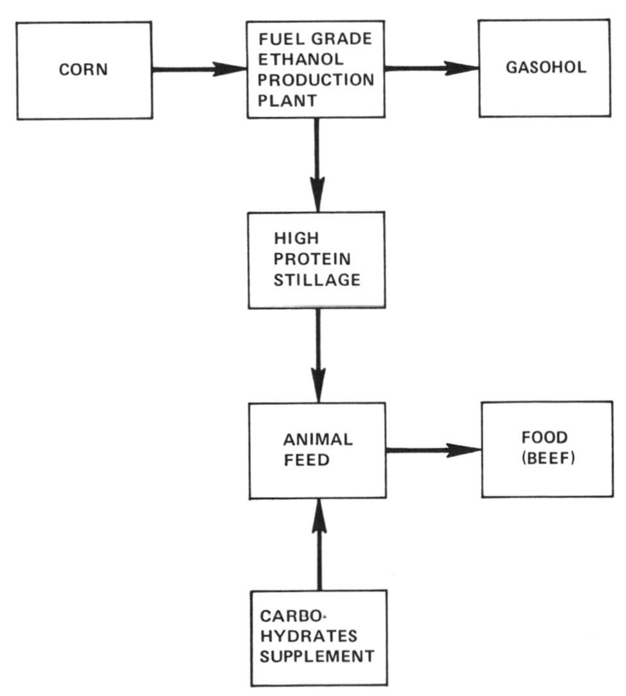

Source: NAFI

Figure 1-19. Use of corn for food and fuel.

starch. In fact, there is economic benefit in utilizing DDG or DDGS rather than corn, when the animal feed has to be shipped considerable distances, particularly for the export market. DDG and DDGS are more-stable materials than the original corn, and can be shipped more readily and stored over long periods of time. Furthermore, because the protein content of DDG and DDGS is approximately three times that of the original corn, freight and storage costs are incurred on only one-third the weight of material, which makes these products much more economical for the feed mixer and ultimate user than grain itself. Also, export of the DDG does not require the same duty payments as export of corn.

In the normal development of yeast to produce alcohol, approximately 9 percent of the sugars produced by saccharification of starch (or available directly as sugars, as in the case of cane or beets) is consumed in growth of yeast cells. This is the well-known Pasteur efficiency. This means that for each gallon of fuel alcohol produced, approximately 0.5 pound of yeast is grown, with a protein content of about 50 percent. The nitrogen content of the protein is provided primarily by fermentation nutrients, such as ammonia. However, where protein-containing grains are saccharified and fermented, some of the nitrogen from the grain is utilized by the yeast cells for their growth.

It is difficult to obtain accurate data from distillery operations to measure the actual protein balance. However, an approximation can be made from analyses of feedstock and DDGS. Corn, weighing 56 pounds per bushel at 15.5 percent moisture content, will contain approximately 10 percent protein on a dry basis, equivalent to 4.73 pounds of protein/bushel. DDGS made from such corn averages 29 percent protein, with 10 percent moisture. With a yield of approximately 17 pounds of DDGS/bushel of corn, the protein content of the DDGS is calculated at 4.93 pounds. Thus, there is a net gain of about 0.2 pound of protein for every bushel of grain processed. The Pasteur efficiency calculation would indicate a yeast protein addition of 0.6 pound per bushel, but this assumes that all of the yeast protein is grown from chemical nutrients, and that no protein nitrogen of the grain is consumed in yeast growth. The rationale of this analysis is that there is no loss in protein from grains processed through a fuel-grade ethanol plant. In fact, there is probably a modest gain in protein content. Furthermore, the yeast protein developed during fermentation involves production of vitamin B complex, which is beneficial to livestock.

There should be little conflict in the marketplace as DDG and DDGS become established products. However, with a national gasohol program which is targeted for a 10 percent blend nationwide by 1990, and which could go beyond that to a 20 percent blend with further economic benefits to the country, there will obviously be a shift in animal feed markets. The primary shift will be that much of the grain, primarily corn and milo which goes to the animal feed market, will first pass through fuel-grade ethanol plants. As indicated above, the full protein content of the grain will then be available in the form of DDG and DDGS, with some slight augmentation due to yeast growth. Should the world demand for grains increase, as appears likely, additional corn will be planted. This may reduce the requirement for soybeans, and particularly soybean meal, since DDG and DDGS sell at a lower price per unit of protein content. Generally, farmers who grow corn also grow soybeans as a rotational crop. They would shift their proportion of production to favor corn. Other high protein feedstuffs, such as meat scraps and brewer's yeast, would be of higher value only in special feeding formulas.

Of particular interest is the potential growth in export markets for DDG and DDGS, which will probably be much lower in cost per unit of protein than grains grown in western Europe and Japan. Fuel-grade ethanol plants which cannot market all of their DDG or DDGS locally could sell surplus material to grain and feed brokers who are active in export sales of animal feeds.

To the extent that grains such as corn are used as feedstocks for ethanol production, there may be potential competition in the long term (1990s) between using crops for food (or feed, fiber, or exports) and using these same crops for fuel-grade ethanol production. However, in reality, the choice will not be to produce food or fuel, but rather to produce fuel-grade ethanol. Most farmers will not consciously choose to produce grain or other products specifically for food or for fuel, although there will be exceptions. They will produce wheat, corn, grain sorghum, or other crops on their farms and sell those products in the open market. What they produce and offer for sale will depend on their judgment of the prices for and the profits available from the various crops that are adapted to their area. Market prices offered by competing users for raw materials with multiple end uses will determine, to a large extent, the relative shares of agricultural land, labor, and capital that will be used for various purposes, and whether or not production of some agricultural commodities will increase materially. Production of agricultural products under contract with ethanol producers is an additional possibility, particularly where specialized processing facilities require an assured local supply or during periods when grain supplies are expected to be very limited.

Concern over the use of agricultural resources to produce fuel is a different issue in the world context than in the U.S. context. Clearly, the preemption of large land areas now producing food for fuel production may cause serious food supply problems in some countries if food imports are not increased. The issue is very different for Brazil or India or low-income developing nations that are net food and fuel importers than for the United States or Canada, which already have high meat diets, large grain exports, and substantial capability for increasing or changing the mix of agricultural production in a few years.

Potential users of agricultural products are always competing for supplies of raw materials. Private choices comparable to the "food vs fuel" dilemma now being posed include those made between using land for soybeans or cotton, crops or woodlands, farms or suburbs. The major policy issues that will have to be addressed sometime during the 1980s, unless grains, cultivated land, and other agricultural resources are again chronically in surplus supply, will be how to manage whatever competition develops among alternative uses of farm products if and when grain supplies are limited, and how to develop an alcohol fuels industry concurrent with the development of other alternative fuels, with an appropriate balance among optional feedstocks, including those that could be produced on lands that are currently not in cultivation. Development of cellulose-production processes and methanol technology demonstration will allow increased alcohol production while not increasing the impact on food production.

Difficult policy choices will arise if the United States must consider limiting its coarse grain exports to Europe, China, and the Soviet Union in order to increase production of alcohol fuels at home, but no serious ethical considerations need to be faced. Importing countries would resist any such limitation on exports of feedstuffs, since it would interfere with their food-consumption policies. They might even retaliate against the United States in some way. People in these countries will not be hungry, however, simply because they may have less meat as a result of increased ethanol production in the United States.

Allocation decisions of this type were being made long before energy or gasohol became a major issue in the United States and other countries. Meat diets in rich and middle-income countries now require large amounts of grain, causing grain prices to be higher than they would be if less grain were fed to livestock, thereby inexorably raising the food bill of developing nations. The issue is not to avoid all possibility of competition between use of agricultural resources for food, fiber, lumber, or fuel, but to design and follow an appropriate strategy for development of an alcohol fuels industry consistent with overall energy policy and with comprehensive food and trade policies as well. Minimizing and managing any serious competition that arises between the energy and food sectors will simply be a new aspect of an old policy problem, not a new problem.

The DOE and the USDA should develop a national policy for the long-term development of fuel-grade ethanol production and export of cereal grains. These decisions must be based on the best interests of the United States. Often our farm policy has stressed balance-of-payment gains rather than our national security and survival of America's small farmer.

AVAILABILITY OF LIGNOCELLULOSE TECHNOLOGY

There is a significant research and development effort nationwide and abroad—in industry, government, and university laboratories—to develop improved technology for conversion of cellulosic wastes to sugar which, in turn, can be fermented to fuel-grade ethanol. In the United States this effort is directed toward annual utilization of hundreds of millions of tons of agricultural and forest residue and wastes, which now have little or no economic value. In addition, these waste raw materials would not compete with human food production or fossil fuel. Therefore, it would be a real advantage to utilize these wastes as a feedstock for fuel-grade ethanol production.

It is anticipated that the technical development of processes utilizing either acid or enzyme hydrolysis, or a combination thereof, can lead to reasonable sugar yields and economic alcohol production from cellulose within the next 5-10 years (Fig. 1-17). The key to commericalization, however, will depend on development of economic methods for collecting the widely diffuse raw materials, processing them for transportation and storage, and developing collection and transportation systems over large areas. Facilities for producing fuel-grade ethanol from grain should consider in their design auxiliary installations which can later be made to process municipal, agricultural, or forest wastes. To retrofit an existing grain fuel-grade ethanol plant would require replacing the grain raw material preparation section by equipment to pretreat and sterilize the cellulosic wastes, and modifying the cooking and saccharification section using purchased enzymes. Fermentation and distillation sections should require little or no modification. Boiler facilities may have to be expanded to accommodate steam requirements for sterilization, provided normal 15 psi steam can be used to sterilize the waste cellulosic feedstocks.

The current issue regarding cellulosic conversion can be resolved by aggressive federal (USDA and DOE) support of research and development efforts to bring cellulosic conversion systems on earlier than 1990.

ENERGY BALANCE

Considerable public interest and debate have been focused on the so-called "energy balance" issue involved in the conversion of biomass materials into ethanol for fuel use. Some people have expressed concern that the production of fuel ethanol requires more energy than is present in the ethanol produced. Although such concern is understandable, it often overlooks some principles applying to all energy conversion processes.

By necessity, any energy conversion process—for example, generation of electricity from coal or refining of gasoline from crude petroleum—reduces the total energy that is eventually available to consumers in a usable form. Thus, a coal-fired power plant that is only 33 percent efficient is considered acceptable because it transforms coal to a more useful form of energy, namely, electricity.

The essential question (Fig. 1-20) that must be asked is, "Does the production of ethanol achieve a net gain in a more desirable form of energy?" Put more simply, can the production of ethanol and its use as a motor fuel or

Source: NAFI

Figure 1-20. Fermentation ethanol provides a net liquid fuel gain.

chemical feedstock reduce the need for imported petroleum in this country? This handbook addresses these questions through calculations of the net gains in premium fuels that can be derived from the production and use of ethanol from biomass, and shows that, for the U.S. alcohol fuel program, "energy balance" need not be a concern because:

- Efficient processes have notably reduced the energy needed to produce ethanol fuel.
- Ethanol fuel used in gasohol can replace more liquid fuel than is consumed in its production.
- Using fuels such as coal or wood in producing ethanol effectively converts these abundant energy sources into premium liquid fuels. In addition, economics and national energy policy both discourage ethanol producers from using oil as a process fuel. Thus, producing ethanol fuel can substantively reduce net U.S. oil imports.

The calculation of a net fuel gain was chosen as the criterion on which to judge the energy efficiency of the production and use of biomass-derived alcohol. By definition, a net fuel gain is achieved when the savings in premium fuels obtained by using alcohol as a fuel or feedstock exceeds the investment of premium fuels in the alcohol production cycle.

Three categories of fuel gain are discussed in this handbook:

- Net petroleum gain
- Net premium fuel gain (petroleum and natural gas)
- Net energy gain (for all fuels)

Savings in premium fuels include the direct savings in reduced use of the premium fuel itself, plus the indirect savings in premium fuels that would be used to produce the conventional premium fuel which has been replaced. Indirect savings in premium fuels may also occur if, for example, fuel economy obtained with gasohol were greater than that for unleaded gasoline.

In this handbook, the investment of energy (in the form of premium fuels) in alcohol production includes all investment from cultivating, harvesting, or gathering the feedstock and raw materials, through conversion of the feedstock to alcohol, to the delivery to the end user.

To determine the fuel gains in ethanol production, six cases (see Table 1-5) encompassing three feedstocks, five process fuels, and three process variations were examined. For each case, two end uses (automotive fuel use and replacement of petrochemical feedstocks) were scrutinized. The end uses were further divided into three variations in fuel economy and two different routes for the production of ethanol from petrochemicals.

Energy requirements calculated for the six process cycles accounted for fuels used directly and indirectly in all stages of alcohol production, from agriculture through distribution of product to the end user. Energy credits were computed for coproducts according to the most appropriate current use. Energy credits for the following conversion process coproducts were accounted for:

- Distiller's dried grains, fusel oil, and ammonium sulfate (derived from the flue gas desulfurization system) when coal is the process fuel (basic Katzen design)
- Distiller's dried grains and fusel oil when residual oil or natural gas is the process fuel (two excursion cases)

It should be noted that energy credits for DDG were calculated on a protein equivalence basis with soybean meal produced in Illinois. Credits for DDG are thus based on energy investment over the complete soybean meal production cycle.

Table 1-5. Options Considered for Production and Use of Ethanol

BIOMASS FEEDSTOCK	CONVERSION PROCESS	PROCESS FUEL	END-USE
Corn	Katzen*	Coal	Gasoline Blending
Corn	Katzen	Coal	Chemical Feedstock
Corn	Katzen	Residual Oil	Gasoline Blending
Corn	Katzen	Residual Oil	Chemical Feedstock
Corn	Katzen	Natural Gas	Gasoline Blending
Corn	Katzen	Natural Gas	Chemical Feedstock
Corn	Katzen	Coal/Coal-derived Ammonia	Gasoline Blending
½ Corn/½ Sweet Sorghum	Katzen	Bagasse	Gasoline Blending
½ Corn/½ Sweet Sorghum	Katzen	Bagasse	Chemical Feedstock
Municipal Solid Waste (Cellulose)	Gulf**	Electricity	Gasoline Blending
Municipal Solid Waste (Cellulose)	Gulf	Electricity	Chemical Feedstock

*An energy-efficient plant design presented in "Grain Motor Fuel Alcohol Technical and Economic Assessment Study," Raphael Katzen Associates, December 1978, Department of Energy Report No. HCP/J6639-01.

**An integrated cellulose conversion process developed by Gulf Oil Chemicals Co.

A commercial-scale, energy-efficient coal-fired plant for the production of motor fuel-grade ethanol from corn (Katzen design) was the basis for all calculations in the base case. This was done because basic economics driven by energy costs and federal tax and loan guarantee requirements ensure that energy-efficient designs will be incorporated into new full-scale ethanol plants. Specifically, the Windfall Profit Tax Act will provide a 10 percent energy tax credit for equipment that converts biomass into alcohol fuel only if the equipment producing the alcohol uses an energy source other than oil, natural gas, or a product of oil or natural gas. Additionally, under the Energy Security Act, biomass energy products using a primary fuel other than petroleum or natural gas will be given priority for financial assistance. In the calculations, the plant was assumed to be located in Illinois, corn was produced locally, and process heat requirements were met by Illinois No. 6 coal. Fuel investments during the agricultural production cycle and for such uses as transportation of feedstocks, fuels, and finished products were thus site-specific. Consequently, comparisons with plants located elsewhere should be made with caution.

Excursions on the base case included two fuel substitutions and one material input change. In the first excursion, residual oil was substituted for coal. In the second, natural gas was substituted for coal. Finally, the base case was modified to include the use of ammonia derived from coal for fertilizer production and process needs.

A plant designed to process corn for one-half of the year and sweet sorghum for the other half was also considered. The primary fuel was the bagasse residue obtained from the sweet sorghum feedstock.

Finally, the production of ethanol from municipal solid waste (primarily cellulose) by the Gulf process was examined. The supplemental fuel, used for process heat requirements in excess of those met by the lignocellulosic process residue, was electricity.

Considering only the fuel investments in the six processes described in this report, and temporarily excluding end use, the cellulose process was found to require the least petroleum fuel investment; the corn/residual oil case required the most petroleum.

Net fuel gains for the six ethanol production cases with the ethanol used as automotive fuel are summarized in Table 1-6. The most noteworthy conclusion to be drawn from the table is that a net energy gain is obtained for all six production cycles even when the worst case fuel economy situation is used.

For the base corn/coal case with ethanol used as a motor fuel, if mileage with gasohol is assumed to be at the middle of the range considered in this handbook ("mileage equal"), the net petroleum gain is equivalent to 0.83 gallon of crude oil per gallon of ethanol produced. The premium fuel gain is 0.81 gallon. The net energy gain for this case is almost one-half gallon of petroleum equivalent for each gallon of ethanol produced. The premium fuel

Table 1-6. Net Fuel Gains Including Credit for Coproducts with Biomass-Derived Ethanol Used as Automotive Fuel Blending Stock* (Gallons of Crude Oil per Gallon of Ethanol)**

PROCESS TYPE / NET FUEL GAINS	NET PETROLEUM GAIN			NET PREMIUM FUEL GAIN (Petroleum & Natural Gas)			NET ENERGY GAIN (All Fuels)		
	GASOHOL MILEAGE 4% LESS	MILEAGE EQUAL	GASOHOL MILEAGE 4% GREATER	GASOHOL MILEAGE 4% LESS	MILEAGE EQUAL	GASOHOL MILEAGE 4% GREATER	GASOHOL MILEAGE 4% LESS	MILEAGE EQUAL	GASOHOL MILEAGE 4% GREATER
PROCESS TYPE									
CORN PROCESS									
— Coal as Process Fuel	0.46	0.83	1.21	0.40	0.81	1.22	0.06	0.49	0.93
— Resid as Process Fuel	0.18	0.56	0.93	0.11	0.52	0.93	0.06	0.50	0.93
— Gas as Process Fuel	0.46	0.83	1.21	0.08	0.49	0.90	0.03	0.47	0.90
— Coal as Process Fuel & Ammonia Derived from Coal	0.46	0.84	1.21	0.51	0.92	1.33	0.06	0.50	0.93
CORN AND SWEET SORGHUM PROCESS									
— Bagasse as Process Fuel	0.42	0.79	1.17	0.34	0.75	1.16	0.30	0.74	1.17
CELLULOSIC PROCESS									
— Electricity as Process Heat	0.52	0.90	1.27	0.57	0.98	1.39	0.25	0.68	1.12

*Gain = (Saving — Investment), with ethanol considered just as a gasoline extender. By-products considered for credits from biomass — ethanol production are fusel oil, DDG and ammonium sulfate as appropriate. Fuel savings at end-use excluding by-product credits were not much different from those shown above.

**Based on conversion of calculated fuel gain in Btu to crude oil equivalent at 138,000 Btu/gal.

gain of 0.98 gallon for the cellulose-to-ethanol route is the most favorable of the six cases considered under the equal mileage assumption.

For the less favorable fuel economy situation, the petroleum gain is approximately one-half gallon for all cases except the corn/residual oil case. Although the net energy gain calculated for all fuels is small (0.06 gallon or less), the magnitude of the petroleum is impressive.

If the fuel economy of gasohol is greater than that of unleaded gasoline, the fuel gains improve dramatically.

For end use as a chemical feedstock, five of the six cases evaluated in the automotive end use were considered, and results are presented in Table 1-7. (The case omitted is that involving coal-derived ammonia.) The net fuel gains are given for ethanol from each process cycle substituted: (1) for ethanol derived from ethane and (2) for ethanol derived from naphtha.

In the least favorable cases (ethanol from corn/residual oil or corn/coal process replacing naphtha-derived ethanol), the net energy gain is approximately one-half gallon for each gallon of biomass-derived ethanol which replaces petroleum ethanol. In several cases, the petroleum gain in crude oil equivalent is actually greater than one gallon for each gallon of fermentation alcohol.

Penetration of ethanol into the U.S. market will be influenced by numerous other factors, regardless of net fuel gain. These factors include feedstock availability and distribution, alternate plant locations, ability of markets to absorb coproducts, overall process economics, plant capacities, regulatory implications, environmental effects, and impacts on the agricultural sector and on the overall economy. From our calculations, we conclude the following:

- Total net energy gain defined to include all energy inputs (low-grade fuels and premium fuels) does not focus attention on the advantages that biomass alcohol processes offer in using low-utility fuels (such as coal and solar energy) to produce premium transportation fuel.
- The definition and calculation of net fuel gains highlights the major objective of a biomass alcohol process—the production of premium fuels from lower-utility fuels or energy to displace petroleum and natural gas.
- For all the specific processes and options considered, ethanol can be produced from biomass with net gains in premium fuels. This conclusion holds even when the ethanol production processes are treated as being premium fuel (petroleum or natural gas) intensive, if the plant utilizes the innovative, energy-efficient designs which are currently available.
- Despite the fact that net premium fuel gains are possible even with alcohol processes which are premium fuel intensive, maximization of available premium fuel for the domestic economy dictates that

Table 1-7. Net Fuel Gains Including Credit for Coproducts with Biomass-Derived Ethanol Used as Chemical Feedstock* (Gallons of Crude Oil per Gallon of Ethanol)**

NET FUEL GAINS / PROCESS TYPE	NET PETROLEUM GAIN		NET PREMIUM FUEL GAIN (Petroleum & Natural Gas)		NET ENERGY GAIN (All Fuels)	
	BIOMASS PROCESS COMPARED TO ETHANE/ETHYLENE ROUTE	BIOMASS PROCESS COMPARED TO NAPHTHA/ETHYLENE ROUTE	BIOMASS PROCESS COMPARED TO ETHANE/ETHYLENE ROUTE	BIOMASS PROCESS COMPARED TO NAPHTHA/ETHYLENE ROUTE	BIOMASS PROCESS COMPARED TO ETHANE/ETHYLENE ROUTE	BIOMASS PROCESS COMPARED TO NAPHTHA/ETHYLENE ROUTE
PROCESS TYPE						
CORN PROCESS						
— Coal as Process Fuel	1.05	0.91	1.01	0.87	0.67	0.52
— Resid as Process Fuel	0.77	0.63	0.73	0.58	0.67	0.52
— Gas as Process Fuel	1.05	0.91	0.70	0.55	0.64	0.49
CORN & SWEET SORGHUM PROCESS						
— Bagasse as Process Fuel	1.01	0.87	0.95	0.81	0.91	0.76
CELLULOSIC PROCESS						
— Electricity as Process Heat	1.[illegible]2	0.98	1.19	1.05	0.86	0.71

*Gain = (Saving — Investment). By-products considered for credits from biomass ethanol production are DDG and ammonium sulfate as appropriate. Fuel savings at end-use excluding by-product credits were not much different from those shown above.
**Based on conversion of calculated fuel gain in Btu to crude oil equivalent at 138,000 Btu/gal.

nonpremium fuels such as coal or wastes should be used as process fuel.

At this time, the energy balance issue is a dead issue.

SUMMARY

Gasohol is currently being marketed over wide areas of the United States with the largest sales in the midwestern states. The gasohol mixture of 10 percent anhydrous fermentation ethanol blended with unleaded fuel has received wide public acceptance, even though gasohol is sold at a 5 cent premium over unleaded regular gasoline.

The primary limitation on the amount of gasohol that could be sold is alcohol availability. Current production and new plants under construction and retrofitted existing facilities could provide 920 million gallons per year production by 1984.

Gasohol initially encountered considerable resistance from the petroleum and automobile industries. However, they now seem to accept the fact that gasohol is viable, that it does perform satisfactorily, and that it can make a real contribution to octane improvement. Consequently, some major petroleum companies are now marketing gasohol, including Texaco, Phillips, Amoco, and Arco. Others are planning to follow.

An intensive effort to construct new alcohol production facilities will have to occur within the next few years in order to meet the expected market demand for gasohol. Government funds available under Public Law 96-126 and Public Law 96-294 will provide significant "pump priming" funds for ethanol production. There is also a considerable amount of private capital which will become available to finance alcohol production facilities when the finance community is assured that there is a long-term role for ethanol production.

Alcohol is not a total solution to our motor fuel energy problem, but can contribute substantially toward ameliorating the situation by improving gasoline performance (octane rating) and reducing the oil imports currently required.

Specifically, gasohol has three important elements working in its favor:

1. Fuel alcohol can be produced from starch- or sugar-containing biomass substrates with existing proven technology. In addition, new cellulosic waste conversion technologies should become commercially viable within the next 3-5 years.
2. The market acceptance of gasohol has exceeded expectations, even though gasohol prices run 5 cents per gallon higher than regular unleaded gasoline, and in spite of the fact that the mileage performance of gasohol has not been accurately established.
3. The addition of 10 percent ethanol to regular unleaded gasoline significantly increases the antiknock value of the fuel mixture from

87 to 90 octane.[6] This improvement in octane rating is equivalent to an extra refinery cost of 3 to 6 cents/gallon gasoline, or a credit of 30 to 60 cents per gallon of fuel alcohol. This octane credit given to gasohol follows from the fact that, to achieve 90 octane in the gasoline pool, additional refining and alkylation would be required, which increases investment and operating costs, and reduces gasoline yield per barrel of oil.

The science and art for production of fuel-grade ethanol by fermentation are well established. Although yields and energy requirements appear to be close to maximum and minimum values, respectively, some improvements may still be possible, arising out of research in areas such as continuous fermentation, development of yeast strains which are more temperature tolerant, and alcohol separation and dehydration methods other than distillation.

As stated, the market acceptance of gasohol has exceeded expectations. The volume currently available does not begin to satisfy present demands. Fuel-grade ethanol for gasohol use is currently sold for as much as $1.80 to $1.85 per gallon, whereas 200-proof high-grade industrial alcohol (synthetic) has been selling (first quarter, 1980) for $2.02 per gallon.

The Energy Security Act of 1980, Public Law 96-294, targets a 10 percent usage of fuel-grade ethanol in gasoline, nationwide, by 1990. This would require the construction of an ethyl alcohol industry that is 22 times as large as our current capacity for production of industrial and potable grades. We can also use a 20 percent fermentation ethanol blend without engine modification. Such a blend will increase octane rating by 5 points, permitting reduction of the refinery gasoline pool to 85 octane, which would decrease refinery costs, increase gasoline yields per barrel of oil, and permit further reduction in petroleum imports.

Current incentives available from federal and state government will encourage production and use of fermentation ethanol in gasohol. The federal government has allowed a federal excise tax exemption of 4 cents per gallon of gasohol,[7] which amounts to 40 cents per gallon of fuel alcohol.

In addition to this exemption, 12 of the 50 states have allowed highway fuel tax exemptions on gasohol, ranging from 1 to 9.5 cents/gallon (10 to 95 cents/gallon of ethyl alcohol).

By way of projection, the price of oil and gasoline should continue to increase at a rate in excess of the general inflation rate, while biomass-based fermentation ethanol should increase only at the general inflation rate. Thus, by 1982, we can reasonably anticipate a gasoline refinery price of $1.50 to $2.00 per gallon. Beyond this point, the ethyl alcohol price should become economically competitive without government subsidy incentives.

[6]Average of research octane and motor octane numbers.

[7]Applied only to fuel alcohol prepared from biomass substrates.

Organization of the Handbook

Alcohol-fuel production can be an attractive agribusiness opportunity provided it is approached in a businesslike fashion and full advantage is taken of the various federal incentives presently available.

The economics and, therefore, the profitability of alcohol-fuel production are sensitive to a number of factors ranging from site-specific feedstock and stillage disposition conditions to national and international policies. Raising the necessary starting capital through conventional financing channels may be difficult because of increased interest rates and the new business field risks associated with alcohol production.

This handbook is designed to lead the potential entrepreneur/ethanol producer through all the steps necessary to develop a business plan and prepare a feasibility analysis for a site-specific fermentation ethanol project. Therefore, emphasis is placed on technical, marketing, financing, management, and incentives information. In this context, it is hoped that the handbook will provide the entrepreneur/ethanol producer with the required background for investment in the fermentation ethanol fuel business.

The introduction has provided an overview of the perspectives and issues affecting the gasohol industry. Chapter 2 discusses factors that affect the decisionmaking process and addresses questions the entrepreneur must face before investing in the gasohol industry. The chapter leads the entrepreneur step-by-step through the series of decisions and choices to be made before reaching the final decision to enter the gasohol business. Simple decision and planning worksheets are provided to aid in the decisionmaking process.

Basic fermentation-ethanol-production processes are described in Chapter 3. The chapter provides detailed information on each stage of the production process, delineates plant design criteria, and presents in significant detail the USDA model ethanol plant process characteristics.

Chapter 4 describes the types of feedstocks available for fermentation-ethanol production, and relates their characteristics and availability to regions within the United States. Trends and fluctuations in the price of the major grain feedstocks, such as corn, are discussed in terms of their potential use and value compared to other feeds. The impacts on agriculture due to the production of fermentation ethanol are also discussed in Chapter 4.

The market potential for fermentation ethanol used as gasohol and its coproducts is fully covered in Chapter 5. This chapter addresses ethanol as a fuel, the price of ethanol, and the public market response to its use as gasohol. Since the stillage coproduct is a fundamental coproduct which must be effectively marketed, a considerable portion of Chapter 5 is devoted to animal feed coproducts, human food coproducts, and the marketing of animal feed stillage coproducts.

While small on-farm fermentation ethanol stills are oriented toward energy self-sufficiency, the larger-scale plants are developed as normal businesses which must be profit-making operations. Chapter 6 describes the economics of ethanol production for small-, intermediate-, and large-scale production systems.

Chapter 7 provides a description of the technology that has been available, the state of the art, and the potential for expanded alcohol-fuel production. It provides a broader scope than just fermentation ethanol in its coverage of cellulose and advanced process technology.

Specific ethanol plant designs are discussed in Chapter 8; a full range of production plants are described. The Solar Energy Research Institute's typical 25-gallon-per-hour ethanol plant has been built by DOE/EG&G, and the new plant's characteristics are discussed in detail. Also, the Katzen 50-million-gallon-per-year plant is analyzed to provide the general technical background and costing data required in analyzing plants of commercial sizes and designs.

The environmental and safety problems associated with ethanol are minor compared to most other energy technologies. However, it is important to understand the impacts on the ethanol plant due to environmental and safety issues. Chapter 9 provides a clear summary of the environmental issues and safety concerns.

It is useless to produce gasohol if a market for the product is not developed. Chapter 10 discusses in detail the suitability of ethanol for fuel, and provides marketing information about gasohol usage. This chapter provides a ten-step guide to carburetor modification so you could use 100 percent ethanol to run your automobile.

The basic elements of a business plan for fermentation ethanol are described in Chapter 11. This material leads to an approach for development of the feasibility analysis in Chapter 12.

The search for synthetic petroleum fuels is not confined to the United States. Brazil is the international leader in developing an alcohol-fueled transportation system. Chapter 13 stresses Brazil's accomplishments and summarizes other international alcohol-fuel efforts.

The appendixes to this handbook were prepared to provide additional specific information for the reader. It is hoped that these support materials will direct you to the right source of additional information. A definitive list of resource people, organizations, and related contacts is provided in Appendix A. The reference sources utilized in this handbook are provided in Appendix B. A bibliography of the basic books or reports on fermentation ethanol and gasohol are listed in Appendix C. A glossary of the unique terms used in this handbook is presented in Appendix D. A series of tables delineating feedstock by states and other tabular reference data are provided in Appendix E.

It is hoped that this comprehensive collection and synthesizing of gasohol information will be of assistance to those interested in learning more about gasohol.

Chapter 2

Decision to Produce Gasohol

Fermentation ethanol-fuel production using biomass feedstocks appears to be an attractive business venture for a wide range of potential entrepreneurs. The production of ethanol fuel from agricultural crops or wastes is strongly supported by the farming community, and the public's initial response to the introduction of gasohol as a substitute automotive fuel has been favorable.

Ethanol production raises numerous issues which may complicate the task of investors when deciding whether to enter the ethanol market. The production of ethanol through fermentation of agricultural feedstocks requires interaction with a variety of producer and consumer groups and local, state, and federal agencies. The feedstock for production originates from the agricultural community and some of the coproducts of ethanol production may impact the same farming community. The production and marketing of ethanol fuel are subjected to regulations similar to the production of alcoholic beverages; simultaneously, the marketing and distribution of ethanol fuel may require the cooperation of gasoline producers and distributors. The economics of ethanol-fuel production benefit from state or federal incentives, and will also be influenced by uncontrollable factors such as climate or world events which impact the price of agricultural feedstocks.

The initial idea of entering the ethanol-fuel business may stem from a variety of motivations, among others: to make a profit; to diversify investments; to make use of idle or surplus facilities or resources; or to fill the needs of a perceived market. Whatever the motivation, the decision to produce ethanol must be based on a careful and thorough analysis of all aspects of the proposed business. In view of the factors impacting on the project, it will be necessary for prospective entrepreneurs to seek and secure the help of individuals or teams familiar with the various aspects of ethanol-fuel production. This groundwork will include the collection of data, analysis of feedstock and production options, and evaluation of financial opportunities which are summarized in the form of the comprehensive feasibility analysis. The conclusion drawn in the feasibility analysis should form the basis upon which the decision to produce will be made. The feasibility analysis is also needed to interest other investors or sources of financing in taking part in the project. It must therefore be recognized that a certain amount of front-end money will be required to reach a substantiated decision to enter the fermentation ethanol business.

The objective of this chapter is to describe the major steps required in deciding to produce ethanol, to discuss the data to be gathered and questions to be answered at each of these steps, and to indicate the kind of effort needed to complete the feasibility analysis on which the final decision will be based.

Overview of Ethanol-Fuel Production

THE OPERATING PLANT AND ITS ENVIRONMENT

Fermentation ethanol-fuel production involves multiple interactions among a variety of "players." Each of these aspects must be addressed during the decisionmaking process in order to avoid overlooking a problem that could be detrimental to the economic viability of the project.

Figure 2-1 presents an overview of an ethanol business, and identifies some of the major elements and entities involved. Four major areas of concern are identified.

Technical Aspects

Ethanol production through fermentation of grain is a well-established technology practiced by the alcoholic beverage industries in the United States. The production of fermentation ethanol fuel is an extension of this technology. The economics involved, however, are quite different. Alcoholic beverage industries are geared toward satisfying a luxury market; therefore, their profitability is not as critically dependent on process efficiency or price of supplies as is that of ethanol-fuel plants which must compete with a still relatively cheap product—gasoline. When evaluating the feasibility of a fermentation ethanol-fuel plant, it will be essential to assess carefully the method of procurement of the feedstock (such as corn) over a long period, the availability of resources such as fuel for the process heat source and water, the sale of the stillage, the efficiency of the ethanol-production process, and other technical aspects which can impact the overall economics of the plant.

As indicated in Fig. 2-1, various government agencies influence the process of fermentation ethanol production. The Department of Agriculture (USDA) may impact feedstock availability and prices, the Bureau of Alcohol, Tobacco, and Firearms (BATF) requires permits for production, the Environmental Protection Agency (EPA) impacts stillage disposal, the Department of Energy (DOE) may encourage the use of renewable process fuels, etc.

Marketing Aspects

Successful marketing of both fermentation ethanol fuel as gasohol and the stillage coproducts of ethanol is essential to ensure economic viability of

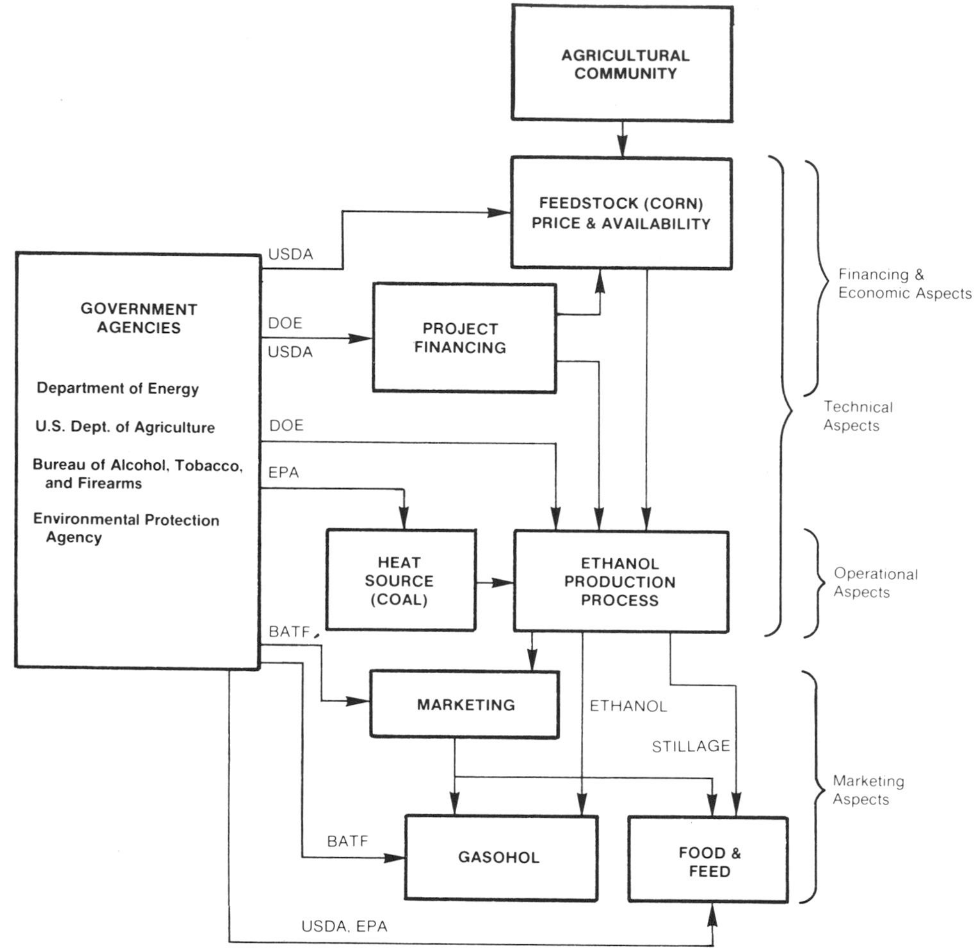

Source: *A Guide to Commercial Scale Ethanol Production and Financing*, SERI, Denver, 1980.

Figure 2-1. Basic organizational interfaces for production of gasohol.

the project. A careful identification and evaluation of the long-term market will be required to assess the economic potential of the production plant. Contractual agreements for the sale of ethanol and stillage, prior to plant operation, will develop a sound basis for the venture.

Various government agencies may have an impact on the marketing process: exemption from the excise tax on fuel blends containing ethanol is controlled by the federal and/or state governments; government policies concerning the export of food and feed products and income tax credits for fuel blenders may modify the marketing plans adopted for the products of the plant.

Financing and Economic Aspects

Obtaining financing from private sources for the proposed project will, to a large extent, depend on the quality and loan thoroughness of the feasibility analysis.

In the course of performing the feasibility analysis, federal loan guarantee or loan programs should be reviewed to determine the eligibility of the fermentation ethanol project being considered. The purpose of the loan guarantee programs is to provide funds to sectors of the economy that otherwise could not obtain credit or could only obtain it under unfavorable conditions through existing commercial financial organizations. Federal agencies having the authority to provide funds to prospective fermentation ethanol-fuel producers include DOE, the Small Business Administration (SBA), the Farmers Home Administration (FmHA) of USDA, the Department of Housing and Urban Development's (HUD) Urban Development Action Grants, and the Economic Development Administration (EDA) of the Department of Commerce. Securing loan guarantees or loans for the project must be considered a priority item for the entrepreneur. Indeed, the terms of these loans are usually attractive and can improve the economics of the project, and loan guarantees will facilitate the task of raising funds from private sources. Securing loans or loan guarantees, however, requires the preparation of documented applications which may delay the decision-making.

Operational Aspects

The investor must address a number of aspects related to the operation of a fermentation ethanol plant. The availability of trained or trainable labor must be considered. Can the regional labor force provide the technical personnel for the plant? Will the creation of a new industry upset the local or regional organizations and therefore risk antagonizing the community?

The above factors, as well as the possible impact of agencies such as the Occupational Safety & Health Administration (OSHA), must be evaluated carefully.

APPROACH TO IMPLEMENTING AN ETHANOL PLANT

The preceding discussion has shown that many factors may affect the profitable operation of an ethanol plant. Before finalizing the decision to start an ethanol plant, the four key aspects of ethanol production will have to be discussed and evaluated to choose the options or compromises which will optimize financial return to the investors.

Figure 2-2 illustrates the decisionmaking and implementation phases which are the major elements involved in bringing a plant to production. The scope and purpose of these elements are described below.

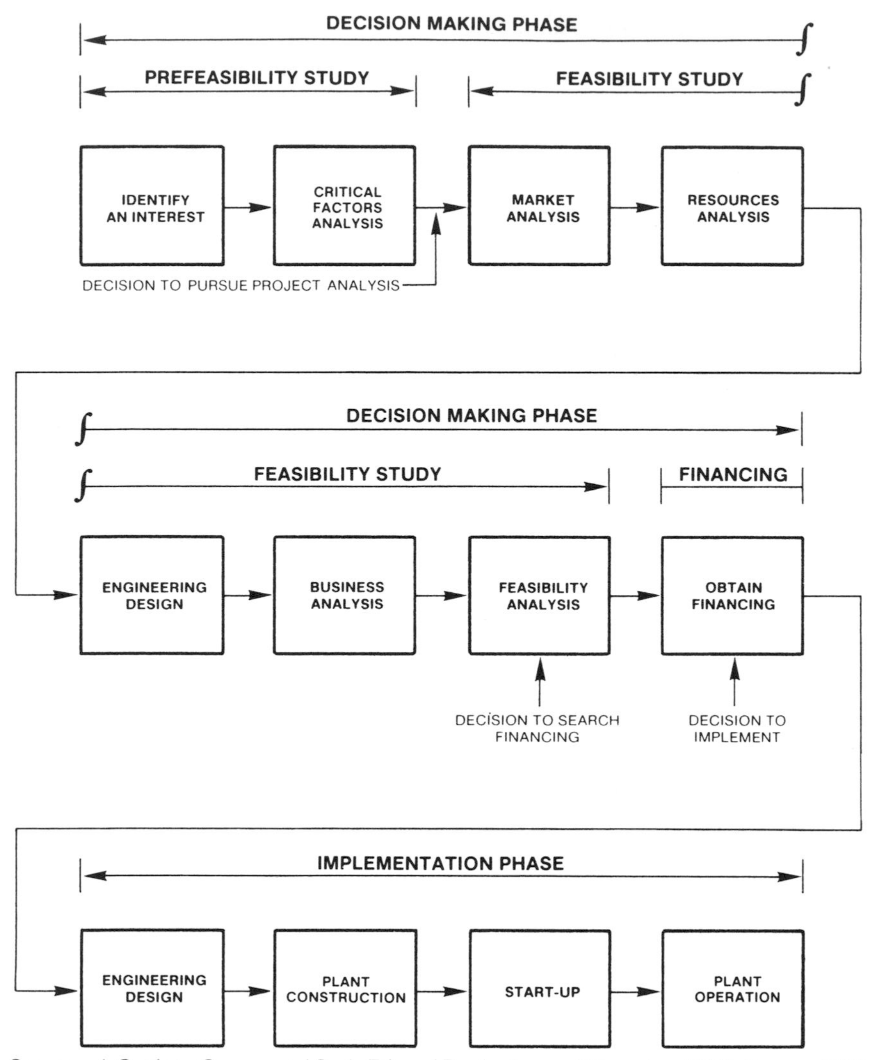

Source: *A Guide to Commercial Scale Ethanol Production and Financing*, SERI, Denver, 1980.

Figure 2-2. Steps involved in bringing a plant to production.

Decisionmaking Phase

The overall purpose is to reach a final decision on whether to become involved in a fermentation ethanol project. This phase involves three major steps:

Prefeasibility study. Having identified an interest in the production of ethanol, an investor and his or her other partners must be convinced that the project is viable, at least in general terms. Some help from consultants or

local/regional sources such as extension agents or agricultural school staff may be needed in discussing some specific aspects of the project. The product of this step is a prefeasibility study which should establish the rationale for pursuing the project. Some of the key areas to be addressed include existence of a market for both ethanol and its stillage coproduct, distribution mechanism for the products, selection and availability of a feedstock, approximate size of the proposed plant, availability of non-petroleum fuels for process steam and process water, general design of the fermentation plant, approximate cost of the plant, business plan, potential methods of financing, and economic viability of the project. At the end of this first step, potential investors in the project should be able to make the decision on whether to pursue the project.

Feasibility study. The objective of this step is to perform detailed technical and financial analyses of the viability of the project. The feasibility study will address four major areas—marketing, technical feasibility, financing and economics, and operation—and will develop a realistic plan. The feasibility analysis will serve three purposes: confirm and refine the estimates of the prefeasibility study; develop a business plan and business planning schedule; and serve as supporting evidence when negotiating the financing of the project.

Performing a detailed feasibility analysis will require contributions from many areas: engineering firms, market specialists, suppliers, environmentalists, law firms, accountants, etc. An adequate budget must be ear-marked for the purpose of generating a credible, well-documented study.

At the end of this step, the potential entrepreneur will be in a position to decide whether the project has sufficient commercial potential for the financial community to be interested.

Financing. The last and most important step is to secure the necessary financing for the project. In these negotiations, the entrepreneur will need the support of law firms, accountants, and other specialists familiar with the alcohol-fuel industry.

When financing has been secured, the final decision to implement the project can be made.

Implementation Phase

This phase involves the same steps followed for any commercial/industrial project: engineering design, construction, start-up, and operation.

The role of the investors is somewhat reduced in this phase, because at this point a management team could become responsible for the project implementation on behalf of the investors. The primary role of the investors is to assist the team in reaching the final decision of implementing the project. To do so, investors must be aware of the questions to be asked, issues to be raised, problems to be solved, and "players" involved in the overall im-

plementation of such a project. Some of these questions, problems, or issues related to the process of deciding to produce ethanol are more complex than those for more conventional ventures. This results from the particular significance of fermentation ethanol in the present energy and international context and because of its intimate relationship with various markets and economic sectors.

The Decisionmaking Process

The decisionmaking process involves addressing and answering specific questions in order to reach a decision concerning project implementation.

The questions to be addressed are similar to those relating to any business venture: Should one get involved? Is it really worth it? What is the market? Is the project technically feasible? Is the project economically viable? Is financing available? The approach to answering these questions resembles that used for other similar businesses: market and engineering studies, business analysis, etc. The purpose of this section is to identify some specific questions and options raised by fermentation ethanol production and to indicate sources of data that could help the investors reach a decision.

THE DECISIONMAKING FLOW DIAGRAM

Figure 2-3 presents a flow diagram of the decisionmaking process. The figure is an expansion of the first phase of the implementation diagram shown earlier (Fig. 2-2).

The top row of the matrix identifies questions to be addressed or decisions to be reached. Each column of the matrix suggests an approach to answering the corresponding questions, the type of answer or output desired, the sources of information and possible options, and an estimate of the front-end cost required to obtain an answer or reach a decision. These costs are approximate and may include such items as consultant fees, data gathering, or trips.

ELEMENTS OF THE DECISIONMAKING PROCESS

The various steps in the decisionmaking process are discussed below.

Initial Expression of Interest by Investors

Should investors get involved at all in the ethanol-fuel business?

Investors will need to identify the existence of interest in ethanol production at the local, regional, or national level. They will need to evaluate the availability of federal incentives to support their project. Their expression of interest could result from contacts with grain producers, agribusinesses, farmers seeking new markets for their products or alternative methods of disposal of stillage or waste products, or distributors trying to secure a long-

term reliable source of substitute fuel. Investors will need to evaluate the true level of interest and the commercial opportunity it generates and then must define their objectives, which could range from producing ethanol fuel as a separate business entity to combining it with current business interests. The alcohol fuel could be marketed through cooperatives for special markets or regional distribution.

In clarifying their objective, investors will not only have to rely on inquiries but also become familiar with state and federal programs to determine if their perceived interest matches national and regional policies. The cost of this first step is approximately $15,000, which includes travel costs, quick inquiries and surveys, and general data and information gathering for a commercial-scale plant.

Attractiveness of the Proposed Project

Is the identified project worth investigating further?

Having established that significant interest exists for an ethanol-fuel project, investors must determine if the basic elements for a successful business are present. A prefeasibility study will provide this answer.

The major aspects to be considered in a prefeasibility study include

- Existence of an identified market for the ethanol fuel
- Availability of a market for the stillage and coproducts
- Availability of sufficient feedstocks such as corn and alternate feedstocks to reduce the project's risks
- Availability of a nonpetroleum heat source such as coal
- Feasibility of the technical process and plant economics

The initial contact with potential investors or financing institutions should be made during this prefeasibility study to determine if the proposed project has a realistic chance of being financed. The product of this study should be a document or a data base sufficient for the investors to determine whether the project appears attractive enough to justify the expenditure of further front-end money. At this time, the investors should also have some knowledge of how their project will be received by the financing community and whether some other investors or the government may be willing to supply some of the seed money for the forthcoming feasibility study. Formation of a corporation may be justified at this time.

Sources to contact in performing this prefeasibility study may include the DOE, USDA, SBA, and EDA through their regional offices or extension services. State agencies, engineering firms, BATF, and local agricultural economists could be contacted to broaden the information base. Some specialized help will be required, the cost of which could reach about $50,000. The timeframe needed for this study may be as short as one or two months but, in most cases, will probably extend over a four- to six-month

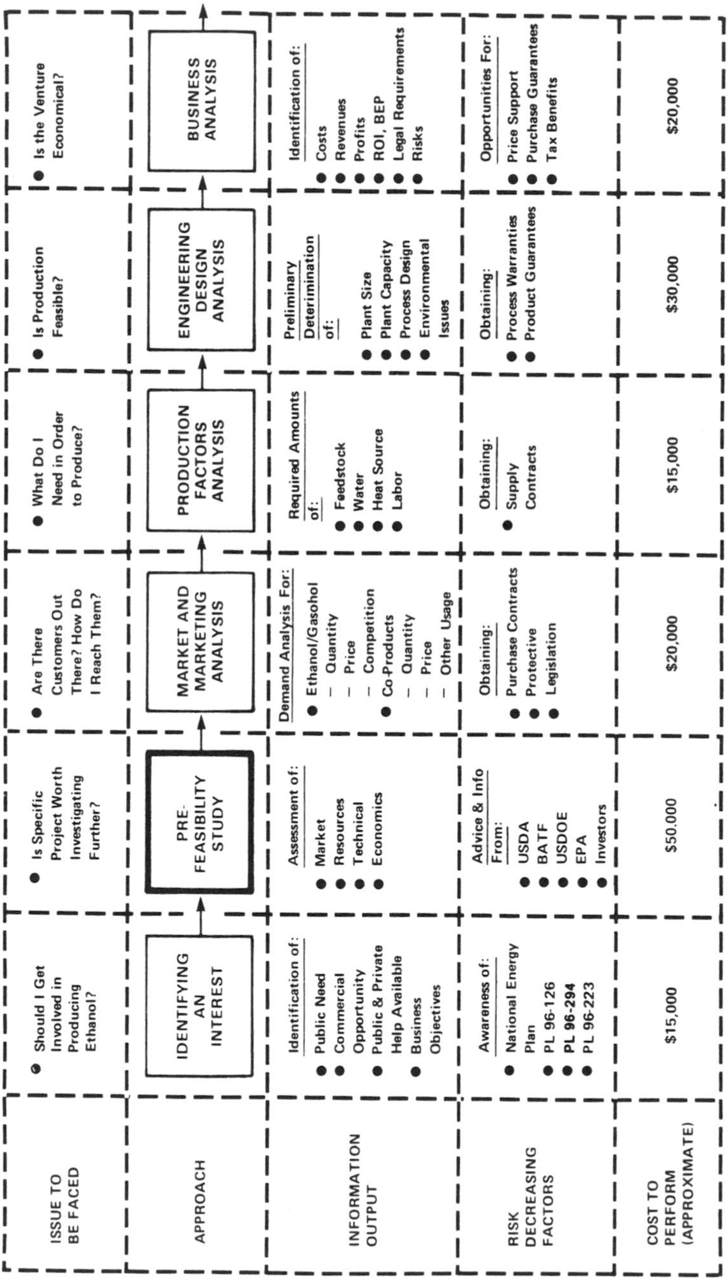

ISSUE TO BE FACED	● Should I Get Involved in Producing Ethanol?	● Is Specific Project Worth Investigating Further?	● Are There Customers Out There? How Do I Reach Them?	● What Do I Need in Order to Produce?	● Is Production Feasible?	● Is the Venture Economical?
APPROACH	IDENTIFYING AN INTEREST →	PRE-FEASIBILITY STUDY →	MARKET AND MARKETING ANALYSIS →	PRODUCTION FACTORS ANALYSIS →	ENGINEERING DESIGN ANALYSIS →	BUSINESS ANALYSIS →
INFORMATION OUTPUT	Identification of: ● Public Need ● Commercial Opportunity ● Public & Private Help Available ● Business Objectives	Assessment of: ● Market ● Resources ● Technical ● Economics	Demand Analysis For: ● Ethanol/Gasohol – Quantity – Price – Competition ● Co-Products – Quantity – Price – Other Usage	Required Amounts of: ● Feedstock ● Water ● Heat Source ● Labor	Preliminary Deterimination of: ● Plant Size ● Plant Capacity ● Process Design ● Environmental Issues	Identification of: ● Costs ● Revenues ● Profits ● ROI, BEP ● Legal Requirements ● Risks
RISK DECREASING FACTORS	Awareness of: ● National Energy Plan ● PL 96-126 ● PL 96-294 ● PL 96-223	Advice & Info From: ● USDA ● BATF ● USDOE ● EPA ● Investors	Obtaining: ● Purchase Contracts ● Protective ● Legislation	Obtaining: ● Supply Contracts	Obtaining: ● Process Warranties ● Product Guarantees	Opportunities For: ● Price Support ● Purchase Guarantees ● Tax Benefits
COST TO PERFORM (APPROXIMATE)	$15,000	$50.000	$20,000	$15,000	$30,000	$20,000

Figure 2-3. Decisionmaking process for entering the ethanol business.

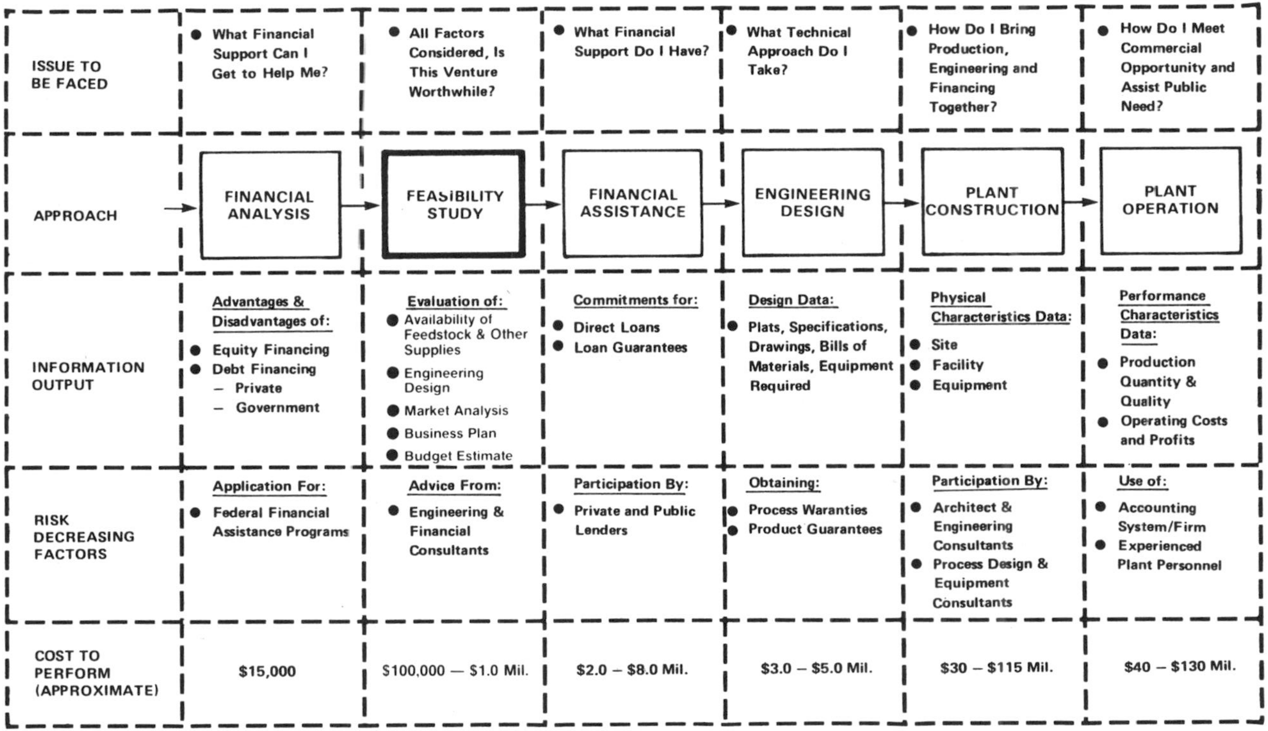

ISSUE TO BE FACED	• What Financial Support Can I Get to Help Me?	• All Factors Considered, Is This Venture Worthwhile?	• What Financial Support Do I Have?	• What Technical Approach Do I Take?	• How Do I Bring Production, Engineering and Financing Together?	• How Do I Meet Commercial Opportunity and Assist Public Need?
APPROACH	→ FINANCIAL ANALYSIS →	FEASIBILITY STUDY →	FINANCIAL ASSISTANCE →	ENGINEERING DESIGN →	PLANT CONSTRUCTION →	PLANT OPERATION
INFORMATION OUTPUT	Advantages & Disadvantages of: • Equity Financing • Debt Financing – Private – Government	Evaluation of: • Availability of Feedstock & Other Supplies • Engineering Design • Market Analysis • Business Plan • Budget Estimate	Commitments for: • Direct Loans • Loan Guarantees	Design Data: • Plats, Specifications, Drawings, Bills of Materials, Equipment Required	Physical Characteristics Data: • Site • Facility • Equipment	Performance Characteristics Data: • Production Quantity & Quality • Operating Costs and Profits
RISK DECREASING FACTORS	Application For: • Federal Financial Assistance Programs	Advice From: • Engineering & Financial Consultants	Participation By: • Private and Public Lenders	Obtaining: • Process Warranties • Product Guarantees	Participation By: • Architect & Engineering Consultants • Process Design & Equipment Consultants	Use of: • Accounting System/Firm • Experienced Plant Personnel
COST TO PERFORM (APPROXIMATE)	$15,000	$100,000 — $1.0 Mil.	$2.0 – $8.0 Mil.	$3.0 – $5.0 Mil.	$30 – $115 Mil.	$40 – $130 Mil.

Source: *A Guide to Commercial Scale Ethanol Production and Financing*, SERI, Denver, 1980.

Figure 2-3. (Continued)

period. The following chapters of this handbook provide the basic data needed to pursue the project, as well as references to information sources.

Having determined that the project is worth pursuing, the investors must analyze in further detail the various aspects of the proposed business. These detailed analyses will result in a feasibility study which will be the basis for the financing and implementation of the project.

Identification of Markets

Are there markets for the ethanol and coproducts of the plant, and how can the markets be reached?

A market analysis is required to determine such elements as the size of the proposed plant, its location, and the marketing structure required. The size of the potential market, market penetration, market price, location of the market, competition, and other relevant factors will be estimated for both ethanol fuel and its coproducts. When feasible, letters of intent, letters of interest, or tentative contractual agreements will be obtained from prospective customers. This market analysis will require expert help from firms specialized in marketing and distribution of both fuel and feed. Background data on these market aspects can be found in Chapter 5. The cost of the market analysis has been estimated at about $20,000.

Production Factors

What are the resources needed to produce?

Having sized the market, the investors can then size the capacity of the proposed plant. It will be necessary to estimate the resources required for plant operation. These resources include feedstocks, a nonpetroleum process heat source, water, land, and labor. A market for stillage-derived coproducts or a facility for stillage disposal is also required. Material balances should be performed to estimate the needed gross quantities of production resources.

Where possible, tentative supply contracts will be drawn for feedstock and stillage, zoning or preliminary zoning permits will be searched, and local environmental issues will be reviewed. The result of this analysis will be the identification of a tentative site or sites and a certain degree of confidence that all required production elements are available.

With technical and financial assistance, it is estimated that this step may cost about $15,000. If permits are to be obtained, the timeframe to finish this task may extend from several months to a year in the worst case.

Technical Feasibility

Is production feasible?

Although conventional ethanol production from sugar-starch feedstock is a well-established technology, site-specific conditions or constraints such as quality of feedstock, source of process heat, and water quality, require modifications to the conventional plant design. A preliminary process and engineering design analysis, which will include comparison of available options, must be performed by a reputable, well-organized engineering firm. This analysis will include an optimized process design, material requirements and energy flow diagrams, plant capacity, expected on-stream factors, and other aspects required in a complete engineering package. Process warranties and product guarantees should also be stated. Availability of electric power, sewage disposal, and water should also be discussed.

Background data relative to the plant design aspects can be found in Chapter 8, but it is stressed that this analysis must be performed by a well-established firm to lend credibility to the study and facilitate financing. Such a study could cost about $50,000 and may extend over several months, depending on the number of options to be analyzed.

Economic Viability

Is the project economically viable?

This is probably the most crucial element of the feasibility study in terms of securing adequate financing. A complete business plan must be proposed, including period of erection of the plant, planned date of operation, legal status (corporation, cooperative, etc.), capital and working capital requirements, and operating costs. These data then will be used to determine estimated return on investment (or price of ethanol required to obtain a proposed return) using discounted cash flow or other methods suitable to the type of business envisioned. The financial method used to estimate the economics of the plant must be flexible enough that sensitivity analyses can be performed to evaluate the potential impact of factors such as fluctuations in feedstock and stillage prices, tax credits, and inflation rate on the economics and return of investment from the plant. It is important that the perceived risks be quantified as much as possible to ensure the development of a viable business plan.

Performing this business analysis will require a data base on feedstocks, chemical and fuel supplies, and market prices of products, as well as appropriate financial data such as inflation rates for fuels, products and labor, prevalent rate of interest, and others. Major accounting and engineering firms have developed computer software for the business analysis discussed here. Legal and financial advice also will be needed in performance of this task. The business analysis could cost $20,000 and extend over several weeks once the necessary data have been collected.

Sources of Financing

What is available?

Possible sources of financing should be reviewed, including private, governmental, or mixed funding of the project. Loans or loan guarantees

Figure 2-4. The written commitment of a supplier to provide feedstock will help establish the viability of the plant.

should be considered. The impact of each of these options must be reviewed in terms of the overall economic viability of the project.

The product of this analysis will be the identification of a preferred method of financing and some alternative methods, if warranted.

Such an analysis will require legal, financial, and analytical inputs from specialized firms or consultants and could cost $20,000 or more.

Overall Feasibility of the Project

Is the venture technically feasible and economically viable?

Having performed the detailed market, resources, technical, and economic analyses, an overall evaluation of the project must now be performed. This feasibility analysis will help in deciding whether or not to seek financing, and in describing the total project and its prospects to future investors.

The feasibility analysis must address all aspects of the proposed business, attempt to answer all foreseeable questions, and establish the credibility of the originator(s) of the project and its supporting team (legal, technical, etc.). Letters of commitment or intent from suppliers, markets, permit and zoning agencies, and others should be an important part of this package. The proposed management team and its credentials should be described.

Performing such a feasibility analysis will require contributions from various teams and individuals and could be quite expensive. An outline for a typical feasibility analysis is described in Chapter 12.

Financing

Prior to completion of the feasibility study, a team must be assembled to prepare a detailed economic and financial analysis of the project and to negotiate and arrange the financing. The qualifications of this financing team are contingent upon the company's internal management team and the complexity of the financing to be arranged, which generally includes specialists in the following areas: economic and financial analysis; negotiation and placement of debt; placement of stock, partnership interest, or bonds; and grants.

Timeframe for the Decisionmaking Process

Developing a credible feasibility analysis and negotiating the required financing, i.e., reaching the decision to implement the plant, may be a time-consuming process.

The prefeasibility study must be performed before engaging in the feasibility study.

While performing the feasibility analysis, several tasks may be conducted in parallel: market evaluation, feedstock supplies, application for permits, negotiation of tentative sales and purchase contracts, etc. This approach could accelerate the process but runs the risk of expending more funds than otherwise necessary if at any point one of the elements of the feasibility package is missing or one of the analyses suggests abandoning the project. For instance, refusal of the necessary permits at a late date or rejection of the project as environmentally unsound could result in the loss of significant amounts of money if all tasks required under the feasibility study are conducted in parallel. The investor will need to decide whether reaching the market at an earlier date is worth risking a large fraction of the front-end investment.

Decision and Planning Worksheets

The previous discussion emphasized the cost and value regarding intermediate- and commercial-scale plant decisions. This section provides decision and planning worksheets oriented toward the on-farm smaller-scale ethanol plant.

The following questions are based on the considerations involved in deciding to proceed with development of a small-scale fermentation ethanol plant. Questions 1-28 are concerned with determining the potential market and production capability; questions 9, 20, and 29-47 examine plant size by comparing proposed income and savings with current earnings; questions 48-53 look at plant costs; questions 54-70 relate to financial and organizational requirements; and questions 71-85 examine financing options.

The final decision to produce ethanol is the result of examining all associated concerns at successively greater levels of detail. Initially a basic determination of feasibility must be made and its results are more a "decision to proceed with further investigation" than an ultimate choice to build a plant or not. This initial evaluation of feasibility is performed by examining: (1) the total market (including on-farm uses and benefits) for the ethanol and coproducts; (2) the actual production potential; (3) the approximate costs for building and operating a plant of the size that appropriately fits the potential market and the production potential; (4) the potential for revenues, savings, or indirect benefits; and (5) personal financial position with respect to the requirements for this plant. There are several points during the course of this evaluation that result in a negative answer. This does not necessarily mean that all approaches are unfeasible. Retracing a few steps and adjusting conditions may establish favorable conditions; however, adjustments must be realistic, not overly optimistic. Similarly, completion of the exercise with a positive answer is no guarantee of success, it is merely a positive preliminary investigation. The real work begins with specifics.

MARKET ASSESSMENT

1. List equipment that run on gasoline and estimate annual consumption for each.

	Equipment	*Fuel Consumption*
a.	____________	____________ gal/yr
b.	____________	____________ gal/yr
c.	____________	____________ gal/yr
d.	____________	____________ gal/yr
e.	____________	____________ gal/yr
	TOTAL	____________ gal/yr

2. List the equipment from Question 1 that you intend to run on a 10 percent EtOH/90 percent gasoline blend.[1]

	Equipment	*Fuel Consumption*	
a.	________________	________________	gal/yr
b.	________________	________________	gal/yr
c.	________________	________________	gal/yr
d.	________________	________________	gal/yr
e.	________________	________________	gal/yr
	TOTAL	________________	gal/yr

3. Take the total from Question 2 and multiply by 10 percent to obtain the quantity of EtOH to supply your own gasohol needs.

________________ × 0.1 = ________________ gal EtOH/yr

4. List the equipment from Question 1 that you are willing to modify for straight EtOH fuel.

	Equipment	*Fuel Consumption*	
a.	________________	________________	gal/yr
b.	________________	________________	gal/yr
c.	________________	________________	gal/yr
d.	________________	________________	gal/yr
e.	________________	________________	gal/yr
	TOTAL	________________	gal/yr

5. Take the total from Question 4 and multiply by 120 percent to obtain the quantity of EtOH for use as straight fuel in spark ignition engines.

________________ gal/yr × 1.2 = ________________ gal EtOH/yr

[1]Throughout these worksheets ethanol is abbreviated EtOH.

6. List your pieces of equipment that operate on diesel fuel.

	Equipment	*Diesel Fuel Consumption*	
a.	____________	____________	gal/yr
b.	____________	____________	gal/yr
c.	____________	____________	gal/yr
d.	____________	____________	gal/yr
e.	____________	____________	gal/yr
	TOTAL	____________	gal/yr

7. List the equipment from Question 6 that you will convert to dual-injection system for 50 percent EtOH/50 percent diesel fuel blend.

	Equipment	*Diesel Fuel Consumption*	
a.	____________	____________	gal/yr
b.	____________	____________	gal/yr
c.	____________	____________	gal/yr
d.	____________	____________	gal/yr
e.	____________	____________	gal/yr
	TOTAL	____________	gal/yr

8. Take the total from Question 7 and multiply by 50 percent to obtain the quantity of EtOH required for dual-injection system equipment.

____________ gal/yr × 0.5 = ____________ gal EtOH/yr

9. Total the answers from Questions 3, 5, and 8 to determine your total annual on-farm EtOH consumption potential.

______ gal EtOH/yr + ______ gal EtOH/yr + ______ gal EtOH/yr = ______ gal EtOH/yr

10. List the number of cattle you own that you intend to feed stillage.

a. ____________ Feeder Calves

____________ Mature Cattle

A mature cow can consume the stillage from 1 gallon of ethanol production in 1 day. A feeder calf can consume from 0.7 gallon of ethanol production in 1 day. Multiply the number of feeder calves by 0.7. Add this product to the number of mature cattle to obtain the daily maximum EtOH production rate for which stillage can be consumed by cattle.

b. ________ Feeder Calves × 0.7 + ________ Mature Cattle = ________ gal/day

11. List the number of cattle that neighbors and/or neighboring feedlots own which they will commit to feed your stillage at full ration.

    ____________________ Feeder Calves

    ____________________ Mature Cattle

    ________ Feeder Calves × 0.7 + ________ Mature Cattle = ________ gal/day

12. Total the answers from Questions 10 and 11 to determine the equivalent daily EtOH production rate for which the stillage can be consumed by cattle.

    ____________ gal/day + ____________ gal/day = ____________ gal/day

13. Determine the number of pigs you own that you can feed stillage.

    a. ____________________ Pigs

    Determine the number of pigs owned by neighbors or nearby pig feeders that can be committed to feeding your stillage at full ration.

    b. ____________________ Neighbors' Pigs

    Total the results from a and b.

    a + b. ______________________________ Total Pigs

14. Multiply total from Question 13 by 0.4 to obtain equivalent daily EtOH production for which stillage can be consumed by pigs.

    ____________________ Pigs × 0.4 = ____________________ gal/day

15. Repeat the exercise in Question 13 for sheep.

    a. ____________________ Sheep Owned

    b. ____________________ Neighbors' Sheep

    a + b. ____________________ Total Sheep

16. Multiply total from Question 15 by the quantity of linseed meal normally fed every day to sheep in order to obtain the equivalent daily EtOH production rate for which stillage can be consumed by sheep.

    ________ Sheep × ________ gal/day = gal/day

17. Repeat the exercise in Question 13 for poultry. Poultry can consume less than 0.05 lb of distiller's dried grains per day. This corresponds to about 0.07 gallon of whole stillage per day. Unless the poultry operation is very large, it is doubtful that this market can make any real contribution to consumption.

    a. ________ Poultry Owned

    b. ________ Neighbors' Poultry

    a + b. ________ Total Poultry

18. Take total from Question 17 and multiply by 0.05 to obtain the equivalent daily EtOH production rate for which stillage can be consumed by poultry.

    ________ Poultry × 0.05 gal/day = ________ gal/day

19. Total the answers from Questions 11, 12, 14, 16, and 18 to obtain the total equivalent daily EtOH production rate for which stillage can be consumed by local livestock.

    __ gal/day + __ gal/day + __ gal/day + __ gal/day + __ gal/day = __ gal/day

20. Multiply the total from Question 19 by 365 to obtain the total annual EtOH production for which the stillage will be consumed.

    ________ gal/day × 365 = ________ gal/yr

Compare the answer from Question 20 to the answer from Question 9. If the answer from Question 20 is larger than the answer from Question 9, all of the stillage produced can be consumed by local livestock. This is the first production-limiting consideration. If the answer to Question 20 is smaller than the answer for Question 11, a choice must be made between limiting production to the number indicated by Question 20 or purchasing stillage-processing equipment.

21. Survey the local EtOH pu purchase market to determine the quantity of EtOH that they will commit to purchase.

	High Proof	*Anhydrous*
a. Dealers	______ gal/yr	______ gal/yr
b. Local Distributors	______ gal/yr	______ gal/yr
c. Regional Distributors	______ gal/yr	______ gal/yr
d. Other Farmers	______ gal/yr	______ gal/yr
e. Transportation Fleets	______ gal/yr	______ gal/yr
f. Fuel Blenders	______ gal/yr	______ gal/yr
TOTAL	______ gal/yr	______ gal/yr

22. Combine the answers from Questions 9 and 21 to determine annual market for EtOH.

______ gal/yr + ______ gal/yr = ______ gal/yr

This is the ethanol market potential.
It is not necessarily an appropriate plant size.

PRODUCTION POTENTIAL

23. Which of the following potential EtOH feedstocks do you now grow?

	Acres	*Yield/Acre*	*Annual Production*
a. Corn	______	______	______ bu/yr
b. Wheat	______	______	______ bu/yr
c. Rye	______	______	______ bu/yr
d. Barley	______	______	______ bu/yr
e. Rice	______	______	______ bu/yr
f. Potatoes	______	______	______ cwt/yr
g. Sugar beets	______	______	______ tons/yr
h. Sugarcane	______	______	______ tons/yr
i. Sweet sorghum	______	______	______ tons/yr

24. Do you have additional uncultivated land on which to plant more of any of these crops?

	Anticipated Acres	*Potential Yield/Acre*	*Additional Annual Production*	
a. Corn	________	________	________	bu/yr
b. Wheat	________	________	________	bu/yr
c. Rye	________	________	________	bu/yr
d. Barley	________	________	________	bu/yr
e. Rice	________	________	________	bu/yr
f. Potatoes	________	________	________	cwt/yr
g. Sugar beets	________	________	________	tons/yr
h. Sugarcane	________	________	________	tons/yr
i. Sweet sorghum	________	________	________	tons/yr

25. Can you shift land from production of any of the crops not mentioned in Question 24 to increase production of one that is? If so, calculate the potential increase as in Question 24.

Crop	*Anticipated Acres*	*Potential Yield/Acre*	*Additional Annual Production*
________	________	________	________
________	________	________	________

26. Add the annual production values separately for each crop from Questions 22, 23, and 24. (This procedure can be used for other crops; however, reliable data for other crops are not available at this time.)

	Ceral Grains (combine totals) bu/yr	*Potatoes cwt/yr*	*Sugar Beets ton/yr*
a.	________	________	________
b.	________	________	________
c.	________	________	________
TOTAL	________ Column I	________ Column II	________ Column III

27. Multiply the Question 26 answers from:

a. Column I by 2.5 to obtain annual potential EtOH production from cereal grains;

__________ bu/yr × 2.5 gal/bu = __________ gal EtOH/yr

b. Column II by 1.4 gal/cwt to obtain annual potential EtOH production from potatoes;

__________ cwt/yr × 1.4 gal/cwt = __________ gal EtOH/yr

c. Column III by 20 gal/ton to obtain annual potential EtOH production for sugar beets.

__________ ton/yr × 20 gal/ton = __________ gal EtOH/yr

If the answer to Question 27 is greater than the answer to Question 22, the *maximum* size of the plant would be the value from Question 22.

PLANT SIZE

Neither the size of the market nor the production potential are sufficient to determine the appropriate plant size although they do not provide an upper limit. A good starting point is to fill your own fuel needs (answer to Question 9) and not exceed local stillage consumption potential (answer to Question 20). Since the latter is usually larger and the equipment for treatment of stillage introduces a significant additional cost, the value from Question 20 is a good starting point. Now the approximate revenues and savings must be compared to current earnings from the proposed ethanol feedstock to determine if there is any gain in value by building an ethanol plant. Assume all feedstock costs are charged to production of EtOH.

FUEL SAVINGS

28. Multiply the total of Questions 3 and 5 by the current price you pay for gasoline in $/gal.

(__________ gal/yr + __________ gal/yr) × __________ $/gal = __________ $/yr

This is the savings from replacing gasoline with EtOH.

29. Multiply the answer from Question 8 by the current price you pay for diesel fuel in $/gal. (Refer to Chapter 10 for ethanol/diesel fuel tradeoffs.)

__________ gal/yr × __________ $/yr = __________ $/yr

This is the savings from replacing diesel fuel with EtOH.

30. Total Questions 29 and 30 to obtain the fuel savings.

__________ $/yr + __________ $/yr = __________ $/yr

FEED SAVINGS

31. Total the answers from Questions 10b, 14, 16, and 18.

_____ gal/day + _____ gal/day + _____ gal/day + _____ gal/day = _____ gal/day

32. Total the answers from Questions 11, 14, 16, and 18.

_____ gal/day + _____ gal/day + _____ gal/day + _____ gal/day = _____ gal/day

33. Multiply the answer to Question 31 by 6.8 to obtain the dry mass of high-protein material represented by the whole stillage fed (if using cereal grain feedstock).

_____ gal/day × 6.8 lb dry mass/gal EtOH = _____ lb dry mass/day

34. Multiply the answer to Question 33 by the protein fraction (e.g., 0.28 for corn) of the stillage on a dry basis.

_____ lb dry mass/day × _____ (protein fraction) = _____ lb/protein/day

35. a. Determine the cost (in $/lb protein) of the next less expensive protein supplement and multiply this number by the answer to Question 34 (answer this question only if you buy protein supplement).

_____ $/lb protein × _____ lb protein/day = _____ $/day

b. Multiply the answer to Question 35 by 365 (or the number of days per year you keep animals on protein supplement) to obtain annual savings in protein supplement.

_____ $/day × 365 days/yr = _____ $/yr

PRODUCTION SAVINGS

36. a. Determine the cost of production of high-protein feeds on your farm in $/lb dry mass and multiply by the protein fraction of each to obtain your actual cost of producing protein for feeding on-farm.

_____ $/lb dry mass × _____ (protein fraction) = _____ $/lb protein

b. Multiply the answer to Question 36a by the answer to Question 34 (or by the amount of protein you actually produce on-farm: quantity in lbs times protein fraction, whichever is smaller) to obtain potential protein.

_____ $/lb protein × _____ lb protein/day = _____ $/day

c. Multiply the answer to Question 36b by the number of days you keep animals on protein supplement during the year up to 365.

_______________ $/day × _______________ days/yr = _______________ $/yr

37. Total the answers from Questions 35b and 36c.

_______________ $/yr + _______________ $/yr = _______________ $/yr

This is the total animal feed savings you will realize each year.

REVENUES

38. a. Multiply the answer from Question 27 by the reasonable market value of the stillage you produce.

_______________ gal/day × _______________ $/gal = _______________ $/day

b. Multiply the answer obtained in Question 38a by the number of days during the year that this quantity of stillage can be marketed, up to 365.

_______________ $/day × _______________ days/yr = _______________ $/yr

This is the total stillage sales you will realize each year.

39. Total the answers from Questions 37 and 38b.

_______________ $/yr + _______________ $/yr = _______________ $/yr

This is the total market value of the stillage you will produce.

40. Subtract the answer to Question 9 from the answer to Question 20 to obtain the EtOH production potential that remains for sale.

_______________ gal/yr − _______________ gal/yr = _______________ gal/yr

41. Multiply the answer from Question 40 by the current market value for ethanol.

_______________ gal/yr × _______________ $/gal = _______________ $/yr

This is the annual ethanol sales potential.

42. Total the answers from Questions 30, 37, 39, and 41 to obtain the total revenues and savings from this production rate.

________ $/yr + ________ $/yr + ________ $/yr + ________ $/yr = ________ $/yr

43. Divide the answer to Question 20 by:

a. 2.5 gal/bu if the feedstock to be used is cereal grain.

_______________ gal/yr ÷ 2.5 gal/bu = _______________ bu/yr

b. 1.4 gal/cwt if the feedstock to be used is potatoes.

_______________ gal/yr ÷ 1.4 gal/cwt = _______________ cwt/yr

c. 20 gal/ton if the feedstock to be used is sugar beets.

_______________ gal/yr ÷ 20 gal/ton = _______________ tons/yr

44. Multiply:

a. The answer from Question 43a by the appropriate market value for cereal grains to obtain the potential earnings for direct marketing with EtOH production;

_______________ bu/yr × _______________ $/bu = _______________ $/yr

b. The answer from Question 43b by the appropriate market value for potatoes;

_______________ cwt/yr × _______________ $/cwt = _______________ $/yr

c. The answer from Question 43 by the appropriate market value for sugar beets.

_______________ tons/yr × _______________ $/ton = _______________ $/yr

45. Total the answers from Questions 44a, 44b, and 44c to obtain the potential earnings from directly marketing crops without making EtOH.

_______________ $/yr + _______________ $/yr + _______________ $/yr = _______________ $/yr

Compare the answers from Questions 45 and 42. If Question 45 is as large or nearly as large as the answer from Question 42, the construction of an ethanol plant of this size cannot be justified on a purely economic basis. Consider scaling down to a size that fills your own fuel needs and recompute Questions 28 through 45. If Question 42 is considerably larger (two to three times) than Question 45, you can consider increasing your plant size within the bounds of the answers to Question 22 (market) and Question 27 (production potential). Care must be taken to assess local competition and market share as you expand plant size.

If a market share exists or if there is good reason to believe that you can acquire a share by superior techniques, the initial plant sizing must accurately reflect this realistic market share.

46. a. Multiply the initial plant production capacity (in gal EtOH/hr) by 16 gal of water per gal EtOH production capacity.

_______________ gal EtOH/hr × 16 gal H_2O/gal EtOH = _______________ gal H_2O/hr

b. Can the answer to Question 46a be realistically achieved in your area? If yes, no adjustment to chosen plant size needs to be made to account for water availability. If no, reduce plant size to realistically reflect available water.

FACTORS DETERMINING COST OF PLANT

The cost of the equipment you choose will be a function of the labor available, the maintenance required, the heat source selected, and the type of operating mode. Select a plant design that accomplishes your determined production rate and fits your production schedule considering each of the following factors.

Labor Requirements

47. a. How much time during the normal farming routine can you dedicate to running the ethanol plant? ________ hr

    b. Do you have any hired help or other adult family members, and if so, how much time can they dedicate to running the ethanol plant? ________ hr

    c. Can you or your family or help dedicate time at periodic intervals to operating the ethanol plant? ________ hr

If labor is limited, a high degree of automatic control is indicated.

Maintenance

48. What are your maintenance capabilities and equipment? ________

Heat Source

49. What is the least expensive heat source available that is practical in your area and does not require the use of petroleum fuel? ________

Cost Calculations

50. List all of the plant components and their costs.

    a. storage bins $ ________________

    b. grinding mill $ ________________

    c. meal hopper $ ________________

    d. cookers $ ________________

    e. fermenters $ ________________

    f. distillation columns $ ________________

    g. storage tanks (product and coproduct) $ ________________

h. pumps $ ______

i. controllers $ ______

j. pipes and valves $ ______

k. metering controls $ ______

l. microprocessors $ ______

m. safety valves $ ______

n. heat exchangers $ ______

o. instrumentation $ ______

p. insulation $ ______

q. boiler $ ______

r. fuel-handling equipment $ ______

s. feedstocks-handling equipment $ ______

t. storage tanks (stillage) $ ______

u. stillage treatment equipment (screen, dryers, etc.) $ ______

v. CO_2 handling equipment $ ______

w. ethanol dehydration equipment $ ______

TOTAL $ ______

51. Determine operating requirements for cost.

Plant capacity = ______ gal anhydrous ethanol/hr

Production = ______ gal/hr × hr operation/yr = ______ gal/yr

Feed materials = production ______ gal/yr ÷ ______ gal/bu = ______ bu/yr

	$/yr	*$/gal*
a. Operating Costs		
Feed materials		
grain ($/bu ÷ gal anhydrous ethanol/bu = $/gal)	______	______

	$/yr	$/gal
or ($/bu × bu/yr = $/yr)	______	______
Supplies		
enzymes	______	______
other	______	______
Fuel for plant operation	______	______
Waste disposal	______	______
Operating labor (operating crew × hours of operation per year × $/hr = $/yr)	______	______
Total operating costs	______	______
b. Maintenance Costs		
Routine scheduled maintenance	______	______
Labor (maintenance crew staff × hrs/yr × $/hr)	______	______
Supplies and replacement parts	______	______
Maintenance equipment rental	______	______
Unscheduled maintenance (estimated)	______	______
Labor	______	______
Supplies	______	______
Maintenance equipment	______	______
Total maintenance costs	______	______
c. Capital or Investment Costs		
Plant equipment costs	______	______
Land	______	______
Inventory	______	______
grain	______	______
supplies	______	______
ethanol	______	______

	$/yr	$/gal
spare parts	________	________
Total	________	________
Taxes	________	________
Insurance	________	________
Depreciation	________	________
Interest on loan or mortgage	________	________
Total capital investment costs	________	________
TOTAL COSTS (Totals of a, b, and c)	________	________

Financial Requirements

52. Capital costs:

Item	*Cost Estimate*	*Considerations*
Real estate	________	________
Buildings	________	________
Equipment	________	________
Business formation	________	________
Equipment installation	________	________
Licensing costs	________	________

53. Operating costs:

Item	*Cost Estimate*	*Considerations*
Labor	________	________
Maintenance	________	________
Taxes	________	________
Supplies	________	Includes raw materials, additives, enzymes, yeast, and water.

Delivery	________	________
Expenses	________	Includes electricity and fuel(s).
Insurance	________	________
Interest on debt	________	Includes interest on long- and short-term loans.
Bonding	________	________

54. Start-up working capital:

Item	*Cost Estimate*	*Considerations*
Mortgage	________	Principal payments only, for first few months.
Cash to carry accounts receivable for 60 days	________	________
Cash to carry a finished goods inventory for 30 days	________	________
Cash to carry a raw material inventory for 30 days	________	________

55. Working capital:

Item	*Cost Estimate*	*Considerations*
Mortgage	________	________

ASSETS (TOTAL NET WORTH)

56. List all items owned by the business entity operating the ethanol plant.

Item	*Value*
________	________
________	________
________	________
________	________
________	________
________	________
________	________
________	________
________	________
________	________

ORGANIZATIONAL FORM

57. Are you willing to assume the costs and risks of running your own EtOH-production facility? ______

58. Are you capable of handling the additional taxes and debts for which you will be personally liable as a single proprietor? (Includes interest on long- and short-term loans.) ______

59. Is your farm operation large enough or are your potential markets solid enough to handle an EtOH-production facility as a single proprietor? ______

60. Is your credit alone sufficient to provide grounds for capitalizing a single proprietorship? ______

61. Will a partner(s) enhance your financial position or supply needed additional skills? ______

62. a. Do you need a partner to get enough feedstock for your EtOH-production facility? ______

    b. Are you willing to assume liabilities for product and partner? ______

63. Is your intended production going to be of such a scale as to far exceed the needs for your own farm or several neighboring farms? ______

64. Do you need to incorporate in order to obtain adequate funding? ______

65. Will incorporation reduce your personal tax burden? ______

66. Do you wish to assume product liability personally? ______

67. How many farmers in your area would want to join a cooperative? ______

68. Do you plan to operate in a centralized location to produce EtOH for all the members? ______

69. Is your main reason for producing EtOH to service the needs of the cooperative members or of others, or to realize a significant profit? ______

FINANCING

If you are considering borrowing money, you should have a clear idea of what your chances will be beforehand. The following questions will tell you whether debt financing is a feasible approach to your funding problem.

70. a. How much money do you already owe? ____________

    b. What are your monthly payments? ____________

71. How much capital will you have to come up with yourself in order to secure a loan? ____________

72. Have you recently been refused credit? ____________

73. a. How high are the interest rates going to be? ____________

    b. Can you cover them with your projected cash flow? ____________

74. If the loan must be secured or collateralized, do you have sufficient assets to cover your debt? ____________

If you are already carrying a heavy debt load and/or your credit rating is low, your chances of obtaining additional debt financing is low and perhaps you should consider some other type of financing. Insufficient collateral, exorbitant interest rates, and low projected cash flow are also negative indicators for debt financing.

The choice between debt and equity financing will be one of the most important decisions you will have to face since it will affect how much control you will ultimately have over your operation. The following questions deal with this issue, as well as the comparative cost of the two major types of financing.

75. How much equity do you already have? ____________

76. Do you want to maintain complete ownership and control of your enterprise? ____________

77. Are you willing to share ownership and/or control if it does not entail more than a minority share? ____________

78. Will the cost of selling the stock (broker's fee, bookkeeper, etc.) be more than the interest you would have to pay on a loan? ____________

If you are reluctant to relinquish any control over your operation, you would probably be better off seeking a loan. On the other hand, if your chances of obtaining a loan are slim, you might have to trade off some personal equity in return for a better borrowing position.

79. Do you have other funds or materials to match with federal funds? (It is usually helpful.) ____________

80. Do you live in a geographical area that qualifies for special funds? ____________

81. Will you need continued federal support at the end of your grant period? ____________

82. Are you going to apply for grant funds as an individual, as a nonprofit corporation, or as a profit corporation? ____________

83. Are you a private nonprofit corporation? ____________

84. Is there something special about your alcohol facility that would make it attractive to certain foundations? ____________

You should now have a good ideas as to where you are going to seek your initial funding. Remember that most new businesses start up with a combination of funding sources. It is important to maintain a balance that will give you not only sufficient funding when you need it, but also the amount of control over your operation that you would like to have.

Completion of these worksheets can lead to an initial decision on the feasibility to proceed. However, this should not be construed as a final decision, but rather a step in that process.

If the financial requirements are greater than the capability to obtain financing, it does not necessarily mean the entire concept will not work. Rather, the organizational form can be reexamined and/or the production base expanded in order to increase financing capability.

Chapter 3

Basic Production Processes

The production of ethanol is an established process that involves some of the knowledge and skill used in normal farm operations, especially the cultivation of plants; it is also a mix of technologies which include microbiology, chemistry, and engineering. Basically, fermentation is a process in which microorganisms such as yeasts convert simple sugars to ethanol and carbon dioxide. Some plants directly yield simple sugars; others produce starch or cellulose that might be converted to sugar. The sugar obtained must be fermented and the beer produced must then be distilled to obtain fuel-grade ethanol. Each step is discussed individually. This chapter provides an introduction to the basic production process and model plant characteristics. A basic flow diagram of ethanol production is shown in Fig. 3-1.

PREPARATION OF FEEDSTOCKS

Feedstocks can be selected from among many plants that either produce simple sugars directly or produce starch and cellulose. This means there is considerable diversity in the initial processing, but some features are universal:

- Simple sugars must be extracted from the plants that directly produce them
- Starch and cellulose must be reduced from their complex form to basic glucose
- Stones and metallic particles must be removed

The last feature must be taken care of first. Destoning equipment and magnetic separators can be used to remove stones and metallic particles. Root crops require other approaches since mechanical harvesters do not differentiate between rocks and potatoes or beets of the same size. Water jets or flumes may be needed to accomplish this.

The simple sugars from such plants as sugarcane, sugar beets, or sorghum can be obtained by crushing or pressing the material. The low sugar bagasse and pulp which remain after pressing can be leached with water to remove residual sugars. The fibrous cellulosic material theoretically could be

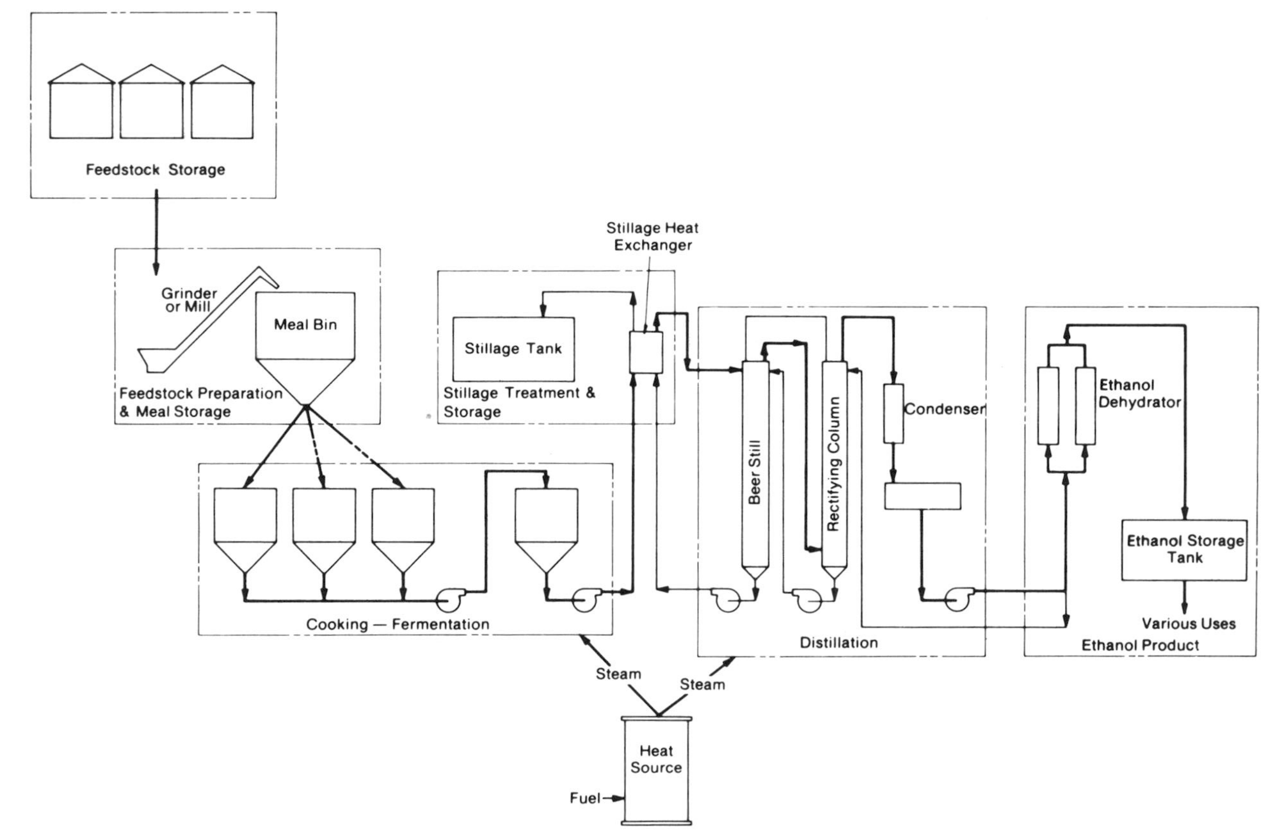

Source: *Fuel from Farms—A Guide to Small-Scale Ethanol Production*, SERI, Denver, 1980.

Figure 3-1. Ethanol-production flow diagram.

treated chemically or enzymatically to produce more sugar. However, no commercially available process currently exists.

Grains and potatoes are commonly used starchy feedstocks. Starch is roughly 20 percent amylose (a water-soluble carbohydrate) and 80 percent amylopectin (which is not soluble in water). These molecules are linked together by means of a bond that can be broken with relative ease. Cellulose, which is also made up of glucose, differs from starch mainly in the bond between glucose units.

Starch must be broken down because yeast can only act on simple sugars to produce ethanol. This process requires that the material be broken mechanically into the smallest practical size by milling or grinding, thereby breaking the starch walls to make all of the material available to the water. From this mixture, a slurry can be prepared, and it can be heated to temperatures high enough to break the cell walls of the starch. This produces complex sugars which can be further reduced by enzymes to the desired sugar product.

CONVERSION OF STARCHES BY ENZYMATIC HYDROLYSIS

Consider the preparation of starch from grain as an example of enzymatic hydrolysis. The intent is to produce a 14-20 percent sugar solution with water and whole grain. Grain is a good source of carbohydrate, but to gain access to the carbohydrate, the grain must be ground. A rule of thumb is to operate grinders so that the resulting mash can pass a 20-mesh screen. This assures that the carbohydrate is accessible and the solids can be removed with a finer screen if desired. If the grain is not ground finely enough, the resultant lumpy material is not readily accessible for enzymatic conversion to sugar. The next step is to prepare a slurry by mixing the meal directly with water. Stirring should be adequate to prevent the formation of lumps and enhance enzyme contact with the starch (thus speeding liquefaction).

High-temperature and high-pressure processes may require a full-time operator, thus making it difficult to integrate into farming operations. Therefore, when deciding which enzyme to purchase, consideration should be given to selecting one that is active at moderate temperature, i.e., 200°F (93°C), near-ambient pressure, and nearly neutral pH. The acidity of the slurry can be adjusted by addition of dilute basic solution (e.g., sodium hydroxide) if the pH is too low, and addition of concentrated sulfuric acid or lactic acid if the pH is too high.

The enzyme should be added to the slurry in the proper proportion to the quantity of starch to be converted. If not, liquefaction is incomplete or takes too long to complete for practical operations. Enzymes vary in activity but thermophyllic bacterial amylases, which are commercially available, can be added at rates slightly greater than ¾ ounce per bushel of meal. Rapid dispersion of the dry enzyme is best accomplished by mixing a premeasured quantity with a small volume of warm water prior to addition to the slurry. Liquefaction should be conducted in the specific temperature range and pH suggested by the supplier of the enzyme used.

After the enzyme is added, the grain mash is heated to break the cell walls of the starch. However, the enzyme must be added before the temperature is raised because once the cell walls rupture, a gel forms and it becomes almost impossible to accomplish good mixing of the enzyme with the starch. The rupture of cell walls, which is caused by heating in hot water, is called gelatinization because the slurry (which is a suspension of basically insoluble material in water) is converted to a high-viscosity solution. Under slow cooking conditions and normal atmospheric pressure, gelatinization can be expected to occur around 140°F (60°C).

The temperature is then raised to the optimal functional range for the enzyme and held for a period of time sufficient to completely convert the starch to soluble dextrins (polymeric sugars). There are commercially available enzymes that are most active around 200°F (93°C) and require a hold time of 2 ½ hours if the proper proportion of enzyme is used. When this step is complete, the slurry has been converted to an aqueous solution of dextrines. Care must be taken to assure that the starch-conversion step is complete because the conditions for the glucose-producing enzyme (glucoamylase), which is introduced in the next step, are significantly different from those for liquefaction.

The next step, saccharification, is the conversion of dextrins to simple sugars, i.e., glucose. The mash temperature is dropped to the active range of the glucoamylase, the enzyme used for saccharification, and the pH of the solution is adjusted to optimize conversion activity. The pH is a critical factor because the enzymatic activity virtually ceases when the pH is above 6.5. Glucoamylase is added in the proportion required to convert the amount of sugar available. Again, depending on the variety selected and its activity, the actual required quantity of enzyme varies.

After the enzyme is added, the temperature of the mash must neither exceed 140°F (60°C) nor drop below 122°F (50°C) during the saccharification step or the enzyme activity will be greatly reduced. The mash, as in the prior step, must be stirred continuously to assure intimate contact of enzyme and dextrin. The mash should be held at the proper temperature and pH until conversion of the dextrin to glucose is complete.

FERMENTATION

Fermentation is the conversion of an organic material from one chemical form to another using enzymes produced by living microorganisms. In general, these bacteria are classified according to their tolerance of oxygen. Those that use oxygen are called aerobic and those that do not are called anaerobic. Those that start with oxygen but continue to thrive after all of the available oxygen is consumed are called facultative organisms. The yeast used to produce ethanol is an example of this type of facultative anaerobe. The breakdown of glucose to ethanol involves a complex sequence of chemical reactions which can be summarized as

$$\underset{\text{(glucose)}}{C_6H_{12}O_6} \longrightarrow \underset{\text{(ethanol)}}{2C_2H_5OH} + \underset{\text{(carbon dioxide)}}{2CO_2} + \text{heat}$$

Actual yields of ethanol generally fall short of predicted theoretical yields because about 5 percent of the sugar is used by the yeast to produce new cells and minor products such as glycerols, acetic acid, lactic acid, and fusel oils.

Yeasts are the microorganisms responsible for producing the enzymes which convert sugar to ethanol. Yeasts are single-cell fungi widely distributed in nature, commonly found in wood, dirt, and plant matter, and on the surface of fruits and flowers. They are spread by wind and insects. Yeasts used in ethanol productions are members of the genus *Saccharomyces*. These yeasts are sensitive to a wide variety of variables that potentially affect ethanol production; however, pH and temperature are the most influential of these variables. *Saccharomyces* are most effective in pH ranges between 3.0 and 5.0 and in temperatures between 80°F (27°C) and 90°F (35°C). The length of time required to convert a mash to ethanol is dependent on the number of yeast cells per quantity of sugar. The greater the number initially added, the faster the job is complete. However, there is a point of diminishing returns.

Yeast strains, nutritional requirements, sugar concentration, temperature, infections, and pH influence yeast efficiency. They are described as follows.

Yeast Strains

Yeasts are divided informally into top and bottom yeasts according to the location in the mash in which most of the fermentation takes place. The top yeasts, *Saccharomyces cerevisiae*, produce carbon dioxide and ethanol vigorously and tend to cluster on the surface of the mash. Producers of distilled spirits generally use top yeasts of high activity to maximize ethanol yield in the shortest time; producers of beer tend to use bottom yeasts which produce lower ethanol yields and require longer times to complete fermentation. Under normal brewing conditions, top yeasts tend to flocculate (aggregate together into clusters) and to separate out from the solution when fermentation is complete. The various strains of yeast differ considerably in their tendency to flocculate. Those strains with an excessive tendency toward premature flocculation tend to cut short fermentation and thus reduce ethanol yield. This phenomenon, however, is not singularly a trait of the yeast. Fermentation conditions can be an influencing factor. The cause of premature flocculation seems to be a function of the pH of the mash and the number of free calcium ions in solution. Hydrated lime, which is sometimes used to adjust pH, contains calcium and may be a contributory factor.

Nutritional Requirements

Yeasts are plants, despite the fact that they contain no chlorophyll. As such, their nutritional requirements must be met or they cannot produce ethanol as fast as desired. Like the other living things that a farmer cultivates and nurtures, an energy source such as carbohydrate must be provided for metabolism. Amino acids must be provided in the proper proportion and major chemical elements such as carbon, nitrogen, phosphorus, and others must be available to promote cell growth. Some species flourish without vitamin supplements, but in most cases cell growth is enhanced when B vitamins are available. Carbon is provided by the many carbonaceous substances in the mash.

The nitrogen requirement varies somewhat with the strain of yeast used. In general, it should be supplied in the form of ammonia, ammonium salts, amino acids, peptides, or urea. Care should be taken to sterilize farm sources of urea to prevent contamination of the mash with undesired microbial strains. Since only a few species of yeasts can assimilate nitrogen from nitrates, this is not a recommended source of nitrogen. Ammonia is usually the preferred nitrogen form, but in its absence, the yeast can break up amino acids to obtain it. The separation of solids from the solution prior to fermentation removes the bulk of the protein and, hence, the amine source would be removed also. If this option is exercised, an ammonia supplement must be provided or yeast populations will not propagate at the desired rates and fermentation will take an excessive amount of time to complete. However, excessive amounts of ammonia in solution must be avoided because it can be lethal to the yeast.

Although the exact mineral requirements of yeasts cannot be specified because of their short-term evolutionary capability, phosphorus and potassium can be identified as elements of prime importance. Care should be taken not to introduce excessive trace minerals, because those which the yeast cannot use increase the osmotic pressure in the system. (Osmotic pressure is due to the physical imbalance in concentration of chemicals on either side of a membrane. Since yeasts are cellular organisms, they are enclosed by a cell wall. An excessively high osmotic pressure can cause the rupture of the cell wall which in turn kills the yeast.)

Sugar Concentration

There are two basic concerns that govern the sugar concentration of the mash: (1) excessively high sugar concentrations can inhibit the growth of yeast cells in the initial stages of fermentation, and (2) high ethanol concentrations are lethal to yeast. If the concentration of ethanol in the solution reaches levels high enough to kill yeast before all the sugar is consumed, the quantity of sugar that remains is wasted. The latter concern is the governing control. Yeast growth problems can be overcome by using large inoculations to start fermentation. *Saccharomyces* strains can utilize effectively all of the sugar in solutions that are 16 percent to 22 percent sugar while producing a beer that ranges from 8 percent to 12 percent ethanol by volume.

Temperature

Fermentation is strongly influenced by temperature, because the yeast performs best in a specific temperature range. The rate of fermentation increases with temperature in the temperature range between 80°F (27°C) and 95°F (35°C). Above 95°F (35°C), the rate of fermentation gradually drops off, and ceases altogether at temperatures above 109°F (43°C). The actual temperature effects vary with different yeast strains and typical operating conditions are generally closer to 80°F (27°C) than 95°F (35°C). This choice is usually made to reduce ethanol losses by evaporation from the beer. For every 9°F (5°C) increase in temperature, the ethanol evaporation rate increases 1.5 times. Since scrubbing equipment is required to recover the ethanol lost by evaporation and the cost justification is minimal on a small scale, the lower fermentation temperature offers advantages of simplicity.

The fermentation reaction gives off energy as it proceeds (about 500 Btu per pound of ethanol produced). There will be a normal heat loss from the fermentation tank as long as the temperature outside the tank is less than the inside. Depending on the location of the plant, this will depend on how much colder the outside air is than the inside air and on the design of the fermenter. In general, this temperature difference will not be sufficient to take away as much heat as is generated by the reaction except during the colder times of the year. Thus, the fermenters must be equipped with active cooling systems, such as cooling coils and external jackets, to circulate air or water for convective cooling.

Infections

Unwanted microbial contaminants can be a major cause of reductions in ethanol yield. Contaminants consume sugar that would otherwise be available for ethanol production and produce enzymes that modify fermentation conditions, thus yielding a drastically different set of products. Although infection must be high before appreciable quantities of sugar are consumed, the rate at which many bacteria multiply exceeds yeast propagation. Therefore, even low initial levels of infection can greatly impair fermentation. In a sense, the start of fermentation is a race among the microorganisms present to see which can consume the most. The objective is the selective culture of a preferred organism. This means providing the conditions that are most favorable to the desired microorganism. As mentioned previously, high initial sugar concentrations inhibit propagation of *Saccharomyces cerevisiae* because it is not an osmophylic yeast (i.e., it cannot stand the high osmotic pressure caused by the high concentration of sugar in the solution). This immediately gives an advantage to any osmophylic bacteria present.

Unwanted microbes can be controlled by using commercially available antiseptics. These antiseptics are the same as those used to control infections in humans, but are less expensive because they are manufactured for industrial use.

DISTILLATION

The purpose of the distillation process is to separate ethanol from the ethanol-water mixture. There are many means of separating liquids comprised of two or more components in solution. In general, for solutions comprised of components of significantly different boiling temperatures, distillation has proved to be the most easily operated and thermally efficient separation technique.

At atmospheric pressure, water boils around 212°F (100°C) and ethanol boils around 172°F (77.7°C). It is this difference in boiling temperature that allows for distillative separation of ethanol-water mixtures. If a pan of an ethanol and water solution is heated on the stove, more ethanol molecules leave the pan than water molecules. If the vapor leaving the pan is caught and condensed, the concentration of ethanol in the condensed liquid will be higher than in the original solution, and the solution remaining in the pan will be lower in ethanol concentration. If the condensate from this step is again heated and the vapors condensed, the concentration of ethanol in the condensate will again be higher. This process could be repeated until most of the ethanol was concentrated in one phase. Unfortunately, a constant boiling mixture (azeotrope) forms at about 96 percent ethanol. This means that when a pan containing a 96-percent-ethanol solution is heated, the ratio of ethanol molecules to water molecules in the condensate remains constant. Therefore, no concentration enhancement is achieved beyond this point by the distillation method.

The system shown in Fig. 3-2 is capable of producing 96-percent-pure ethanol, but the amount of final product will be quite small. At the same time there will be a large number of products of intermediate ethanol-water compositions that have not been brought to the required product purity. If, instead of discarding all the intermediate concentrations of ethanol and water, they were recycled to a point in the system where the concentration was the same, we could retain all the ethanol in the system. Then, if all of these steps were incorporated into one vessel, the result would be a distillation column. The advantages of this system are that no intermediate product is discarded

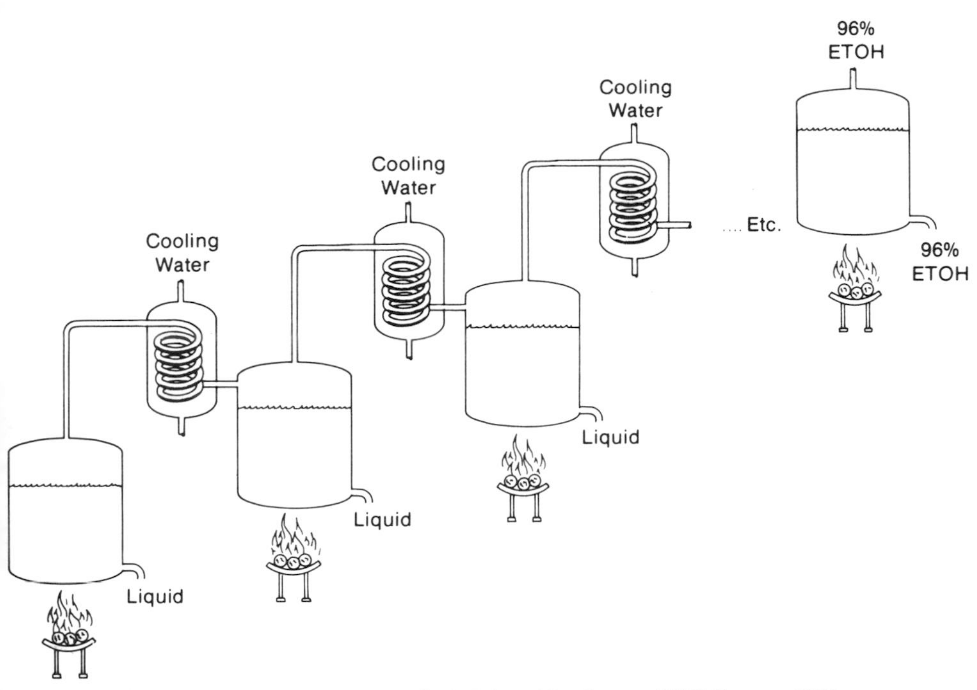

Source: *Fuel from Farms—A Guide to Small-Scale Ethanol Production*, SERI, Denver, 1980.

Figure 3-2. Basic process of successive distillation to increase concentration of ethanol.

and only one external heating and one external cooling device are required. Condensation at one stage is affected when vapors contact a cooler stage above it, and evaporation is affected when liquid contacts a heating stage below. Heat for the system is provided at the bottom of the distillation column; cooling is provided by a condenser at the top where the condensed product is returned in a process called reflux. It is important to note that without this reflux the system would return to a composition similar to the mixture in the first pan that was heated on the stove.

The example distillation sieve tray column given in Fig. 3-3 is the most common single-vessel device for carrying out distillation. The liquid flows down the tower under the force of gravity while the vapor flows upward under the force of slight pressure drop.

The portion of the column above the feed is called the rectifying or enrichment section. The upper section serves primarily to remove the component with the lower vapor pressure (water) from the upflowing vapor, thereby enriching the ethanol concentration. The portion of the column below the feed, called the stripping section, serves primarily to remove or strip the ethanol from the down-flowing liquid.

Figure 3-4 is an enlarged illustration of a sieve tray. In order to achieve good mixing between phases and to provide the necessary disengagement of

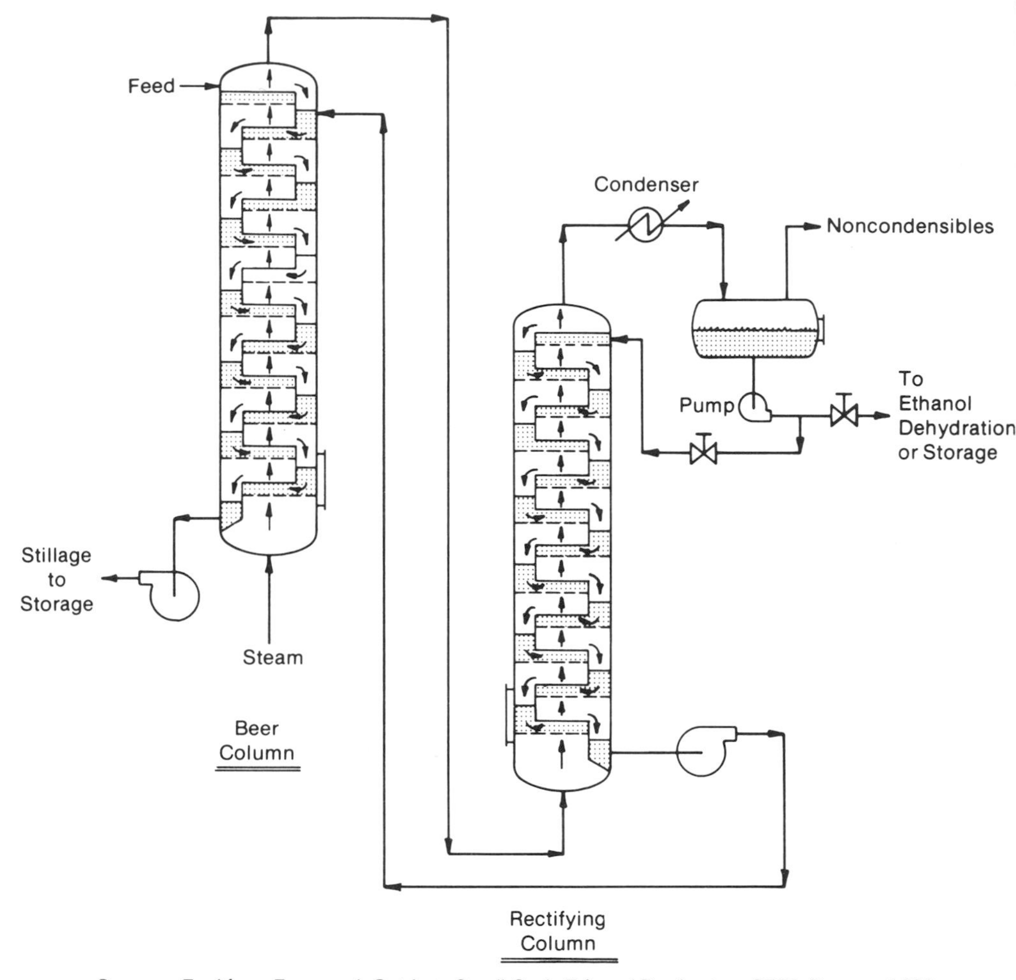

Source: *Fuel from Farms—A Guide to Small-Scale Ethanol Production*, SERI, Denver, 1980.

Figure 3-3. Schematic diagram of sieve tray distillation of ethanol.

vapor and liquid between stages, the liquid is retained on each plate by a weir (a dam that regulates flow) over which the solution flows. The effluent liquid then flows down the downcomer to the next stage. The downcomer provides sufficient volume and residence time to allow the vapor-liquid separation.

It is possible to use several devices other than sieve tray columns to achieve the countercurrent flow required for ethanol-water distillation. A packed column is frequently used to effect the necessary vapor-liquid contacting. The packed column is filled with solid material shaped to provide a large surface area for contact. Countercurrent liquid and vapor flows proceed in the same way described for the sieve tray column.

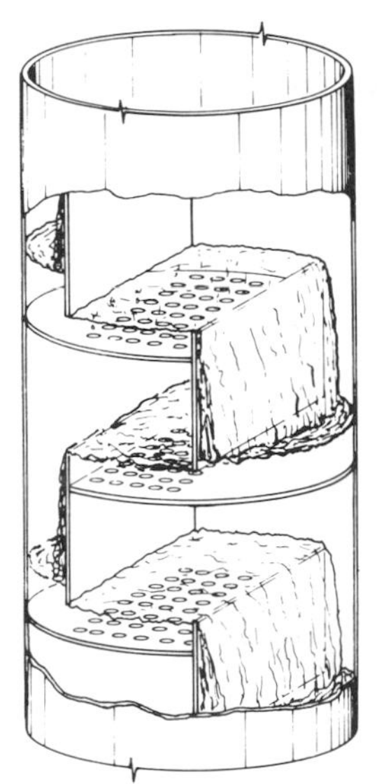

Source: *Fuel from Farms—A Guide to Small-Scale Ethanol Production*, SERI, Denver, 1980.

Figure 3-4. Enlarged illustration of sieve tray.

Fermentation is effected by a variety of conditions. The more care used in producing optimum conditions, the greater the ethanol yield. Distillation can range from the simple to the complex. Fortunately, the middle line works quite satisfactorily for on-farm ethanol production.

Plant Design Criteria

The criteria affecting the decision to produce ethanol and to establish a production facility can be categorized into two groups: fixed and variable. The fixed criteria are basically how much ethanol and coproducts can be produced and sold. This chapter is concerned with the second set of variable criteria and their effect on plant design.

Plant design is delineated through established procedures which are complex and interrelated. The essential elements, however, are described here.

The first step is to define a set of criteria which affect plant design. These criteria (not necessarily in order of importance) are

- Amount of labor that can be dedicated to operating a plant
- Size of initial investment and operating cost that can be managed in relation to the specific financial situation and/or business organization
- Ability to maintain equipment both in terms of time to do it and anticipated expense

- Federal, state, and local regulations on environmental discharges, transportation of product, licensing, etc.
- Intended use (on-farm use and/or sales) of chemicals
- Desired form of coproducts
- Safety factors
- Availability and expense of heat source
- Desired flexibility in operation and feedstocks

The second step is to relate these criteria to the plant as a whole in order to set up a framework or context for plant operations. The third step is complex and involves relating the individual systems or components of production to this framework and to other connected systems within the plant. Finally, once the major systems have been defined, process control systems can be integrated where necessary. This design process leads to specifying equipment for the individual systems and process control.

After the process is discussed from overall plant considerations through individual system considerations to process control, representative ethanol model plants are described.

OVERALL PLANT CONSIDERATIONS

Before individual systems and their resulting equipment specifications are examined, the criteria listed above are examined in relation to the overall plant. This establishes a set of constraints against which individual systems can be correlated.

Required labor. The expense the operation can bear for labor must be considered. To some extent the latter concern is modified by the size of plant selected (the expense for labor is less per gallon the more gallons produced). If it is possible to accomplish the required tasks within the context of daily farming activities, additional outside labor will not be required. A plant operated primarily by one person should, in general, require attention only twice or at most three times a day. If possible, the time required at each visit should not exceed 3 hours. The labor availability directly affects the amount and type of control and instrumentation that the plant requires, but it is not the sole defining criteria for plant specification.

Maintenance. The plant should be relatively easy to maintain and not require extensive expertise or expensive equipment.

Feedstocks. The process should use crop material in the form in which it is usually or most economically stored (e.g., forage crops should be stored as ensilage).

Use. The choice of whether to produce anhydrous or lower-proof ethanol depends on the intended use or market and may also have seasonal dependencies. Use of lower-proof ethanols in spark ignition tractors and trucks poses no major problem during summertime (or other periods of moderate ambient temperature). Any engine equipped for dual injection does not require anhydrous ethanol during moderate seasons (or in moderate climates). If the ethanol is to be sold to blenders for use as gasohol, the capability to produce anhydrous ethanol may be mandatory.

Heat source. Agricultural residues, coal, waste wood, municipal waste, producer gas, geothermal water, solar, and wind are the preferred possibilities for heat sources. Examples of these considerations are shown in Table 3-1. Each poses separate requirements on the boiler selected, the type and amount of instrumentation necessary to fulfill tending (labor) criteria, and

Table 3-1. Heat-Source Selection Considerations

Heat Source	Heating Value (dry basis)	Form	Special Equipment Req'd	Boiler Types	Source	Particular Advantages	Particular Disadvantages
Agriculture Residuals	3,000-8,000 Btu/lb	Solid	Handling and feeding equipment: collection equipment	Batch burner-fire tube; fluidized bed	Farm	Inexpensive; produced on-farm	Low bulk density; requires very large storage area
Coal	9,000-12,000 Btu/lb	Solid	High sulfur coal requires stack scrubber	Conventional grate-fire tube; fluidized bed	Mines	Widely available demonstrated technology for combustion	Potentially expensive; no assured availability; pollution problems
Waste Wood	5,000-12,000 Btu/lb	Solid	Chipper or log feeder	Conventional fluidized bed	Forests	Clean burning; inexpensive where available	Not uniformly available
Municipal Solid Waste	8,000 Btu/lb	Solid	Sorting equipment	Fluidized bed or conventional fire tube	Cities	Inexpensive	Not widely available in rural areas
Pyrolysis Gas		Gas	Pyrolyzer-fluidized bed	Conventional gas-boiler	Carbonaceous materials	Can use conventional gas-fired boilers	Requires additional piece of equipment
Geothermal	N.A.	Steam/ hot water	Heat exchanger	Heat exchanger water tube	Geothermal source	Fuel cost is low	Capital costs for well and heat exchanger can be extremely high
Solar	N.A.	Radiation	Collectors, concentrators, storage batteries, or systems	Water tube	Sun	Fuel cost is low	Capital costs can be high for required equipment
Wind	N.A.	Kinetic energy	Turbines, storage batteries, or systems	Electric	Indirect solar	Fuel cost is low	Capital costs can be high for required equipment

Source: *Fuel From Farms – A Guide to Small-Scale Ethanol Production,* SERI, Denver, 1980.

the cash flow necessary to purchase the necessary quantity (if not produced on-farm). This last consideration is modified by approaches that minimize the total plant energy demand.

Safety. An ethanol plant poses several specific hazards. Some of these are enumerated in Table 3-2 along with options for properly addressing them. Detailed safety information is provided in Chapter 9.

Coproduct form and generation. Sale or use of the coproducts of ethanol production is an important factor in overall profitability. Markets must be carefully weighed to assure that competitive influences do not diminish the value of the coproduct that results from the selected system. In some areas, it is conceivable that the local demand can consume the coproduct produced by many closely located small plants; in other areas, the local market may only be able to absorb the coproducts from one plant. If the latter situation occurs, this either depresses the local coproduct market value or encourages the purchase of equipment to modify coproduct form or type so that it can be transported to different markets.

Flexibility in operation and feedstocks. Plant profitability should not hinge on the basis of theoretical maximum capacity. Over a period of time, any of a myriad of unforeseen possibilities can interrupt operations and depress yield. Market (or redundant commodity) variables or farm operation considerations may indicate a need to switch feedstocks. Therefore, the equipment for preparation and conversion should be capable of handling cereal grain and at least one of the following:

- Ensiled forage material
- Starchy roots and tubers
- Sugar beets, or other storable, high-sugar-content plant parts

Initial investment and operating costs. All of the preceding criteria impact capital or operating costs. Each criterion can influence production rates which, in turn, change the income potential of the plant. An optimum investment situation is reached only through repeated iterations to balance equipment requirements against cost in order to achieve favorable earnings.

INDIVIDUAL SYSTEM CONSIDERATIONS

Design considerations define separate specific jobs which require different tools or equipment. Each step depends on the criteria involved and influences related steps. Each of the components and systems of the plant must be examined with respect to these criteria. Figure 3-5 diagrams anhydrous ethanol production. The typical plant that produces anhydrous ethanol contains the following systems and/or components: feedstock handling and storage, conversion of carbohydrates to simple sugars, fermentation, distillation, drying ethanol, and stillage processing.

Table 3-2. Ethanol Plant Hazards

Hazards	Precautions
1. Overpressurization; explosion of boiler	• Regularly maintained/checked safety boiler "pop" valves set to relieve when pressure exceeds the maximum safe pressure of the boiler or delivery lines.
	• Strict adherence to boiler manufacturer's operating procedure.
	• If boiler pressure exceeds 20 psi, acquire ASME boiler operator certification. Continuous operator attendance required during boiler operation.
2. Scalding from steam gasket leaks	• Place baffles around flanges to direct steam jets away from operating areas.
	• (Option) Use welded joints in all steam delivery lines.
3. Contact burns from steam lines	• Insulate all steam delivery lines.
4. Ignition of ethanol leaks/fumes or grain dust	• If electric pump motors are used, use fully enclosed explosion-proof motors.
	• (Option) Use hydraulic pump drives; main hydraulic pump and reservoir should be physically isolated from ethanol tanks, dehydration section, distillation columns, condenser.
	• Fully ground all equipment to prevent static electricity build-up.
	• Never smoke or strike matches around ethanol tanks, dehydration section, distillation columns, condenser.
	• Never use metal grinders, cutting torches, welders, etc. around systems, or equipment containing ethanol. Flush and vent all vessels prior to performing any of these operations.
5. Handling acids/bases	• Never breathe the fumes of concentrated acids or bases.
	• Never store concentrated acids in carbon steel containers.
	• Mix or dilute acids and bases slowly—allow heat of mixing to dissipate.
	• Immediately flush skin exposed to acid or base with copious quantities of water.
	• Wear goggles whenever handling concentrated acids or bases; flush eyes with water and immediately call physician if any gets in eyes.
	• Do not store acids or bases overhead work areas or equipment.
	• Do not carry acids or bases in open buckets.
	• Select proper materials of construction for all acid or base storage containers, delivery aides, valves, etc.
6. Suffocation	• Never enter the fermenters, beer well, or stillage tank unless they are properly vented.

Source: *Fuel From Farms – A Guide to Small-Scale Ethanol Production,* SERI, Denver, 1980.

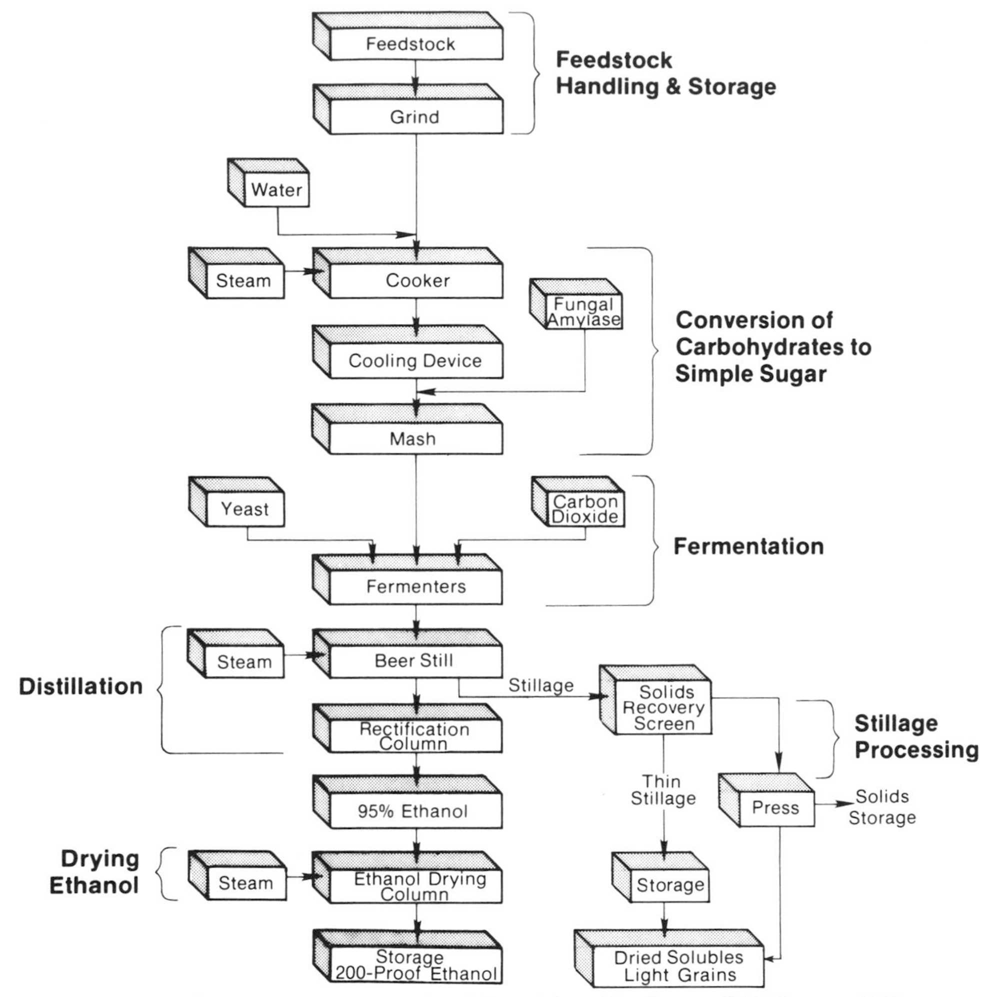

Source: *Fuel from Farms—A Guide to Small-Scale Ethanol Production*. SERI, Denver, 1980.

Figure 3-5. Anhydrous-ethanol-production flow chart.

Feedstock Handling and Storage

Grain. A small plant should be able to use cereal grains. Since grains are commonly stored on farms in large quantity, and since grain-growing farms have the basic equipment for moving the grain out of storage, handling should not be excessively time-consuming. The increasing popularity of storing grain at high moisture content provides advantages since harvesting can be done earlier and grain drying can be avoided. When stored as whole grain, the handling requirements are identical to those of dry grain. If the grain is ground and stored in a bunker, the handling involves additional labor since it must be removed from the bunker and loaded into a grainery from which it can be fed by an auger into the cooker. This operation probably could be performed once each week, so the grains need not be ground daily as with whole grain for a small-scale plant.

Roots and tubers. Potatoes, sugar beets, fodder beets, and Jerusalem artichokes are generally stored whole in cool, dry locations to inhibit spontaneous fermentation by the bacteria present. The juice from the last three can be extracted but it can only be stored for long periods of time at very high sugar concentrations. This requires expensive evaporation equipment and large storage tanks.

Belt conveyers will suffice for handling these root crops and tubers. Cleaning equipment should be provided to prevent dirt and rocks from building up in the fermentation plant.

Sugar crops. Stalks from sugarcane, sweet sorghum, and Jerusalem artichokes cannot be stored for long periods of time at high moisture content. Drying generally causes some loss of sugar. Field drying has not been successful in warm climates for sugarcane and sweet sorghum. Work is being conducted in field drying for sweet sorghum in cooler climates; results are encouraging though no conclusions can be drawn yet.

Canes and stalks are generally baled and the cut ends and cuts from leaf stripping are seared to prevent loss of juice.

A large volume of material is required to produce a relatively small amount of sugar, thus a large amount of storage space is necessary. Handling is accomplished with loaders or bale movers.

Conversion of Carbohydrates to Simple Sugars

Processing options available for converting carbohydrates to simple sugars are

- Enzymatic vs acid hydrolysis
- High-temperature vs low-temperature cooking
- Continuous vs batch processing
- Separation vs nonseparation of fermentable nonsolids

Enzymatic vs acid hydrolysis. Enzymatic hydrolysis of the starch to sugar is carried out while cooling the cooked meal to fermentation temperature. The saccharifying enzyme is added at about 130°F (54°C), and this temperature is maintained for about 30 minutes to allow nearly complete hydrolysis following which the mash is cooled to fermentation temperature. A high-activity enzyme is added prior to cooking so that the starch is quickly converted to soluble polymeric sugars. The saccharifying enzyme reduces these sugars to monomeric sugars. Temperature and pH must be controlled within specific limits or enzyme activity decreases and cooking time is lengthened. Thus equipment for heating and cooling and addition of acid or base are necessary.

Acid hydrolysis of starch is accomplished by directly contacting starch with dilute acid to break the polymer bonds. This process hydrolyzes the starch very rapidly at cooking temperatures and reduces the time needed for cooking. Since the resulting pH is lower than desired for fermentation, it may be increased after fermentation is complete by neutralizing some of the acid with either powder limestone or ammonium hydroxide. It also may be desirable to add a small amount of glycoamylase enzyme after pH correction in order to convert the remaining dextrins.

High-temperature vs low-temperature cooking. Grain must be cooked to rupture the starch granules and to make the starch accessible to the hydrolysis agent. Cooking time and temperature are related in an inverse ratio; high temperatures shorten cooking time. Industry practice is to heat the meal-water mixture by injecting steam directly rather than by heat transfer through the walls of the vessel. The latter procedure runs the risk of causing the meal to stick to the walls; the subsequent scorching or burning would necessitate a shutdown to clean the surface.

High-temperature cooking implies a high-pressure boiler. Because regulations may require an operator in constant attendance for a high-pressure boiler operation, the actual production gain attributable to the high temperature must be weighed against the cost of the operator. If there are other supporting rationale for having the operator, the entire cost does not have to be offset by the production gain.

Continuous vs batch processes. Cooking can be accomplished with continuous or batch processes. Batch cooking can be done in the fermenter itself or in a separate vessel. When cooking is done in the fermenter, less pumping is needed and the fermenter is automatically sterilized before fermenting each batch. There is one less vessel, but the fermenters are slightly larger than those used when cooking is done in a separate vessel. It is necessary to have cooling coils and an agitator in each fermenter. If cooking is done in a separate vessel, there are advantages to selecting a continuous cooker. The continuous cooker is smaller than the fermenter, and continuous cooking and hydrolysis lend themselves very well to automatic, unattended operation. Energy consumption is less because it is easier to use counterflow heat exchangers to heat the water for mixing the meal while cooling the cooked meal. The load on the boiler with a continuous cooker is constant. Constant boiler load can be achieved with a batch cooker by having a separate vessel for preheating the water, but this increases the cost when using enzymes.

Continuous cooking offers a high-speed, high-yield choice that does not require constant attention. Cooking at atmospheric pressure with a temperature a little over 200°F yields a good conversion ratio of starch to sugar, and no high-pressure piping or pumps are required.

Separation vs nonseparation of nonfermentable solids. The hydrolyzed mash contains solids and dissolved proteins as well as sugar. There are some advantages to separating the solids before fermenting the mash, and such a step is necessary for continuous fermentation. Batch fermentation requires separation of the solids if the yeast is to be recycled. If the solids are separated at this point, the beer column will require cleaning much less frequently, thus increasing the feasibility of a packed beer column rather than plates. The sugars that cling to the solids are removed with the solids. If not recovered, the sugar contained on the solids would represent a loss of 20 percent of the ethanol. Washing the solids with the mash water is one way of recovering most of the sugar.

Fermentation

Continuous fermentation. The advantage of continuous fermentation of clarified beer is the ability to use high concentrations of yeast (this is possible because the yeast does not leave the fermenter). The high concentration of yeast results in rapid fermentation and, correspondingly, a smaller fermenter can be used. However, infection with undesired microorganisms can be troublesome because large volumes of mash can be ruined before the problem becomes apparent.

Batch fermentation. Fermentation time periods similar to those possible with continuous processes can be attained by using high concentrations of yeast in batch fermentation. The high yeast concentrations are economically feasible when the yeast is recycled. Batch fermentations of unclarified mash are routinely accomplished in less than 30 hours. High conversion efficiency is attained as sugar is converted to 10-percent-alcohol beer without yeast recycled. Further reductions in fermentation require very large quantities of yeast. The increases attained in ethanol production must be weighed against the additional costs of the equipment and time to culture large yeast populations for inoculation.

Specifications of the fermentation tank. The configuration of the fermentation tank has very little influence on system performance. In general, the proportions of the tank should not be extreme. Commonly, tanks are upright cylinders with the height somewhat greater than the diameter. The bottom may be flat (but sloped for drainage) or conical. The construction materials may be carbon steel (commonplace), stainless steel, copper, wood, fiberglass, reinforced plastic, or concrete coated on the inside with sprayed-on vinyl. Usually, the tanks are covered to permit collection of the CO_2 evolved during fermentation so that the ethanol which evaporates with it can be recovered.

Many potential feedstocks are characterized by relatively large amounts of fibrous material. Fermentation of sugar-rich material such as sugar beets, sweet sorghum, Jerusalem artichokes, and sugarcane as chips is not a demonstrated technology and it has many inherent problems. Typically, the weight of the nonfermentable solids is equal or somewhat greater than the weight of fermentable material. This is in contrast to grain mashers which contain as much fermentable material as nonfermentable material in the mash. The volume occupied by the nonfermentable solids reduces the effective capacity of the fermenter. This means that larger fermenters must be constructed to equal the production rates from grain fermenters. Furthermore, the high volume of nonfermentable material limits sugar concentrations and, hence, the beer produced is generally lower in concentration (6 vs 10 percent) than that obtained from grain mashes. This fact increases the energy spent in distillation.

Since the nonfermentable solid chips are of larger size, it is unlikely that the beer containing the solids could be run through the beer column. It may be necessary to separate the solids from the beer after fermentation because of the potential for plugging the still. The separation can be easily accomplished, but a significant proportion of the ethanol (about 20 percent) would be carried away by the dewatering solids. If recovery is attempted by "washing out," the ethanol will be much more dilute than the beer. Since much less water is added to these feedstocks than to grain (the feedstock contains large amounts of water), only part of the dilute ethanol solution from the washing out can be recycled through the fermenter. The rest would be mixed with the beer, reducing the concentration of ethanol in the beer which, in turn, increases the energy required for distillation. Another approach is to evaporate the ethanol from the residue. By indirectly heating the residue, the resulting ethanol-water vapor mixture can be introduced into the beer column at the appropriate point. This results in a slight increase in energy consumption for distillation.

The fermenter for high-bulk feedstocks differs somewhat from that used for mash. The large volume of insoluble residue increases the demands on the removal pump and pipe plugging is more probable. Agitators must be sized to be self-cleaning and must prevent massive settling. High-speed and high-power agitators must be used to accomplish this.

The equipment for separating the fibrous residue from the beer when fermenting sugar crops could be used also to clarify the grain mash prior to fermentation. This would make possible yeast recycle in batch fermentation of grain.

Temperature control. Since there is some heat generated during fermentation, care must be taken to ensure that the temperature does not rise too high and kill the yeast. In fermenters the size of those for on-farm plants, the heat loss through the metal fermenter walls is sufficient to keep the temperature from rising too high when the outside air is cooler than the fermenter. Active cooling must be provided during the periods when the temperature differential cannot remove the heat that is generated. The maximum heat generation and heat loss must be estimated for the particular fermenter to assure that water-cooling provisions are adequate.

Distillation

Preheater. The beer is preheated by the hot stillage from the bottom of the beer column before being introduced into the top of the beer column. This requires a heat exchanger. The stillage is acidic and hot so copper or stainless-steel tubing should be used to minimize corrosion to ensure a reasonable life. Because the solids are proteinaceous, the same protein buildup that plugs the beer still over a period of time can be expected on the stillage side of the heat exchanger. This mandates accessibility for cleaning.

Beer column requirements. The beer column must accept a beer with a high solids content if the beer is not clarified. Not only are there solids in suspension, but also some of the protein tends to build up a rather rubbery coating on all internal surfaces. Plate columns offer the advantage of relatively greater cleaning ease when compared to packed columns. Even if the beer is clarified, there will be a gradual buildup of protein on the inner surfaces. This coating must be removed periodically. If the plates can be removed easily, this cleaning may be done outside of the column. Otherwise, a caustic solution run through the column will clean it.

The relatively low pH and high temperature of the beer column will corrode mild steel internals, and the use of stainless steel or copper will greatly prolong the life expectancy of the plates in particular. Nevertheless, many on-farm plants are being constructed with mild steel plates and columns in the interest of low first cost and ease of fabrication with limited shop equipment. Only experience will indicate the life expectancy of mild steel beer columns.

Introducing steam into the bottom of the beer column rather than condensing steam in an indirect heat exchanger in the base of the column is a common practice. The latter procedure is inherently less efficient but does not increase the total volume of water in the stillage as does the former. Indirect heating coils also tend to suffer from scale buildup.

Rectifying column. The rectifying column does not have to handle liquids with high solids content and there is no protein buildup, thus a packed column suffers no inherent disadvantage and enjoys the advantage in operating stability. The packing can be a noncorroding material such as ceramic or glass.

General considerations. Plate spacing in the large column of commercial distilleries is large enough to permit access to clean the column. The small columns of on-farm plants do not require such large spacing. The shorter columns can be installed in farm buildings of standard eave height and are much easier to work on.

All items of equipment and lines which are at a significantly higher temperature than ambient should be insulated, including the preheated beer line, the columns, and the stillage line. Such insulation is more significant for energy conservation in small plants than for large plants.

Drying Ethanol

Addition of a third liquid to the azeotrope. Ethanol can be dehydrated by adding a third liquid such as gasoline to the 190-proof constant boiling azeotrope. This liquid changes the boiling characteristics of the mixture and further separation to anhydrous ethanol can be accomplished in a reflux still. Benzene is used in industry as a third liquid, but it is very hazardous for on-farm use. Gasoline is a suitable alternative liquid and does not pose the same health hazards as benzene, but it fractionates in a distillation column because gasoline is a mixture of many organic substances. This is potentially an expensive way to break the azeotrope unless the internal reflux is very high, thereby minimizing the loss of gasoline from the column. Whatever is chosen for the third liquid, it is basically recirculated continually in the reflux section of the drying column, and thus only very small fractions of makeup are required. The additional expense for equipment and energy must be weighed carefully against alternative drying methods or product value in uses that do not require anhydrous ethanol.

Molecular sieve. The removal of the final 4-6 percent water has also been accomplished on a limited basis using a desiccant (such as synthetic zeolite) commonly known as a molecular sieve. A molecular sieve selectively absorbs water because the pores of the material are smaller than the ethanol molecules but larger than the water molecules. The sieve material is packed into two columns. The ethanol—in either vapor or liquid form—is passed through one column until the material in that column can no longer absorb water. Then the flow is switched to the second column, while hot [450°F (232°C)] and preferably nonoxidizing gas is passed through the first column to evaporate the water. Carbon dioxide from the fermenters would be suitable for this. Then the flow is automatically switched back to the other column. The total energy requirement for regeneration may be significant (the heat of absorption for some synthetic zeolites is as high as 2500 Btu/lb). Sieve material is available from the molecular sieve manufacturers listed in Appendix A, but columns of the size required must be fabricated. The molecular sieve material will probably serve for 2000 cycles or more before significant deterioration occurs.

Selective absorption. Another very promising (though undemonstrated) approach to dehydration of ethanol has been suggested by Dr. Ladisch of Purdue University. Various forms of starch (including cracked corn) and cellulose selectively absorb water from ethanol-water vapor. In the case of grains, this opens the possibility that the feedstock could be used to dehydrate the ethanol and, consequently, regeneration would not be required. More investigation and development of this approach is needed.

Stillage Processing

The stillage is a valuable coproduct of ethanol production. The stillage from cereal grains can be used as a high-protein component in animal feed rations, particularly for ruminants such as steers or dairy cows. Small on-farm plants may be able to directly use the whole stillage as it is produced since the number of cattle needed to consume the stillage is about one head per gallon of ethanol production per day.

Solids separation. The solids can be separated from the water to reduce volume (and hence shipping charges) and to increase storage life. Because the solids contain residual sugars, microbial contaminants rapidly spoil stillage if it is stored wet in warm surroundings. The separation of the solids can be done easily by flowing the stillage over an inclined, curved screen consisting of a number of closely spaced transverse bars. The solids slide down the surface of the screen, and the liquid flows through the spaces between the bars. The solids come off the screen with about 85 percent water content, dripping wet. They can drop off the screen into the hopper of a dewatering press which they leave at about 65 percent water content. Although the solids are still damp, no more water can be extracted easily.

Transporting solids. The liquid from the screen and dewatering press still contains a significant proportion of dissolved proteins and carbohydrates. If these damp solids are packed in airtight containers in a CO_2 atmosphere, they may be shipped moderate distances and stored for a short time before microbes cause major spoilage. This treatment would enable the solids from most small plants to reach an adequate market. While the solids may easily be separated and dewatered, concentrating the liquid (thin stillage) is not simple. It can be concentrated by evaporation, but the energy consumption is high unless multiple-effect evaporators are used. These evaporators are large and expensive, and may need careful management with such proteinaceous liquids as thin stillage.

Stillage from aflatoxin-contaminated grains or those treated with antibiotics are prohibited from use as animal feed.

Distiller's solubles, which is the low-concentration (3-4 percent solids) solution remaining after the solids are dewatered, must be concentrated to a syrup of about 25 percent solids before it can be economically shipped moderate distances or stored for short times. In this form it can be sold as a liquid protein to be used in mixed feed or it can be dried along with the damp distiller's grains.

Disposing of thin stillage. If the distance from markets for the ethanol coproduct necessitates separating and dewatering the stillage from plant, and if the concentration of the stillage for shipment is not feasible, then the thin stillage must be processed so that it will not be a pollutant when discharged. Thin stillage can be anaerobically fermented to produce methane. Conventional flow-through-type digesters are dependent on so many variables that they cannot be considered commercially feasible for on-farm use. Experimental work with packed-bed digesters is encouraging because of the inherent stability observed.

Another way to dispose of the thin stillage is to apply it to the soil with a sprinkler irrigation system. Trials are necessary to evaluate the various processes for handling the thin stillage. Because the stillage is acidic, care must be taken to assure that soil acidity is not adversely affected by this procedure.

PROCESS CONTROL

Smooth, stable, and trouble-free operation of the whole plant is essential to efficient conversion of the crop material. Such operation is, perhaps, more important to the small ethanol plant than to a larger plant, because the latter can achieve efficiency by dependence on powerful control systems and constant attention from skilled operators. Process control begins with equipment characteristics and the integration of equipment. There is an effect on every part of the process if the conditions are changed at any point. A good design will minimize negative effects of such interactions and will prevent any negative disturbance in the system from growing. Noncontinuous processes (e.g., batch fermentation) tend to minimize interactions and to block such disturbances. The basic components requiring process control in a small-scale ethanol plant are cooking and hydrolysis, fermentation, distillation, ethanol-drying system, pumps and drives, and heat source.

Control of Cooking and Hydrolysis

Input control. All inputs to the process must be controlled closely enough so that the departures from the desired values have inconsequential effects. The batch process has inherently wider tolerances than the continuous process. A variation of 0.5 percent in ethanol content will not seriously disturb the system. This corresponds to about 3 percent tolerance on weight or volume measure. Meal measurement should be made by weight, since the weight of meal filling a measured volume will be sensitive to many things, such as grain moisture content and atmospheric humidity. Volume measurement of water is quite accurate and easier than weighing. Similarly, volume measurement of enzymes in liquid form is within system tolerances. Powdered enzymes ideally should be measured by weight but, in fact, the tolerance on the proportion of the enzymes is broad enough so that volume measure also is adequate.

Temperature, pH, and enzyme control. The temperature, pH, and enzyme addition must also be controlled. The allowed variation of several degrees means that measurement of temperature to a more than adequate precision can be accomplished easily with calibrated, fast-response indicators and readouts. The time dependence brings in other factors for volume and mass. A temperature measurement should be representative of the whole volume of the cooker; however, this may not be possible because, as the whole mass is heating, not all parts are receiving the same heat input at a given moment since some parts are physically far removed from the heat source. This affects not only the accuracy of the temperature reading but also the cooking time and the action of the enzyme. Uniformity of temperature and of enzyme concentration throughout the mass of cooking mash is desired and may be attained by mixing the mass at a high rate. Thus, agitation is needed for the cooker. The temperature during the specific phases of cooking and hydrolysis must be controlled by regulating steam and cooling water flow rates based on temperature set points.

Automatic controls. An automatic controller could be used to turn steam and cooling water on and off. The flow of meal, water, enzymes, and yeast could be turned on and off by the same device. Therefore, the loading and preparation of a batch cooker or fermenter could easily be carried out automatically. Safety can be ensured by measuring limiting values of such quantities as temperature, water level, and pH, and shutting down the process if these were not satisfied. Any commercial boiler used in a small plant would be equipped with simple, automatic controls including automatic shutdown in case certain conditions are not met. There is a need for an operator to check on the system to assure that nothing goes wrong. For example, the mash can set up during cooking, and it is better to have an operator exercise judgment in this case than to leave it entirely to the controls. Since cooking is the step in which there is the greatest probability of something going wrong, an operator should be present during the early, critical stages of batch cooking. If continuous cooking is used, unattended operation requires that the process be well-enough controlled so that there is a small probability of problems arising.

Control of Fermentation

Temperature and pH control. Batch fermentation does not need direct feedback control except to maintain temperature as long as the initial conditions are within acceptable limits. For the small plant, these limits are not very tight. The most significant factors are pH and temperature; of the two, temperature is most critical. It is very unlikely that the change in pH will be great enough to seriously affect the capacity of the yeast to convert the sugar. Fermentation generates some heat, so the temperature of the fermenter tends to rise. Active cooling must be available to assure that summertime operations are not drastically slowed because of high-temperature yeast retardation.

The temperature of the fermenter can be measured and, if the upper limit is exceeded, cooling can be initiated. It is possible to achieve continuous control of the fermenter temperature through modulation of the cooling rate of the contents. Such a provision may be necessary for very fast fermentation.

Automatic control. Continuous fermentation, like continuous cooking, should have continuous, automatic control if constant attendance by an operator is to be avoided. The feasibility of continuous, unattended fermentation in on-farm plants has not been demonstrated, although it is a possibility.

Control with attention at intervals only. The feasibility of batch fermentation with attention at intervals has been established. After initiating the cooking and hydrolysis steps, the operator could evaluate the progress of fermentation at the end of the primary phase and make any adjustments necessary to assure successful completion of the fermentation. This interval between the points requiring operator attention can vary widely, but is usually from 8 to 12 hours. Fermentation can be very fast—as short as 6 hours—but the conditions and procedures for reliably carrying out such fast fermentations have not yet been completely identified and demonstrated. The schedule for attending the plant should allow about 15 percent additional time over that expected for completion of the fermentation process. This permits the operator to maintain a routine in spite of inevitable variations in fermentation time.

Controls for Distillation

Continuous control of the distillation process is not mandatory because the inputs to the columns can easily be established and maintained essentially constant. These inputs include the flow rate of beer, the flow rate of steam, and the reflux flow rate. These are the only independent variables. Many other factors influence column operation, but they are fixed by geometry or are effectively constant. Once the distillation system is stabilized, only changes in ambient temperature might affect the flow balance as long as the beer is of constant ethanol content. Sensitivity to ambient temperature might affect the flow balance as long as the beer is of constant ethanol content. Sensitivity to ambient temperature can be minimized by the use of insulation on all elements of the distillation equipment and by installing the equipment in an insulated building. Occasional operator attention will suffice to correct the inevitable slow drift away from set values. The system also must be adjusted for changes in ethanol content from batch to batch.

Distillation column design can aid in achieving stable operation. Packed columns are somewhat more stable than plate columns, particularly as compared to simple sieve plates.

Starting up the distillation system after shutdown is not difficult and can be accomplished either manually or automatically. An actual sequence of events is portrayed in the representative plant described at the end of this chapter. The process is quite insensitive to the rate of change of inputs, so the demands made on the operator are not great. It is important that the proper sequence be followed and that the operator know what settings are desired for steady-state operation.

Control of Ethanol-Drying System

Operation of a molecular sieve is a batch process. As such, it depends on the capacity of the desiccant to ensure completion of drying. No control is necessary except to switch ethanol flow to a regenerated column when the active column becomes water saturated. Water saturation of the sieve can be detected by a rise in temperature at the discharge of the column. This temperature rise signals the switching of flow to the other column, and regeneration of the inactive column is started immediately. The regeneration gas, probably CO_2 from the fermenter, is heated by flue gas from the boiler. The control consists of initiating flow and setting the temperature. The controller performs two functions: it indicates the flow and sets the temperature of the gas. Two levels of temperature are necessary: the first [about 150°F (66°C)] is necessary while alcohol clinging to the molecular sieve material is being evaporated; the second (about 450°F) is necessary to evaporate the absorbed water. Here, again, the completion of each phase of the regeneration cycle is signaled by a temperature change at the outlet from the column. Finally, the column is cooled by passing cool CO_2 through it until another outlet temperature change indicates completion of the regeneration cycle. The controls required for a dehydration distillation column are essentially the same as those required for the rectification column.

Controls for Pumps and Drives

The pumps used in this plant can be either centrifugal or any one of a number of forms of positive displacement pumps. The selection of the pump for mash or beer needs to take into account the heavy solids loading (nearly 25 percent for mash), the low pH (down to 3.5 for the beer), and the milk abrasive action on the mash.

The pumps might be powered by any of a number of different motors. The most probable would be either electric or hydraulic. If electric motors are used, they should be explosion proof. Constant-speed electric motors and pumps are much less expensive than variable-speed motors. Control of the volume flow for the beer pump, the two reflux pumps, and the product pump would involve either throttling with a valve, recirculation of part of the flow through a valve, or variable-speed pumps. Hydraulic drive permits the installation of the one motor driving the pumps to be located in another part of the building, thereby eliminating a potential ignition source. It also provides inexpensive, reliable, infinitely variable speed control for each motor. Hydraulic drives could also be used for the augers, and the agitators for the cookers and fermenters. Since hydraulics are used universally in farm equipment, their management and maintenance is familiar to farmers.

Heat Source Controls

There are basically two processes within the ethanol-production system that require heat: the cooking and the distillation steps. Fortunately, this energy can be supplied in low-grade heat [less than 250°F (121°C)]. Potential sources of heat include coal, agriculture residues, solar, wood wastes, municipal wastes, and others. Their physical properties, bulk density, calorific value, moisture content, and chemical constituency vary widely. This, in turn, requires a greater diversity in equipment for handling the fuel and controls for operating the boilers. Agriculture residues vary in bulk density from 15 to 30 lb/ft^3 and the calorific value of oven-dry material is generally around 8000 Btu/lb. This means that a large volume of fuel must be fed to the boiler continuously. For example, a burner has been developed that accepts large, round bales of stover or straw. The boiler feed rate will vary in direct proportion to the demand for steam. This in turn is a function of the distillation rate, the demand for heat for cooking (which varies in relation to the type of cooker used—batch or continuous).

Emissions. Emissions' control on the boiler stack are probably minimal, relying instead on efficient burner operation to minimize particulate emissions. If exhaust gas scrubbers or filters are required equipment, they in turn require feedback control. Filters must be changed on the basis of pressure drop across them, which indicates the degree of loading (plugging). Scrubbers require control of liquid flow rate and control of critical chemical parameters.

Boiler safety features. Safety features associated with the boiler are often connected to the control scheme to protect the boiler from high-pressure rupture and to prevent burnout of the heat expander tubes. Alarm systems can be automated and have devices to alert an operator that attention is needed. For instance, critical control alarm can activate a radio transmitter, or "beeper," that can be worn while performing other normal work routine.

Model Plants

Modern ethanol-fuel production plants should be designed to achieve maximum production with minimum energy. In order that a variety of types and sizes of production plants would be included in this chapter, six model ethanol plants were selected.

GENERAL CRITERIA

Facilities for producing ethanol by fermentation may be grouped according to process type and annual production capacity. Some typical classifications are shown in Table 3-3.

The farmer-built, one-of-a-kind still is expected generally to operate intermittently to produce 160-190-proof alcohol. The wet stillage produced would be used in a farmer's own or nearby operation. The plant would not lend itself readily to project-type financing and predictable production quantities or quality. At the other extreme is the custom-designed, manufactured, and erected large commercial-scale ethanol plant producing

Table 3-3. Ethanol Plant Segments by Size and Type of Manufacturing

Type of manufacturing	Small (<1 million gallons 160-190 proof ethanol & wet stillage)	Medium (1-5 million gal. 190-200 proof ethanol with wet stillage or DDGS)	Intermediate (5-15 million gal. 200 proof ethanol and DDGS)	Large (>15 million gal. 200 proof ethanol and DDGS)
Farmer built one of a kind	2-40 GPH intermittent operation	NA	NA	NA
Factory built equipment package repetitively produced and sold as a unit	20-50 GPH intermittent or continuous	150-700 GPH continuous operation	NA	NA
Custom design manufacture and erection	NA	300-700 GPH continuous operation	700-6,000 GPH continuous operation	Designed Production Capacity

NA - not applicable
GPH - gallons of ethanol per hour
Source: *Small-Scale Fuel Alcohol Production,* USDA, Washington, D.C., 1980.

over 15 million gallons of 200-proof ethanol and distiller's dried byproduct, annually.

As shown in Table 3-3, the factory-built equipment plant uses a standardized design. It is manufactured as an integrated equipment package and is repetitively produced. The sizes and types of these plants range from one of intermittent production of less than 20 gallons per hour to one of continuous production of up to 5 million gallons per year of 200-proof ethanol. Plants in this classification are most applicable to farm- and community-level operations.

The characteristics of these typical plants differ. For example, some use mild steel but a few specify such variations as stainless steel, steel-epoxy and fiberglass tanks and stainless steel and steel-copper stills.

A number of companies (Katzen, Vulcan, PEDCo, etc.) market an ethanol plant or process of the intermediate and large commercial-scale type.

With two exceptions, the model plants were designed to use corn as the feedstock: one plant can use both corn and potatoes, and another, both corn and molasses. In practice, probably all could use other small grains if their special processing characteristics (e.g., the increased foaming of wheat gluten) are understood when specifying operating procedures. Until operating procedures are finalized and actual operating data are acquired, no conclusion can be drawn as to the relative merits of the various plant designs.

Three features have been identified, however, that affect energy inputs and the efficiency of ethanol-fuel production: (a) the design and cost differences necessitated by feedstock variation, (b) types of small boilers available, and (c) plant operating skills required.

Variations with Feedstocks

The basic principles involved in ethanol production are the same for all feedstocks; however, as noted earlier in this chapter, the specific feedstocks used affect methods of storage and handling, the hydrolysis of starch or the extraction of sugar, and the processing of the coproducts. These differences result in production-cost variations.

In using feedstocks other than corn, a number of cost factors need to be considered aside from those for feedstock and the value of the coproduct.

Table 3-4 shows the considerations associated with the equipment differences identified with feedstock processing. Corn has been considered the primary feedstock; a representative number of the potential agricultural products and coproducts commonly mentioned as feedstocks are compared to corn. The three sugar crops—sweet sorghum, sugarcane, and sugar beets—have special characteristics that require significantly different equipment.

These sugar crops do not require the hydrolyzing step, for they can be fermented directly once the sugar has been extracted by pressure (in sugarcane milling) or by diffusion (in sugar beet processing). These crops have a short harvest season, and once harvested they deteriorate rapidly. To avoid a loss of their sugar content, prompt processing is required in warm climates. In northern areas, sugar beets are usually stored during the winter in a cold or frozen condition and the processing is completed before spring. Plants processing these sugar crops should be designed to handle multiple crops (i.e., sugars and starches) and, thus, operate the year around. Coproduct starch and molasses are relatively easy to process, but their availability and supply are limited. Potatoes can be stored for longer periods than can sugar crops, but the cost of storage is greater. Little cost information regarding cooking and enzyme treatment is available.

Table 3-5 is a subjective comparison of the relative investment costs for processing various ethanol feedstocks. All comparisons are for plants having equal annual capacity and compare costs to those for processing corn.

The investment costs and operating costs vary with the characteristics of the feedstocks. In Table 3-6 a similar display indicates the magnitude of operating cost by item for these same feedstocks relative to corn. These variable costs are given on a per-gallon basis.

Biomass Boilers

Ethanol production has a higher energy requirement compared to the energy content of ethanol. If ethanol is to make a significant net contribution to the liquid-fuel supply, it is imperative that fuels other than fuel oil or natural gas be used for process heat energy.

Because development of small-scale ethanol plants is envisioned for on-farm operation, it has been widely assumed that crop residues (stover) and

Table 3-4. Comparison of Process Factors of Other Feedstocks to Corn

	Storage length of supply	Feedstock preparation	Sugar extraction	Cooking hydrolysis	Fermentation	Distillation	Dehydration	Coproduct processing
Corn	Grain bins, 12 months	Grind	None	Heat and hydrolyze	With solids	Two columns	Benzene extraction	DDGS or wet stillage
Wheat	Grain bins, 12 months	Grind	None	Heat and hydrolyze	With solids	Two columns	Benzene extraction	DDGS or wet stillage
Milo/sorghum	Grain bins, 12 months	Grind	None	Heat and hydrolyze	With solids	Two columns	Benzene extraction	DDGS or wet stillage
Sweet sorghum	Piles, 3 months	Crush in roller mill	Separate sugar & fiber	None	Liquid only	Larger size	Liquid only	Low value fiber
Sugarcane	Piles, 3 months	Crush in roller mill	Separate sugar & fiber	None	Liquid only	Larger size	Liquid only	Low value fiber
Sugar beets	Piles, 4 months	Slice	Extensive extraction	None	Liquid only	Larger size	Liquid only	Beet pulp dry or wet
Potatoes	Controlled atmosphere, 12 months	Grind	None	Heat and hydrolyze	Liquid only	Somewhat larger size	Liquid only	Potato pulp dry or wet
Starch	Bins, 12 months	None	None	Heat and hydrolyze	Liquid only	Somewhat larger size	Liquid only	Yeast only
Molasses	Tanks, 12 months	None	None	None	Liquid only	Somewhat larger size	Liquid only	Yeast only

Source: *Small-Scale Fuel Alcohol Production,* USDA, Washington, D.C., 1980.

Table 3-5. Comparison Cost of Investment of Other Feedstocks with Corn for Same Ethanol Capacity

	Feedstock storage	Feedstock preparation	Hydrolysis or sugar extraction	Distillation	Dehydration	Coproduct drying	Total investment
Corn	1.0	1.0	1.0	1.0	1.0	1.0	1.0
Wheat	1.0	1.0	1.0	1.0	1.0	1.0	1.0
Milo/sorghum	1.0	1.0	1.0	1.0	1.0	1.0	1.0
Sweet sorghum[a]	0.3	3.0	0.1	3.0	3.0	0.1	2.5
Sugarcane[a]	0.3	3.0	0.1	3.0	3.0	0.1	2.5
Sugar beets[a]	0.3	4.0	2.0	3.0	3.0	1.5	3.0
Potatoes	1.5	1.2	1.0	1.2	1.0	1.0	1.2
Starch	0.5	0.1	0.8	1.0	1.0	0.1	0.8
Molasses	0.5	0.1	0.1	1.0	1.0	0.1	0.7

[a]Assumes that plant with some total annual capacity will have to have three times capacity per month because of approximately four month season.

Source: *Small-Scale Fuel Alcohol Production,* USDA, Washington, D.C., 1980.

Table 3-6. Comparison of Operating Costs[a] Per Gallon of Processing Other Feedstocks with Corn for Plants of Same Ethanol Capacity

	Total costs					Indirect costs				Direct & Indirect Total
	Labor	Fuel	Electricity	Other	Total	Running & Main.	Ins. & Tax	General & Admin.	Total	
Corn	1.0	1.0	1.0	1.0	1.0	1.0	1.0	1.0	1.0	1.0
Wheat	1.0	1.0	1.0	1.0	1.0	1.0	1.0	1.0	1.0	1.0
Milo/sorghum	1.0	1.0	1.0	1.0	1.0	1.0	1.0	1.0	1.0	1.0
Sweet sorghum	1.2	0.8	1.0	0.8	1.0	1.3	2.0	1.0	1.4	1.2
Sugarcane	1.2	0.8	1.1	0.8	0.9	1.3	2.0	1.0	1.5	1.3
Sugar beets	1.3	1.2	1.2	0.8	1.2	1.5	2.5	1.0	1.6	1.5
Potatoes	1.1	1.3	1.0	1.0	1.2	1.2	1.0	1.0	1.1	1.1
Starch	0.7	1.0	0.8	0.9	0.8	0.8	0.8	0.9	0.8	0.8
Molasses	0.6	0.7	0.7	0.8	0.7	0.6	0.7	0.9	0.7	0.7

[a]Costs of operation do not include feedstock cost or coproduct credit.

Source: *Small-Scale Fuel Alcohol Production,* USDA, Washington, D.C., 1980.

wood would be used as the boiler fuel. Two factors may negate this assumption, however. The first factor is that, with the exception of one manufacturer, all present developmental work uses natural gas or oil-fired boilers; no commercially available boilers, in sizes required for these plants, burn solid fuels other than coal or wood. Secondly, plant designers are developing plants with automated features to minimize operator attention; indeed, an automatically controlled boiler is a high-priority requirement. Small boilers burning solid fuels, especially crop residue, are difficult to control unless hand fired, a significant deterrent to automated operation. Additionally, biomass-fueled boilers are still developmental. Early indications are, however, that biomass boilers will cost two to three times as much as gas-fired boilers.

Farmers themselves could develop such equipment. They could combine tub grinders (capable of grinding large round and conventional rectangular bales) with various stoker and blower devices to automatically feed the fire box of a variety of standard boilers.

USDA MODEL ETHANOL PLANTS

This section describes the six model small-scale ethanol plants delineated by the USDA, which encompass the range of types and sizes of plants currently being marketed. These model plants were designed to be characteristic of plants projected for on-farm and rural community operation in the near-term. These six plants vary by ethanol processing, capacity, product proof, style of operation, and the form of the coproduct stillage.

Table 3-7 outlines the following six distinctive small-scale plants:

- Pot still, intermittent production of alcohol; wet stillage
- Small on-farm, intermittent production of alcohol; wet stillage
- Large on-farm, continuous production of alcohol; wet stillage
- Small community, continuous production of alcohol; wet stillage
- Small community, continuous production of alcohol; DDGS
- Large community, continuous production of alcohol; DDGS

In addition, a central "topping-cycle" dehydration plant was anticipated. This plant would operate as an addition to the community-sized plants and have an excess dehydrating capacity. For the latter portion of the operation, lower-proof alcohol would be collected from nearby small stills for processing to 200 proof.

In this handbook, the on-farm, small-scale discussions refer to the six USDA model plants. The general characteristics of each production plant are described in the following paragraphs. Key information for each plant includes type and size of operation, location, feedstock, type and use of products, investment costs, and operating costs.

Table 3-7. Model Small-Scale Ethanol-Plant Characteristics

Model plant designation	Size	Type	Distillation rate (GPH)	Proof	Operation schedule		Annual production (000 gal/yr)	Coproduct type	Plant construction type
					(hrs/day)	(days/yr)			
Pot still	Small	Intermittent	20	160–190	8	100	16	Wet stillage	Pot still package
Small on-farm	Small	Intermittent	25	190	8	300	60	Wet stillage	Package plant
Large on-farm	Medium	Continuous	50	190	24	300	360	Wet stillage	Package plant
Small community, wet	Large	Continuous	150	200	24	300	1,000	Wet stillage	Custom built
Small community, DDGS	Large	Continuous	150	200	24	300	1,000	DDGS	Custom built
Large community, DDGS	Large	Continuous	300	200	24	300	2,000	DDGS	Custom built

Source: *Small-Scale Fuel Alcohol Production,* USDA, Washington, D.C., 1980.

Pot Still

This model is the simplest of all ethanol plants and represents a commercial model of the legendary moonshiner's pot still. From this model, a farmer or other operator can easily learn the techniques of operation and acquire the experience required to proceed to a larger continuous still that will be a more-sophisticated and more-efficient fuel-producing unit.

Type and size of operation. A pot still usually will be operated by one person who will prepare the mash, cook, and ferment in one tank on a 3-day cycle. On the first day, the mash will be prepared by cooking and treating with enzymes and then left to ferment, largely unattended, for 2 or 3 days. On the third day, when the mash reaches an alcohol level of 7-10 percent, the operator will distill the mash in a single- or double-column still to produce 160-190-proof ethanol and wet stillage. Thus, every third to fourth day or twice a week, the operation will produce ethanol. The operator will then recharge the cook tank for the next batch. A typical, commercially available still will produce about 160 gallons of alcohol per batch. If operated regularly and year-round, the annual production would be 16,000 gallons.

Location. This model plant probably will be located on a farm in the open or in a building, depending on the climate and weather protection required. Its process and some of its cooling water will come from a well or similar supply. Some cooling water may come from a stream or pond.

Feedstock. The mash will be prepared from shelled corn or other grain or from sample-grade grains that may be obtained on farm or locally at below market prices.

Type and use of products. The primary product, ethanol, at 160-190-proof, will be used on the farm as fuel for vehicles (properly converted for alcohol-fuel use), space heating, cooking, grain drying, and other farm liquid-fuel requirements. Some excess product may be sold "as is" to neighbors or to a community plant capable of further processing the alcohol to 200 proof for use in the gasohol market.

The secondary product, wet stillage, because of the probably intermittent production of a pot still, will be available about twice a week for feed use for low-production animals. If such feed uses are not available, the wet stillage may be distributed on the land for its value as fertilizer.

Investment costs. Some pot stills may be farmer built from miscellaneous materials with a cost that may vary from a few thousand to $10,000. However, either skid or trailer-mounted commercial plants are expected to cost about $15,000 for the major parts; an additional $10,000 will be required for storage tanks (grain and alcohol) and other test and auxiliary equipment. The total model plant is estimated at $25,000 and has an expected life of 5 years.

Operating costs. Operating costs are the direct costs and the indirect costs of converting corn to 190-proof ethanol. Costs do not include grain costs or the value of the coproduct stillage.

The direct costs are those directly related to the amount of ethanol produced and include labor, fuel, electricity, and supplies of yeast, enzymes, and other chemicals.

For the pot still, it is assumed that one part-time person will operate the plant while carrying out other farm operations. Although some operators may choose to declare this intermittent labor to have no value, it has been assumed that 3 hours of labor will be required for each batch of 160 gallons or 0.05 man-hours per gallon of ethanol. A minimal, low hourly rate of $3.00 per hour has also been assumed that the labor per gallon is calculated at $0.15.

A hand-fired coal or wood-burning boiler has been assumed; the operator will stoke and regulate the steam pressure while the cooking and distillation operations are underway. The energy required to produce steam for cooking and distilling is estimated as follows:

Operation	*Btu/gal 190 proof*
Cooking	4,000
Distillation	30,000
Miscellaneous	9,000
Total	43,000

This amount of steam requires a 25-horsepower boiler (80 percent efficient). If coal has a heat content of 11,000 Btu per pound and costs $40 per ton, the fuel cost for this model plant will be $0.10 per gallon.

The electrical energy needed to operate the grinder, augers, mixers, and pumps for the complete plant has been estimated to be 0.5 kWhr per gallon. Assuming an electrical rate of $0.05/kWhr, the electricity cost per gallon is $0.03.

An alpha-amylase is used in the cooking process to gelatinize the starch. At a price of $1.75 per pound for this enzyme, the cost per bushel of grain will be $0.0595. The glucoamylase enzyme used to convert starch to sugar is priced at $2.63 per liter and costs $0.08925 per bushel. A commercial dry yeast for fermentation is priced at $0.90 per pound and costs $0.054 per bushel. These three supplies total $0.20 per bushel of grain; thus, if the theoretical yield of alcohol is 2.6 gallons per bushel, the per-gallon cost of the items is $0.08. An additional $0.02-per-gallon cost has been assumed to cover the miscellaneous chemicals needed for cleaning, for sterilizing, and for adjusting the pH of the mash and treating the boiler feedwater. The total cost for supplies is $0.09 per gallon.

Indirect costs are annual costs not closely related to production. These include maintenance, general and administrative, property taxes, insurance, tax bond, and depreciation costs. For the pot still, it has been assumed that general and administrative costs are part of the farm operation and no specific cost for this function has been assigned.

Maintenance costs for this operation have been assumed to be 3 percent of the investment cost. Prorated over 16,000 gallons of product, the maintenance cost is $0.05 per gallon.

Four types of insurance should be carried by a small-still operator. The rates and costs shown below will vary from state to state and will also vary among insurance companies. These costs are estimated to be:

Insurance type	*Rate*	*Annual premium*
General liability $500,000 coverage	$0.65/$100 payroll	$ 15.60
Product liability	$1.00/$1,000 sales	12.90
Workmen's compensation (State of Kansas)	$4.93/$100 payroll	118.30
Fire and extended coverage	$0.80/$100 valuation	200.00
Total annual premium		$346.80

$346.80/16,000 gallons = $0.02/gal

The property taxes will vary by county and by state. For purposes of this handbook, two assumptions have been made: the tax valuation will be based on 20 percent of the fair market value (investment cost of equipment) and the tax rate will be 100 mills per dollar of valuation. Thus, for this model plant, the valuation is $5,000 and the annual tax is $500 or $0.03 per gallon of ethanol.

The holder of a Bureau of Alcohol, Tobacco, and Firearms (BATF) commercial permit must post a bond equivalent to the tax of $10.50 per proof gallon (100 proof) on 15 operating days of alcohol production. For this plant, 15 days of production is 640 gallons of 190 proof or 1216-proof gallons. At $10.50 per proof gallon, the bond required is approximately $13,000. The bond rate quoted now is $12.00/$1000 of tax liability or $156 per year, and the per-gallon cost is $0.01 per gallon.

The depreciation cost is important only for calculating income tax and, therefore, in the financial analysis shown in the preceding portion of this handbook, it is part of capital recovery. However, the depreciation figure used for calculating the cost of converting corn to alcohol is based on the estimated life of the equipment. For this model plant, 5 years is the estimated life and the depreciation rate is 20 percent per year. For this plant then, depreciation is $5,000 divided by 16,000 gallons or $0.31 per gallon of ethanol.

Conversion costs can be summarized as follows:

Item	*Annual ($000)*	*Unit ($/gal)*
Direct		
Labor	2.4	0.15
Fuel	1.6	0.10
Electricity	0.4	0.03
Supplies	1.4	0.09
Total	5.8	0.37

Indirect		
Maintenance	0.8	0.05
General and administrative	—	—
Taxes	0.5	0.03
Insurance	0.3	0.02
Tax Bond	0.2	0.01
Total	1.8	0.11

Small On-Farm Ethanol Plant

The key characteristic of the small on-farm model plant is its multiple fermentation tanks which permit the continuous fermentation and distillation of ethanol for 6 days per week. The plant equipment will be purchased as a package unit for small-scale, on-the-farm ethanol production.

Type and size of operation. This plant will feature a continuous batch fermentation process that will result in six 8-hour days of distillation at the rate of 25 gallons of 190-proof ethanol per hour. The operation of the plant will require one part-time operator 6 days per week and, when operated for 300 days per year, it will result in approximately 60,000 gallons of ethanol product. The distillation will be carried out on one shift only; fermentation will continue around the clock unattended for 16 hours per day for 6 days and for 24 hours on the seventh day.

Location. A building is desired to house this type and size of plant. Cooling and process water will be drawn from existing wells and ponds.

Feedstock. Feedstock will be locally grown and stored corn or similar grain, unless there is an opportunity to obtain sample-grade or damaged grains nearby.

Type and use of product. It is assumed that approximately one-half of this plant's 60,000 annual gallons of 190-proof ethanol will be used on the farm and one-half will be sold locally in a 190-proof market or to a commercial plant for upgrading to 200 proof.

The coproduct, wet stillage, will be integrated into a continuous feeding program of high-production livestock, and none will be dried or sold. Good management of coproduct use will be required in order to have a supply available on the day when there is no distillation.

Investment costs. The investment costs for this on-the-farm ethanol plant have been made with the following assumptions:

No land cost—by using land of existing farmstead

No grain storage cost—by using existing storage

No feedstock-preparation equipment—by using feed grinder now on farm

No water supply cost—by using existing wells and/or ponds and pumps in place

No building cost—by using an existing building

The other costs that are included as investment costs are estimated from the tentative prices of packaged ethanol plants offered for sale by current manufacturers. The summary of investment costs are:

Site preparation	$ 10,000
Supply storage	1,000
Boiler (gas-fired)	30,000
Cooking equipment	2,000
Fermentation tanks (6)	12,000
Instrumentation	5,000
Stripping column (12″ × 20′)	15,000
Rectification column (12″ × 20′)	15,000
Denaturization equipment	2,000
Alcohol storage (5,000 gallons)	2,000
Stillage storage	2,000
Miscellaneous	15,000
Subtotal	111,000
Erection and installation (25 percent)	29,000
Total investment cost	$140,000

The above costs assume that the cooking, fermentation, storage tanks, and columns will be fabricated from mild carbon steel.

Operating costs. Operating costs are those direct and indirect costs to convert corn to 190-proof alcohol at the rate of 25 gallons per hour. These costs do not include the feedstock (corn) costs nor the value of the wet stillage for livestock feeding.

The direct operating costs are those related to the number of gallons of ethanol produced and are labor, fuel, electricity, and supplies.

This plant will require one person working up to 8 hours per day for 6 days a week. On a per gallon basis, 0.04 man-hours will be required at an assumed rate per hour of $5.00 per hour or $0.20 per gallon.

Energy use is estimated to be 43,000 Btu per gallon; with a 60 horsepower natural gas boiler. With a gas price of $2 per thousand cubic feet (mcf), the cost of 1000 Btu, assuming an 80 percent efficiency, will be $0.0023. This fuel cost is $0.10 per gallon of ethanol produced. The heat required has been derived from these calculations:

Operation	*Btu/gal (190 proof)*
Cooking	4,000
Distillation	30,000
Miscellaneous	9,000
Total	43,000

The electrical energy needed to operate the grinder, augers, mixers, and pumps for this plant is estimated at 0.5 kWhr per gallon. The assumed electrical rate per kWhr is $0.05; the electrical cost is $0.03 per gallon.

The alpha-amylase enzyme used in cooking starch is priced at $1.75 per pound and 0.03 pound is used per bushel; hence, the cost is $0.0595 per bushel of grain. The glucoamylase enzyme used to convert starch to sugar is valued at $0.08925 per bushel based on a commercial price of $2.63 per liter. Dry, powered yeast for the fermentation process is priced at $0.90 per pound or $0.054 per bushel of corn. These supplies total $0.20 per bushel of corn and, with an estimated yield of 2.6 gallons per bushel, the per-gallon cost is $0.08. An additional $0.01 per gallon cost of acid, base, and other chemicals make the total supply cost $0.09 per gallon.

Indirect costs are annual costs and are relatively independent of ethanol produced: maintenance, general and administrative expense, property taxes, insurance, tax bond cost, and depreciation. For this plant, the general and administrative expenses are considered a part of the total farm operation. No cost has been assumed for these functions.

The annual cost for maintenance has been calculated to be 3 percent of the equipment investment cost for the first 5 years and 5 percent for the next 5 years.

Four types of insurance are needed. The insurance rates will vary from company to company and from state to state; however, for estimation purposes, the following rates are used:

Insurance type	*Rate*	*Annual premium*
General liability $500,000 coverage	$0.65/$100 payroll	$ 78.00
Product liability	$1.00/$100 sales	48.60
Workmen's compensation	$4.93/$100 payroll	591.60
Fire and extended coverage	$0.80/$100 sales	1,120.00
Total annual premium		$1,838.20

$1,838.20/60,000 gallons = $0.03 per gallon

Although taxes will vary by county and by state, two simplifying assumptions have been made for this handbook in order to estimate the tax cost. First, it has been assumed that tax valuation will be based on 20 percent of fair market value (i.e., equipment investment cost). The second assumption is that the tax will be 100 mills on each dollar of valuation. For this model plant, the tax valuation will be $28,000 and the tax, $2800 per year, or $0.05 per gallon of ethanol.

According to current regulations, a holder of a BATF commercial permit must post a bond equivalent to the tax of $10.50 per proof gallon (100 proof) on 15 days' alcohol production. For this plant, a 15-day production in proof gallons is 6316 gallons with a tax liability of $66,000. The cost of a tax bond in this amount is $792 per year (based on a premium of $12.00/$1000 bond) or $0.01 per gallon of production.

The calculation of this cost is important only for income tax calculation. From preceding financial analysis data, depreciation is shown as a part of the capital recovery. However, the expected life of the equipment is a key determinant in calculating depreciation and capital recovery. For this plant, a 10-year life has been assumed or a depreciation rate of 10 percent. With the investment cost of $140,000, the depreciation is $14,000 annually or $0.23 per gallon.

In summary, conversion costs are as follows:

Item	Annual ($000)	Unit ($/gal)
Direct		
Labor	12.0	0.20
Fuel	6.0	0.10
Electricity	1.5	0.03
Supplies	5.4	0.09
Total	24.9	0.42
Indirect		
Maintenance	3.9	0.07
General and administrative	—	—
Taxes	2.8	0.05
Insurance	1.8	0.03
Tax bond	0.8	0.01
Total	9.3	0.16

Large On-Farm Ethanol Plant

This model is a farm-located, continuous fermentation and distillation process. It is the largest farm ethanol plant illustrated by USDA and produces 190-proof ethanol and wet stillage.

Type and size of operation. Operating continuously, 24 hours per day, 7 days per week, and 300 days per year at a rate of 50 gallons per hour, this plant is capable of producing 360,000 gallons annually. Four shifts of operators will be required to man this continuously operating plant.

Location. The plant will probably be located on a farm; however, it could be considered a small community or cooperative still. Relatively elaborate site preparation, building, and storage facilities will be required, and the availability of an adequate water supply will be a key location consideration. Access to wastewater sewage disposal is required.

Feedstock. The feedstock will be locally grown and stored corn or similar grain unless purchasable sample-grade or damaged gain is nearby.

Type and use of products. Of the 360,000 gallons of 190-proof ethanol produced by this model plant, it is assumed that one-half will be consumed in other farm activities and one-half will be sold in local markets. The wet stillage will be fed to high-production livestock on a continuous basis. The third product, carbon dioxide, is presumed to have no local use and will be vented.

Investment costs. No land cost is included on the assumption that this ethanol plant will be part of the farmstead. Storage, feedstock-preparation equipment, and suitable building costs are included in the cost estimated. The equipment for fermentation and distillation plus the instrumentation for control of these processes will be purchased as a package from the manufacturer. Design, engineering, and erection costs will be considerably less than that for custom-designed and built equipment. Investment costs for 1979 can be summarized as follows:

Items	*Costs ($000)*
Site preparation	20
Building	20
Receiving and storage	20
Supply storage	2
Feedstock preparation	15
Water supply	5
Boiler	35
Cooking	5
Fermentation	36
Instrumentation	15
Distillation	25
Rectification	25
Denaturization	2
Alcohol storage	3
Stillage storage	2
Miscellaneous	15
Subtotal	245
Engineering and erection	120
Total investment cost	365

Operating costs. Operating costs are considered to be those direct costs and indirect costs to convert corn to 190-proof alcohol at the rate of 50 gallons per hour. These costs do not include the feedstock (corn) costs or the value of the wet stillage for livestock feeding, for these costs are similar for all model plants and are evaluated elsewhere in the report.

The direct costs are those directly related to the number of gallons of ethanol produced and are labor, fuel, electricity, and supplies.

This plant requires one full-time person working 8 hours per shift for four shifts. On a per gallon basis, 0.02 man-hours will be required at an assumed rate per hour of $6.00 or $0.12 per gallon.

To calculate the fuel cost for the boiler, the model assumes a 43,000 Btu per gallon heat requirement for a 50-horsepower natural gas boiler. Such a boiler will require $0.0023 worth of fuel per 1000 Btu, assuming an 80 percent efficiency. This fuel cost is $0.09 per gallon for ethanol produced. The heat required has been derived from these calculations:

Operation	*Btu/gal (190 proof)*
Cooking	4,000
Distillation	30,000
Miscellaneous	9,000
Total	43,000

The electrical energy needed to operate the grinder, augers, mixers, and pumps for this plant is estimated at 0.5 kWhr per gallon. The assumed electrical rate per kWhr is 0.05 or rounded to $0.03 per gallon.

The alpha-amylase enzyme used in cooking starch is currently priced at $1.75 per pound; at a rate of 0.03 pound per bushel, cost is $0.0595 per bushel of corn. The glucoamylase enzyme used to convert starch to sugar costs $0.08925 per bushel based on its commercial price of $2.63 per liter. Dry, powered yeast for the fermentation process is priced at $0.90 per pound or $0.054 per bushel of corn. These supplies total $0.20 per bushel of corn and, with an estimated yield of 2.6 gallons per bushel, the per-gallon cost is $0.08. An additional $0.01 per-gallon cost of acid, base, and other chemicals makes the total supply cost $0.09 per gallon.

Indirect costs are annual costs relatively independent of production: maintenance, general and administrative expense, property taxes, insurance, tax bond cost, and depreciation.

The annual cost for maintenance has been estimated at 3 percent of the equipment investment cost for the first 5 years and 5 percent thereafter.

Four types of insurance are needed. The insurance rates will vary from insurance company to company and from state to state. However, for estimation purposes, rates quoted by a midwestern firm have been used for the following calculations:

Insurance type	*Rate*	*Annual premium*
General liability $500,000 coverage	$0.65/$100 payroll	$ 304.20
Product liability	$1.00/$100 sales	583.20
Workmen's compensation	$4.93/$100 payroll	2,307.00
Fire and extended coverage	$0.80/$100 sales	2,920.00
Total annual premium		$6,114.40

$6,114.40/360,000 gallons = $0.02 per gallon

Although taxes will vary by county and by state, two simplifying assumptions have been made for this handbook in order to estimate the tax cost. First, it has been assumed that tax valuation will be based on 20 percent of fair market value (i.e., equipment investment cost). Second, the tax will be 100 mills on each dollar of valuation. For this model plant, the tax valuation will be $73,000 and the tax $7300 per year or $0.02 per gallon of ethanol.

The general and administrative expense for this plant is $15,000, which is estimated to cover the cost of one person's time for managing the plant and marketing the products.

According to current regulations, a holder of BATF commercial permit must post a bond equivalent to the tax of $10.50 per proof gallon (100 proof) and 15 days' alcohol production. For this plant, 15 days' production in proof gallons is 31,580 gallons with a tax liability of the maximum $200,000. The cost of a tax bond in this amount is $2,400.00 per year based on a premium of $12.00/$1000 bond. On a cost-per-gallon basis, the tax bond is $0.007/gallon of production.

The calculation of this cost is important only for income tax calculations. From preceding financial analysis sections, depreciation is shown as a part of the capital recovery. However, the expected life of the equipment is a key determinant in calculating depreciation and capital recovery. For the plant shown in the model size, 10 years has been assumed as the life of the plant, a depreciation rate of 10 percent. With the investment cost of $365,000, the depreciation is $36,500 annually or $0.10 per gallon.

In summary, the conversion costs are as follows:

Item	*Annual ($000)*	*Unit ($/gal)*
Direct		
Labor	43.2	0.12
Fuel	34.2	0.09
Electricity	9.0	0.03
Supplies	32.4	0.09
Total	118.8	0.33
Indirect		
Maintenance	10.0	0.03
General and administrative	15.0	0.04
Taxes	7.3	0.02
Insurance	6.1	0.02
Tax bond	2.4	0.00
Total	40.8	0.11

Small Community, Wet Stillage

This model plant is assumed to be a community- or co-op-operated plant not part of a farm enterprise. This model produces about 1 million gallons annually of anhydrous ethanol (200 proof).

Type and size of operation. On a basis of a 24-hour-per-day, 300-day-per-year operation at the rate of 150 gallons per hour, this plant has capacity for about 1 million gallons of 200-proof ethanol. This size of operation is considered a commercial or industrial plant with hired management and has custom designed and installed equipment.

Location. It is expected that a plant of this size and type of operation will be located on the edge of a rural community, very likely near a feedlot, a petroleum processor or distributor, or both. Water availability will be a prime consideration in locating this plant and, therefore, the availability of adequate water, surface and/or groundwater, is essential. Adequate wastewater sewage disposal is also required.

Feedstock. The feedstock will be purchased from a local grain elevator with the possibility that there may be an occasional opportunity to obtain sample-grade or damaged grains nearby.

Type and use of product. The major product, 1 million gallons of 200-proof ethanol, will be sold in the gasohol market by contract with a nearby petroleum refiner or distributor. The secondary product, stillage, will likewise be contracted to nearby feedlots for cattle.

Investment costs. A modest land cost of $10,000 is included in the investment cost for this model plant on the assumption that the plant will be located in conjunction with either a feedlot or a petroleum processing or distribution operation. Costs for other equipment have been estimated on the basis of custom design fabrication and installation as a turnkey project. Investment costs can be summarized as follows:

Item	*Cost* ($000)
Land	10.0
Site preparation	10.0
Building	54.0
Receiving and storage	71.0
Supply storage	5.0
Feedstock preparation	11.0
Water supply	12.5
Boiler (coal-fired)	175.0
Cooking	21.0
Fermentation	111.0
Instrumentation	40.0
Distillation	44.0
Rectification	31.0
Dehydration	78.0
Denaturization	3.0
Alcohol storage	4.0
Stillage storage	85.0
Wastewater treatment	10.0
Miscellaneous	20.0
Subtotal	795.5
Engineering and erection	404.5
Total investment cost	1,200.0

Operating costs. Operating costs are considered to be those direct and indirect costs to convert corn to 200-proof alcohol at a continuous rate of 150 gallons per hour. These costs do not include the feedstock (corn) costs or the value of the concentrated wet stillage for livestock feeding.

The direct costs are those directly related to the number of gallons of ethanol produced and are labor, fuel, electricity, and supplies.

This plant will require two full-time persons working 8 hours per day for two shifts and one person each for the other two shifts per week. On a per-gallon basis, 0.010 man-hours will be required at an assumed rate per hour of $6.00 per hour, a $0.06 per-gallon rate.

An estimated 61,000 Btu per gallon heat is required in a 300-horsepower coal (or combination coal, wood or natural gas) boiler. Such a boiler will require $0.0023 worth of fuel per 1000 Btu, assuming an 80 percent efficiency and using 11,000-Btu coal delivered at the site for $40.00 per ton. This fuel cost is $0.14 per gallon of ethanol produced. The heat required has been derived from the following calculations:

Operation	*(000) Btu/gal (190 proof)*
Cooking	3.6
Distillation	28.0
Dehydration	20.0
Miscellaneous	9.4
Total	61.0

The electrical energy needed to operate the grinder, augers, mixers, and pumps for this plant is estimated at 0.05 kWhr per gallon. The assumed electrical rate per kWhr is 0.05 or rounded to $0.03 per gallon.

The alpha-amylase enzyme used in cooking starch is priced at $1.75 per pound at current commercial rates or $0.03 lbs per bushel or $0.0595 per bushel of corn. The glucoamylase enzyme used to convert starch to sugar is valued at $0.08925 per bushel based on a commercial price of $2.63 per liter. Dry, powdered yeast for the fermentation process is priced at $0.84 per pound or $0.05 per bushel of corn. These supplies total $0.20 per bushel of corn and, with an estimated yield of 2.6 gallons per bushel, the per-gallon cost is $0.08. An additional $0.01 per-gallon cost of acid, base, and other chemicals makes a total supply cost of $0.09 per gallon.

Indirect costs are annual costs which are relatively independent of production. They include maintenance, general and administrative expenses, property taxes, insurance, tax bond, and depreciation costs.

The annual cost for maintenance has been estimated to be 3 percent of the equipment investment cost for the first 5 years and 5 percent thereafter.

The estimated cost for management and marketing for this plant (general and administration) is based on a cost of two full-time persons, an annual cost of $30,000 per year total or $0.03 per gallon.

There are four types of insurance to protect the financial interest of the enterprise and, therefore, these insurance costs are a part of the cost of ethanol production. For estimation purposes, insurance rates quoted for a midwestern location have been used for the following calculations:

Insurance type	*Rate*	*Annual premium*
General liability $500,000 coverage	$0.65/$100 payroll	$ 390.00
Product liability	$1.00/$100 sales	1,740.00
Workmen's compensation	$4.93/$100 payroll	2,958.00
Fire and extended coverage	$0.80/$100 valuation	9,600.00
Total annual premium		$14,688.00

$$\$14{,}688/1{,}000{,}000 \text{ gallons} = \$0.01 \text{ per gallon}$$

Although taxes will vary, two simplifying assumptions have been made for this handbook in order to estimate the tax cost. First, it has been assumed that tax valuation will be based on 20 percent of fair market value (i.e., equipment investment cost). Second, the tax will be 100 mills on each dollar of valuation. For this model plant, the tax valuation will be $240,000 and the tax will be $24,000 per year or $0.02 per gallon of ethanol.

According to current regulations, a holder of a BATF commercial permit must post a bond equivalent to the tax of $10.50 per proof gallon (100 proof) on 15 days' alcohol production. For this plant, 15 days' production in proof gallons is 108,000 gallons with a tax liability in excess of $200,000, the maximum tax bond required. The cost of a tax bond in this amount is $2400 per year based on a premium of $12.00/$1000 or $0.002 per gallon of production.

The calculation of this cost is important only for income tax calculations. From preceding financial analysis data, depreciation is shown as a part of the capital recovery. However, the expected life of the equipment is a key determinant in calculating depreciation and capital recovery. For the plant shown in this model size, 20 years has been assumed as the life of the plant which makes the depreciation rate 5 percent. With the investment cost $1,200,000, the depreciation is $60,000 annually or $0.06 per gallon.

Conversion costs can be summarized as follows:

Item	*Annual ($000)*	*Unit ($/gal)*
Direct		
Labor	60.0	0.06
Fuel	141.2	0.14
Electricity	25.0	0.03
Supplies	90.0	0.09
Total	316.2	0.32
Indirect		
Maintenance	33.8	0.03
General and administrative	30.0	0.03
Taxes	24.0	0.02
Insurance	14.7	0.01
Tax bond	2.4	0.01
Total	104.9	0.10

Small Community, DDGS

This model is a variation of the last previous model and was developed to show the change in operating costs of a plant when the additional cost of drying stillage (DDGS) is added to the cost of ethanol production. This plant is identical except that the costs both in investment and in operations associated with drying the stillage are added to, and reflected in, the ethanol cost per gallon.

Type and size of operation. On a 24-hour-per-day, 300-day-per-year operation, at the rate of 150 gallons per hour, this plant has a capacity for about 1 million gallons of 200-proof ethanol. This size operation is considered a commercial or industrial plant with hired management and custom-designed and installed equipment.

Location. It is expected that a plant of this size and type operation will be located on the edge of a rural community, very likely near a feedlot and a petroleum processor or distributor or both. Water availability will be a prime consideration in locating this plant and, therefore, the availability of adequate surface and/or groundwater is essential.

Feedstock. The feedstock will be purchased from local grain elevators. Occasionally sample-grade or damaged grains may be available nearby.

Type and use of product. The major product will be 200-proof ethanol. The coproduct, DDGS, will have a broad market and, because it is a dried product, will be of a type that can be shipped over great distances to reach national or world markets.

Investment costs. In addition to the investment of equipment shown in the small community, wet model, added costs for the dryer and a larger boiler (including installation) will amount to $375,000. This addition makes the total investment cost $1,575,000.

Operating costs. The operating costs for this model plant will remain the same as the previous model plant with the exception of those items listed below.

This plant will require two full-time persons working eight hours per day for each of the four weekly shifts. On a per-gallon basis, 0.0139 man-hours will be required at an assumed rate of $6.00 per hour or $0.08 per gallon.

Fuel costs for the stillage drying will add 21,000 Btu per gallon to the fuel requirements of the previous model. This additional fuel will add $0.05 per gallon to those costs and will make the total fuel cost for this model plant $0.19 per gallon for the total 82,000 Btu per gallon. The boiler will be a 400-horsepower boiler.

The cost of maintenance is assumed to be 3 percent of the equipment investment cost for the first 5 years and 5 percent thereafter.

The overhead costs for this plant are based on the requirement of two full-time persons at a total of $40,000 per year or $0.04 per gallon of ethanol output.

Four types of insurance are needed. The rates and costs shown below will vary from one insurance company to another, and from one state to another; however, for estimation purposes, rates quoted for a midwestern location have been used for the following calculations:

Insurance type	*Rate*	*Annual premium*
General liability $500,000 coverage	$0.65/$100 payroll	$ 520.00
Product liability	$1.00/$100 sales	1,740.00
Workmen's compensation	$4.93/$100 payroll	3,944.00
Fire and extended coverage	$0.80/$1000 valuation	12,600.00
Total annual premium		$18,804.00

$18,804/1,000,000 gallons = $0.02 gallon

The tax valuation is $315,000 for this plant so that the tax based on 100 mills per dollar of value will be $31,500 or rounded to $0.03 per gallon.

According to current regulations, a holder of a BATF commercial permit must post a bond equivalent to the tax of $10.50 per proof gallon (100 proof) on 15 days' alcohol production. For this plant, 15 days' production in proof gallons is 108,000 gallons with a tax liability in excess of $200,000, the maximum tax bond required. The cost of a tax bond in this amount is $2400 per year based on a premium of $12.00/$1000 or $0.002 per gallon of production.

With the investment cost at $1,575,000 for a 20-year life, the depreciation per gallon for this plant is $0.08.

Conversion costs can be summarized as follows:

Item	*Annual ($000)*	*Unit ($/gal)*
Direct		
Labor	84.0	0.08
Fuel	190.0	0.19
Electricity	25.0	0.03
Supplies	90.0	0.09
Total	389.0	0.39
Indirect		
Maintenance	44.9	0.04
General and administrative	40.0	0.04
Taxes	31.5	0.03
Insurance	18.8	0.02
Tax bond	2.4	0.00
Total	137.6	0.13

Large Community, DDGS

This model plant is a community or co-op-operated plant and produces 2 million gallons annually of anhydrous ethanol (200 proof). This model also assumes the sale of a dried byproduct, DDGS.

Type and size of operation. On a 24-hour-per-day, 300-day-per-year operation at the rate of 300 gallons per hour, this plant has the capacity for 2 million gallons of 200-proof ethanol. This operation is considered a commercial or industrial plant with hired management and custom-designed and installed equipment.

Location. It is expected that a plant of this size and type operation will be located on the edge of a rural community, very likely near a feedlot and a petroleum processor or distributor or both. Water availability will be a prime consideration in locating this plant; therefore, the availability of adequate water from rainfall, surface, or groundwater is essential.

Feedstock. The feedstock for the mash will be purchased from local grain elevator storage. Occasionally damaged grains may be obtained locally.

Type and use of product. The major product, 2 million gallons of 200-proof ethanol, will be sold into the gasohol market by contract with a nearby petroleum refiner or distributor. The secondary product, DDGS, will be sold as a commodity on the national market.

Investment cost. Land cost is included in the investment cost for this model plant on the assumption that the plant will be located separately but near either a feedlot or a petroleum processing or distribution operation. Costs for other equipment have been estimated on the basis of custom-designed fabrication and installation as a turnkey project.

A summary of these investment costs are as follows:

Item	*Cost* *($000)*
Land	90.0
Site preparation	45.0
Building	135.0
Receiving and storage	90.0
Supply storage	10.00
Feedstock preparation	50.00
Water supply	40.00
Boiler (coal-fired)	295.0
Cooking	40.0
Fermentation	200.0
Instrumentation	70.0
Distillation	50.0
Rectification	40.0
Dehydration	100.0
Denaturization	5.0
Alcohol storage	8.0
Stillage storage	100.0
Drying	400.0
Wastewater treatment	20.0
Miscellaneous	40.0
Subtotal (Equipment)	1,828.0
Engineering and erection	922.0
Total investment cost	2,750.0

Operating costs. Operating costs are considered to be those direct costs and indirect costs needed to convert corn to 200-proof alcohol at a continuous rate of 300 gallons per hour. These costs do not include the feedstock (corn) costs or the value of the concentrated wet stillage for livestock feeding.

The direct costs are those directly related to the number of gallons of ethanol produced and are labor, fuel, electricity, and supplies.

This plant will require two full-time persons working eight hours per day on each of four weekly shifts operating continuously at the rated output of this plant. On a per gallon basis, 0.0069 man-hours will be required at an assumed rate per hour of $6.00 per hour or $0.04 per gallon.

An estimated 82,000 Btu per gallon heat is required in an 800-horsepower coal- (or combination coal, wood, or natural gas) fired boiler. Such a boiler will require $0.0023 worth of fuel per 1000 Btu, assuming an 80 percent efficiency and using 11,000-Btu coal delivered at the site for $40.00 per ton. This fuel cost is $0.19 per gallon of ethanol produced. The heat required has been derived from these calculations:

Operation	*(000) Btu/gal (200 proof)*
Cooking	3.6
Distillation	28.0
Dehydration	20.0
Drying	21.0
Miscellaneous	9.4
Total	82.0

The electrical energy needed to operate the grinder, augers, mixers, and pumps for this plant is estimated at $0.5 kWhr per gallon. The assumed electrical rate per kWhr is $0.05 or rounded to $0.03 per gallon.

The alpha-amylase enzyme used in cooking starch is priced at $1.75 per pound at current commercial rates, which amounts to 0.03 pound per bushel or $0.0595 per bushel of corn. The glucoamylase enzyme used to convert starch to sugar is valued at $0.08925 per bushel based on a commercial price of $2.63 per liter. Dry, powdered yeast for the fermentation process is priced at $0.84 per pound or $0.05 per bushel of corn. These supplies total $0.20 per bushel of corn and, with an estimated yield of 2.5 gallons per bushel, the per-gallon cost is $0.08. An additional $0.01 per-gallon cost of acid, base, and other chemicals makes the total supply cost $0.09 per gallon.

Indirect costs are annual costs and are independent of the number of gallons of ethanol produced. These are maintenance, general, and administrative expenses, property taxes, insurance, tax bond, and depreciation costs.

The annual cost for maintenance has been estimated to be 3 percent of the equipment investment cost for the first 5 years and 5 percent thereafter.

The estimated cost for management and marketing for this plant (general and administration) is based on a cost of two full-time persons with an annual cost of $50,000 per year total or $0.03 per gallon.

Four types of insurance are needed. Insurance rates vary from insurance company to company and from state to state; however, for estimation purposes, rates quoted for a midwestern location have been used for the following calculations:

Insurance type	*Rate*	*Annual premium*
General liability $500,000 coverage	$0.65/$100 payroll	$ 538
Product liability	$1.00/$100 sales	3,480
Workmen's compensation	$4.93/$100 payroll	4,930
Fire and extended coverage	$0.80/$100 valuation	22,000
Total annual premium		$30,948

$30,948/2,000,000 gallons = $0.01 per gallon

Although taxes will vary by county and by state, two simplifying assumptions have been made for this handbook in order to estimate the tax cost. First, it has been assumed that tax valuation will be based on 20 percent of fair market value of the equipment investment cost. Second, the tax will be 100 mills on each dollar of valuation. For this model plant, the tax valuation will be $550,000 and the tax $55,000 per year or $0.03 per gallon of ethanol.

According to current regulations, a holder of a BATF commercial permit must post a bond equivalent to the tax of $10.50 per proof gallon (100 proof) on 15 days' alcohol production. For this plant, 15 days' production in proof gallons is over the maximum tax liability of $200,000. Therefore, the cost of a tax bond for $200,000 is $2400 per year based on a premium of $12.00/$1000 bond or $0.001 per gallon of production.

The calculation of this cost is important only for income tax calculations. From preceding financial analysis sections, depreciation is shown as a part of the capital recovery. However, the expected life of the equipment is a key determinant in calculating depreciation and capital recovery. For the plant shown in this model size, 20 years has been assumed to be the life of the plant, a period which makes the depreciation rate 5 percent. With the investment cost of $2,750,000, the depreciation is $137,500 annually or $0.07 per gallon.

Conversion costs can be summarized as follows:

Item	*Annual* ($000)	*Unit* ($/gal)
Direct		
Labor	84.0	0.04
Fuel	380.0	0.19
Electricity	50.0	0.03
Supplies	180.0	0.09
Total	694.0	0.35
Indirect		
Maintenance	74.3	0.04
General and administrative	50.0	0.03
Taxes	55.0	0.03
Insurance	30.9	0.01
Tax bond	2.4	0.00
Total	212.6	0.11

Central "Topping-Cycle" Dehydration Plant

A variation of the model plants previously described will be the operation of a centralized dehydration process. This alternative model plant will operate a collection service and process lower-proof ethanol, produced elsewhere, into anhydrous or 200-proof ethanol.

This plant will operate as an addition to a community plant which produces its own anhydrous product via fermentation and has extra dehydration capacity. Such a plant will probably not operate as a stand-alone plant because of the economies possible when it is attached to an ethanol-production plant. If designed originally with oversized dehydration capacities, the equipment will be sized accordingly. However, for purposes of this analysis, it is assumed that the dehydration equipment is duplicated and operated as one unit.

Type and size of operation. This plant will be designed as an addition to the small community model plants and will produce 1 million gallons of anhydrous ethanol annually with dehydration equipment capacity to process an additional 1 million gallons of 190-proof ethanol. The additional dehydration unit will operate 300 days per year at the rate of 150 gallons per hour.

Location. The plant will be located as an integral part of the small community model plants.

Feedstock. The feedstock for the plant will be 190-proof ethanol collected by tank truck from a variety of on-farm ethanol plants within an average 10-mile radius of the central plant.

Type and use of product. The final and only product will be 200-proof ethanol. This will be combined with the 200-proof ethanol produced at the plant and marketed through its distribution system.

Investment cost. The following equipment list represents the additional equipment to dehydrate the additional ethanol. The equipment for producing the 190-proof feedstock has already been described in models for the on-farm model plants.

Item	*Cost ($000)*
2,500-gallon tank truck	45
25,000-gallon storage tank	10
Unloading and metering equipment	2
Dehydration column, 24″ × 30″ plates	25
Condenser/heater	3
Product cooler	4
Benzene cooler	2
Benzene-column calandria	5
Instrumentation	10
Ethanol storage tank, 25,000 gallons	10
Additional 100-horsepower boiler	36
Subtotal	152
Engineering and installation	50
Total investment cost	202
Additional working capital to finance 15-day feedstock operating costs, and final product	$150,000

Operating costs. The operating costs for this plant are the direct costs associated with the number of gallons or product processed and the indirect costs, which are usually annual costs not related directly to production.

The direct costs are the costs of collecting lower-proof ethanol from the farm, transporting the product to the central plant, and dehydrating it. They include labor, fuel, electricity, supplies, and trucking costs.

One man with a truck can transport 2500 gallons of 190-proof ethanol from the farm to the central plant, can run two trips per day, and load and unload the product while working an eight-hour day, 5-day week. An assumed rate of $6.00 per hour computes to $48 per day, $240 per week, or $12,500 per year. Based on 1 million annual gallons, the per-gallon cost is $0.01.

The estimated energy requirement for the dehydration process is 20,000 Btu per gallon. Assuming a coal-fired boiler using $40 per ton coal (80 percent efficiency), the fuel cost per gallon is $0.043 rounded to $0.05.

The electrical power estimate is less than 0.2 kWhr per gallon for the pumping requirements on a per-gallon basis. Assuming electrical costs at $0.05 per kWhr, the electrical cost is $0.01 per gallon.

The truck costs are based on the operation of a single truck that will make two trips per day, hauling 2500 gallons per trip, for a five-day week, 43-week year. This schedule will meet the requirements of the plant. The collection area is estimated to lie within a 10-mile radius with the average round-trip length of 20 miles. Thus, the annual mileage is 8600 miles with an assumed per mile cost of $1.00. Adjusting the annual cost of $8600 to a per-gallon cost results in $0.0086 per gallon or $0.01.

Indirect costs include maintenance, general and administration, insurance, taxes, and depreciation.

Costs for maintenance have been estimated at 3 percent of the equipment cost for the first five years and 5 percent for the remaining 15 years of plant life.

To coordinate and schedule the pick up of lower-proof ethanol and to keep records of all the transactions of the collection system, one person has been designated as the administrator. The annual cost is $15,000, which rounds to $0.02 per gallon for the general and administration cost.

The property tax cost for the additional property valuation for this plant is based on 20 percent of the $202,000 investment cost of the equipment times the tax rate of 100 mills or a $4500 per year tax cost. The tax bond cost will require another tax bond to cover transportation functions and will be $2,400 per year. The insurance cost is as follows:

Insurance type	*Rate*	*Annual premium*
General liability	$0.65/$100 payroll	$ 179
Product liability	$1.00/$1000 sales	1,500
Workmen's compensation	$4.93/$100 payroll	616
Fire and extended coverage	$0.80/$1000 valuation	1,616
Total annual premium		$3,911

These annual costs total $10,811 or $0.01 per gallon of ethanol processed.

Depreciation as shown here is a memo cost only as it appears as part of the capital recovery in the financial analysis. For purposes of this analysis, depreciation is 5 percent per year, $10,000, or $0.01 per gallon.

Conversion costs can be summarized as follows:

Item	Annual ($000)	Unit ($/gal)
Direct		
Labor	12.5	0.01
Fuel	46.3	0.05
Electricity	10.0	0.01
Truck costs	8.6	0.01
Total	77.4	0.08
Indirect		
Maintenance	6.0	0.01
General and administrative	15.0	0.02
Taxes and insurance	10.8	0.01
Total	31.8	0.04

Chapter 4

Fundamentals of Feedstocks

Selection of the feedstock is one of the most critical elements in assessing the feasibility of an ethanol-production facility. Closely linked with this decision are an evaluation of the quantities of the selected feedstock that can be secured annually or, on a long-term basis, that can support a plant, and an estimation of the cost of the resource. This first step is essential as it will provide an order of magnitude for the size of the project considered and for the market required to absorb the ethanol fuel and its marketable coproducts. The selection of a feedstock will affect the design and characteristics of the projected plant and the economics of the overall project.

Many feedstocks can be used to produce fermentation ethanol, as shown in Table 4-1. Despite the fact that corn presently offers the largest potential for U.S. grain production, other crops or residuals could be considered. Particular attention should be given to site- or region-specific conditions that make particular crops or their residues especially attractive as feedstock material. Disposal of food-processing residues, damaged crops, or crop residues through ethanol production could result in a credit or attractive feedstock prices for the ethanol producer. Such favorable situations may in some cases make small commercial plants more economically attractive than larger ones using more conventional feedstocks such as grain or sugarcane.

The prospective producer should therefore approach the problem of feedstock selection with an open mind. Also, he should attempt to integrate his operation in the context of the region considered, in order to take advantage of local or regional conditions that could be beneficial. The producer should also consider the existence of other plants in the same area and their feedstock requirements.

Selection of a particular feedstock will have a direct bearing on the design of the plant. Therefore, this chapter discusses some feedstock characteristics, which are essential for the prospective producer to know before he makes any decisions.

The average composition of some of the possible feedstocks is shown in Table 4-2. Only the carbohydrate component of the feedstock is used by the fermentation process; all other components remain in the distiller's coproduct.

Table 4-1. Summary of Feedstock Characteristics

TYPE OF FEEDSTOCK	PROCESSING NEEDED PRIOR TO FERMENTATION	PRINCIPAL ADVANTAGES	PRINCIPAL DISADVANTAGES
SUGAR CROPS Sugar Beets Sweet Sorghum Sugar Cane Fodder Beets Jerusalem Artichoke	Milling to extract sugar.	• Preparation is minimal. • High yields of ethanol per acre. • Crop coproducts have value as fuel, livestock feed, or soil amendment.	• Storage may result in loss of sugar. • Cultivation practices vary widely especially with "nonconventional" crops.
STARCH CROPS Grains: Corn Wheat Sorghum Barley Tubers: Culled Potatoes Potatoes	Milling, liquefaction, and saccharification.	• Storage techniques well developed. • Cultivation practices are widespread with grains. • Livestock coproduct is relatively high in protein.	• Preparation involves additional equipment, labor, and energy costs. • DDG from aflatoxin-contaminated grain is not suitable as animal feed.
CELLULOSIC Crop Residues: Corn Stover Wheat Straw Forages: Alfalfa Sudan Grass Forage	Milling and hydrolysis of the linkages	• Use involves no integration with the livestock feed market. • Availability is widespread.	• No commercially cost-effective process exists for hydrolysis of the cellulosic linkages.

Source: *A Guide to Commercial Scale Ethanol Production and Financing,* SERI, Denver, 1980.

Four products result from the fermentation of the raw materials—ethanol, distiller's coproduct, carbon dioxide, and water. The relationships of these substances to that of the starting materials have been calculated and are shown in Table 4-3. Differences in the initial starch (or sugar) and moisture content of the raw materials will result in variations in the quantities of products produced.

The coproduct associated with each feedstock is a viable commercial material, usually livestock feed or fertilizer. When feed grains are used as the feedstock, the resulting coproduct stillage is a high protein feed. For example, spent grains and distiller's grains have been marketed for a number of years by the beverage industry. The market for the other stillage coproducts of the feedstock is less well established.

Types of Feedstocks

The biological production of ethanol is accomplished through the fermentation of six-carbon sugar units (principally glucose) in the presence of yeast. All agricultural crops and crop residues contain six-carbon sugars or compounds of these sugars, and can be used to produce ethanol, provided that the six-carbon sugars they contain are accessible for fermentation. Agricultural crops and residues can be subdivided into three broad classes: sugar crops, starch crops, and cellulosic material, as shown in Table 4-1. In

Table 4-2. Average Composition of Possible Feedstocks for Ethanol Production[a]

				Carbohydrates						
	Water	Protein	Fat	Fiber	N-free extract	Mineral matter	Ca	P	N	K
Sugar beet	83.6	1.6	0.1	1.0	12.6	1.1	0.04	0.04	0.26	0.25
Molasses, beet	19.5	8.4	0	0	62.0	10.0	0.05	0.02	1.34	4.77
Artichoke tubers	79.5	2.0	0.1	0.8	15.9	1.7	—	0.06	0.32	0.41
Cassava roots	67.4	1.1	0.3	1.4	28.8	1.0	—	0.04	0.18	0.33
dried	5.6	2.8	0.5	5.0	84.1	2.0	—	—	0.45	—
Potatoes, tuber	78.8	2.2	0.1	0.4	17.4	1.1	0.01	0.05	0.35	0.48
Sugarcane	76.8	1.0	0.8	6.8	13.4	1.2	—	0.04	0.16	0.37
molasses cane or blackstrap	26.6	3.0	0	0	61.7	8.6	0.66	0.08	0.48	3.67
Sweet potatoes	68.2	1.6	0.4	1.9	26.7	1.2	0.03	0.04	0.26	0.38
Corn, dent no. 3	16.5	8.9	3.8	2.0	67.5	1.3	0.02	0.26	1.42	0.28
Milo	11.0	10.9	3.0	2.3	70.7	2.1	0.03	0.28	1.74	0.35
Rice	12.2	9.1	2.0	1.1	74.5	1.1	0.04	0.25	1.46	—
Rye	10.5	12.6	1.7	2.4	70.9	1.9	0.10	0.33	2.02	0.47
Wheat, hard winter southern plains	10.6	13.5	1.8	2.8	69.2	2.1	0.05	0.42	2.16	—
Raisins, cull	15.2	3.4	0.9	4.4	73.1	3.0	—	—	0.54	—

[a]All figures given in this table are percentages.

Source: Frank B. Morrison, *Feeds and Feeding,* Abridged, 9th ed., The Morrison Publishing Co., 1961.

Table 4-3. Input–Output Relationships for Converting Selected Raw Materials to Ethanol

Item	Corn	Grain sorghum	Wheat	Potatoes	Sugar beets
Input					
Unit	bu	bu	bu	cwt	ton
Weight per unit	56	56	60	100	2,000
Moisture content (percent)	13	13	13	78	75
Output (yield) per unit in percent					
Ethanol	30.7	31.8	30.2	8.6	8.2
Distiller's coproduct	33.0	30.9	33.8	6.9	8.9
Carbon dioxide	29.3	30.3	28.8	8.2	7.9
Water	7.0	7.0	7.2	76.3	75.0
Total	100.0	100.0	100.0	100.0	100.0
Output (yield) per unit in weight (lbs)					
Ethanol	17.2	17.8	18.1	8.6	164.3
Distiller's coproduct (dry)	18.5	17.3	20.3	6.9	178.5
Carbon dioxide	16.4	17.0	17.3	8.2	157.2
Water	3.9	3.9	4.3	76.3	1,500.0
Total	56.0	56.0	60.0	100.0	2,000.0
Conversion rate, gallons ethanol per unit	2.6	2.7	2.74	1.3	24.9

Source: *Small-Scale Fuel Alcohol Production,* USDA, Washington, D.C., 1980.

Source: *A Guide to Commercial Scale Ethanol Production and Financing,* SERI, Denver, 1980.

Corn offers the largest potential for ethanol production in the United States.

sugar crops, the six-carbon sugars or fermentable sugars occur individually or in bonded pairs. Minimal mechanical treatment will release the fermentable sugars. In starch crops, the six-carbon sugars are linked in long, branched chains (called starch). These chains must be broken down into individual or pairs of six-carbon sugars before yeast can use the sugars to produce ethanol. These crops, therefore, will require additional treatment (mechanical, chemical, or biological) before fermentation can occur. In cellulosic materials, the six-carbon sugars are linked in extremely long chains involving strong chemical bonding. Releasing the six-carbon sugars for fermentation requires extensive pretreatment. The optimum method for recovering the fermentable sugars from cellulosic materials has not been commercially demonstrated, and various research programs are being pursued to improve the process.

Specifics regarding crop selection and its impact on the project are given below.

SUGAR CROPS

Processing and Storage Requirements

As previously mentioned, fermentable sugars are easily released from sugar crops. Preparation of the crop for fermentation involves milling or crushing and sugar extraction. Variations of the sugar recovery process are used for different crops, i.e., slicing of sugar beets followed by sugar extraction, pressing of citrus crops, etc. The sugar recovery process involves relatively low capital, labor, and energy costs.

The ease of recovery of fermentable sugars from sugar crops is counterbalanced by a significant disadvantage. The high moisture content of these easily accessible sugars makes them very susceptible to infestation by microorganisms, resulting in crop spoilage during storage. Crop spoilage in turn will result in reduced ethanol production. Sugar loss during storage can be reduced or eliminated by treatment of the extracted sugar solution. Two processes for treating the extracted sugars may be used—pasturization of the solution or evaporation of part of the water to obtain a concentrated sugar solution. Both processes, however, are costly in terms of equipment and energy.

Potential Sugar Crops

The two major sugar crops that have been cultivated at a significant commercial level of production are sugarcane and sugar beets. Other alternative sugar crops that are or could be cultivated in the United States include sweet sorghum, Jerusalem artichokes, fodder beets, and fruits.

Sugarcane. Sugarcane is considered an attractive feedstock because of its high yield of sugar per acre and a correspondingly high yield of residue known as bagasse, which can be used as fuel to generate process heat. The major drawback of this feedstock is the limited availability of land suitable for economical production of the crop. The potential for expansion of the production of sugarcane in the United States to support a large ethanol industry appears limited owing to specific geographic conditions necessary to its cultivation.

Sweet sorghum. The name sweet sorghum refers specifically to varieties of sorghum bicolor. Sweet sorghum is grown on a small scale for syrup and silage. Other sorghums are grown for grain. Sweet sorghum is a potentially attractive feedstock because of its high yield of ethanol per acre and its adaptability to a wide range of climates and soils. It could make a significant contribution to the feedstock resource for future ethanol production. Although this feedstock is easier to store than sugarcane, it deteriorates rapidly in storage. The juice from sweet sorghum and milo once extracted could be concentrated and then preserved for later fermentation. Systems for doing this have only been tested at the laboratory level at this time. However, given the initial success of these systems and the development of an appropriate processing infrastructure, sweet sorghum and milo could readily become a principal biomass resource for ethanol production. The full potential of the crop as a feedstock has not yet been realized because genetic improvements and improvements in crop management have not been implemented.

Fodder beets. Fodder beets are a high-yielding forage crop obtained by crossing two other beet species, sugar beets and mangolds. Fodder beets have higher sugar yields per acre, better storage characteristics, and are less demanding to grow than sugar beets. When fully developed, fodder beets could contribute significantly to the feedstock resource base for ethanol production.

Sugar beets. Sugar beets tolerate a wide range of climatic and soil conditions and therefore offer a possibility of expanded production to support ethanol-production facilities. Widespread expansion of sugar beet cultivation, however, is limited to some extent by the necessity to rotate this with nonroot crops on a 3-year basis. Sugar beets are considered an attractive feedstock because of their high yield of sugar per acre and correspondingly high yield of beet pulp and beet top coproducts.

Jerusalem artichokes. The Jerusalem artichoke has shown potential as an alternative sugar crop. It is well adapted to northern climates and a variety of soils, and is not demanding of soil fertility. With expanded production, this crop could make a significant contribution to the feedstock resource for future ethanol production.

Fruit crops. Fruit crops are not likely to be used as direct feedstock for ethanol-fuel production because of their high market value for direct human consumption. The coproducts of processing fruit crops (citrus molasses, for example) could be used as feedstocks because fermentation is an economical method for reducing the potential environmental impact of disposal of untreated wastes containing fermentable sugars. Although prime fruit crops are

Source: *A Guide to Commercial Scale Ethanol Production and Financing*, SERI, Denver, 1980.
Fermentable sugars can be recovered from sugar beets.

too valuable to use, distressed fruit crops are an excellent feedstock for ethanol production.

STARCH CROPS

The starch crop used as feedstock for ethanol production includes grains (corn, wheat, barley, grain, sorghum, etc.) and tubers (potatoes and sweet potatoes).

Processing and Storage Requirements

Yeasts cannot directly use starch, the long-branched chains of six-carbon sugars, to produce ethanol. Before fermentation, the starch chains must be broken down into individual or pairs of six-carbon sugar units. This step involves the reaction of the starch-containing material with water (hydrolysis) in the presence of enzymes to produce a simple sugar solution. In the case of grain feedstocks, the process includes milling of the grains to a fine meal to expose the starch and slurrying the meal with water to form a mash, followed by hydrolysis. Hydrolysis involves the liquefaction of the mash into a solution of high-molecular-weight sugars (dextrins) followed by their conversion to fermentable sugars. Both steps are conducted in the presence of enzymes (protein catalysts) under controlled temperature. The preparation process

Source: *A Guide to Commercial Scale Ethanol Production and Financing*, SERI, Denver, 1980.
The starch in grains can be converted to fermentable sugar in a simple two-step process.

prior to fermentation may include several variations of the general procedure just described. The grains may be prepared under dry or wet milling conditions resulting in the production of various coproducts (germs, oil, hulls, etc.). These process variations will be discussed further.

The conversion of potatoes or, more generally, tubers to ethanol is similar to that of grains, with minor modifications to the process. The potatoes are washed to remove dirt and soil microbes, sliced, and cooked before the hydrolysis step.

A distinct advantage of starch crops is the relative ease with which these crops can be stored with minimal loss of the fermentable portion. Ease of storage is related to the fact that a conversion step is needed before fermentation. Many microorganisms including yeasts can use individual or small groups of sugar units but not long chains. Some microorganisms present in the environment produce the enzymes needed to break up the chains, but, unless certain conditions such as moisture, temperature, and acidity are just right, the rate of conversion during storage is very low. When crops and other feeds are dried to about 12-percent moisture (the percentage below which most microorganisms are not active), the deterioration of starch and other valuable components such as proteins and fats is minimal. Grains are routinely dried before storage and, therefore, little risk of loss is expected for these feedstocks. Potatoes can usually be stored about 6 months before significant losses occur.

POTENTIAL STARCH CROPS

Grains

Potential grain feedstocks include rye, wheat, milo, rice, barley, and corn. Corn is currently the largest potential feedstock supply. The following discussion, nevertheless, is applicable to other grains with relatively minor variations of the process described for corn.

Corn kernels are composed of three major constituents: the germ (2-3 percent of the kernel) which is rich in oil and proteins; the hull and bran layers (13-17 percent of the kernel) which are rich in protein, cellulose, and minerals; and the endosperm (the remainder of the kernel) which is mostly starch fixed in a matrix of protein. Various food or feed products may be extracted from the kernels through different processing methods. The simplest way of processing corn kernels prior to fermentation to alcohol is shown in Fig. 4-1. Whole kernels are ground and the resulting meal is treated through slurrying and hydrolyzing to convert the starch contained in the grain to fermentable sugars. The nonfermentable part of the grain or spent grain

Source: *A Guide to Commercial Scale Ethanol Production and Financing*, SERI, Denver, 1980.

Milo is a popular grain crop in the southwest.

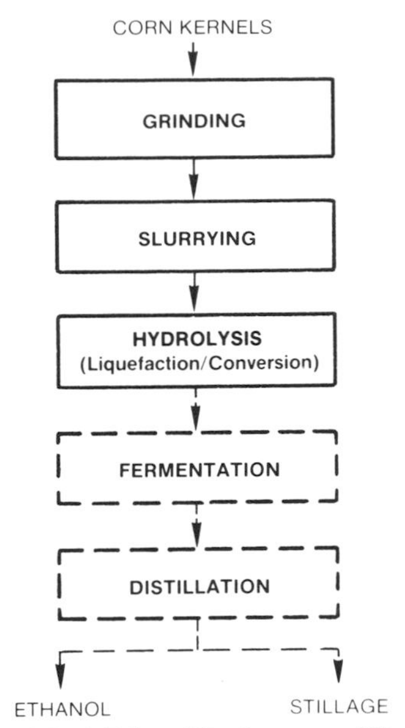

Source: *A Guide to Commercial Scale Ethanol Production and Financing*, SERI, Denver, 1980.

Figure 4-1. Simple processing of corn prior to fermentation.

contains most of the nonstarch nutritive elements originally present in the kernels. This coproduct of alcohol production may be retrieved after distillation of the alcohol, as shown in Fig. 4-1, or at various other steps in the alcohol-production chain, such as after hydrolysis or fermentation. More intricate approaches to converting corn into food or feed are the dry- and wet-milling processes.

In some modern dry-milling processes, the germ is first removed and treated to produce oil and germ meal that is used for food or feed (see Fig. 4-2). After grinding of the degerminated kernel, the hull is separated and used for feed or food additives. The remaining endosperm fraction of the kernel is then milled and used for making a variety of food products, such as corn meal, grits, and corn flakes.

A part of the endosperm fraction of the kernel may be used as feedstock for an ethanol-production unit. The residue of ethanol production, i.e., the stillage itself, is a valuable coproduct which can be used as animal feed or feed supplement. Figure 4-2, therefore, indicates that ethanol production can be integrated into a corn dry-milling process having food or feed production as its main objective. The advantage of dry milling over the simple processing described in Fig. 4-1 is that a variety of high-valued coproducts (oil, food additives, feed, or feed supplements) is produced in addition to ethanol. This advantage, however, is counterbalanced by both the higher capital required for an integrated dry-milling/ethanol-production unit and the need for sophisticated marketing to dispose of the coproducts.

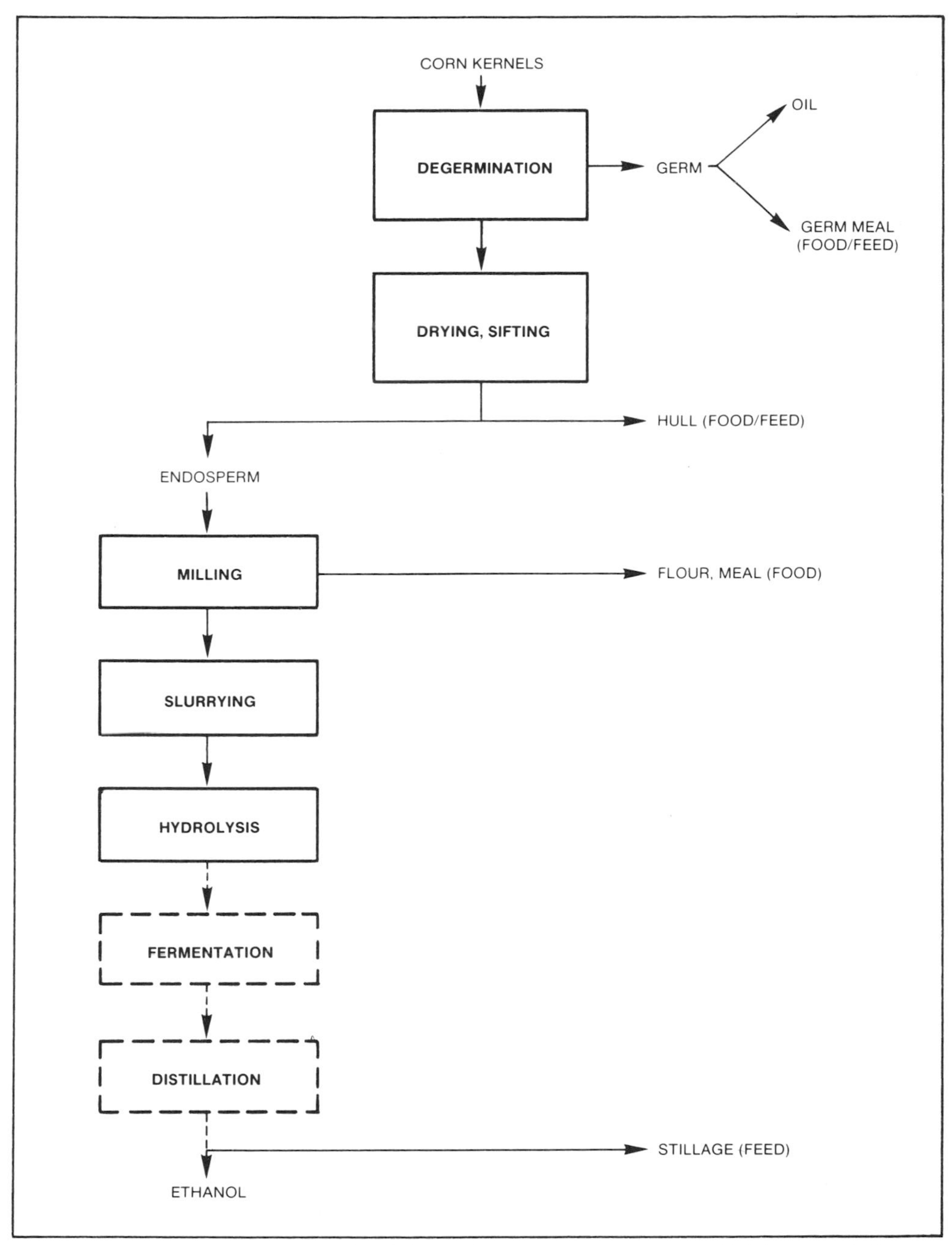

Source: *A Guide to Commercial Scale Ethanol Production and Financing*, SERI, Denver, 1980.

Figure 4-2. Dry-milling process with ethanol production.

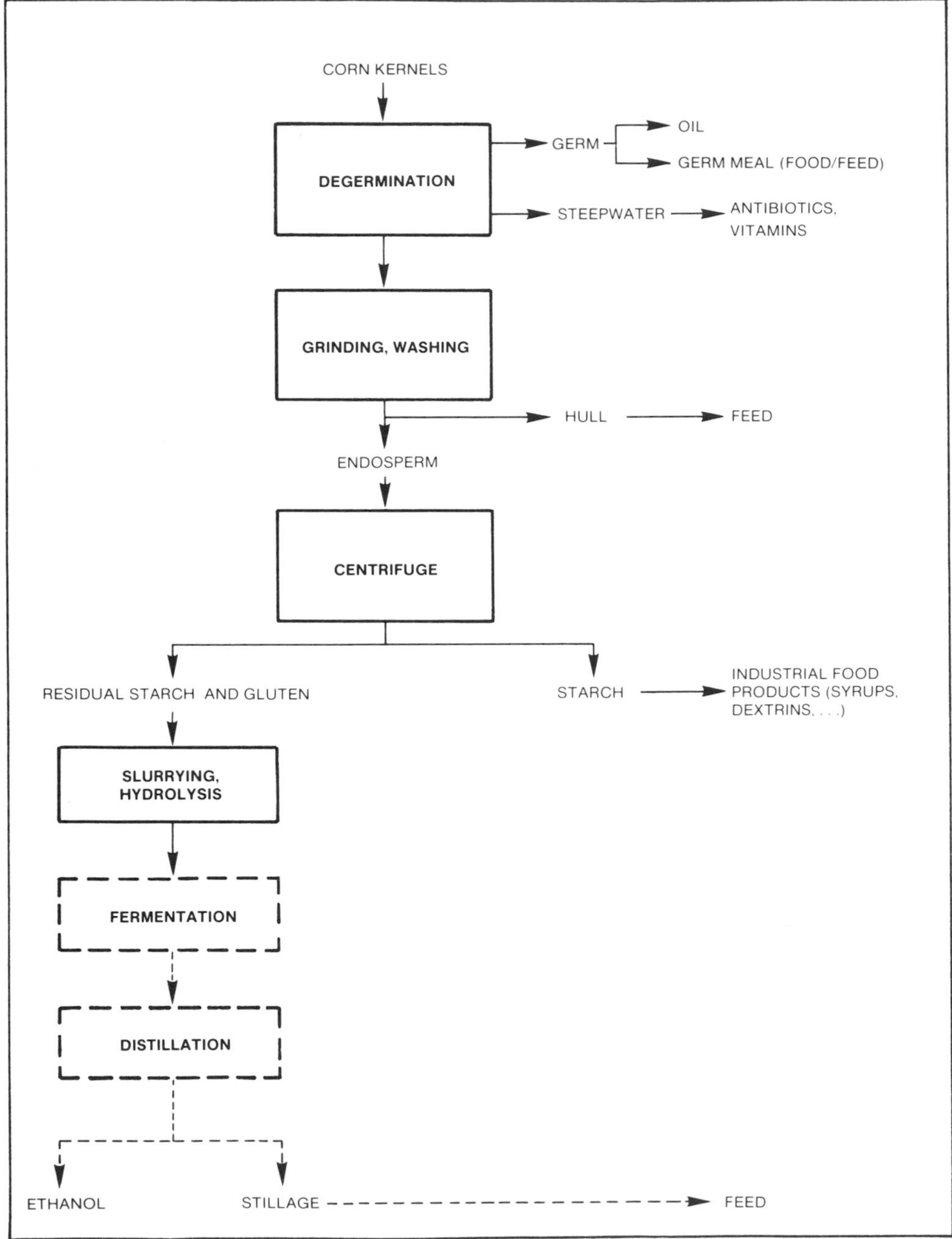

Source: *A Guide to Commercial Scale Ethanol Production and Financing*, SERI, Denver, 1980.

Figure 4-3. Wet-milling process with ethanol production.

In the wet-milling process, the starch contained in the endosperm is further separated from the protein matrix (gluten) in which it is embedded. Figure 4-3 illustrates a wet-milling process in which an alcohol-production line using part of the starch has been incorporated. Several plants of this type are either in operation or in the planning stages. Multiple high-value coproducts are generated during ethanol production via both the wet- and dry-milling processes. The largest traditional wet-milling plants process between 150,000 and 209,000 bushels of corn per day. This would correspond to a production capacity of 380,000 to 620,000 gallons per day for a plant of simplified design.

In a combined wet-milling/fermentation ethanol process, however, emphasis is placed on food or feed production, and the ethanol output is only 70 to 80 percent that of a plant of simplified design with the same corn input devoted primarily to ethanol production. The capital cost of the combined wet-milling/fermentation ethanol plant will be higher than that for a simple ethanol plant, and a careful analysis of the potential benefits of generating valuable feed or food coproducts will be required before selecting the corn wet-milling processing route.

Tubers

Potential tuber feedstocks include potatoes or potato wastes from food-processing plants, sweet potatoes, and other starchy tubers. The grinding, slurrying, and hydrolysis operations prior to fermentation are similar to those for grains, but modified to account for the size and moisture content of the tuber feedstock.

CELLULOSIC FEEDSTOCKS

Enormous amounts of cellulosic materials are potentially available for ethanol production. Although no practical process for converting cellulosic materials to ethanol on an industrial scale has been demonstrated, research is underway. Cellulosic materials require extensive preprocessing to release the fermentable sugars. Several processes which may include mechanical, chemical, and biological treatments are presently being investigated. This technology, therefore, must be considered a future method of ethanol production. It is currently impossible to project its real potential or economic feasibility. It should be noted, however, that available data suggest that the preprocessing of cellulose prior to fermentation will be different from that practiced for starch feedstocks. Modification of a plant from starch to a cellulosic crop would require the addition of an appropriate front-end processing area to the original plant and very little if any modifications to the existing starch-processing front-end. Moreover, that plant then would have the capability of processing both starch/sugar and cellulosic feedstocks, which could be an advantage. Storage of cellulosic crops is relatively simple.

Source: *A Guide to Commercial Scale Ethanol Production and Financing*, SERI, Denver, 1980.
Stover is a potential feedstock for cellulosic conversion technologies presently under development.

Potential feedstocks include such materials as corn stover, straw, sugarcane bagasse, mill residues, forest residues, and industrial and urban wastes.

Yields of Products

ETHANOL

The potential conversion rate of feedstocks to ethanol is an important element when selecting a feedstock. The absolute conversion rate expressed in gallons of ethanol per unit quantity of feedstock, such as gallons per bushel of grain, will establish a relationship between the amount of feedstock required and the size of the plant contemplated or, conversely, will determine the size of the plant which can be supported by the resources available in a certain region. Also, when several feedstocks could be available, all other factors being equal, the feedstock having the highest conversion rate and requiring the least handling and logistics for procurement of bulky raw material will be preferred. The relative conversion rate expressed in gallons of ethanol per acre of land devoted to a given crop is also an important aspect of crop selection. This yield per unit land area provides an estimate of the geographic crop area required to support a plant of given capacity, and, therefore, of the potential impact of an ethanol plant on other users of the crop of interest in the region. A comparison of land requirements for feedstocks, when a choice is available, will indicate which feedstock is most appropriate to the farming patterns of the region and which feedstock has the highest potential for long-term availability.

Table 4-4 summarizes conversion yield data (quoted as average values) for various feedstocks. Fluctuations around these values will occur due to such factors as the degree of preprocessing of the feedstock prior to fermentation and the quality of the feedstock. As an example, conversion yields as high as 2.7 gallons of ethanol per bushel for corn and other grains, and yields ranging from 75.0 to 60.3 gallons of ethanol per ton for sorghum and citrus molasses, respectively, have been quoted. The conversion yields quoted in the table assume that the main objective of the conversion plant is ethanol production. Small-scale plants due to design/operational limitations may have smaller yields than the average. The yields of ethanol fuel per unit feedstock will be lower when ethanol production is integrated with food- or feed-production processes. (This lower yield of ethanol per unit feedstock does not result from lower sugar to ethanol conversion efficiency, but rather from a different utilization of the components of the feedstock. In these

Table 4-4. Probable Commercial Yields of 200-Proof Ethanol From Various Feedstocks

FEEDSTOCK	GALLONS PER BUSHEL*	GALLONS PER TON	GALLONS PER ACRE**
Corn	2.5	89	228
Grain Sorghum	2.4	86	135
Wheat	2.4	80	74
Rye	2.2	79	54
Oats	1.0	64	57
Barley	2.1	88	92
Rice	1.8	80	175
Potatoes	0.69	23	299
Potato Wastes	—	13	—
Sweet Potatoes	0.94	34	190
Yams	0.75	27	NA
Jerusalem Artichokes	0.60	20	NA
Sugar Beets	—	22	412
Sugar Cane	—	15	555
Sweet Sorghum	NA	NA	NA
Apples	0.35	14	NA
Peaches	0.28	12	NA
Molasses	—	68	NA

*Average Yields
**Based on Average 1977 Crop Yields

Source: U.S. Department of Agriculture, "Small-Scale Fuel Alcohol Production," prepared with the assistance of Development Planning and Research Associates, Inc., Manhattan, KS, 66502, March 1980, Washington, D.C. 20250.

food/feed/ethanol integrated processes, a fraction of the starch—the source of fermentable sugars—is recovered for food and feed uses and, therefore, is not available for ethanol production.)

The data in Table 4-4 also show that the wastes of feedstocks or feedstocks of lesser quality (distressed grains, for instance) may have yields lower than those mentioned in the table. For instance, potato wastes from french-fry-processing plants produce about one-half of the fermentation ethanol obtained from whole potatoes for equal feedstock inputs. These lower yields per unit input and the associated costs of handling larger quantities of feedstock must be compared to the lower feedstock costs for waste materials. In some cases, the use of wastes may result in a credit because ethanol production may provide a method for disposal of the wastes from food-processing plants. The ethanol-production rates per acre are based on average U.S. yields of the various crops per acre. Significant regional differences in yields have been recorded. As an example, in 1977, while the United States average yield of corn for grain was 90.8 bushels per acre, yields of 29 bushels per acre were reported for Alabama and yields of 116 and 105 bushels per acre were reported for Colorado and Ohio, respectively.

On the basis of the data from Table 4-4, a 50-million-gallon-per-year plant would require about 21.3 million bushels of corn per year, the crop harvested from about 233,000 acres of farm land having a productivity of 91 bushels per acre. Assuming that the conversion plant is located in the center of a corn-production area and that the crop land is about 50 percent of all land, the ethanol plant would consume all the grain produced within a circle of about 30.4 miles in diameter. For the range of productivities mentioned above (i.e., 30-130 bushels per acre-year) the diameter of the land area required to sustain the ethanol plant could range from about 13 to 27 miles. These simplified considerations give an order of magnitude for the crop area which will be affected by the installation of an ethanol plant, and should be considered when siting the plant to avoid direct competition with other users of the feedstock.

Table 4-4 also shows that sugar crops, such as sugarcane grown at present in the United States, have a higher ethanol productivity per acre than the starch crops presently produced. This yield advantage of sugar crops, however, is counterbalanced by the fact that sugar crops spoil more quickly in storage.

Of the starchy feedstocks listed in the table, corn and potatoes have the highest potential for fermentation ethanol production per unit land area.

COPRODUCTS

The fermentation process resulting in the production of ethanol also yields several coproducts, including carbon dioxide, fusel oil, yeast, and stillage. Other coproducts such as food or feed components also may be generated when ethanol production is integrated in a food processing chain such as the dry or wet milling of grains described earlier. The coproducts of ethanol production may have a beneficial effect on the overall economics of ethanol production if they can be recovered economically in significant quantities, and if a commercial market exists to absorb these coproducts. The amount, quality, and, therefore, market value of the coproducts vary widely and depend on the feedstock and the processing steps used in producing ethanol.

Carbon Dioxide

Carbon dioxide is used in carbonated beverages, in fire extinguishers, in the manufacture of dry ice, and in food preservation. The recovery of carbon dioxide is likely to be practical only for large-scale plants and only when a local market is readily available. In most cases, carbon dioxide recovery will not be economically justifiable and, except for special conditions, no credit can be expected from carbon dioxide.

Fusel Oil

Fusel oil is a poisonous liquid mixture of alcohols consisting mostly of normal amyl and iso-amyl alcohol. Where corn is the feedstock for ethanol production, less than 1 percent of the total amount of alcohol produced is fusel oil. Fusel oil can be used as a denaturing agent for the ethanol produced.

Yeast

Yeast present in the fermentation medium may be recycled or recovered for commercial uses. In either case, its value as a coproduct should be determined. If yeast is not recovered, it will contribute to the high protein content of the stillage.

Stillage

The stillage from fermentation contains fibrous carbohydrate material, high-protein yeast, proteins from the original feedstock, and nonfermentable solids and solubles including various minerals and other nutrients. The actual nutritional value of the stillage will vary among feedstocks and with the ethanol-production process used. Whole thin stillage usually contains about 90 percent water. Although whole stillage is currently used as animal feed or feed supplement, its usefulness is limited by the high water content which prevents animals from consuming large quantities of it. Moreover, whole stillage is very susceptible to microbial degradation and must be delivered and consumed within a short time after removal from the distillery. In summer, it is recommended that the stillage be consumed within 24 hours after recovery from the still. This can result in complex and expensive logistic problems for commercial-scale operations generating large quantities of stillage daily. Stillage may be dried to avoid or reduce the storage, marketing, and distribution problems; however, this increases the production cost.

Four types of coproducts are presently recovered from the stillage of distilleries using grain as feedstock: distiller's dried solubles (DDS), obtained by evaporating and drying the thin stillage to recover the minerals and nutrients in solution; condensed distiller's solubles, obtained by concentrating the thin stillage to a semi-solid form; distiller's dried grains (DDG), obtained by separating the coarse grains from the whole stillage and drying the solid fraction recovered; and distiller's dried grains with solubles (DDGS), which result from the blending of DDG and DDS prior to drying. The paths of utilization of stillage and its coproducts are illustrated in Fig. 4-4. DDG or DDGS is generally marketed at 10 percent moisture and has a protein content of the order of 25-28 percent dry basis, i.e., lower than soybean meal (about 45 percent) but significantly higher than corn (about 9 percent). The dried coproducts recovered from stillage are easier to store, transport, and market than fresh stillage.

Table 4-5 summarizes the production rates of stillage and stillage-derived coproducts for some of the major feedstocks currently available for ethanol production. A 50-million-gallon-per-year ethanol plant using corn feedstock will generate about 500 million gallons of fresh stillage annually or about 1.3 million gallons of stillage daily. The same plant will produce about 163,000 tons of dried distiller's products annually or about 450 tons daily. This comparison suggests that treatment of the stillage to dried distiller's products will simplify the logistics of marketing the coproducts of a commercial-size plant.

The output of coproducts shown in Table 4-5 refers to a simple fermentation ethanol plant, i.e., a plant designed primarily for ethanol production. The outputs from an ethanol-production plant integrated with a grain-processing plant (dry- or wet-milling process) will be quite different. As an example, the treatment of one bushel (56 pounds) of corn through a wet-milling process will result in the production of about 9.2 pounds of feed at about 21 percent protein; 2.7 pounds of gluten meal at about 60 percent protein; 3.5 pounds of germ at about 50 percent oil; and 31.5 pounds of starch. This starch can be partially used for food or totally converted to ethanol. In the latter case, about 2.5 to 2.7 gallons of ethanol would be produced per bushel of corn treated. As most of the food or feed components of the grain has been removed through the wet-milling process, only about 3.2 pounds of residual distiller's coproducts would be recovered from the stillage. The above-mentioned outputs of a wet-milling process are only indicative and will vary according to the design of the system.

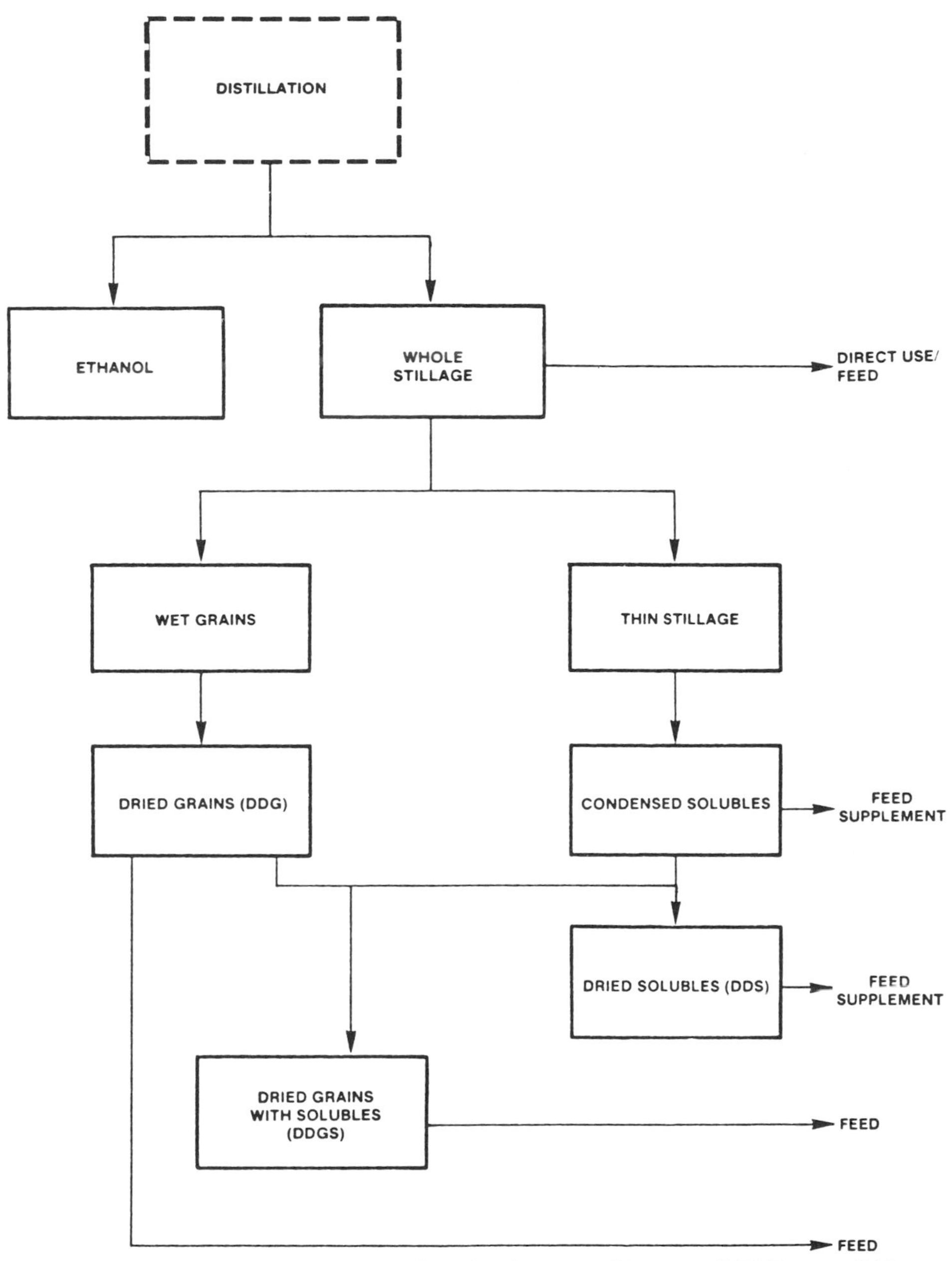

Source: *A Guide to Commercial Scale Ethanol Production and Financing*, SERI, Denver, 1980.

Figure 4-4. Paths of utilization of residual stillage from a grain feedstock distillery.

Table 4-5. Production Rates of Stillage and Stillage-Derived Products for Ethanol-Production Feedstocks (per Gallon of Ethanol)

FEEDSTOCK	UNIT INPUT	WEIGHT INPUT[1] (LB)	VOLUME OF STILLAGE[2] (GAL.)	WEIGHT OF STILLAGE[2] (LB)	WEIGHT OF DISTILLERS COPRODUCTS[3] (LB)
Corn	0.38 bu	21.5	10.4	92.7	6.5
Wheat	0.38 bu	23.1	10.4	94.2	8.0
Grain Sorghum	0.38 bu	21.5	10.4	92.7	6.5
Potatoes	0.71 cwt	71.4	7.0	58.4	10.6
Sugar Beets	0.049 ton	98.5	10.3	85.6	13.0
Molasses	2.5 gal.	29.3	4.8	39.0	–

[1] These yields are slightly different from average yields and reflect expected variations in ethanol production for various feedstocks.

[2] Volume and weight of stillage will depend on amount of water used in fermentation mash and amount of water recycled to process after distillation.

[3] 10% moisture for grain feedstocks; 75% moisture for sugar feedstocks.

Source: M.L. David, G.S. Hammaker, R.J. Buzenberg, and J.P. Wagner, "Gasohol Economic Feasibility Study", Report prepared for Energy Research and Development Center, University of Nebraska, Development Planning and Research, Inc., p. 261, July 1978, Manhattan, KS, 66502. U.S. Department of Agriculture, "Small Scale Fuel Alcohol Production", prepared with the assistance of Development Planning and Research Associated, Manhattan, KS 66502, March 1980, Washington, D.C. 20250.

The major market for stillage or stillage-derived coproducts is animal feed or animal feed supplement. The quantity of stillage or stillage-derived products, which can usefully be consumed by animals, is a function of a number of factors, such as the type of animal, diet needs within its life cycle, and ability to digest and absorb the products. Considerable research is in progress to optimize the use of distiller's residues in formulating animal diets. Therefore, some uncertainty exists as to the exact amount of stillage or its derivatives which can be tolerated by various types of animals.

Some typical quantities of coproducts which could be fed to various types of animals in combination with other feeds required to supply a balanced ration are shown in Table 4-6. The animal population required to absorb the distiller's coproducts generated by ethanol plants of various sizes is reflected in the data of Tables 4-5 and 4-6. Some typical values are shown in Table 4-7.

The table shows that the animal population required to absorb the distiller's coproducts from commercial-size plants is large. The problem of marketing these coproducts will have to be carefully evaluated before choosing an ethanol plant site.

The available markets for stillage will probably be expanded severalfold in the next few years by technological developments which have already been proven at the laboratory level and are going into pilot testing at this time. Membrane processes show promise for the low-cost denaturing of stillage together with the capability of reducing its salt content. Denaturing costs by membrane processes will be about a few dollars per thousand gallons as

Table 4-6. Animal Consumption of Distiller's Coproducts

ANIMALS	STILLAGE 10% SOLIDS (GAL/DAY/ANIMAL)	STILLAGE 20% SOLIDS (GAL/DAY/ANIMAL)	DRIED DISTILLERS GRAINS WITH SOLUBLES (LB/DAY/ANIMAL)
Calf (550 lb)	6.3	3.1	5.8
Steer (770 lb)	9.2	4.6	8.5
Cow (1,300 lb)	7.2	3.6	6.6
Pig (60 lb)	1.2	0.6	1.1
Pullet (3.7 weeks)	–	–	0.13
Pullet (7.5 weeks)	–	–	0.22

Source: U.S. Department of Agriculture, "Small-Scale Fuel Alcohol Production," prepared with the assistance of Development Planning and Research Associates, Inc., Manhattan, KS, 66502, March 1980, Washington, D.C. 20250.

Table 4-7. Animal Population (in Thousands) Required to Utilize the Distiller's Coproducts from Various-Sized Ethanol Plants

PLANT SIZE IN MM GAL/YR	5		15		25		50		100	
ANIMAL / COPRODUCT	STILLAGE[1]	DDGS	STILLAGE	DDGS	STILLAGE	DDGS	STILLAGE	DDGS	STILLAGE	DDGS
Calf (550 lb)	25	17	75	51	125	85	250	170	500	340
Steer (770 lb)	17	12	51	86	85	60	170	120	340	240
Cow (1,300 lb)	22	15	66	45	110	75	220	150	440	300
Pig (60 lb)	131	90	393	270	655	450	1,310	900	2,620	1,800
Pullets (3.7 Weeks)	NA	758	NA	2,274	NA	3,790	NA	7,580	NA	15,160
Pullets (7.5 Weeks)	NA	448	NA	1,344	NA	2,240	NA	4,480	NA	8,960

Assumes corn feedstock and 330 days production per year.

[1]Ten percent solids.

Source: *A Guide to Commercial Scale Ethanol Production and Financing,* SERI, Denver, 1980.

contrasted with more than 20 dollars when accomplished by evaporators. These membrane denaturing plants can also be quite small as contrasted to evaporators which are efficient only in relatively large plants. Membrane processes remove salts, and it has been shown that the salt content is one of the elements which limits the amount of distiller's coproducts which can be fed to animals. There is evidence that stillage without excess salt can be fed to animals as readily as any animal feed material.

Availability of Feedstocks

A major consideration in evaluating the potential for ethanol production from agricultural products is the present availability of feedstocks. With the exception of very small amounts of the feedstocks being utilized by the beverage industry, very little of the potential feedstocks are used in alcohol production. U.S. production and dispersal of the most frequently mentioned ethanol feedstocks in 1976 have been compiled and are shown in Table 4-8. Much of the current production of the two major small grains, corn and wheat, are used for feed, food, industrial use, and exports.

Consider, for example, the quantities of ethanol that could be produced if all the 1976 (the latest year for which complete data are available) ending stocks of all small grains were converted to ethanol. Assuming 2.5 gallons per bushel, the 2,380 million bushels of small grains (excluding rice) would convert to a total of 6.0 billion gallons of 200-proof ethanol. For comparison, the U.S. domestic demand for motor gasoline in the month of January 1976, was 8.3 billion gallons. However, it should be pointed out that there are tremendous year-to-year variations in ending stocks. The 1975 ending stocks were about one-half those of 1976, while those for 1974 were only one-third of that for 1976. For 1974 then, the quantity of alcohol that could have been produced from these stocks would have been reduced to 2.0 billion gallons.

The major grain used on farms where it is produced is corn. In 1976, over one-third of the corn produced was used on the farms where it was grown. If one-half of this quantity or 1,110 million bushels were to be converted to ethanol, at a rate of 2.5 gallons per bushel, then 2.8 billion gallons of 200-proof ethanol could potentially be produced. For comparison, 3.5 billion gallons of gasoline were used for crop production and livestock operations in 1978.

Assessment of potential feedstock by examining the total U.S. production only can be misleading, as there are major regional variations in the production of the potential feedstocks. This has definite implication for the quantities of ethanol that could be produced in an area and also for the type of ethanol-production facilities that would be appropriate. In Table 4-9, the production of major feedstocks by region is shown.

Nearly 90 percent of the small grains is produced in the Cornbelt, Lake, and Great Plains states. Potatoes are grown primarily in the western states and to a lesser extent in the Lake and New England states. Sugarcane production is limited essentially to the southeastern region and Hawaii. Sugar beet production is somewhat more widespread with 52 percent of the production in the western states and 27 percent in the Lake states. Sweet sorghum is grown to a very limited extent at present, but the feasibility of growing it at a variety of sites is being encouraged, since its alcohol production per acre of land is potentially greater than that from the grain crops.

Table 4-8. Feedstock Supply and Disappearance, 1976[1]

Feedstock	Supply				Used on farms where produced	Disappearance					Ending stocks	Put under supp.	
	Base stocks	Pro-duction	Impacts	Total		Food Seed	Industrial feed	Total	Exports	Total disap-pearance		Quantity	Percent of Production
Small Grains													
Corn (bu)	399	6,266	3	6,668	2,219	513	3,587	4,100	1,684	5,784	884	276	4.4
Grain													
Sorghum (bu)	51	720	—	771	207	6	428	434	246	680	91	11.8	1.6
Wheat (bu)	665	2,142	3	2,810	104	553 92	103	748	950	1,698	1,112	495	23.0
Rye (bu)	4.4	15	0.2	19.6	3.6	3.7 4.7	6.7	15.1	0.04	15.2	4.4	0.1	1.0
Oats (bu)	205	546	1	752	353	88	489	577	10	587	165	4.5	0.8
Barley (bu)	128	372	11	511	98	158	161	319	66	385	126	18.7	5.0
Rice (cwt)	36.9	115.4	0.1	152.6	641	29.2	13.5	42.7	65.6	108.3	40.5	23.4	20.2
Potatoes		357.7		6.6[2]									
White (cwt)		13.4		1.3[3]									
Sweet Sorghum[4]	In 1973–1975, the area harvested was less than 1,620 acres with yields of 84--209 tons per acre												
Sugar beets (tons)		29.4											
Sugarcane		28.1				26.9 1.2							

[1]All figures are in million units, except for percent of production.

[2]1975 data: Used on farms where produced for seed, feed, and household use. Shrinkage and loss 22.2 million cwt.

[3]1975 data: Used on farms where produced. Shrinkage and loss of 0.7 million cwt.

[4]No official statistics.

Source: United States Department of Agriculture, *Agricultural Statistics,* U.S. Government Printing Office. Washington, 1977 and 1978.

Table 4-9. Regional Production of Potential Feedstocks, 1977

	Corn[1]	Milo[1]	Wheat[1]	Oats[1]	Barley[1]	Rice[2]	Potatoes[2]				Sweet potatoes[3]	Sugar beets[3]	Sugarcane[3]
							Winter	Spring	Summer	Fall			
New England	0	0	2	2	0	0	0	0	0	31	0	0	0
Midatlantic	208	0	22	35	12	0	0	0	3	18	1	0	0
Southeast	373	22	73	18	9	56	1	8	5	0	9	0	33
Cornbelt	3,462	68	259	142	2	1	0	0	2	4	0	1	0
Lake states	1,077	0	166	257	56	0	0	0	3	40	0	7	0
Great Plains	1,090	662	1,044	270	137	23	0	1	2	23	1	5	2
Western States	136	36	462	20	199	18	1	14	6	187	1	13	0
Hawaii	0	0	0	0	0	0	0	0	0	0	0	0	20
Alaska	0	0	0	0	0	0	0	0	0	0	0	0	0
United States	6,346	788	2,026	744	415	98	2	23	21	303	12	26	55

[1]Units: 10^6 bushels.

[2]Units: 10^6 hundredweight.

[3]Units: 10^6 tons.

Source: U.S. Department of Agriculture, *Agricultural Statistics,* U.S. Government Printing Office, Washington, 1978.

Potato production is classified by season, and over 88 percent of the production is classified as fall production. The western states lead the nation with 60 percent of the total production. The Lake states and New England produce another 12 percent and 9 percent, respectively. Only in the southeastern region is the contribution of spring potatoes high with respect to the total produced in that region. Of the total sweet potato production of about 12 million hundredweight, nearly 80 percent was grown in the southeastern region.

OVERVIEW OF FEEDSTOCK AVAILABILITY

Near-Term Availability

Table 4-10 summarizes estimates of the quantities of feedstocks potentially available at present or in the near term in the United States. These estimates assume that none of the feedstocks is diverted from its present food or feed uses (i.e., the amounts shown in the table are essentially surplus production) and that present patterns of agriculture are not changed. The table shows that a goal of over two billion gallons of ethanol annually is a realistic objective in the near future. The table also shows that corn, grain sorghum, and other resources which include grains other than corn constitute the backbone of our resources. Citrus wastes, although significant in quantity, are available only on a seasonal basis, which limits their economic attractiveness.

Long-Term Availability

If the goal of 1.8 billion gallons of ethanol per year as proposed by President Carter is to be reached, a variety of feedstocks must be utilized. Traditionally, U.S. agriculture has shown a great degree of flexibility in adjusting to new market demand and technologies. Increasing demand for corn

Table 4-10. Quantities of Feedstocks for Ethanol Production Potentially Available in the Near Term and Their Ethanol Equivalent

FEEDSTOCKS	QUANTITY (MILLION DRY TONS/YEAR)	ETHANOL EQUIVALENT (MILLION GALLONS/YEAR)
Corn	3.6	360
Grain Sorghum	0.3	30
Citrus Waste	1.9	210
Whey	0.9	90
Others	1.7	150
TOTAL		840

Source: U.S. Department of Energy, Assistant Secretary For Policy Evaluation, "Report of the Alcohol Fuels Policy Review," DOE/PE-0012. Washington, D.C., June 1979.

feedstock for ethanol production accompanied by increased availability of distiller's grains feed supplement would result in a gradual shift from soybean and grain sorghum production to corn or other grains. Similarly, the existence of a steady and significant market for ethanol feedstocks may induce farmers to develop marginally used land areas.

Recent estimates by the U.S. Department of Agriculture (USDA)[1] indicate that about 78 million acres presently not used for crops have a high potential for cropland development. The development of these areas into cropland could translate to a 33 percent increase of corn production or a 43 percent increase of soybean production in the Northern Plains and Southern Plains, respectively, and a 350 percent increase of soybean production in the southeast. Despite increasing national and foreign demand for farm products, the potential exists to increase the ethanol-feedstock resources significantly. It should be mentioned, however, that some development problems such as drought, erosion, periodic flooding, and rocky soil may have to be considered when bringing some of the potential areas into cropland use.

The preceding discussion indicates that, on a national level, sufficient resources exist to justify an ethanol program in the near term, and prospects for maintaining or even expanding an ethanol program in the long term appear favorable.

Regional Availability of Feedstocks

The availability of feedstocks varies from region to region, as climatic and soil characteristics result in specialized crop production. Also, while some crops such as corn are grown in many states, some regions or states will register a surplus of production and others, although producers, will import corn. Surplus regions or the proximity of surplus regions where competition for the feedstock with other users is reduced and prices are less sensitive to changes in demand should obviously be preferred when siting a plant.

Figure 4-5 shows the distribution of potential ethanol-feedstock production by USDA farm-production regions. (The production indicated in the notes to the table are the totals.) The quantity of feedstock available for ethanol production is only a fraction of the total production, i.e., surplus, distressed crops, or wastes from processing. The data show that corn, wheat, and grain sorghum are the major potential feedstocks and that one-half or more of these crops are produced in the Northern Plains, Lake states, and Cornbelt regions. Sugar beet production is spread over many regions but the amounts potentially available are small. Sugarcane production is geographically very limited and its feedstock potential is very small. Potatoes (surplus, distressed crops, or wastes from processing) and wastes from food

[1]L.K. Lee, *A Perspective on Cropland Availability*, U.S. Department of Agriculture—Economics, Statistics, and Cooperative Services, Agricultural Economic Report No. 406, Washington, D.C., 1978.

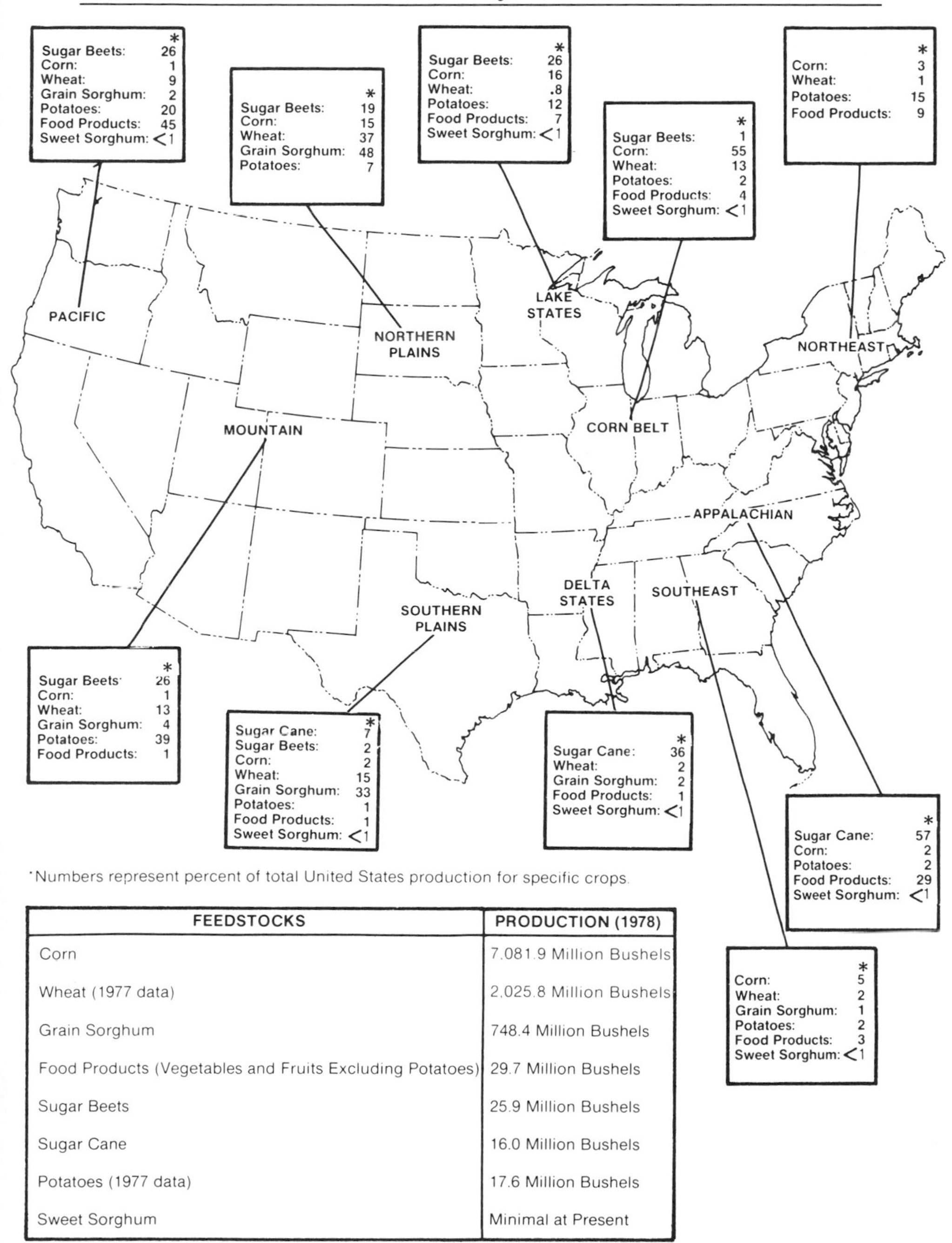

*Numbers represent percent of total United States production for specific crops.

FEEDSTOCKS	PRODUCTION (1978)
Corn	7,081.9 Million Bushels
Wheat (1977 data)	2,025.8 Million Bushels
Grain Sorghum	748.4 Million Bushels
Food Products (Vegetables and Fruits Excluding Potatoes)	29.7 Million Bushels
Sugar Beets	25.9 Million Bushels
Sugar Cane	16.0 Million Bushels
Potatoes (1977 data)	17.6 Million Bushels
Sweet Sorghum	Minimal at Present

Source: *A Guide to Commercial Scale Ethanol Production and Financing*, SERI, Denver, 1980.

Figure 4-5. Potential ethanol feedstocks and their availability by regions.

products processing could supply the feedstock needed in a variety of regions although the total potential resource is limited. Food processing wastes also have the drawback that they are often seasonal feedstocks. As a result, plants relying on these feedstocks could be idle for part of the year if no alternative feedstock is available to fill the idle periods.

FACTORS AFFECTING FEEDSTOCK AVAILABILITY

Climate and Productivity

Increased soil management and particularly increased usage of nitrogen fertilizers have resulted in significant increases in average productivity of grains, especially corn. This increased soil management, however, has also resulted in increasing the amplitude of fluctuations in grain yields due to climate variations because positive response to fertilizer and other management techniques depend on favorable weather conditions.

Figure 4-6 shows historical trends of corn yield per acre during the 1963-1978 period. During the 1970s, yields varied between 72 and 100 bushels per acre, i.e., a variation of about 11-26 percent below and above

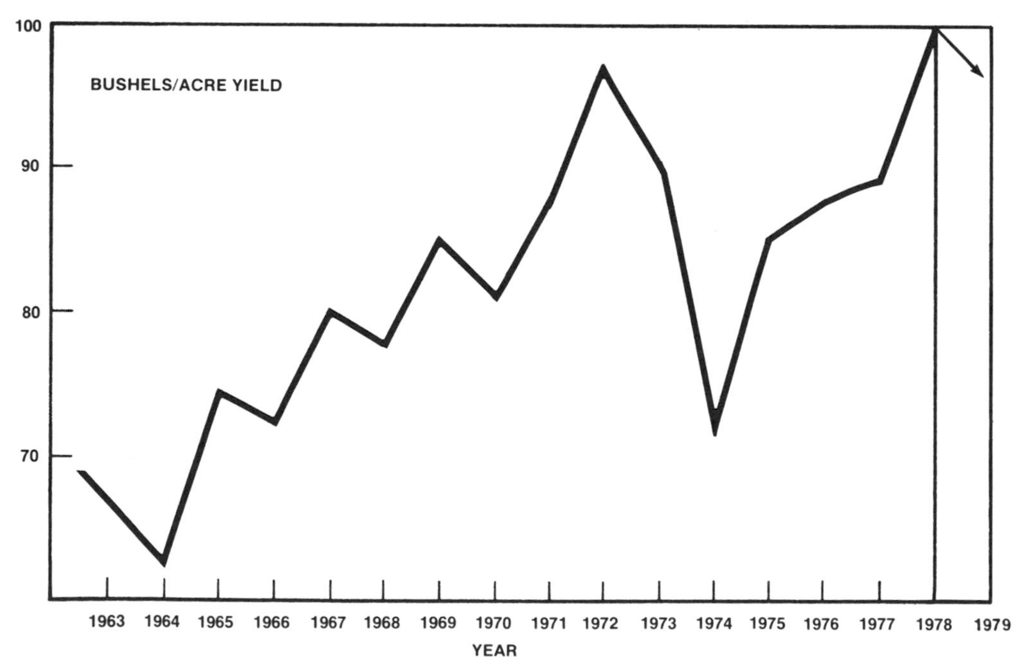

Source: U.S. Department of Agriculture, "Statement of Bob Bergland. Secretary of Agriculture before the Committee on Science and Technology. Subcommittee on Energy Development and Applications. Honorable Richard Ottinger. Chairman, House of Representatives, USDA 1032-79, May 4, 1979.

Figure 4-6. U.S. corn crop yields, 1963-1978.

the average productivity of 81 bushels per acre. Such fluctuations in productivity and production will have an impact on both food and fermentation ethanol-feedstock prices and availability.

Market Demand

Markets for grains include the domestic and foreign food and feed markets and ethanol production. These markets are defined and predictable to a certain extent. However, unforeseen variations resulting from foreign crop deficits, cancellation of foreign deliveries as a result of political decisions, emergency situations, or climate may change the yearly supply and demand balance for grains in the United States. It should be noted that once a significant ethanol industry is established, it will provide a predictable market for feedstocks which can be included in the projections of future U.S. markets.

Government Policies

As a result of the shift toward capital- and energy-intensive agricultural practices, U.S. agriculture has shown periods of overproduction. The federal price support and stabilization policy designed to offset this overproduction has included long-term land retirement programs to remove cropland from intensive cultivation and renewable yearly set-aside programs. Little or no grain cropland is removed from production under either the long-term diversion or yearly set-aside program at present.

HISTORIC TRENDS OF FEEDSTOCK SUPPLIES

Table 4-11 shows trends of corn production, utilization, and carryover stocks from 1965 to 1978. During the 1970s, an average of about 65 million acres was harvested and the data of Table 4-11 reflect the variations in productivity reported in Fig. 4-6. During the 1972-1974 period, a sharp decline in productivity and total production was recorded and foreign demand increased significantly. As a result, stocks, i.e., one of the elements through which fluctuations in supplies and prices can be damped, declined sharply. Concurrently, as discussed below, prices rose by over 60 percent between 1972 and 1974. The supply and price of other feedstocks could undergo similar fluctuations.

These fluctuations in feedstock supplies must be faced by ethanol producers. Food-processing plants such as those for coffee face some of the same uncertainties in supplies, fluctuations due to variations in productivity, and changes in policies by producers of the raw material. Potential ethanol producers must recognize this and explore ways of reducing the impact of these unavoidable fluctuations on the operation (and profitability) of their plant.

Table 4-11. Historical Trends of Corn Production, Utilization, Stocks, and Costs

YEAR	PRODUCTION (MM BUSHELS)	UTILIZATION (MM BUSHELS)	STOCKS[1] (MM BUSHELS)	PRICE[2] ($/BUSHEL)
1960	3,907	3,678	1,787	–
1965	4,103	4,409	842	1.17
1970	4,152	4,494	667	1.33
1971	5,646	5,188	1,127	1.08
1972	5,580	6,000	708	1.57
1973	5,671	5,896	484	2.55
1974	4,701	4,826	361	3.02
1975	5,829	5,793	399	2.54
1976	6,266	5,784	884	2.15
1977	6,425	6,208	1,104	2.02
1978	7,082	–	–	2.11

[1] Carryover Stocks, Sept. 30.

[2] Season Average Price

Source: USDA, *Agricultural Statistics 1979,* U.S. Government Printing Office, Washington, D.C., 1979.

Cost of Feedstock

HISTORIC PERSPECTIVE OF FEEDSTOCK COSTS

Table 4-12 shows wholesale prices for corn, soybean meal, and distiller's grains on the Chicago market for the 1970-1979 period for January and July of each year. Also included is the approximate price for stillage, derived from the historical relationship between stillage price and those for soybean meal and corn. The data of the table show that significant fluctuations in prices of corn and the coproducts of distillation occur on a yearly basis. These have been related earlier to fluctuations in production as well as changes in demand (national and foreign). Seasonal variations also occur as shown by comparing the data for January and July; in many cases, summer prices are higher than winter prices.

Another important fact apparent from the data is the correlation between DDG or stillage prices and those of corn and soybean meal: at constant soybean meal price, DDG or stillage prices increase with that of corn and vice versa. At constant corn price, the price of DDG or stillage increases and falls with that of soybean meal. This is an important relationship since, under certain circumstances, an increase in corn feedstock price resulting from the expected fluctuations in price for this commodity could be partly compensated by the increased value of the coproduct from ethanol production.

As an example, the data in Table 4-12 for January 1973 and 1974 show that the per-gallon cost of raw material for ethanol production would have jumped from $0.62 in 1973 to $1.09 in 1974, assuming a conversion rate of 2.5 gallons per bushel. The value of DDG, i.e., a credit toward feedstock price, simultaneously would have increased per gallon of ethanol from $0.33 in 1973 to $0.45 in 1974 on the basis of a production of 6.5 pounds of DDG

Table 4-12. Wholesale Prices of Corn and Other Commodities on the Chicago Market

YEAR	JANUARY 1					JULY 1				
	#2 YELLOW CORN		SOYBEAN MEAL	DISTILLERS GRAINS	STILLAGE	#2 YELLOW CORN		SOYBEAN MEAL	DISTILLERS GRAINS	STILLAGE
	$/BU[1]	$/TON	$/TON	$/TON	$/1000 GAL[2]	$/BU	$/TON	$/TON	$/TON	$/1000 GAL
1979	2.24	80.00	194.60	141.00	45.76	3.04	108.60	218.30	151.10	53.03
1978	2.18	77.90	182.60	130.00	43.64	2.40	85.70	186.10	120.00	45.15
1977	2.48	88.60	209.70	140.00	49.09	2.20	78.60	184.70	140.00	43.64
1976	2.58	92.10	136.00	105.00	40.61	3.05	108.90	232.70	130.00	55.45
1975	3.32	118.70	139.80	107.50	42.73	2.81	100.00	124.30	103.00	40.61
1974	2.72	97.10	174.80	138.00	46.67	3.06	109.30	107.80	92.50	39.41
1973	1.54	55.00	189.10	100.00		2.38	85.00	306.60	138.00	
1972	1.20	42.90	87.50	64.00		1.26	45.00	106.20	71.00	
1971	1.58	56.40	85.90	70.00		1.52	54.30	88.00	67.00	
1970	1.23	44.00	84.50	64.00		1.39	49.60	85.70	61.00	

[1] Assumes 56 lb per bushel.

[2] Approximate price derived from historical relationship with soybean meal and corn prices.

Source: USDA, *Small-Scale Fuel Alcohol Production,* Washington, D.C., March 1980.

per gallon of ethanol. The net price of the raw material (feedstock price minus credit) would therefore have been $0.29 and $0.64 per gallon in 1973 and 1974, respectively. This is a large increase in feedstock price, i.e., 121 percent, but, nevertheless, smaller than the increase which would have been incurred had the price of DDG been kept at its 1973 level, i.e., 162 percent.

These related feedstock and coproduct prices also could be advantageous to the producer. Between 1975 and 1976 (January prices, Table 4-12), the price of corn decreased by about 22 percent while that of DDG remained essentially constant. As a result, the net price per gallon for the feedstock (feedstock minus credit) decreased from $0.98 to $0.69 over the same period, i.e., a decrease of about 30 percent.

These two examples indicate that the impact of fluctuations in feedstock prices on alcohol-production costs, although potentially very significant, must be evaluated in the context of the overall market situation for the grains and coproducts. The discussion also suggests that the profitability of the ethanol plant should be examined on a plant-life basis, i.e., over a period long enough that the fluctuations in feedstock and coproduct prices compensate each other. This somewhat unpredictable economic performance may be difficult to make acceptable to potential investors. To a certain extent, the investor can anticipate the fluctuations in feedstock prices around a generally predictable trend. (As an example, see Fig. 4-7 for corn average selling prices.)

METHODS OF PROCUREMENT OF THE FEEDSTOCK

As indicated in the previous section, it is necessary for the ethanol produce to explore methods of procurement of feedstocks which will ensure a

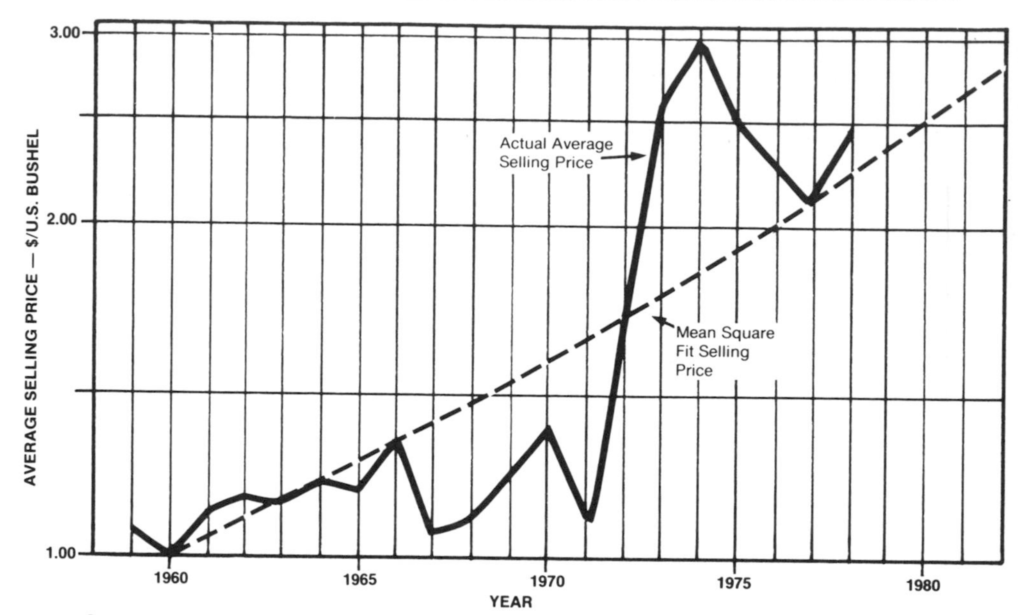

Source: Raphael Katzen Associates. "Grain Motor Fuel Alcohol Technical and Economic Assessment Study," Department of Energy Contract No. EJ-78-C-01-6639, December 1978.

Figure 4-7. The selling price of corn.

continuous flow of raw material despite expected fluctuations in production of the crop.

Feedstock purchase on the commodity market is the most direct method of raw material procurement. This option, however, puts the ethanol producer in direct competition with other users of the commodity and, therefore, the commodity price will be extremely sensitive to supply and demand relationships.

Another option is to secure long-term contracts with farmers for the supply of a fraction of the needed raw material, and rely on the commodity market for the remainder. In this case, purchases on the commodity market can be scheduled to benefit from downward price trends. Long-term contracts with farmers may have to include indexing clauses to account for inflationary trends. Because of the high energy inputs required by intensive farming (fuels and fertilizers), specific indexing agreements addressing these energy costs may have to be negotiated. Also, direct supply contracts with producers may raise problems of storage of the delivered raw material.

Other options which address specific or special circumstances may be considered. A cooperative ethanol-production facility including farmer members may require delivery from the farmer members of a specified amount of raw material over a certain time as part of the cooperative charter. Agreements could be negotiated between an ethanol producer and food processors whereby a long-term supply of wastes is guaranteed in exchange for the service of disposing of the wastes through ethanol production.

Procurement of feedstocks for a commercial-size plant is a critical function which impacts drastically on the economics of ethanol production. The organizational chart of the proposed plant should therefore identify this function and the needed specialized personnel.

FEEDSTOCK COST BASIS

Table 4-13 shows average feedstock costs per gallon of alcohol for some of the major potential feedstocks presently considered. A 15-year average (1963-1977) of prices paid to farmers for the commodities, converted to 1979 dollars by using the GNP price deflator, was used, as were average yields of ethanol per unit of feedstock.

Table 4-13 shows that corn, grain sorghum, and rye are the least expensive feedstocks. Sugarcane and sugar beets compare to wheat and barley. Fruit and potato wastes have been assumed to be available at $1/cwt (hundred pounds) or $20/ton. On that basis, the cost of these feedstocks is comparable to that of sugar crop and wheat.

FACTORS AFFECTING THE COST OF FEEDSTOCKS

Productivity or production, supply and demand relations, and, to a certain extent, government policies such as the agricultural stabilization program will affect feedstock availability and therefore feedstock prices. These have been discussed earlier. Two other factors may also affect the cost of feedstocks or the overall cost of ethanol production: government incentives

Table 4-13. Average Feedstock Cost Per Gallon of Ethanol

FEEDSTOCK	AVERAGE YIELD (GAL./TON)	AVERAGE 1963 TO 1977 PRICE PAID TO FARMERS (1979 $)	$/GAL.
Rye	78.8	2.36/bu	1.07
Grain Sorghum	79.5	2.40/bu	1.08
Corn	84.0	2.69/bu	1.14
Barley	79.2	2.35/bu	1.24
Wheat	85.0	3.46/bu	1.36
Oats	63.6	1.46/bu	1.43
Sugar Beets	22.1	31.58/ton	1.43
Fruit Wastes	13.0	20.00/ton	1.54
Sugar Cane	15.2	23.68/ton	1.56
Potato Wastes	12.5	20.00/ton	1.60
Rice	79.5	229.00/ton	2.88
Potatoes	22.9	99.60/ton	4.35
Sweet Potatoes	34.2	195.60/ton	5.72

Source: USDA, *Small-Scale Fuel Alcohol Production,* Washington, D.C., March 1980.

promoting the use of wastes and the cost of collection of the feedstock for large-scale operations.

Government Incentives

Prior to the Windfall Profits Tax Act, federal law permitted the financing of solid waste disposal facilities through tax-exempt industrial development bonds. The act expands the definition of solid waste disposal facilities to include property used primarily to process solid waste to alcohol. To be qualified as an alcohol producer under this provision, a facility must satisfy three requirements:

- The primary product obtained from the facility must be alcohol (there is no minimum proof requirement).
- More than half of the feedstock used in the production of alcohol must be solid waste or a feedstock derived from solid waste.
- Substantially all of the solid waste-derived feedstock must be produced at a facility, located at or adjacent to the site of the alcohol-producing facility, and both facilities must be owned and operated by the same person (ownership is meant for tax purposes).

The Windfall Profits Tax Act specifies that such bonds will not be tax exempt if they are guaranteed by the federal government or if any payment of the principal or interest is made with funds from a federal, state, or local energy program. The Windfall Profits Tax Act also expands the definition of a solid waste disposal facility that can be financed through tax-exempt industrial development bonds to include solid waste disposal facilities that produce steam or electricity.

These provisions of the Windfall Profits Tax Act, although not directly affecting the feedstock cost, nonetheless encourage the development of ethanol producing plants integrated with solid waste-producing operations. The overall economics of an ethanol plant using wastes from food-processing plants could therefore be more attractive than suggested by the feedstock costs shown in Table 4-13. The provisions of the act can also favor a plant using wastes to produce the process steam and/or electricity required by an ethanol plant. The use of bagasse, i.e., cellulosic residue from sugarcane processing, to fuel process steam boilers would benefit from the provisions of the act, for instance.

Grain Assembly Costs

A commercial 50-million-gallon-per-year plant will require about 20 million bushels of grain annually. The cost of gathering this quantity of feedstock will depend on the density of production of grain per unit land area (bushels-per-acre or tons-per-square-mile) and on the fraction of the grain which can be purchased for ethanol production.

The estimated crop production areas required to support a 50-million-gallon-per-year plan (corn feedstock) are shown in Fig. 4-8 for the states of Illinois, Indiana, and Ohio. The density of corn production per unit land area was estimated for each state by dividing the corn grain production for 1978 (USDA, *Agricultural Statistics 1979,* Washington, D.C., 1979) by the area of each state. Production densities ranged from about 21,000 bushels per square mile for Illinois to about 9,000 bushels per square mile for Ohio. To derive the data in the figure, it is further assumed that 10 percent of the corn produced is available for feedstock use. The average hauling distances (i.e., radius of the circle supplying one-half of the feedstock required) range from 40 miles for Illinois to 61 miles for Ohio. On the basis of the data shown in Fig. 4-8, corn-hauling costs would add about 4 and 5 cents per gallon to the cost of ethanol produced in Illinois and Ohio, respectively, for the corn-assembly scenario used in the present example. Although these supplementary costs are small in absolute value, they may become significant in terms of overall alcohol-production economics.

These transportation costs will become more significant as the capacity of the ethanol plant increases. It will therefore be necessary to carefully evaluate the optimum plant size for which the economy of scale associated with large plants is overriden by the supplementary cost for assembling the feedstock and disposing of the coproducts and effluents.

The cost of assembling the feedstock will also have to be weighed against the cost of marketing the products of the plant: proximity to the markets for

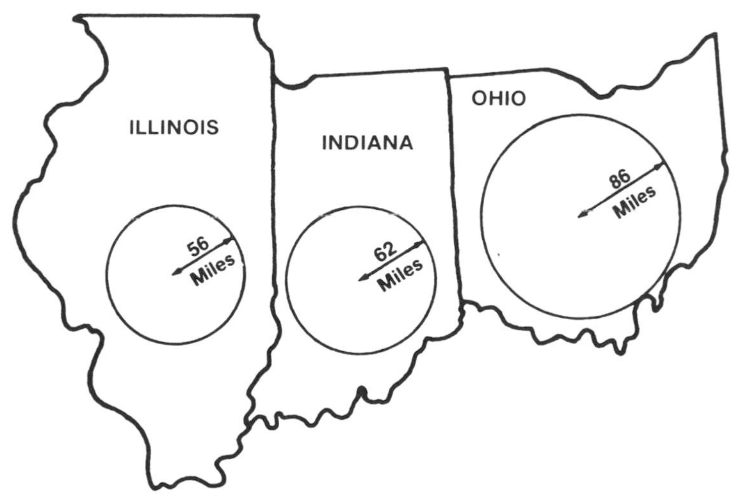

*Assumptions: 1978 productivities; uniform distribution of crop-producing land within the state; 10 percent of crop available for ethanol production.

Source: *A Guide to Commercial Scale Ethanol Production and Financing*, SERI, Denver, 1980.

Figure 4-8. Corn production area required to support a 50-million-gallon-per-year plant (corn feedstock).*

the products, i.e., ethanol, and coproducts may be economically more attractive than proximity to the feedstock source.

TRENDS AND FLUCTUATIONS IN PRICE

The data from Table 4-12—corn prices for January 1 and July 1, 1970 to 1979, on the Chicago Market—are plotted on Fig. 4-9. The solid line shows the trend in prices as obtained by regression of the set of data points. The slope of the line suggests a general price increase at a rate of about 8.4

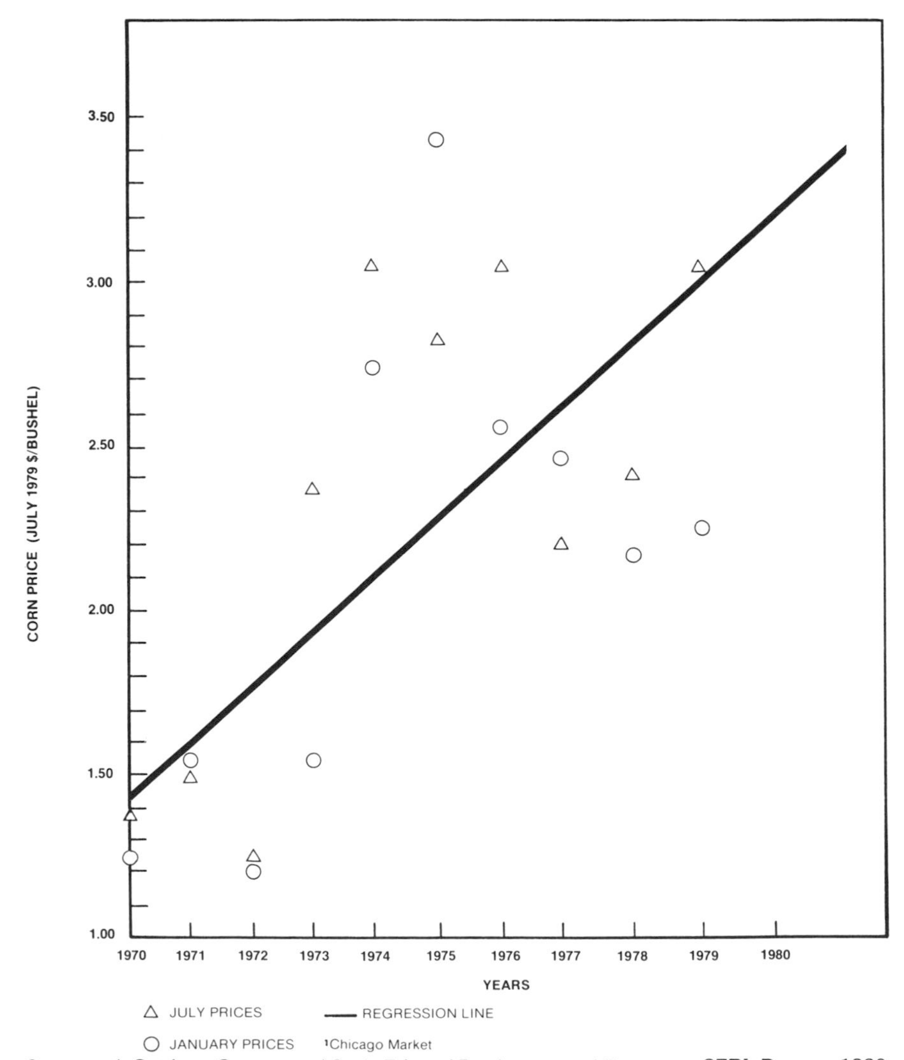

Source: *A Guide to Commercial Scale Ethanol Production and Financing*, SERI, Denver, 1980.

Figure 4-9. Trends and fluctuations in corn prices.[1]

percent per year, i.e., a rate comparable to the average inflation rate during the period considered.

During the period, wide fluctuations around the general trend line were observed. On the basis of a yield of 2.6 to 2.7 gallons per bushel, a fluctuation of 10 cents per bushel around the price trend line corresponds to a fluctuation of 4 cents in the cost of ethanol. The data in the table show that corn purchased in the winter of 1975 would have been about $1.00 over the expected (trend) price line and, therefore, that the cost of ethanol would have been boosted by about $0.40 per gallon during that period. Similarly, corn purchased in January, 1979, would have cost about $0.80 less than the price expected from the general trend line, resulting in a production cost for ethanol of $0.32 per gallon lower than projected. It is therefore apparent that, over a period of time, a certain amount of compensation for the fluctuations in feedstock costs could take place, resulting in relatively constant feedstock costs over that period.

As indicated before, this burden could be alleviated if a sizable fraction of the feedstock supply could be obtained under long-term contracts at predictable prices. Corn prices usually rise sharply when there is a poor crop and the supply of grain for food and exports becomes tight.

Alcohol-Production Impacts on Agriculture

This section discusses the economic and related impacts on agriculture which result from incurred production of fermentation ethanol.[2]

SUPPLY AND DEMAND IMPACTS ON AGRICULTURE

Alcohol production from corn has broad supply/demand implications for agricultural commodities. The following discussion focuses on the general supply and demand responses under conditions where corn is the alcohol feedstock. The numbered boxes refer to those in Fig. 4-10.

Supply Responses

Alcohol production from corn will affect the agricultural sector by increasing the price of corn. The first supply response from increased corn prices would be a reduction in the amount of corn-fed beef and an increase in range-fed beef (Box 1 in Fig. 4-10). The higher corn price would make range-fed beef more competitive at the margin. Higher corn prices would also be likely to reduce the amount of corn used in feeding beef and could possibly reduce total beef production.

Higher beef prices would likely lead to changes in the use of corn for dairy, pork, poultry, and other animal feeds (Box 2 in Fig. 4-10). We cannot

[2]*Impacts of Biomass Energy and Alcohol Fuels Loans and Guarantees,* from Title II-A of the Energy Security Act of 1980, November 8, 1980, USDA-OBBE.

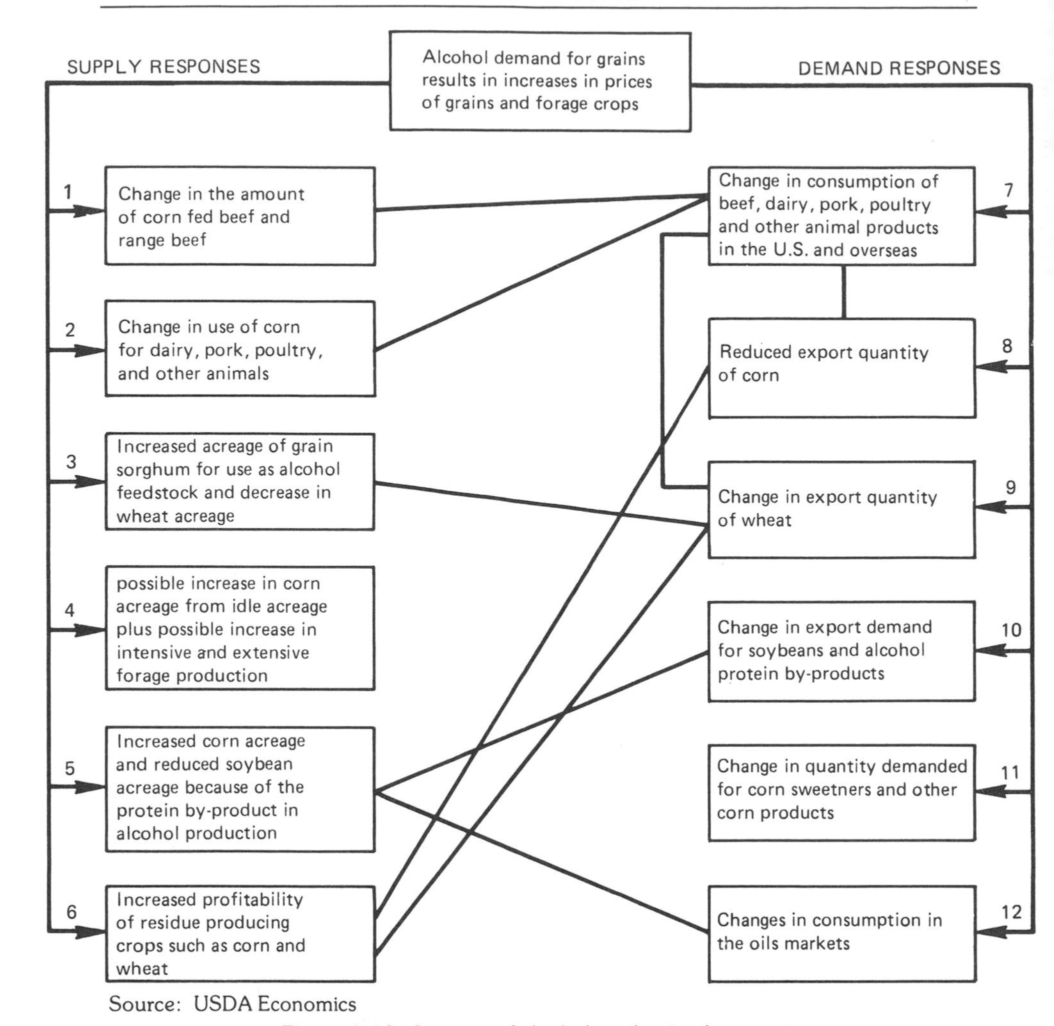

Source: USDA Economics

Figure 4-10. Impacts of alcohol production from grains.

say with certainty whether the total use of corn for animal feeding would increase or decrease, however. Conceivably, greater quantities of corn might be fed to animals that are more efficient converters of corn than are beef cattle. Production of these animals might increase. But the effects on corn usage are uncertain. They would depend on the magnitude of the changes and conversion efficiency in the whole animal system and possible changes in consumer demand for meat and poultry products. More information is needed on the interaction of grain and coproduct prices and the levels of use by the various livestock subsectors.

Another supply response would be an increase in the acreage of grain sorghum for use as an alcohol feedstock (Box 3 in Fig. 4-10) and a corresponding decrease in wheat acreage. Wheat is not likely to be used as an alcohol feedstock because of its high price relative to other grains. However, grain sorghum can be substituted for corn in alcohol production and in animal feeding. Grain sorghum and wheat are substitutes in production in the Great Plains—an area of variable rainfall. Corn can be grown on some of this acreage but it is not a drought tolerant as grain sorghum, so the normal substitution is between grain sorghum and wheat. As grain sorghum prices rise, grain sorghum acreage will increase and cause a corresponding decrease in wheat acreage. This decrease in wheat acreage will reduce the quantity of wheat produced and tend to raise the price of wheat.

An increase in corn acreage from idle land and from conversion of pasture and hayland to row crops would also be expected (Box 4 in Fig. 4-10). The higher corn price would bring in land which had previously been set aside in some years. Also, it would tend to bring into row crop production land which in prior years had been used for pasture or growing hay.

Corn acreage would also increase because soybean acreage would be reduced due to the production of protein coproducts from alcohol production (Box 5 in Fig. 4-10). For each three bushels of corn used in alcohol production, about a bushel of distiller's dried grains (DDGS), the high-protein animal feed, is produced. This coproduct will tend to reduce the demand for soybean meal, which, in turn, will lower the acreage of soybeans.

Demand Responses

On the demand side, the first response (Box 7 in Fig. 4-10) would be a change in consumption of beef, dairy, pork, poultry, and other animal products in the United States and overseas. Higher corn prices would result in higher final product prices in these animal subsectors. As indicated on the supply side, it is likely that the sector most effected would be beef and that substitution of other animal products for beef would occur.

Higher corn prices would reduce export demand for corn (Box 8 in Fig. 4 10). Also, wheat and other grains might be substituted for animal products overseas if the prices of animal products increased significantly (Box 9 in Fig. 4-10). The ultimate direction of the change in export quantity for wheat would depend on the extent to which wheat prices increased because of reduced production (Box 3 on the supply side), and the increased export demand for wheat induced by changes in dietary patterns among grain importing countries.

The next demand response (Box 10 in Fig. 4-10) would be a change in the export demand for soybeans and alcohol protein coproducts such as DDGS and corn gluten meal. One would expect, at least temporarily, a drop in soybean meal prices which would increase the quantity demanded, but over the longer term one cannot be certain what direction the changes would take. Soybean acreage likely would decrease, which would tend to firm up soybean prices. Also, export potential for increased alcohol protein coproducts is uncertain.

Another area affected by higher corn prices would be the corn sweetener market (Box 11 in Fig. 4-10). Higher corn prices would reduce the quantity demanded for corn for use in sweeteners and other products. The extent to which this change would occur depends on the world sugar market.

The last major change on the demand side would be a change in the consumption of edible oils (Box 12 in Fig. 4-10). To the extent that less soybean oil is produced, more oils of other types such as corn and sunflower oil may be produced. The prices of oils might also rise in response to the reduced supply.

Clearly, many of these supply and demand responses are interrelated. The lines drawn in Fig. 4-10 highlight the major areas where impacts are expected from increased grain alcohol production.

The magnitude of the impacts is directly related to the level of corn used in alcohol production. Low production levels of alcohol would not be expected to elicit strong responses in either demand or supply. Most of the impacts described above would be significant at alcohol production levels greater than 500 million gallons annually, if corn is used as the feedstock.

In order to meet the 1982 national alcohol-fuel-production goal, 6.9 million metric tons of corn would be needed to produce ethanol. About 1.7 million metric tons of high-protein feed supplement would be produced as a coproduct and could replace up to 1.3 million metric tons of oilseed protein feeds.

In 1979, U.S. farmers planted 201 million acres to feed grains and oilseed crops. In 1980, they planted 203 million acres. Total utilization of these commodities during the 1979-1980 marketing years will be about 317 million metric tons, with 213 million of this total used in domestic markets. Total usage will be down slightly in 1980-1981, according to current indications, due to the tighter supply and higher price prospects. The tighter supply is caused by decreased yields of most food and feed grain commodities resulting from the widespread drought conditions during the 1980 growing season.

The net impact of the DOE/USDA 1982 alcohol biomass program would be small. There would be a net increase of 0.5 million acres planted to feed grains and oilseeds—corn acreage increasing by 0.9 million acres and soybean acreage decreasing by 0.4 million acres—and a net increase of 3.6 million metric tons in total feed grain supplies. The net acreage increase represents about 0.2 percent of the total planted in 1979. The additional supply of feed grain and protein supplement is equal to 1.1 percent of the total feed grain and oilseed utilization during the 1979-1980 season. The impacts of the DOE/USDA 1982 alcohol biomass program would be gradual in any case, since most of the alcohol conversion capacity would not be in place until mid-1982 or later.

PRICE AND PRODUCTION EFFECTS

The effects of price and production on feedstocks and related issues which change due to the Energy Security Act of 1980, Alcohol from Biomass Program for 1982, Title II-A, are discussed in this section.

Corn

Increases in corn demand resulting from expanded ethanol production will have small but noticeable near-term effects on corn prices, supply, and demand. However, owing to the lag time in getting additional alcohol-production capacity on line, the largest share of the DOE/USDA alcohol biomass program's effects would show up in the 1982-1983 crop year, beginning October 1982.

Case I—preexisting alcohol-fuel program. Using USDA's October 1980 crop-production estimates and assuming average conditions in 1981 through 1982, seasonal average real farm prices for corn are expected to increase less than 1 percent for 1981 and 1982 and 2 percent in 1983 from base projections as shown in Table 4-14. Total utilization is expected to increase only about 0.4 percent in 1983, with shifts occurring between domestic and export uses. Domestic demand increases about 1 percent. Exports decrease each crop year, although the declines are minimal.

Case II—Energy Security Act impact. Seasonal average corn prices are expected to show moderate increases due to the higher demand and reduced supplies caused by the Title II-A program. Real price increases would be 0.9 percent over the base estimate in 1981, 2.8 percent above the base in 1982, and 8.2 percent above the base estimate in 1983. The 8.2 percent differential is the most significant change caused by the Title II-A program.

Total corn utilization shows average increases from base projections of 0.8 percent—0.3 percent in 1980-1981, 0.8 percent in 1981-1982, and 1.4 percent in 1982-1983. Domestic use increases above the base by 1.5, 2.1, and 4.2 percent, respectively. Exports decrease from the base each year.

Soybeans

The production of substantial amounts of high-protein animal feed coproducts from corn used in ethanol production dampens supply, demand, and prices for soybeans. The long-term disequilibrium effects of market imbalance are small.

Case I—preexisting alcohol-fuel program. Using USDA's October 1980 crop-production estimates and assuming average conditions in 1981-1983, seasonal average farm soybean prices are expected to show negligible effects from ethanol-induced demand and supply shifts. Average real prices over the 1980 to 1983 crop years indicate virtually no measurable effect from the current ethanol program (Table 4-14). Differences from base projections are

Table 4-14. Effects of Title II-A of the Energy Security Act on Corn and Soybean Utilization and Prices, 1980-1982

Activity	Percent change from base situation[a]		
	1980/81	1981/82[b]	1982/83[b]
Corn			
Case I (Current ethanol program)			
Total supply	c	c	c
Total use	0.1	0.1	0.4
Domestic	1.1	0.5	1.2
Export	-2.0	-2.8	-1.1
Price	0.6	0.6	2.0
Case II (Title II-A program)			
Total supply	c	c	c
Total use	0.3	0.8	1.4
Domestic	1.5	2.1	4.2
Export	-2.2	-1.7	-4.2
Price	0.9	2.8	8.2
Soybeans			
Case I (Current ethanol program)			
Total supply	c	-2.0	-0.1
Total use	2.2	-0.2	-0.2
Domestic	0.4	-0.4	-1.0
Export	4.7	0.3	0.9
Price	-0.3	-0.1	-0.6
Case II (Title II-A program)			
Total supply	c	-2.1	-0.4
Total use	2.2	0.3	0.6
Domestic	0.4	-1.2	-3.4
Export	5.0	1.7	3.5
Price	0.5	-1.1	-2.0

[a]Base case data: Agricultural Supply and Demand Estimates, USDA, World Food and Agriculture Outlook and Situation Board, August 12, 1980.

[b]Estimates for 1981/82 and 1982/83 are based on output from a feed simulation model (FEED-SIM), Purdue Station Bull. No. 521, March, 1979.

[c]Less than 0.1 percent.

Source: USDA Economics

0.6 percent or less each year. Impacts on total supply and utilization are also negligible, with the effects being 1 percent or less from the base during 1981-1982 and 1982-1983.

Case II—Energy Security Act impact. Using USDA's October 1980 crop-production estimates and assuming average weather conditions in 1981 through 1983, seasonal average farm prices for soybeans are expected to show minimal effects from ethanol-induced demand and supply shifts. Average prices would decline 1-2 percent in 1981-1982 and 1982-1983 from base estimates; total utilization decreases from the base by 0.3 and 0.6 percent during these years. Domestic demand drops about 3.4 percent from the base as ethanol coproduct feeds replace soybean animal feed. However, average soybean exports increase above the base by 1.7 and 3.5 percent, respectively, for 1982 and 1983.

Other Crops

The current program impacts of ethanol production on all other major crops are negligible. Only for barley, cotton, and grain sorghum is there any measurable impact. Given current ethanol-production efforts, shifts to corn production and the increased availability of ethanol coproduct feeds would cause barley production in 1982-1983 to fall below the base estimate by about 0.5 percent. Cotton production would be about 0.5 percent below the base and grain sorghum about 0.5 percent below. Implementing the Energy Security Act Title II-A program causes production decreases of about twice the magnitude that would otherwise be expected—barley, down 1 percent from the base; cotton, down 1.3 percent; and grain sorghum, down 0.8 percent.

Livestock

Ethanol-production impacts on the beef subsector are negligible for the existing program and for the Energy Security Act Title II-A program. Increases in corn feed prices put slight downward pressure on demand as feed input prices filter into output prices. The effect is to decrease production, domestic use, and stocks. Average changes in beef production are less than 0.3 percent with the Energy Security Act Title II-A program.

Land

The impact on farmland of increasing ethanol production to meet the objectives of the Energy Security Act Title II-A is negligible. Much of the increased acreage for corn feedstocks comes from reduced soybean acreage. Preliminary analysis indicates that about 0.5 million acres of additional cropland planted to feed grain would be needed to meet the Energy Security Act Title II-A goal. In 1980, farmers had 354.5 million acres in principal crops out of a total 460.5 million acres of cropland. The additional acres represent only 0.5 percent of the fallow and idle land and land devoted to minor crops.

Gross Farm Receipts

Ethanol-production impacts are expected to increase farm receipts from corn sales and decrease receipts from soybean sales. Current program impacts are moderate for corn receipts (2.5 percent) while soybean receipts decline slightly (−1 percent). The Energy Security Act Title II-A program impacts indicate corn receipts increasing about 9.7 percent from base projections by 1982-1983.

Receipts from soybean sales, however, decline 2.6 percent from base. Changes in gross receipts for all other crops are negligible. Thus, the Energy Security Act Title II-A program effects increase corn receipts about 7.2 percent and decrease soybean receipts about 1.6 percent from the level anticipated with the current ethanol program.

Food Prices

The Energy Security Act Title II-A program effects on food prices are expected to be minimal. To the extent that any effects are noticeable, they will occur in higher beef prices as a result of higher corn feed costs. These higher feed costs could be moderated by increased use of coproduct feeds, though the extent to which this might happen is unknown. Only 3 percent of total corn disappearance is due to human consumption. Increases in corn product prices would likely be offset by reduced consumption.

Chapter 5

Markets for Ethanol and Coproducts

Establishing an economically viable ethanol plant requires four key elements of availability: a reliable, sufficient feedstock supply; the support resources such as process water, process energy, and labor; an adequate market for ethanol and its coproducts; and transportation for feedstock and ethanol. These elements impact directly the selection of a site for the proposed plant and the overall economics of the project. Compromises in the following areas may need to be made to approach an ideal situation: proximity to the feedstock supply; access to plentiful water resources and renewable process energy sources; and direct access to large enough markets for ethanol and its coproducts to absorb the plant's output.

The problem of availability of the required feedstock resources has been addressed in the preceding chapter. This chapter discusses some of the characteristics of the markets for ethanol and its coproducts, and examines some of the issues to be addressed by a potential entrepreneur.

Ethanol Markets

ETHANOL AS A FUEL AND AS A CHEMICAL

Ethanol may be used in various forms for fuel:

As a blend with gasoline in various proportions

As hydrated lower-proof ethanol

As neat anhydrous ethanol

As a fuel supplement in dual-carbureted diesel engines

Ethanol also is used as a chemical in such industries as pharmaceuticals and perfumes. Each of these potential markets for ethanol has some constraints which bear on the decisionmaking process of producing ethanol. Some of these constraints result from the differences in the properties of fuel ethanol compared to petroleum-based fuels now used.

Fuel Properties of Ethanol

Table 5-1 summarizes some of the properties of ethanol and other fuels. Blending ethanol with other fuels results in modifications to the properties of

Table 5-1. Summary of Ethanol and Other Fuel Properties

PROPERTY	GASOLINE	ETHANOL	NO. 1 DIESEL
Molecular Weight	126	46	170
Heating Value			
Higher (Btu/lb)	20,260	12,800	19,240
Lower (Btu/lb)	18,900	11,500	18,250
Lower (Btu/gal)	116,485	76,152	133,332
Latent Heat of Vaporization (Btu/lb)	142	361	115
Research Octane	85-94	106	
Motor Octane	77-86	89	10-30
Stoichiometric Air/Fuel Ratio	14.7	9.0	
Flammability Limits (Volume Percent)	1.4 to 7.6	3.3 to 19	

Source: ***Fuel From Farms – A Guide to Small-Scale Ethanol Production,*** **SERI, Denver, 1980.**

the original fuel. Adding 10 percent ethanol to gasoline results in a lowered energy content (about 112,000 Btu per gallon for gasohol versus about 116,000 Btu per gallon for gasoline) but a higher octane number. The stoichiometric air/fuel ratio for the blend also will be quite different from that of gasoline as the percentage of ethanol in the blend is increased. Therefore, as more ethanol is added to the blend, the air/fuel mixture of a carburetor set for gasoline becomes less favorable, and the driveability of the vehicle may be affected. Comparative fuel economy for gasohol- or gasoline-operated vehicles is still a matter of controversy. In some instances, improved mileage has been claimed for gasohol-operated vehicles, but, in most cases, mileage has not changed. The direct use of ethanol in diesel engines is difficult without major engine modifications. Under present engine designs, ethanol does not meet the fuel specifications of diesel engine manufacturers. Therefore, ethanol cannot presently be substituted for diesel fuel. Ethanol-diesel fuel mixtures have unfavorable self-ignition characteristics due to the low self-ignition tendency of ethanol, and the performance of blend-fed engines is strongly affected. The best approach at present appears to be carbureting ethanol in the diesel engines. This, however, requires engine modifications and separate ethanol and diesel fuel tanks. Refer to Chapter 10 for more information on gasohol utilization.

If the market for diesel fuel warrants it, there are well-established alternative fermentation technologies which can produce diesel fuels from biomass. For example, butanol is readily produced by fermentation and is an excellent diesel fuel. If the producers wished to switch from ethanol to a diesel fuel, this would require only a change from yeast to another biocatalyst and some modifications in distillation equipment.

Ethanol-Gasoline Blends

Because of the phase-separation problems occurring when hydrated ethanol (less than 200-proof ethanol) is mixed with gasoline, ethanol-gasoline blends usually include anhydrous (200-proof) ethanol. Such blends have been used for many years in various countries. A blend of 90 percent unleaded gasoline and 10 percent anhydrous ethanol is marketed at present in many states under the name gasohol as noted in the Introduction.

Despite the controversy concerning the fuel efficiency of gasohol versus gasoline, gasohol has been well received by the public and is generally accepted as a substitute for gasoline. The present political climate and the desire to reduce the nation's dependence on foreign energy supplies may be some of the motivating factors behind the adoption of gasohol. If the present trends are maintained, i.e., if the use of gasohol continues to be encouraged by state and federal governments and if the public continues to accept gasohol as a substitute for unleaded gasoline, the potential market for fermentation ethanol is about 4-5 billion gallons per year if a 10 percent blend is used.

Hydrated Ethanol

Hydrated ethanol (less than 200 proof) can be burned efficiently in internal combustion engine with minor engine modifications. The carburetor jet size must be enlarged slightly when converting from gasoline to ethanol because the ethanol component contains less useful energy per unit volume than gasoline (see Table 5-1). With some engines, it is desirable to modify the intake manifold to ensure proper vaporization of the ethanol so that all cylinders will be operated with the same air/fuel ratios.

The use of hydrated ethanol fuel would have major advantages for the producer. The last step in the usual process of ethanol refining, that is, the dehydratation of the azeotrope, could be eliminated. The most likely hydrated ethanol fuel would be one of about 186 proof, which can easily be obtained at relatively low-energy consumption directly by single distillation of the fermenter beer. Major savings both in equipment and in energy consumption would result with significant reduction in cost for this product. The use of 186 proof as 93 percent ethanol is likely because this is the aim of the Brazilian effort, and both General Motors and Volkswagen are already producing or will produce automobiles for the Brazilian market which are designed to make use of 186-proof fuel. Also, the engine modifications required to make appropriate use of this fuel are modest compared to those required for the use of 160-proof fuel.

The use of hydrated ethanol has certain drawbacks. There will be a reduced mileage range per tank as compared to gasoline. Problems due to the accumulation of water are possible but are easily circumvented by appropriate designs and, with this fuel, there is no danger of freezing at low temperatures. Thus, while the direct hydrated ethanol market may have only a limited attractiveness for a period of time for commercial-sized ethanol plants, a major conversion to the use of this fuel rather than anhydrous ethanol in gasohol could be possible.

Anhydrous Ethanol Fuel

Anhydrous ethanol can be burned directly in spark-ignition engines using essentially the same modifications discussed above for the use of hydrated ethanol. Ethanol may also be used in furnaces, boilers, or gas turbines. In the latter case, efficiencies slightly higher than those obtained with hydrocarbons have been recorded. At present, however, the market for anhydrous ethanol fuel appears limited to blending with gasoline.

Diesel Fuel Supplement

As indicated above, diesel engines can operate on separately carbureted anhydrous ethanol and diesel fuel. Engine modifications and separate tanks for the two fuels are required. Thus, the market for ethanol in this application probably will be limited.

Industrial Applications of Ethanol

The chemical industry consumes large quantities of industrial ethanol as either feeddstock or solvent. In the latter case, one of the major consumers is the pharmaceutical industry, requiring extremely pure ethanol. Industrial ethanol also can be the feedstock to produce two important industrial chemicals: acetic acid and ethylene. Acetic acid can be obtained directly from ethanol by fermentation; ethylene may also be derived from ethanol.

At present, most industrial ethanol is produced from petroleum- or natural gas-derived ethylene. The cost of industrial ethanol therefore is directly related to those of petroleum and natural gas. As petroleum-derived industrial ethanol costs continue to climb, paralleling the cost of petroleum and natural gas, fermentation-derived ethanol will become more attractive as an industrial feedstock or chemical. The potential annual market for industrial ethanol alone is on the order of 200 million gallons.

Impact of Ethanol Fuel on Petroleum Import Requirements

The previous discussion suggests that the most attractive market for commercial-sized fermentation ethanol plants is ethanol to be used as a blend with gasoline, i.e., gasohol. Despite the slightly lower thermal value per unit volume of gasohol compared to gasoline, no significant fuel mileage decrease has been recorded when gasoline is replaced by gasohol. Ethanol, therefore, can displace a quantity of gasoline equivalent to the proportion of ethanol in the gasohol blend.

The addition of ethanol to gasoline increases the octane rating of the blend because anhydrous ethanol is a higher octane fuel. In the past, the octane rating of fuels was increased by adding tetraethyl lead. Because of the adverse effects of lead compounds on humans, the conversion to unleaded gasoline was mandated some years ago. The changes in refinery operations required to produce fuel of a given octane without lead additives reduce the quantity of fuel produced from a barrel of crude oil. The octane-boosting process requires additional energy in the refining process, energy lost from every barrel processed. The addition of ethanol to gasoline gives the required octane boost without the supplementary energy expenditure in the refining process. Therefore, every barrel of ethanol produced decreases the crude oil demand not only by the quantity of gasoline directly replaced by the ethanol but also by the crude oil saved as a result of the value of ethanol as an octane enhancer.

Problems of Storage, Handling, Blending, and Distribution

Some amounts of water in the ethanol will cause the separation of ethanol-gasoline blends in two layers, water-alcohol and gasoline. This separation will result in poor engine performance. It is therefore necessary that ethanol be kept anhydrous during transportation prior to blending with gasoline. The points at which blending of ethanol with gasoline could occur are at the refinery during loading of the trucks, the pipeline terminal as the trucks are loaded, or the retail station by means of a blending pump. In view of the bulk quantities involved when dealing with the output of a commercial operation, the first two blending options are preferred. Hydrated ethanol, on the other hand, offers very few problems and is readily handled by commercial equipment.

Blending shortly before use will have an impact on the economics of ethanol fuel. The costs of transportation of ethanol from the production facility to the point of blending and the storage of ethanol at the site of blending will have to be added to the retail price of ethanol.

The cost of transportation has been estimated to be about $0.008 per gallon for the first 20 miles plus about $0.003 per gallon for each additional 20 miles traveling distance (1980 dollars). The cost of storage facilities is estimated to be about $0.10 per gallon capacity for a 50-million-gallon-per-year facility. With adequate maintenance, the life of storage facilities may exceed 40 years. The proximity of a pipeline terminal or refinery may therefore be a desirable feature when siting an ethanol plant.

MAJOR ETHANOL PRODUCERS AND DISTRIBUTORS

The principal promoters of gasohol to date have been independent oil companies. Texaco, Inc. has taken a lead role and is currently retailing gasohol at most of its outlets. Besides Texaco, prominent distributors of gasohol include Atlantic-Richfield Oil Company and Standard Oil Company (division of Amoco Oil Company). The major companies as a group have shown reserve in entering the market. Nevertheless, some major oil companies are test marketing gasohol. These firms include Cities Service Company, Phillips Petroleum Company, Standard Oil Company of Indiana, and Diamond Shamrock.

In addition, some major oil companies are discussing joint-venture arrangements with large food processors. Three of these joint-venture arrangements are

Texaco and CPC International

Chevron and American Maize

Ashland Oil and Publicker Industries

Some of the major ethanol producers are listed in Table 5-2 and fuel-ethanol plants, which have been projected for the near term, are listed in Table 5-3. Neither the total national production capacity including small producers nor the amount of ethanol sold for fuel is well defined at present.

MARKET FOR ETHANOL

U.S. Market for Gasohol

In 1979, approximately 50 million gallons of ethanol were used to produce gasohol in the United States. The gasohol was sold through almost 2000 retail outlets. Most of the gasohol has been made with unleaded gasoline so that it could compete against premium unleaded fuels. However, some gasohol used regular gasoline. The latter mixture primarily served to extend gasoline supplies during shortages.

Table 5-2. Major Ethanol-Fuel Producers

PRODUCERS	CAPACITY (MILLION GALLONS/YEAR)
Eastman	25
USI	66
Publicker	60
Union Carbide	120
Archer Daniels Midland Company	260
Georgia Pacific	4.6
Midwest Solvents	15

Source: *A Guide to Commercial Scale Ethanol Production and Financing*, SERI, Denver, 1980.

Table 5-3. Potential Ethanol-Fuel Plants—Near Term (Received Grants Under P.L. 96-126)

STATE AND COMPANY	FEEDSTOCK	CAPACITY (Million/ Gallons/ Year)
Alabama		
Grasp, Inc.	Corn and milo	40
Arizona		
Arizona Grain, Inc.	Corn, grain, barley	12.5
Arkansas		
Arkansas Grain Fuels Inc.	Milo	35
California		
Ultrasystems, Inc.	Sugar beets, potatoes, wheat, grain	20
Western Concentrates, Inc.	Corn	132
Colorado		
Grand American, Inc.	Grain	10
Georgia		
Cafpro, Inc.	Corn	2
Nuclear Assurance Corp.	Wood	25
Hawaii		
Hilo Coast Processing	Molasses	23.5
Idaho		
Clearwater Palouse Energy Co-op	Wheat, barley	20
Illinois		
Rochell Energy Developers	Corn	1.5
Indiana		
Agri Answer, Inc.	Corn	Unknown
New Energy Corp.	Corn	50
Iowa		
Agri Grain Power Inc.	Corn	50
Kansas		
Planning, Design & Dev. Inc.	Grain	10
Louisiana		
Apex Oil Inc.	Various	33
Independence Energy, Inc.	Sugar cane, sorghum, molasses, corn	20
Maine		
D.W. Small & Sons	Corn, potatoes	25
Maryland		
Americol Ltd.	Local feedstocks	10
Massachusetts		
Belcher New England, Inc.	Corn, grain, potatoes, cellulose	25
Michigan		
U.S. Ethanol Industries	Corn	40

STATE AND COMPANY	FEEDSTOCK	CAPACITY (Million Gallons Year)
Minnesota		
CBA, Inc.	Corn	24.5
Mississippi		
Alcohol Fuels of Miss. Inc.	Wood chips and dust	1
Missouri		
Missouri Farmers Assoc.	Corn, wood	5
Montana		
Infinity Oil Co. Inc.	Wheat, barley	5
Nebraska		
Nebraska Alcohol Fuels Corporation	Corn	50
Nevada		
Geothermal Food Processors, Inc.	Corn	5
New York		
Andco Environmental Processes	Corn	15
North Carolina		
Diversified Fuels, Inc.	Corn	30-50
North Dakota		
Dawn Enterprise, Inc.	Wheat, potatoes	50
Oklahoma		
Fulton Energy Corp.	Corn, milo	25
Oregon		
Morrow Ag Energy Corp.	Corn, wheat, sugar beet, potatoes	20
Pennsylvania		
Lavco, Inc.	Corn	20
South Carolina		
Energy Conversion Corp.	Corn	20
South Dakota		
Sodak Resources, Ltd.	Corn	20
Vermont		
Alternative Concepts of Energy	Cheese whey	1.5
Virginia		
A. Smith Bowman	Corn	20-40
Washington		
Omega Fuels	Corn	50
Wisconsin		
Dvorak Farms	Corn	2
Wisconsin Agri Energy Corp.	Corn	20
TOTAL PROJECTED PRODUCTION		950-1030 Gallons/Year

Source: *A Guide to Commercial Scale Ethanol Production and Financing,* SERI, Denver, 1980.

Consumer reaction to gasohol has been very favorable, especially in the farming community. Generally, motorists rate the product high in terms of engine performance. It is no surprise that gasohol acceptance appears to be increasing, as is the number of companies selling gasohol.

Gasohol test market results show penetration rates of 8-30 percent of overall gasoline sales. In areas such as the midwest, where gasohol has been promoted extensively, a near-term penetration rate of 20 percent of gasoline sales appears reasonable. In less-developed markets, a conservative estimate of 10 percent penetration may be more realistic in the short term.

The near-term gasohol penetration rate could increase significantly under conditions of gasoline shortage or gasoline price rises relative to ethanol. Many motorists contend that gasohol must overcome its price disadvantage relative to gasoline before its use becomes more widespread.

A factor which once constrained gasohol sales was the difficulty for retailers to get unleaded gasoline allocations for blending. However, gasoline supplies currently appear available. To the extent that supplies remain adequate, this constraint will be moderate. In addition, the Department of Energy (DOE) issued changes in gasoline allocation rules to assign automatically unleaded gasoline to blenders of gasohol. The oil companies are expected to oppose these changes, however, because such gasoline assignments would take gasoline away from existing customers.

In the future, the sales potential of gasohol as an unleaded fuel may increase. If for no other reason, the percentage of gasoline sales comprised of unleaded fuel is increasing as new vehicles (using only unleaded fuels) replace older ones.

REGIONAL MARKETS FOR ETHANOL FUEL

One important consideration in the siting and planning of an ethanol plant is the proximity of a market for the fuel produced and of blending sites. Figure 5-1 shows the percentage of the total number of vehicles registered by U.S. Department of Agriculture (USDA) regions. Also plotted on the figure are approximate locations of refineries and pipeline terminals. A few facts are apparent from the figure:

> The preferred sites for blending of ethanol with gasoline, i.e., refineries and terminals, are widely available within most regions and in particular in the Lake States, Northern Plains, and Cornbelt regions where the largest fraction of the grain feedstocks is produced.
>
> Only about 27 percent of the vehicular fleet is registered in the three major grain-producing regions. Marketing of ethanol produced in those regions will therefore have to include shipment of ethanol or gasohol to large vehicular markets such as the Northeast, Appalachian, and Southeast regions (combined fleet: 42 percent of the national total). A compromise between siting of the plant in the

vicinity of the feedstock resource and close to the market for ethanol may have to be reached.

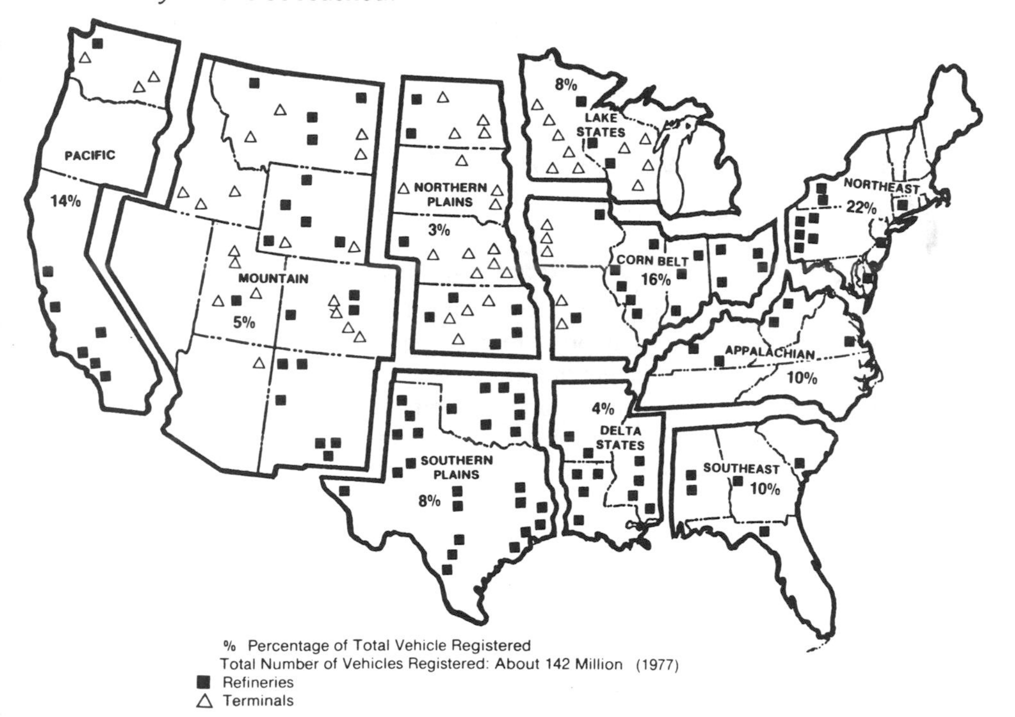

Source: U.S. Department of Commerce, "Statistical Abstract of the United States, 1978," 99th Annual Edition, Washington, D.C., 1978.
M. L. David, G. S. Hammaker, R. J. Buzenberg, and J. P. Wagner, "Gasohol Economic Feasibility Study." Report prepared by Energy Research and Development Center, University of Nebraska, Development Planning and Research Associates, Inc., Manhattan, Kansas, p. 261, July 1978.

Figure 5-1. Percentage of motor vehicles registered by USDA farm regions and location of refineries and terminals.

Government Policies and the Market for Ethanol

The major impact of government policies on the market for ethanol will be through enhancing the economic attractiveness of ethanol over fossil fuels. The excise tax exemption, as well as a 10 percent additional investment tax credit for facilities that convert alternate feedstocks to liquid fuels, are the two major federal policies encouraging the penetration of ethanol in the fuels market. A number of states have also eliminated the state fuel tax for ethanol blend fuels or gasohol.

Expected Market Fluctuations

If present price and incentive policies are maintained, tending to make gasohol competitive with gasoline, and if the present reception of gasohol by the public is sustained, no drastic market fluctuations for fuel ethanol are expected. In the long run, the improved average mileage required under federal law will somewhat tend to decrease the demand for automotive fuel. The impact on the ethanol market will, however, be small.

Price of Ethanol

HISTORICAL REVIEW OF PRICES

Industrial ethanol is manufactured from ethylene. As prices of petroleum products increase, so do those of ethanol. The price of anhydrous industrial ethanol is about $2.02 per gallon (May, 1980, wholesale, f.o.b. plant).

The price of ethanol produced from agricultural feedstocks is dependent on factors such as type and price of the feedstock, plant operating life, financing terms, value of the credit for coproducts, and other factors. Recent estimates suggest market prices ranging from $1.30 to $1.65 per gallon (1980 dollars) for ethanol from corn and wheat feedstocks under various plant-financing conditions. These market prices assume that a credit is received for the coproducts. On an energy-content basis ($/Btu), ethanol from grain is still more expensive than gasoline. However, if one accepts equivalence in performance of gasoline and ethanol in a gasohol blend, the price of ethanol is approximately equivalent to that of gasoline. At present, gasohol retail prices (even with the tax exemptions) are equal to or higher than those for unleaded gasoline. As prices of petroleum increase, the price position of ethanol versus gasoline will further improve.

FACTORS AFFECTING THE PRICE OF ETHANOL

As earlier indicated, one of the provisions of the Windfall Profits Tax is to exempt ethanol from renewable feedstocks from the federal excise tax. This amounts to an exemption of $0.40 per gallon of ethanol or $0.04 per gallon of gasohol. Similarly, many states exempt ethanol or gasohol from state taxes. The amount of the exemption varies.

Transportation and distribution costs of ethanol increase the cost of ethanol used in gasohol blends. This is particularly true for markets far from ethanol-producing plants. As an example, the price of gasohol in Virginia ($0.829 in May 1979) included 1.7 cents per gallon for shipping the ethanol from Illinois.

PROJECTED PRICES AND UNCERTAINTY OF PRICE

The market price of gasohol will follow closely that of gasoline, i.e., increase at a rate slightly higher than inflation. A 10-cent rise in price for unleaded gasoline raises that for gasohol by 9 cents. The present and near-term projected production capacities for ethanol suggest that full production will not glut the market and depress the market price of ethanol.

The production cost of ethanol is very sensitive to net feedstock costs (feedstock cost minus credit for coproducts). As was shown in Chapter 4, wide fluctuations in net feedstock costs have occurred and must be expected. Over the years, these fluctuations tend to smooth out. Their impact on fermentation ethanol prices has a relatively slight effect on gasohol prices—a 10-cent price rise in ethanol results in only a 1-cent rise in gasohol price.

There is a certain risk involved when entering the ethanol-fuel business. The risk is probably related to the uncertainty of the feedstock cost rather than projected changes in demand for fuel ethanol. To minimize the risk, emphasis must be placed on securing long-term reliable feedstock supplies.

Iowa Development Commission Gasohol Acceptance Study

The Iowa Development Commission prepared an analysis titled *Gasohol Acceptance in Established Markets,* dated April 19, 1979.

Since gasohol was first introduced to Iowans in June, 1978, the marketing of the 10 percent alcohol/90 percent unleaded gasoline gasohol mixture has grown at a tremendous rate. During the Iowa Development Commission's consumer acceptance study from June 15 to October 31, 1978, approximately 1,090,000 gallons of gasohol were sold in Iowa. In the month of November, 67 stations sold 600,000 gallons of the new product. During December, the product's popularity rose sharply, with 120 stations retailing 2.6 million gallons of the alcohol-gasoline motor fuel blend.

After the Federal Tax Exemption went into effect on January 1, 1979, the consumption figures increased to 250 stations pumping 4 million gallons of gasohol. In February, gasohol consumption continued to increase to the 5 million gallon level.

March figures show 400 stations marketing gasohol and monthly sales figures running in the neighborhood of 5.6 million gallons of gasohol a month as shown in Fig. 5-2.

In an attempt to obtain a more precise indication of gasohol's market acceptance, the Iowa gasohol acceptance study was formulated where three distinct markets representing a rural, intermediate, and large market area were identified, having a sufficient number of gasohol stations to consider the fuel generally available.

The Iowa communities having these characteristics were the Gowrie (rural), Ames (intermediate), and Cedar Rapids (large urban) markets. The number of gasoline stations selling gasohol for each market were: Gowrie, one; Ames, six; and Cedar Rapids, nine.

A telephone survey was designed by the Iowa Development Commission, and an independent telephone surveying company was hired to conduct the survey. A sample size was determined to yield a 95 percent

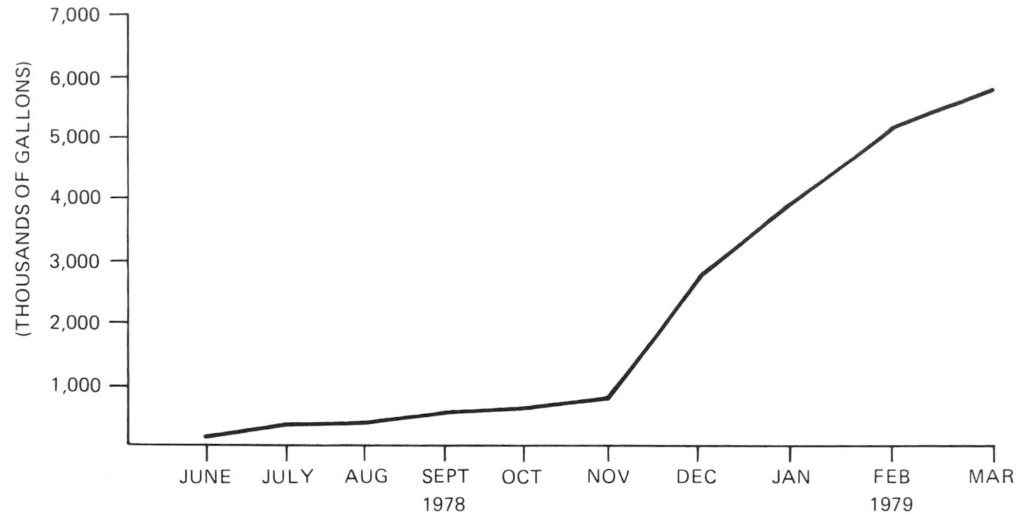

Source: Iowa Development Commission, *Gasohol Acceptance in Established Markets*, April 19, 1979.

Figure 5-2. Iowa's monthly consumption of gasohol based on ethanol.

confidence level that the mean of the sample data would be within, plus or minus, 5 percent of the true mean for the entire market area.

The sample size was apportioned to each respective market area, and telephone numbers were randomly selected from the telephone directories of each community. The survey commenced on March 17, 1979, and was completed on March 27, 1979.

Only the head of household or the spouse of the head of household was interviewed. Although a bias may exist because more female than male interviews were conducted, the Iowa Development Commission believes that this potential source of bias would not significantly skew the results. The age of each respondent was obtained to verify that each age group was adequately represented in the sample.

ACCEPTANCE OF GASOHOL

In the rural market of Gowrie, 30 percent have used or are using gasohol. In Cedar Rapids, where five of the nine gasohol stations had been offering gasohol less than 6 weeks prior to the study, nearly 30 percent of those surveyed had purchased gasohol. In Ames, where most outlets had been offering gasohol for 5 months, the percentage using gasohol was 19 percent.

As shown in Table 5-4, 15 percent of the Iowans interviewed had used gasohol. Heaviest use then was in small rural markets, with the "city" markets of Spencer and Fort Dodge reporting 16 and 10 percent gasohol users, respectively. Currently, the larger city markets of Ames and Cedar Rapids are at considerably higher levels.

Table 5-4. Number of People Using Gasohol

September, 1978	Usage (percent)	March, 1979	Usage (percent)
Four Rural Stations	22	Gowrie, Iowa	30
Fort Dodge, Iowa	10	Ames, Iowa	19
Spencer, Iowa	16	Cedar Rapids, Iowa	30
Total "Yes"	15	Total "Yes"	27

(Sample Count: 500) (Sample Count: 514)

Source: Iowa Development Commission, *Gasohol Acceptance in Established Markets,* April 19, 1979.

PURCHASING FREQUENCY

When asked what percentage of their fuel purchases were gasohol, as shown in Table 5-5, 15 percent of the gasohol users in Gowrie were spending 50-75 percent of their gasoline money for gasohol, and 22 percent were buying nothing but gasohol.

In Cedar Rapids, 8 percent were using gasohol between 30 and 50 percent of the time, and 16 percent were buying gasohol 50-100 percent of the time, which included 10 percent using gasohol all the time.

In Ames, 13 percent of the households surveyed reported a use rate of 25-50 percent of the time, and no household reported a use rate larger than 65 percent.

LIMITING FACTOR: AVAILABILITY

Respondents indicated that their use of gasohol would be greatly increased if the blend were available at most stations. Inconvenience and inability to buy gasohol at their favorite stations are major deterrents to popular use.

Of those who have used gasohol, 75 percent reportedly would use it regularly or occasionally if readily available; 45 percent of the nonusers said they would buy gasohol regularly or occasionally if offered at most stations.

Table 5-5. Gasohol's Percentages of Fuel Purchased for All Family Vehicles

Market	1-24 percent		25-49 percent		50-100 percent		Don't know		Total users
	No.	(percent)	No.	(percent)	No.	(percent)	No.	(percent)	
Gowrie	10	(55)	0	(0)	7	(39)	1	(6)	18
Ames	21	(70)	1	(3)	4	(13)	4	(13)	30
Cedar Rapids	64	(72)	8	(9)	14	(16)	3	(3)	89
All user averages (percent)		69		7		18		6	

Source: Iowa Development Commission, *Gasohol Acceptance in Established Markets,* April 19, 1979.

The initial study determined that 21 percent of gasohol users had switched retail stations to buy the product. Currently, 26 percent have changed stations to get gasohol.

GASOHOL BENEFITS

Asked to list the most important benefit of gasohol (see Table 5-6), the leading answer, from about 30 percent of all responses, was "extending oil supplies." About 20 percent listed "improved auto performance"; 13 percent said "important market for corn." Other benefits mentioned occasionally were "helps economy," "reduced pollution," and "motorists like it."

Only 6 percent believe that gasohol has no special benefits, and 19 percent gave a "don't know" response.

Reasons for nonusers. Nonusers were asked for the "main reason" they have not tried gasohol. Major reasons listed in a total of 375 answers were: "not available/not at my station," 107; "too expensive," 79; "haven't gotten to it," 63; "concerned about effects on car," 26; "use leaded gas," 22.

Premium fuel. About 48 percent of all respondents believed that gasohol is a "premium quality" fuel, 23 percent said "don't know," and 29 percent said "not a premium product."

This response represents a considerable increase over results from the September 1978 study, when only 19 percent said that gasohol was "equal to or better than" unleaded regular, and 3 percent considered it less desirable than the competing gasoline product.

Performance related to weather. Respondents were asked to separately rate the performance of gasohol for warm weather, cold weather, and year-around driving. Only 3 percent of the total sample felt gasohol would be unacceptable in warm weather, 5 percent said unacceptable for cold weather, and 4 percent reported "unacceptable for year-around use." On the acceptable side, half the respondents felt gasohol was an acceptable fuel for all

Table 5-6. Most Important Benifits of Gasohol

	Extending oil supply		Improved performance		Market for corn		Other[a]		None		Don't know	
Market	No.	(percent)	No.	(percent)	No.	(percent)	No.	(percent)	No.	(percent)	No.	(percent)
Ames	56	(37)	11	(7)	18	(12)	18	(12)	13	(9)	35	(23)
Cedar Rapids	88	(29)	77	(26)	31	(10)	41	(14)	15	(5)	47	(16)
Gowrie	15	(21)	14	(19)	21	(29)	7	(10)	1	(1)	15	(21)
Total averages (percent)		30		20		13		13		6		19

[a]Other benefits mentioned included: Renewable resource, less pollution, prevents gasoline freeze, cost, good for economy, benefits oil/alcohol companies.

Source: Iowa Development Commission, *Gasohol Acceptance in Established Markets,* April 19, 1979.

seasons. (The remaining responses to this question were "no opinion" or "don't know.")

Gas station change. The initial study determined that 21 percent of gasohol users had switched retail stations to buy the product. Currently, 26 percent of the users have changed stations to buy gasohol.

Few engine problems. Gasohol apparently causes very few engine problems. Only 15 percent of the gasohol users reported any problems. Of the total sample of 514 respondents, only 21 complaints were mentioned. "Plugged dirty fuel filter" and "rougher running" were mentioned most often, because gasohol cleans out fuel systems.

Aware of gasohol. In an earlier survey, 84 percent of the respondents could locate a gasohol outlet. Currently, 90 percent of the respondents know where gasohol can be purchased.

WILL GASOHOL USAGE GROW?

Asked about their opinions on the future growth of gasohol consumption, 74 percent of all respondents, users and nonusers, said they expect the blend to be more widely used.

The "will grow" side. Those who predicted future growth were asked "why." "Extending petroleum supplies" was the highest-ranking answer, with 41 percent. As usage grows, 16 percent said that prices will become more competitive. "Superior product" answers came from 11 percent, with 9 percent saying that gasohol's popularity will cause growth. "Benefits to farmers," a high-ranking reason given last summer, was mentioned by only 6 percent in this study.

The "won't grow" side. Of the 16 percent predicting no growth for gasohol, 35 cited "cost" as their reason, followed by "apathy," "happy with gasoline," "not an alternative," "needs improvement," "not enough alcohol," and "using food as fuel."

WHO BENEFITS FROM GASOHOL?

In an earlier study, Iowans were asked "Who benefits the most from a national gasohol program?" The responses, from predominately small-town rural areas, favored "extending energy for everyone" (39 percent) over "helps agriculture" (31 percent).

In a current study, an open-ended question was used: "Who do you see benefiting the most from gasohol?" Responses are shown in Table 5-7.

WOULD PEOPLE PAY A PREMIUM?

Respondents were asked how often they would buy gasohol if tax exemptions were removed and gasohol cost 10 cents more than unleaded

Table 5-7. Who Benefits From Gasohol[a]

Market	The public		Farmers		Government		Oil/alcohol companies		No benefit or don't know	
	No.	(percent)	No.	(percent)	No.	(percent)	No.	(percent)	No.	(percent)
Ames	52	(35)	46	(30)	5	(3)	16	(10)	26	(17)
Gowrie	20	(33)	29	(48)	1	(2)	4	(7)	8	(13)
Cedar Rapids	137	(46)	94	(31)	7	(2)	32	(11)	25	(8)
Total averages (percent)		41		33		3		10		12

[a]Other responses mentioned one to three times included: economy, environment, people on low incomes, future generations, gas stations.

Source: Iowa Development Commission, *Gasohol Acceptance in Established Markets,* April 19, 1979.

regular. Their responses: At a 10-cent premium price, 9 percent of all families would use it regularly; 33 percent would use it occasionally; 46 percent would not buy gasohol; and 12 percent responded with "don't know."

Most people who have used gasohol would pay a 10-cent premium. Responses from users showed that 59 percent would still use gasohol regularly or occasionally. (A 10-cent price spread would mean that 36 percent of the current users would not buy gasohol. The remaining 5 percent didn't know.) In an earlier study, only 11 percent of the respondents said they would pay as much as a 5-cent premium for gasohol on a regular or occasional usage basis.

Favor state promotion. Should the State of Iowa continue to promote the development of gasohol? The overall response favored such promotion: 90 percent saying "yes," 6 percent saying "no," and 5 percent "undecided."

In a related question, people were asked if they favor or oppose the use of state fuel tax money to subsidize gasohol. The panel voted 69 percent in favor of the subsidy, 22 percent opposed, and 10 percent said "don't know." (Of those using gasohol, 89 percent favor the gas tax exemption.)

STATION MANAGER INTERVIEWS

In a concurrent study, interviews were completed with the managers/owners of all retail stations offering gasohol in the Ames, Cedar Rapids, and Gowrie markets, as well as some gasoline stations not handling gasohol.

Since only 16 stations were supplying gasohol in these market areas, the responses are not presented on a statistical basis, but as simply a summary of information.

Increased sales. All gas stations gained new customers by providing gasohol. About one-half said the new-customer totals were "many," the other half reported "important increases." (Stations with heavily transient trade said they had no accurate way to determine the special "pull" of gasohol, other than increases in pump volume.)

Asked why customers are buying gasohol, the typical answer was along these lines: "The novelty and the publicity probably caused motorists to try gasohol. Customers have stayed with the blend because of its performance, because it extends petroleum supplies, because it is price competitive, and because it helps farmers."

Eight of the 16 retailers said that "nearly all" of their gasohol customers continue to use the product; and seven others said that "most" customers are repeaters. Only one station reported "few repeaters."

Farmer influence in this study was minor. Except for the farm cooperative station in Gowrie, the city stations reported an average of 4-5 percent farmer customers at their pumps.

Gasohol sales increases. Although one-third of the stations interviewed had been selling gasohol for less than six weeks, the alcohol-gasoline blend was highly competitive with its all-gasoline counterpart, regular unleaded. Six stations reported gasohol having less than 30 percent of their unleaded business, adding that "gasohol's share is growing daily": three stations were nearing the 50-50 point; three stations were selling from 50-59 percent of their unleaded gasohol; two reported gasohol above 60 percent; and five stations have discontinued unleaded regular, offering only unleaded gasohol. (Most operators reported that their breakout of leaded versus unleaded fuels is close to the 50-50 level. As new cars replace those now using leaded fuel, the percentages will continue to move up for unleaded.)

All 16 operators expect gasohol to move into wide usage. Their reasons for this prediction were fairly evenly divided among four factors: it extends available petroleum supplies; it is a better product with its higher octane level; the price is "competitive on a value level"; and "people want it."

Price hikes: 9-15 cents. Operators were asked to predict the retail price of gasoline 6 months in advance, in September, 1979. Operators in Ames are looking for increases averaging about 15 cents per gallon. In Cedar Rapids, price hike predictions averaged 9 cents. Two operators believed that there would be little if any increase, and one said gasoline prices could drop during the period in question.

Premium prices. Would customers pay a premium for gasohol over that of unleaded gasoline? The answers provided an interesting marketing point. In Cedar Rapids, where gasohol and unleaded regular sell at virtually the same price, retailers generally felt that a price spread of 2-3 cents would be "about all the drivers would pay."

In Ames, where gasohol was priced at 3-4 cents higher than unleaded gasoline, station managers thought that a 4-5-cent spread would be the most motorists would pay. (Their customers, surveyed at the same time, would pay 10 cents more for gasohol.)

Although operators recognize that gasohol is a premium fuel and therefore is worth more than unleaded gasoline, they generally said that customers are not yet really aware of increased octane levels and other premium benefits. One operator suggested that "if a couple of major oil companies were to explain gasohol in their national advertising, drivers would expect to pay a premium price for it."

Complaints. After 6-8 months station spokesmen say they are having virtually no complaints about gasohol's performance in their customers' cars. An occasional clogged fuel filter was reported, with operators generally saying "the alcohol cleans out the tank and fuel systems, and most drivers should replace those filters more often anyway."

Other stations would accept gasohol. With all "gasohol stations" reporting volume increases since offering the alcohol blend, many saying the gains have been substantial, the interviewer listed nearby stations not now selling gasohol.

The "gasoline-only" stations generally reported business at normal levels, with some increase in their volume. These operators believe that the switch to gasohol is not cutting heavily into their volume. One operator explained: "Only a handful of stations are getting all of the gasohol business, so impact on the dozens of stations not selling the blend is minor." He added that "there have been enough stations closed in this market that we have all picked up some additional business."

Would the "straight-gasoline" stations offer gasohol? The general answer was, "Yes, if supplies of alcohol were available and the public wants to buy gasohol." Most said they were glad that gasohol is being tested in the open market and that they welcome a new fuel source that "might keep gas tanks full." None of the gasoline-only operators displayed negative attitudes about gasohol.

In summary, the managers of nongasohol stations said that, "If alcohol is available and people want it, we'd be happy to put it in our tanks."

SUMMARY OF IOWA SURVEY

Gasohol acceptance is steadily growing. In the first study in September, 1978, 15 percent of the Iowans interviewed had used gasohol. The second phase analysis showed that 27 percent of those surveyed have used gasohol with acceptance in individual markets as follows: Gowrie, 30 percent; Cedar Rapids, 30 percent; and Ames, 19 percent.

The most important benefit of gasohol mentioned by all respondents was "extending oil supplies" (30 percent); "improved auto performance" (20 percent); and "importance to the corn market" (13 percent).

Availability remains the primary reason for nonusers not having tried gasohol, with 29 percent giving "not available/not at my station" as their reasons. Also, 21 percent felt it was too expensive and 17 percent had not gotten around to purchasing the product.

Gasohol continues to receive a high engine-performance rating, with only 21 out of 514 users surveyed reporting problems. Rougher running engine and plugged fuel filters, which are not unusual problems in initial tankfuls of gasohol, were experienced by 10 of 21 users.

If gasohol were available at "most service stations," 54 percent of the respondents said they would use it regularly or occasionally. This is a sizable increase over the 11 percent who responded similarly in the first Iowa Development Commission's consumer acceptance study. An overwhelming 74 percent of those surveyed felt gasohol use would become more widespread. A primary reason for this expected growth includes gasohol's petroleum extending ability (40 percent). Also, many thought that its price would eventually come down (16 percent).

Consumers outranked farmers as benefactors of gasohol—41 percent to 33 percent. This is a shift from the first study which listed energy and agriculture as equal benefactors.

Although price continues to be the primary problem associated with gasohol's growth, its value as a premium fuel was recognized by 59 percent of the users and 36 percent of the nonusers who indicated a willingness regularly or occasionally to pay a 10-cent premium for the fuel. (The remaining 5 percent did not know.)

Of those surveyed, 89 percent felt the state should continue to promote gasohol, and 68 percent favored the continued exemption from the state fuel tax.

After 6-8 months, station spokesmen who were interviewed in the tested markets said they had "virtually no" complaints about gasohol's performance in their customers' cars.

This survey has been provided in this handbook to provide some information for the marketing problems associated with gasohol and recent experience in our nation's heartland. Introducing gasohol in a new market at the time may be representative of new gasohol markets throughout the country.

Coproducts of Ethanol Manufacture

The two primary coproducts obtained during the production of ethanol by fermentation of agricultural products are the residue, commonly used as an animal feed, and carbon dioxide.

ANIMAL FEED COPRODUCT

Forms of Grain Coproducts

Distillery coproducts from grain are a high-quality ingredient for animal feed. Four products are commercially available and have been defined by the American Feed Control Officials.

Distiller's dried solubles are obtained after the removal of ethyl alcohol by distillation from the yeast fermentation of a grain or a grain mixture by condensing the thin stillage fraction and drying it by methods employed in the grain distilling industry. The predominating grain must be declared as the first word in the name.

Distiller's dried grains are obtained after the removal of ethyl alcohol by distillation from the yeast fermentation of a grain or a grain mixture by separating the resultant coarse grain fraction of the whole stillage and drying it by methods employed in the grain distilling industry. The predominating grain must be declared as the first word in the name.

Distiller's dried grains with solubles are obtained after the removal of ethyl alcohol by distillation from the yeast fermentation of a grain or a grain mixture by condensing and drying at least three-quarters of the solids of the resultant whole stillage by methods employed in the grain distilling industry. The predominating grain must be declared as the first word in the name.

Condensed distiller's solubles are obtained after the removal of ethyl alcohol by distillation from the yeast fermentation of a grain or a grain mixture by condensing the thin stillage fraction to a semisolid. The predominating grain must be declared as the first word in the name.

Because of the high energy requirements for drying products such as distiller's and brewer's coproducts, there have been continuing efforts to use undried products. Various brewers have marketed wet grains within reasonable distances of the brewery, primarily for feeding ruminants. The untreated distiller's coproduct (stillage) has presented greater difficulties because of its higher liquid content; distiller's stillage is approximately 90 percent water compared to 70 percent water in brewer's wet grains. The reason for this difference is that brewers remove as much of the liquid as possible from the solid feedstock after enzyme treatment and this liquid (wort) is then fermented. Distillers simply distill the alcohol out of the fermented solid-liquid mixture.

Distillers could remove the solids fraction from the stillage and sell a wet grain product, but a problem of disposing of the liquid would remain. A significant loss of nutrients (particularly soluble protein) in the liquid fraction would occur.

The equipment and operating costs for drying stillage is high but, for large distillers, the advantages of marketing a dry product (or condensed solubles) are enough to overcome the high production costs. The major advantages of dry products include the following:

- Microbial decomposition is prevented in dry products. Wet products decompose rapidly and must be fed within one to two days in warm weather. Some preservatives might be used to increase the allowable storage life, but insufficient research has been done to determine the types of preservatives and their overall economics. Refrigeration could be used to extend storage time, but operating costs would be high; however, refrigeration costs would be lower than drying costs.

- The transportation costs for marketing dry products are reduced by eliminating the transportation cost for water.
- Feed manufacturers are usually better equipped to handle dry ingredients.
- Farmers are usually equipped to feed dry feeds rather than slurries.

Quantity of Animal Feed Coproduct

About 16 pounds of coproduct (dry weight) are produced from each bushel of corn. About 6 pounds are realized per gallon of 200-proof alcohol.

If the coproduct is not dried but used as stillage, then the same equivalent solid materials will be produced. The concentration of solids in the stillage will depend on the operation of the cooking and fermenting process. In normal operations, about 30 gallons of mash will be used per bushel of grain. After removal of the alcohol, there will be about 27 gallons (224 pounds) of stillage containing about 16 pounds of solids. The quantity will be reduced slightly if some of the liquid from the stillage is set back for use in a succeeding batch. The amount of liquid which may be set back has not been accurately determined but may be as much as 50 percent of the liquid. A stillage can probably be produced which will be more concentrated than the normal stillage. With maximum (50 percent) set back, there would be 14 gallons (116 pounds) of stillage with 12 percent solids per bushel of corn.

A large increase in fuel-alcohol production would lead to a change in the supply of vegetable protein supplement available for animal feeds. A major increase in alcohol production, say to the point of using 1 billion bushels of corn per year would increase the DDGS supply from the present 450,000 tons per year to 8,000,000 tons per year. Obviously this would have a significant effect on the supply and price of vegetable protein supplements. However, this diversion of corn would be equivalent to about 10,000,000 acres of corn production at about average yields. Such a diversion of corn supplies could change the relative corn-soybean acreages to reduce soybean supplies, and overall protein supplies might not be affected greatly.

Table 5-8 shows present overall vegetable protein supplement supplies from major sources. An additional 8,000,000 tons of DDGS would represent a very significant increase in that supply but a much smaller percentage of the overall supply.

Stillage as a Feed Ingredient

In spite of the advantages of dry products, stillage is probably the preferred form of disposing of coproducts from small stills. The quantity of

Table 5-8. Vegetable Protein Meal Production

Material	Calendar or crop year	U.S. Production		
		Domestic consumption (1,000 T)	Export (1,000 T)	Total (1,000 T)
Corn gluten meal and feed	1978-1979	1,038	2,031	3,069
Cottonseed meal	1978	1,996	26	2,022
Distiller's dried grain	1978-1979	496	–	496
Linseed meal	1978	81	38	119
Peanut meal	1978	89	–	89
Soybean meal	1978	16,498	6,520	23,018
Sunflower meal	1978-1979	198	–	198
TOTAL	–	20,396	8,615	29,011

Source: USDA, ESCS, Fats and Oils Situation.

stillage should be small enough to allow it to be fed within a reasonable distance from the point of production.

Since stillage feeding has not been widely practiced, there is limited research or documented use data on which to base recommendations for feeding from either nutritional, animal performance, or technical (handling) standpoints.

The Kentucky Agricultural Experiment Station reported some research on stillage feeding during the 1940s. This research was directed toward maximum usage of stillage rather than a most efficient use of its nutrient elements.

In general, the Kentucky recommendations were that adult beef animals should be limited to 40 gallons of stillage per day and swine to 4 gallons per day. Beyond these limits, urinary and other animal growth problems arose. When stillage is fed at these levels, animals are being forced to consume water at an above normal level. In this case, protein is supplied in excessive amounts. In the discussion which follows later, it will be assumed that stillage will be fed at a level which will make the best economic usage of its nutritional (chemical) elements; and, under these conditions, the normal animal intake of water will usually not be exceeded.

Since water is normally considered a "free good," feeding trials do not normally report water consumption. A few reports have been found from which rough estimates of water consumption may be made.

Dairy cattle require a high water intake which is related to the level of milk production. Most research reports were made before present production levels were reached. In 1954, Brody reported water consumption of 16.5 to 19.7 gallons per day at a temperature of 50°F (10°C) for Holstein cows producing 34.3 to 45.5 pounds milk per day; at an air temperature of 17°F (−8°C), water intake varied from 15.8 to 18.6 gallons per day with production of 30.4 to 35.7 pounds of milk per day. Total dry matter intake was not reported (only total digestible nutrients), but the water-feed ratio was probably about 4:1.

Stillage and wet grains must be used in a relatively short time to avoid excessive microbial decomposition. Little research has been reported, but a 1979 publication by Stechley reports on the effect of storage on brewer's yeast slurry. Slurry was stored at 4°, 21°, and 30°C (39°, 70°, and 86°F) for 35 days. Storage at 4°C (39°F) showed small changes in chemical composition during the 5-week period. Storage at 21°C (70°F) showed significant changes in the first seven days: 15 percent of the dry matter was lost, and true protein dropped from about 30 percent to 20 percent (a 30 percent loss) on a dry-matter basis. At 30°C (86°F), the dry matter loss was about 16 percent and protein dropped from about 30 to 13 percent. Total crude protein increased slightly as dry matter decomposed, but the true protein was converted to ammoniacal form. The protein conversion would have little effect on ruminants, but it would be undesirable for monogastrics.

The conclusion to be reached is that stillage and/or wet grains must be fed within one or two days to avoid significant nutrient loss or be given special treatment, probably refrigeration. Although this problem is not insurmountable, it will be a significant problem, particularly for the smaller farm-size still which may not distill a batch every day.

Stillage or wet grains have a distinct odor and cannot be added to or removed from rations randomly without an adverse effect on feed consumption and animal performance. This will be a very important factor with dairy cattle. Also, with dairy cattle, the material may affect milk flavor; hence, it is probably best to feed just after, rather than shortly before, milking.

Possible physical forms of coproducts which appear practical for small plants include:

- Stillage which is probably the only practical form for disposing of coproducts from farm stills
- Wet grains from medium-size operations
- Condensed solubles from medium or large operations
- Distiller's dried grains or distiller's dried grains with solubles from medium or large operations

HUMAN FOOD COPRODUCTS

As discussed in Chapter 4, an ethanol production unit can be integrated with a dry- or wet-milling grain plant. Dry- and wet-milling operations are specialized industries requiring unique marketing efforts to dispose of the variety of products generated and to adjust to the changes in demand for food/feed products.

The most common case of integration of an ethanol plant with such operations will probably result from the addition of an ethanol unit to an existing mill rather than the creation of an entirely new milling/ethanol complex. The former case is being implemented at the Archer Daniels Midland Company, the nation's largest ethanol producer for use as gasohol. In this instance, ethanol is really a coproduct of the major products (food and feed) of the plant and, as such, is a manifestation of a desire to diversify by taking advantage of an emerging market. The other approach, i.e., creation of a totally new integrated milling/ethanol complex, has some drawbacks for the prospective investor. The capital cost for an integrated milling/ethanol plant is much higher than that for an ethanol plant; the problems of marketing are increased because of the variety of products ranging from animal and human food/feed and pharmaceutical derivatives to fuel. Each component of the integrated project, i.e., milling and ethanol production, is a venture in itself including its own risks. Integrating the two components may provide some hedge against these risks but also may result in accumulated problems and risks unattractive to the investor unfamiliar with these industries.

The remainder of this section will therefore describe coproducts in terms of a proposed plant that is primarily producing ethanol fuel and that the major coproduct is animal feed (DDG or its equivalent).

MARKETING ANIMAL FEED COPRODUCTS

Stillage, the residue of fermentation and distillation in the production of ethanol, contains many nutritive elements as noted earlier. This is particularly true of grain stillage, which has been used as animal feed or feed supplement over the years. Marketing these coproducts is essential to the economics of ethanol production. The present section focuses on coproducts resulting from the production of ethanol from grain. Some of the characteristics of these coproducts have already been discussed in Chapter 4 but will be briefly summarized as needed here.

Marketing Options for the Coproducts

Fresh stillage is the mixture of various nutrients dissolved or suspended in water. It can be fed directly to animals but is not tolerated in large quantities because of the limited capacity for water intake by cattle and other animals. Fresh stillage also degrades rapidly, particularly in warm climates; therefore, disposal of the stillage output of a commercial plant will result in a complex distribution problem. Fresh stillage can be concentrated or dried. In this form, the product can be stored and shipped, making marketing an easier, more predictable task. This approach, however, requires a supplementary investment in drying equipment and storage facilities for the coproducts. In the forthcoming sections it is assumed that stillage is marketed as a dried product—DDG or DDGS.

Table 5-9. Market for Some Commercial Feeds in the United States

YEAR	IN THOUSAND TONS							
	CORN	SOYBEANS	WHEAT MILL FEEDS	GLUTEN FEED MEAL	BREWERS DRIED GRAINS	DISTILLERS DRIED GRAINS	DRIED/MOLASSES BEET PULP	TOTAL COMMERCIAL FEEDS
1964	82,800	9,236	4,716	1,165	295	409	1,289	99,910
1965	94,100	10,274	4,612	1,135	304	426	1,153	119,170
1966	93,200	10,820	4,499	1,193	324	425	1,129	118,998
1967	98,200	10,753	4,490	1,053	336	447	1,130	123,724
1968	100,200	11,525	4,469	963	333	437	1,523	127,226
1969	106,300	13,582	4,633	1,000	361	428	1,675	135,006
1970	100,300	13,467	4,499	1,236	361	382	1,509	128,905
1971	111,400	13,173	4,364	1,067	369	404	1,570	139,788
1972	120,700	11,972	4,397	1,262	361	428	1,566	148,175
1973	117,700	13,854	4,465	1,361	348	456	1,375	146,505
1974	90,300	12,552	4,693	1,502	346	339	1,325	117,872
1975	100,600	15,613	4,933	1,477	321	400	1,860	131,640
1976	100,400	14,056	4,797	1,038	297	374	1,760	129,180
1977[1]	103,800	16,277	4,970	1,223	282	403	1,500	128,455
1978[1]	112,000	17,400	–[2]	–	–	–	–	–

[1]Preliminary.

[2]Not available as yet.

Source: U.S. Department of Agriculture, *Agricultural Statistics, 1979,* United States Government Printing Office, Washington, 1979.

Market for Coproducts

Table 5-9 shows the market for selected animal feeds and the total U.S. market for the years 1963 to 1976. The total market includes oilseed, animal protein, and other mill products. As a point of reference, a 50-million-gallon ethanol plant will produce about 177,000 tons of DDG per year, i.e., about 45 percent of the 1976 market for that commodity. A 600-million-gallon ethanol program—a near-term objective in the United States—will produce almost 2 million tons of DDG (or the equivalent) or about 7 percent of the total 1976 feedmarket and about five times the amount of DDG sold in the United States in 1976. In the 1963 to 1976 period, the domestic market for all animal feeds increased at an average rate of about 1.2 percent annually. The market for soybean meal (with which DDG is more directly competing) expanded at a rate of about 3.5 percent over that period. This expansion is not quite sufficient to absorb the expected expansion in production of DDG as a result of ethanol-fuel production. Some substitution between DDG and soybean may take place. As a result, some readjustment of agricultural production patterns will take place as the demand for corn feedstock increases.

Exports of corn and soybeans have expanded at an average rate of about 11 and 9 percent, respectively, over the 1960-1977 period. Part of these exports are used for animal feed; this foreign market could absorb some coproducts such as DDG if the buyers are willing to substitute DDG for the feeds presently used. U.S. exports amounted to about 47 and 17 million tons of corn and soybeans, respectively, in 1976. The potential of the foreign markets should be investigated seriously by ethanol producers.

Geographic Distribution of the Market

As was discussed in Chapter 4, the major markets for DDG and DDGS are cattle and milk cows where DDG and DDGS are a protein supplement. Other farm stock such as pigs and chickens provide a significantly smaller market. Figure 5-3 shows the distribution of the cattle population and large feedlots (larger than 4,000 heads) by regions. Over 60 percent of the potential market is located in the Northern and Southern Plains, Mountain, and Cornbelt regions. Over 80 percent of the large feedlots are located in the Mountain and Northern and Southern Plains regions. As a point of reference, a 50-million-gallon-per-year ethanol plant produces enough DDG and DDGS to feed about 120,000 steers or about thirty 4,000-head feedlots. As an example, the total cattle population of the Northern Plains region in 1978,

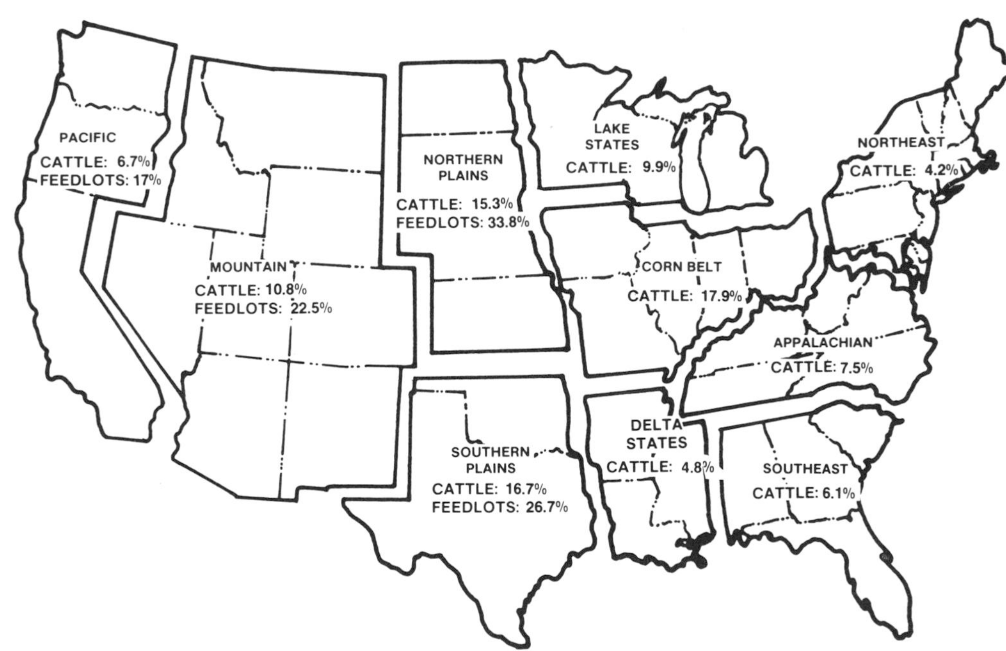

CATTLE: Cattle and milk cows — Total Population: About 125 million (1978). Number is percentage of total population in the region.

FEEDLOTS: Percentage of the number of large feedlots (larger than 4,000 head) in the region.
Total feedlot number: 551 - Animals marketed in 1976: about 13 million

Source (cattle): U.S. Department of Agriculture, "Agricultural Statistics 1979", United States Government Printing Office, Washington, D.C. 1979.

Source (feedlots): Schooley, F., et al., "Mission Analysis for the Federal Fuels for Biomass Program", SRI International, Menlo Park, Cal., Dec. 1978.

Figure 5-3. Distribution of cattle and milk cow population and large feedlots by regions.

i.e., about 18 million head, could have absorbed the DDG and DDGS production of about 120 50-million-gallon-per-year plants if DDG and DDGS were the only sources of feed. The impact of such an approach on other sources of feed would, however, be quite dramatic.

The data presented indicate that a large potential market for DDG is available, a large fraction of the market is in the major grain-producing or neighboring regions, and a large national ethanol program may induce significant changes in the production patterns of feed-related agricultural products.

Price of Coproducts

Table 5-10 shows the historical trends in the prices of corn, stillage, DDG, and soybean meal. The discussion relating to the data indicates the interdependence of the prices of these commodities as well as the range of price fluctuations to be expected.

The price of the coproducts will be influenced by factors such as processing and transportation. Processing costs will be addressed in a later section. Transportation costs are an important item because of the relative geographical distribution of the markets for ethanol and its coproducts and of the feedstock supply regions.

Figure 5-4 compares shipping rates for the three commodities involved in a marketing/production effort: grains, feed, and ethanol. Rates are expressed per gallon of ethanol and therefore estimate the impact of market location versus production location directly on the price of ethanol delivered. It must be stressed that the shipping rates shown in the graph are susceptible to significant variance among states. The data show that proximity of the

Table 5-10. Wholesale Prices in Chicago (Dollars Per Ton, Bulk)

Year	January 1			July 1		
	#2 yellow corn	Soybean meal	Distiller's grains	#2 yellow corn	Soybean meal	Distiller's grains
1979	80.00	194.60	141.00	108.60	218.30	151.10
1978	77.90	182.60	130.00	85.70	186.10	120.00
1977	88.60	209.70	140.00	78.60	184.70	140.00
1976	92.10	136.00	105.00	108.90	232.70	130.00
1975	118.70	139.80	107.50	100.30	124.30	103.00
1974	97.10	174.80	138.00	109.30	107.80	92.50
1973	55.00	189.10	100.00	85.00	306.60	138.00
1972	42.90	87.50	64.00	45.00	106.20	71.00
1971	56.40	85.90	70.00	54.30	88.00	67.00
1970	44.00	84.50	64.00	49.60	85.70	61.00

Source: *Feed Market News,* USDA, AMS.

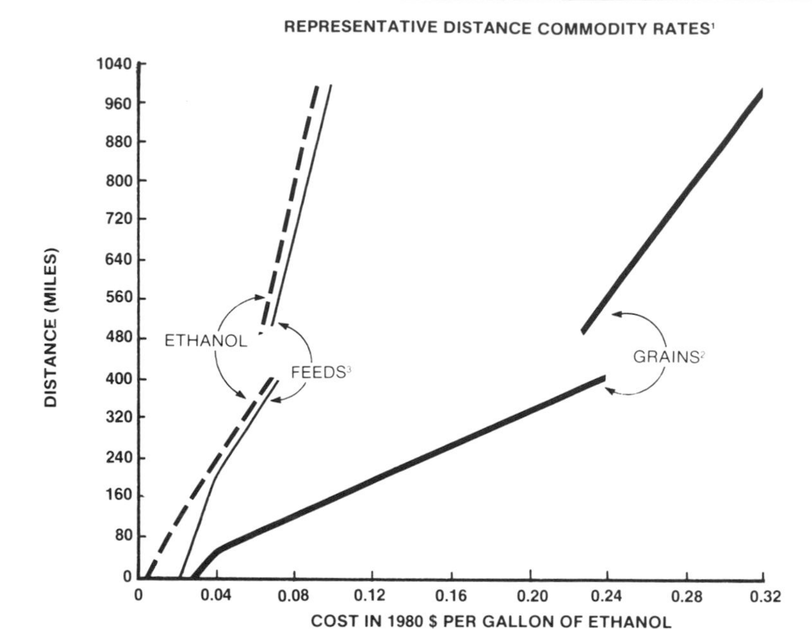

[1]Rates for 400 miles or less are for 45,000 lb. hauling trucks.
Rates for more than 400 miles are for railroad carloads.
[2]Assumes 56 lb. per bushel, 2.5 gallons ethanol per bushel.
[3] Assumes 6.6 lb. DDG per gallon ethanol

Source: *A Guide to Commercial Scale Ethanol Production and Financing*, SERI, Denver, 1980.

Figure 5-4. Representative distance commodity rates (in 1980 dollars per gallon of ethanol).

feedstock supply is an important factor, i.e., hauling the grain rather than the products is less favorable, especially where hauling distances reach 100 miles and over. For distances less than 100 miles, it is preferable to look for proximity of the feed market rather than proximity of the ethanol market. Above that distance, the shipping costs of both commodities are comparable.

As an example, assume that the ethanol market for a plant is the East Coast of the United States, about 500 miles away from the feedstock supply, and that the DDG will be shipped to European markets. A plant located on the East Coast, at a terminal by a shipping harbor, would incur grain shipping costs of $0.232 per gallon of ethanol. A plant located in the feedstock region would incur shipping costs of only $0.135 to deliver the ethanol and DDG to their markets. Siting of the plant therefore must be carefully evaluated, as it may have a major impact on overall plant economics.

Value of Distiller's Coproducts

Historical prices for distiller's dried grains exist, for they have been used for many years by commercial feed manufacturers. Table 5-10 shows January 1 and July 1 prices of corn, soybean meal, and distiller's grains at Chicago over a 10-year period.

A linear regression analysis of the Chicago data yielded the following relationship of distiller's grains price to the price of corn and soybean meal:

Distiller's grain price = 16.28 + 0.439 (corn price) + 0.356 (soybean meal price)

All prices are in dollars per ton. The standard error of estimate for the three constants are 16.28 ± 10.5, 0.439 ± 0.14, and 0.356 ± 0.06. Figure 5-5 relates distiller's grains and corn and soybean meal prices based on this relationship.

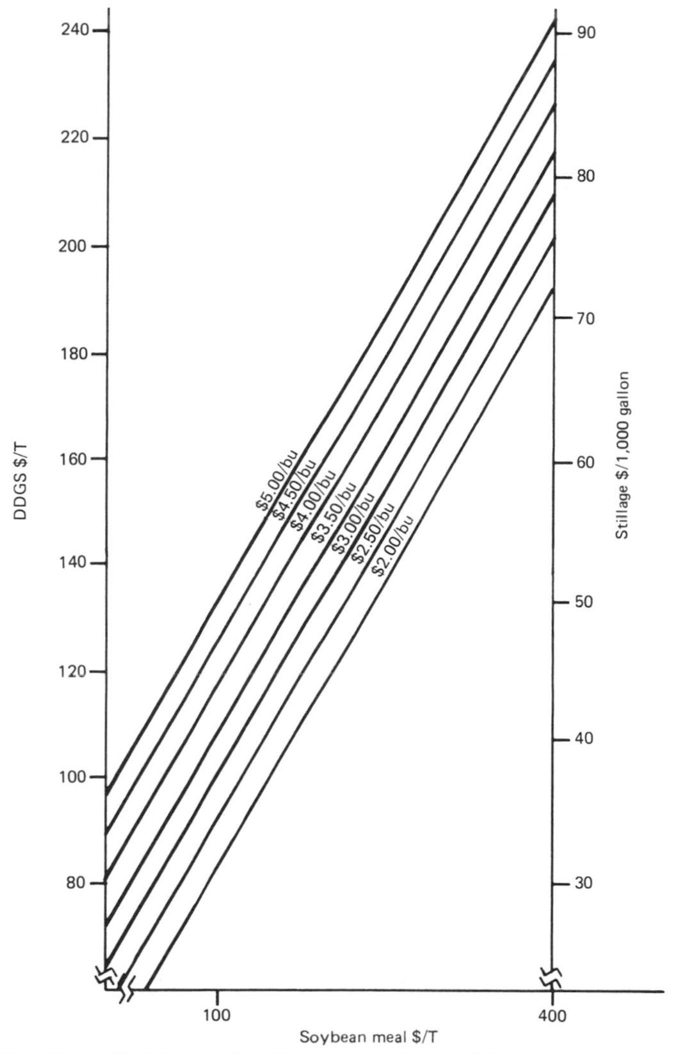

Source: USDA, Small-Scale Fuel Alcohol Production, March 1980.

Figure 5-5. Price relationships of DDGS and stillage to corn and soybean meal (based on Chicago historical prices).

Stillage prices have not been reported and published. Some beverage distiller's sold stillage at prices which are probably just about sufficient to cover the cost of handling and which would encourage farmers to use the product on a regular basis.

Owing to the constraints of storage and palatability noted earlier (if stillage is not kept constantly in the rations of high-production animals), small farm stills may have to dispose of the coproduct by spreading it on land or feeding it, when available, to low-production animals such as dry cows. Under these circumstances, the value of the coproduct may be nominal.

When a plant becomes large enough to distill every day, then the constant supply of stillage provides the opportunity to take advantage of its full nutritional value. If it can be delivered and fed every day with a holding time not exceeding 24 hours, then little nutrient loss will occur, and including it in the ration every day should eliminate its palatability problem.

The nutrient composition of stillage has not been extensively reported. For this handbook it was assumed that for a given amount of solid material the composition of stillage will be the same as DDGS. Table 5-11 compares some important properties of DDGS with corn and soybean meal.

Table 5-11 shows energy to be about equal for all three ingredients. DDGS has a lower protein content than soybean meal but would be considered a protein feed rather than an energy feed such as corn. The major deficiency of DDGS is its very low lysine content, but pure lysine, commercially available at rather reasonable cost, can overcome this deficiency. The relatively high fiber content of DDGS will limit somewhat its use in

Table 5-11. Average Nutrient Composition of DDGS, Corn, and Soybean Meal

Nutrient	DDGS	Corn	Soybean meal
Dry matter (percent)	93	89	89
Digestible energy (kcal/kg)	3568	3525	3350
Protein (percent)	27.2	8.8	44.0
Lysine	0.6	0.24	2.93
Methionine	0.6	0.2	0.7
Cystine	0.3	0.2	0.7
Crude fiber (percent)	9.1	2.2	7.3
Calcium (percent)	0.35	0.02	0.29
Phosphorous (percent)	0.95	0.28	0.65
Niacin (mg/kg)	80	34	60
Pantothenic acid (mg/kg)	11.0	7.5	13.3
Riboflavin (mg/kg)	8.6	1.0	2.9

Source: *Nutrient Requirements of Swine,* National Research Council, 1979.

poultry broiler and layer rations where very high energy levels are used to achieve maximum performance. The level of major minerals, calcium and phosphorus, is higher in DDGS than in soybean meal. Limiting and important B-vitamin levels in DDGS compare favorably with soybean meal.

Ration Formulation and Use of Coproducts

National Research Council (NRC) nutrient requirements were used to formulate rations for several important animal classes:

- Beef steers of about 550 pounds
- Beef steers of about 750 pounds
- Dairy cows
- Swine on a growing ration
- Pullets on a developer ration (only DDGS was considered)

The ration formulation used a least-cost linear program to determine the maximum price at which DDGS would replace soybean meal as the protein source. A total liquid constraint was imposed which would limit the moisture content of the ration to 4 pounds of water per pound of dry feed (solids). Nutrient factors for other constraints were selected to include those now being used by a medium-sized Kansas commercial feed mill (Blair Million Co. of Atchison, Kansas). The same feedmill furnished current prices for all ingredients except roughage, corn, soybean meal, and DDGS.

Selection of prices for roughages and corn was made based on 1979 farm prices. The price of soybean meal was taken as 50 percent higher than the current Kansas City bulk wholesale carlot price. Large farmers buying carloads of soybean meal would pay less than 50 percent over wholesale prices. Smaller farmers who purchase commercial protein supplements would pay about 50 percent over wholesale for the soybean meal component of the protein supplement. Other ingredients required to balance the ration were similarly priced at 50 percent over wholesale. Table 5-12 shows the prices used for all ingredients except DDGS.

After the linear programming model had established the level and price for DDGS and the major ingredients, the formula was checked against total nutrient requirements and a premix was formulated to supply the required levels of other ingredients such as vitamins A and D and salt. The retail price of this premix was calculated for each ration.

Table 5-13 shows the nutritional constraints used in the linear program for each ration formulated. Selection of the type of energy constraint was arbitrary and based on the practice of the Kansas feed mill. Urea was allowed only in the steer ration, although some urea might be allowed in calf and dairy rations by some nutritionists. Stillage was not thought to be appropriate in the pullet rations because of handling and mixing problems. Ingredients con-

Table 5-12. Ingredient Prices Used for Formulation[a]

Ingredient	$/kg	$/2,000 lb	Other
Brome hay	0.055	50	
Corn stover	0.0275	25	
Corn	0.0982	89	$2.50/bu
Milo	0.0869	79	$3.95/cwt
Soybean meal	0.3135	285	
Meat and bone meal	0.396	360	
17 percent dehydrated alfalfa	0.181	164	
Urea	0.280	254	
Calcium	0.0363	33	
Dicalcium phosphate	0.3383	307	
Salt	0.0825	75	
Trace mineral mix	1.30		
Lysine	6.60		$3/lb
DL-Methionine	5.35		$2.43/lb
Vitamin mix	2.20		$1/lb

[a]Nonfarm ingredients priced at 50 percent over wholesale to reflect manufacturing and marketing costs of typical feed supplements.

Source: USDA, *Small-Scale Fuel Alcohol Production,* U.S. Government Printing Office, Washington, DC, 1980.

sidered in each formulation were those considered as appropriate and likely to be used in a practical operating situation in the midwest.

Table 5-14 shows the rations which were formulated for the ruminants including those components which would be incorporated in a supplement to be used with the farm-supplied ingredients to meet the NRC nutritional requirements. The steer ration which allowed some nonprotein nitrogen (urea) to enter the ration does not need a natural protein supplement; hence, stillage can only be forced into the ration by lowering the price until it becomes competitive with corn as an energy source.

The replacement pullet ration uses some soybean meal but could be forced out entirely by lowering the price of DDGS even further.

The swine ration was stillage for all of the protein supplementation. Lysine supplementation is required, but the cost is not great. Because of the free fatty acid components of corn oil, high levels of DDGS or stillage in swine finishing rations would cause the same soft pork problem as has been observed with peanut meal supplements. It may be necessary to reduce the level of stillage in the later finishing phases, but the exact time and level will have to be determined from feeding trials. These formulas were restricted to the prefinishing phase.

Table 5-13. Nutrient Constraints Used in Least-Cost Formulation

Item	Units per ton dry solids	Ration				
		Calf[a]	Steer[b]	Dairy[c]	Swine[d]	Pullet[e]
TDN[f]	kg	>720	>770			
Net energy	Mcal			>1520		
Metabolizable energy	Mcal					>2900
Digestive energy	Mcal				>3380	
Fiber	kg			>170		
Protein	kg	110–130	110–130	140–160	>160	
NPN protein[g]	kg	0	<15	0	0	0
Lysine	kg				>7	>11
Methionine	kg				>2.3	>4
Methionine + Cystine	kg				>4.5	>7.5
Thiamine	g				>1100	
Riboflavin	g				>2600	
Pantothenic acid	g					>10,000
Niacin	g					>27,000
Calcium	kg	3.5–4.5	2.9–3.9	4.8–5.8	6–8	8–18
Phosphorous	kg	>3.1	>2.6	>3.4	>5	>4
Water		<4000	<4000	<4000	<4000	Used DDGS

[a] 250 kg calf graining 0.9 kg/day.
[b] 350 kg steer graining 1.3 kg/day.
[c] 600 kg cow producing 23 kg/day.
[d] 20–35 kg hog gaining 600 g/day.
[e] 0–6-week-old replacement pullets.
[f] TDN = total digestible nutrients.
[g] NPN = nonprotein nitrogen.

Source: USDA, *Small-Scale Fuel Alcohol Production,* U.S. Government Printing Office, Washington, D.C., 1980.

Potential for Utilization of Stillage

The potential for stillage usage is very site specific. Transportation costs will be a major factor in assigning a value at the plant since the value at the farm is the one calculated in the preceding formulation analysis. If the plant is owned by the farmer, then no marketing costs would be incurred. If stillage incurred marketing costs, other than transportation, then the effective price is further reduced. Marketing and processing costs, excluding transportation, are generally about 33 percent of the retail price for manufactured feeds. Retailer margins are, of course, less. A marketing margin of 10 percent should be reasonable for a plant which sold stillage at retail.

Transportation costs will be a major factor in determining the radius over which stillage might be utilized economically.

Table 5-14. Composition of Rations Formulated and Price of Stillage or DDGS Quantity of Ingredient Per Metric Ton of Dry Solids Fed

Ingredient	Calf	Calf	Steer	Dairy	Swine	Pullet
Brome hay	732		360	432		
Corn stover (kg)		779				
Corn (kg)		121	263	471	723	
Stillage (kg)	4267	1833	4375	1786	3273	
Stillage (gal)	1131	485	1159	473	867	
DDGS (kg, equivalent)	474	203	486	198	364	965
Soybean meal (kg)						21
Dicalcium phosphate (kg)		7				
Limestone (kg)			2	9		12
Salt (kg)	2.5	2.5	2.5	2.5	2.5	2.5
Trace minerals (kg)	0.55	0.55	0.55	0.55	0.56	0.55
Lysine (kg)					4.23	4.67
Choline chloride (kg)					0.25	
Vitamin premix (kg)	0.55	0.55	0.55	0.55	0.25	1.39
Stillage price ($/kg)	0.174	0.022	0.010	0.021	0.016	—
Stillage price ($/gal)	0.066	0.083	0.04	0.079	0.060	
DDGS price ($/kg)	0.174	0.22	0.10	0.21	0.16	0.11
($/2000 lb)	158	200	110	191	145	100

Source: USDA, *Small-Scale Fuel Alcohol Production,* U.S. Government Printing Office, Washington, D.C., 1980.

Table 5-15. Transportation Costs for a Typical Feed Mill Operating 20-Ton Capacity Trucks in 1978

Cost item	$/mile	$/mile
Fixed expenses		0.11
Labor		0.50
Variables		
Maintenance and repair	0.14	
Tires	0.03	
Fuel, oil, grease	0.12	
Other	0.04	
Total variables		0.33
Total		0.94

Source: USDA, *Small-Scale Fuel Alcohol Production,* U.S. Government Printing Office, Washington, D.C., 1980.

Stillage would have to be delivered by tank truck, and the quantities would be variable depending on the size of customer. Loading and unloading times would be similar to that of other feed. Truck operating costs were obtained from two large feed manufacturers.

Table 5-15 shows major costs incurred by one feed manufacturer for a small fleet of trucks operating in Alabama in 1978. The miles per trip would be greater than for stillage since dry feed was delivered. Another midwest feed manufacturer estimates current truck costs at $0.95 to $1.00 per mile.

If truck operating costs are $1.00 per mile and a 20-ton (4,800-gal) load is assumed, then the cost per mile per gallon of stillage is $0.0002 per mile per gallon or $0.0004 per delivery mile (the truck must travel out and back). If stillage has a value at the farm of $0.075 per gallon, then the plant value drops to zero for a 188-mile delivery radius. (For a delivery distance of 18 miles, the value of the stillage is reduced by 10 percent to $0.068.) If the marketing cost, exclusive of transportation, is 10 percent, then the maximum delivery radius is reduced to 170 miles.

Stillage Drying

Although the drying of stillage to DDGS requires large amounts of energy, it is necessary for large plants which have little chance to dispose of large amounts of stillage within a reasonable radius.

Small plants which might consider using relatively simple drum dryers, such as those used in alfalfa dehydration, would use about 1,400 Btu per pound of water evaporated. If stillage were 10 percent solids, this would require about 75,000 Btu per gallon of alcohol. Obviously, this would be generally infeasible.

Larger plants, in the range of upward from 1 million gallons per year, can justify the investment in more efficient drying techniques such as vapor recompression evaporators. Their energy efficiency is improved sufficiently to allow dehydration to be feasible.

Coproduct Consumption and Animal Populations

The amount of marketable coproducts, particularly stillage, will depend on the local numbers and mix of animals. As shown in Table 5-14, ruminants, particularly calves and dairy cows, represent the most attractive outlet. Exact quantities used depend on numbers of animals and their production levels. Larger amounts of stillage could be used in some cases by using stillage as an energy rather than a protein source, but its value will drop rapidly to compete with grain rather than protein sources such as soybean meal.

Some typical quantities of coproducts which might be used by various kinds of animals are shown in Table 5-16.

Table 5-16. Animal Consumption of Distiller's Coproducts

Type of animal	Production rate	Feed			Stillage, gal/day	
		kg feed/day	lb feed/day	DDGS (lb/day)	10 percent solids	20 percent solids
550-lb calf (hay ration)	2 lb/day, gain	6.2	13.7	5.8	6.3	3.1
770-lb steer	2.8 lb/day, gain	8.8	19.4	8.5	9.2	4.6
1300-lb cow	50 lb/day, 3.5 percent milk	16.8	37.0	6.6	7.2	3.6
60-lb pig	1.3 lb/day, gain	1.5	3.3	1.1	1.2	0.6
Pullet, age 3.7 weeks		0.057	0.13	0.13	NA[a]	NA
Pullet, age 7.5 weeks		0.1	0.22	0.22	NA	NA

[a]NA = not applicable

Source: USDA, *Small-Scale Fuel Alcohol Production,* U.S. Government Printing Office, Washington, D.C. 1980.

Table 5-17. Number of Animals Required to Use Stillage[a]

Type of animal	Size of still	
	60,000 gal/yr	360,000 gal/yr
550-lb calf	230[b]	1365[b]
770-lb steer	155	931
1300-lb cow	200	1200
60-lb pig	1,200	7200

[a]Based on 300 days/year production.

[b]Number of animals.

Source: USDA, *Small-Scale Fuel Alcohol Production,* U.S. Government Printing Office, Washington, D.C., 1980.

Table 5-17 shows the number of animals which would be required to consume the stillage produced by 60,000- and 360,000-gallon-per-year stills. The number of animals required for the larger still, those ranging from 7200 pigs of 60-pound weight to 931 steers of 770-pound weight, do not appear to present great problems for stillage disposal. Many large cattle feedlots could even use the stillage from a 1-million-gallon-per-year still.

Some Unique Problems of Stillage Usage

Stillage will produce rations of unique physical form, odor, and palatability. High-production animals must be kept constantly on such rations to maintain the feed consumption levels required. Frequent interruption of supply would drastically reduce the value of the stillage; consequently, if frequent interruptions occur, perhaps due to shut down for maintenance, then it might be necessary to maintain refrigerated supplies for emergency use.

Rapid loss of nutrients due to microbial action will require that stillage holding time be limited; it should be fed within about 24 hours during summer months.

Handling and feeding stillage during very cold weather will present some new and unique problems because of freezing. Twice-a-day feeding will be necessary under many conditions where feed bunks are located outside in northern regions.

Feed bunks will have to be capable of holding feeds containing a large amount of liquid. Concrete bunks would probably be adequate with some sealing at joints. The dry feed could be placed in the trough and then the stillage placed on top. It may be possible to mix roughage and stillage in the normal mixer-feeder wagons.

Slurry feeding equipment has been developed for swine feeding in Europe and should be available with little modification.

Stillage and wet distiller's grains have not been defined by the American Feed Control Officials. This will be necessary before they can be widely marketed.

One potential use of condensed solubles was considered but not investigated. Liquid feed supplements for beef are becoming increasingly popular. Usually they are molasses-urea mixtures, but some natural protein would be preferred by many nutritionists. A few feed manufacturers are using substantial quantities of condensed solubles in their liquid feed formulations.

Other Coproducts

The foregoing discussion focused on grain distiller's coproducts. Although certain characteristics of the coproducts produced from other feedstock differ, the analysis of their usage would be analogous. Distinct properties of potato and sugar crop coproducts are delineated in the following material.

Potato dry coproduct. The coproduct of potatoes should contain a higher ash content by a factor of at least 5 as compared to that of corn distiller's grains. The protein content should be slightly less and the fiber content about equal to those for corn distiller's grains. Amino acid data are not readily available. B-vitamin composition should be about the same since this originates from the yeast. Table 5-18 shows an estimated composition. Because of the lower protein, the material would have a lower value than DDGS.

Sugar beet coproduct. Sugar beet pulp is a well-defined commercial feed ingredient which has been priced in Chicago at $130 per ton compared to $145 per ton for distiller's dried grains. If the solubles were condensed, the price should be about the same as brewer's dried yeast which is priced at about $400 per ton basis Chicago; hence, the overall value of these coproducts should be greater than that for grain coproducts.

Sugarcane coproducts. The plant residue from sugar manufacture (bagasse) is of relatively low value; some is used for the manufacture of

Table 5-18. Estimated Composition of Potato Distiller's Dried Coproduct

Element	Composition (percent)
Water	10
Ash	9
Crude fiber	5
Ether extract	1
Nitrogen-free extract	53
Crude protein	22

Source: USDA, *Small-Scale Fuel Alcohol Production,* U.S. Government Printing Office, Washington, D.C., 1980.

building board, but most is used for boiler fuel. Its yeast residue could be recovered and should have the same value as other dried yeast.

Sorghum coproducts. The plant residue may have a slightly higher feeding value than sugarcane but no nutritional data have been found. The yeast coproduct would be valuable.

CARBON DIOXIDE

Carbon dioxide is produced in about equal weights with alcohol during the fermentation process. One pound of carbon dioxide has a volume of 8.1 cubic feet per pound. A million-gallon-per-year still would, therefore, produce about 21,000 pounds of carbon dioxide per day having a volume of 170,000 cubic feet.

The gas from the fermentors would be relatively pure carbon dioxide (and water vapor) if the fermentors were relatively tight and would be suitable for many uses. Keeping a slight positive pressure inside the fermentor would exclude air.

As the gas is removed from the fermentor, it could be dried, compressed for storage, and used as a gas or frozen to form dry ice.

Carbon dioxide is frequently used to preserve the quality, including color, of many agricultural products. Fresh fruits and vegetables are preserved in improved condition in limited oxygen atmospheres. The color of meat can be maintained in an inert gas atmosphere.

In the past, many beverage distilleries produced dry ice, but the advent of mechanical refrigeration equipment for transport equipment reduced the market for the product.

Today, most distillers vent the carbon dioxide to the atmosphere because it is uneconomical to process it. The use of carbon dioxide would be very site specific. The value of the product would probably be about equal to the costs of recovery. Prices of $3 to $5 per ton have been reported for uncleaned, uncompressed, raw gas. In most cases, the problem in marketing the carbon dioxide relates to plant size and cost of handling. Many banks will not give value for this coproduct, unless the plant is very large.

Relationship between the Ethanol and Coproduct Markets and the Feedstock Supplies

Figure 5-6 summarizes the data concerning the respective locations of the ethanol and coproduct markets and the feedstock supplies.

The major grain feedstock regions are the Northern Plains and Cornbelt. The major market regions for DDG are the same, as well as regions to the north and west. Major markets for ethanol are the East and West Coast regions. In view of the discussion on the relative transportation costs of the commodities involved, siting a plant in or close to the grain-producing regions gives direct access to the grain supply and coproduct markets and limits the marketing problem to that of distribution of ethanol in the major market areas. As was discussed, blending points for gasohol are available in all potential market areas for ethanol.

As was discussed in Chapter 4, some local feedstock resources exist in other areas such as potatoes in the Mountain region. Figure 5-6 suggests that a plant located in Idaho, close to a potato or potato waste supply, would have good access to both ethanol and coproduct markets.

The conclusions derived from the data in Fig. 5-6 are site specific and will have to be refined to account for specific factors such as water and auxiliary fuel resources.

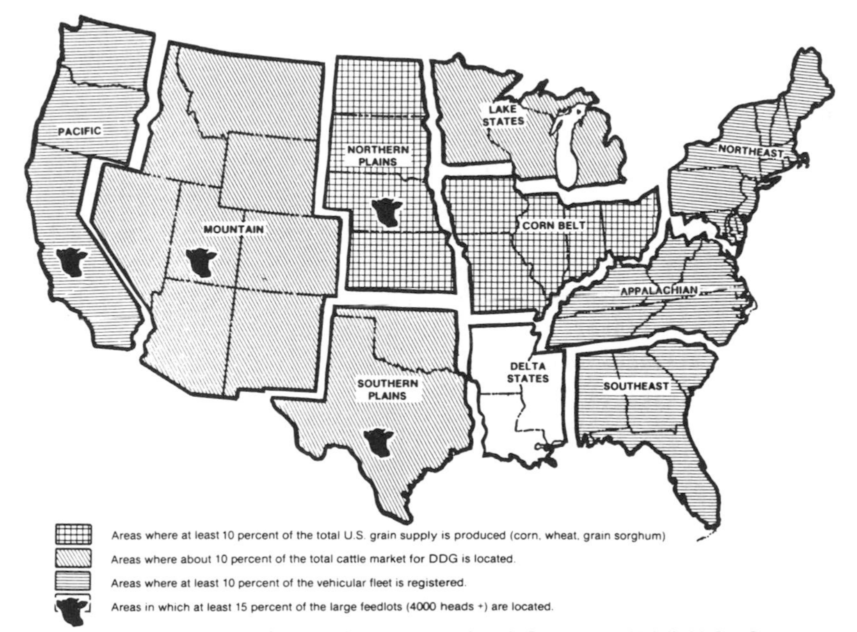

Source: U.S. Department of Agriculture "Agricultural Statistics 1979," U.S. Government Printing Office, Washington, D.C., 1979. U.S. Department of Commerce "Statistical Abstract of the United States 1978", 99th Annual Edition, Washington, D.C., 1978.

Figure 5-6. Location of major grain feedstock supplies and potential markets for ethanol and coproducts.

Chapter 6

Economics of Ethanol Production

This chapter presents the estimated costs of ethanol production; it brings together feedstock and coproduct prices and operating, maintenance, and capital costs into an estimated production cost for the various types of ethanol plants discussed in Chapter 3. It should be recognized that, while these cost estimates are based on the best information now available, there is not sufficient operating experience with small-scale ethanol plants from which actual operating characteristics and costs can reliably be derived. As operating experience and records become available, these cost estimates can be refined.

The chapter first establishes a set of base conditions which reflect the current best estimates of plant-operating characteristics and costs. Then, sensitivity analyses are presented to show the effect of such different operating characteristics and cost levels. This type of analysis not only dramatically shows the important cost elements, but also provides a basis for users of this book to adjust the base case results to their particular situations.

In making the base cost estimates, input values were determined by estimating their potential worth to investors. For example, for the small on-farm pot still, it was assumed that land and certain grain-handling and -grinding equipment were already owned by the farmer. Corn was valued at the long-term price received by farmers on the basis that it could alternatively be sold rather than processed into ethanol. Labor was valued as if it were hired. However, a specific entrepreneur (particularly at the farm level) may have a unique circumstance regarding the availability of land or grain-handling equipment or may perceive the value of the corn or labor differently. In such instances, the sensitivity analysis can be used to estimate these potential input values for specific situations.

The latter part of this chapter examines these estimated costs in relation to ethanol-fuel values for selected applications.

Cost Elements

The economic cost elements can be broadly arranged into four major groups—feedstock costs, coproduct credits, operating costs, and capital costs. Within each of these are subelements involving input-output rates and prices. The following presents the values used to make the base cost estimates.

Because of the scarcity of information on plants using nongrain feedstocks, the cost estimates presented herein are based on grain. Although all grains can be used as a feedstock, corn currently commands the most interest and is used as the basis of estimating feedstock cost.

FEEDSTOCK

Corn prices vary by location and reflect local supply-demand conditions, transportation facilities, and markets, as discussed in Chapter 4. For purposes of these cost estimates, the long-term corn price received by farmers was used for the on-farm plants on the assumption that this price reflects farm-level prices. For the large community plants, a higher price reflecting the long-term average corn price at major markets was used to reflect, in turn, the additional transportation and competitive pressures which a larger plant would face. Specifically, a farm price of $2.50 per bushel (1979 dollars) and an off-farm price of $2.75 per bushel (1979 dollars) were used. It is noted that the average farm price of grain sorghum is about $2.25 per bushel (1979 dollars) and the major market price is $2.60-$2.90 per bushel (1979 dollars) for Kansas City and Fort Worth, respectively.

In addition to differing price levels, various theoretical ethanol yield rates were used. The base case estimates were based on the following yields:

Type	*Proof*	*Yield (gal/bu)*
Pot	190	2.4
Small on-farm	190	2.5
Large on-farm	190	2.5
Small community, wet	200	2.5
Small community, DDGS	200	2.5
Large community, DDGS	200	2.5

To date, several operators have reported ethanol yields, and these have been lower than those shown above. However, under proper operation, particularly with regard to cooking, these yields are believed to be reasonably achievable. The pot still yield was reduced below the other on-farm units to reflect less than optimal operating conditions for a very small unit of this type.

COPRODUCT CREDITS

As discussed in Chapter 5, the feeding value of wet stillage appears to vary by class of livestock and ranges from about $110 to $200 per ton of DDGS equivalent. For cost estimating purposes, a single value was selected as representative of each model plant's stillage and marketing situations.

The pot still was estimated to yield 21.4 gallons per bushel of wet stillage with 9 percent dry matter. The other two on-farm and small community stills were estimated to yield 15.7 gallons per bushel of wet stillage composed of 12.3 percent dry matter. The two community stills producing DDGS were estimated to produce 17.8 pounds of DDGS at 10 percent moisture.

Because the small pot still would not produce a daily supply of wet stillage, it was assumed that the stillage would be fed to low production livestock and would have a value equivalent to hay. The other stills analyzed were based on continuous operations so that a daily stillage supply would be available. The long-term (15-year average) DDGS wholesale price at Cincinnati has been $130 per ton (1979 dollars). As prices paid by farmers for DDGS are not reported, the difference in wholesale and prices paid by farmers for soybean meal was used for estimating purposes. On a long-term basis, the prices paid by farmers for soybean meal has been about 28 percent above its Decatur wholesale price. This suggests that prices for on-farm or near-farm use of wet stillage should be higher than the equivalent wholesale price for DDGS.

Because the small pot still would not produce a daily supply of wet stillage, it was assumed that the stillage would be fed to low-production livestock and would have a value equivalent to hay. The other stills analyzed were based on continuous operations so that a daily stillage supply would be available. The long-term (15-year average) DDGS wholesale price at Cincinnati has been $130 per ton (1979 dollars). Prices paid by farmers for DDGS are not reported. On a long-term basis, this price has been about 28 percent above the Decatur wholesale price. This suggests that prices for on- or near-farm use of wet stillage should be higher than the equivalent wholesale price for DDGS.

The wet stillage material will require more effort for handling and feeding than does soybean meal. For purposes of these estimates, DDGS prices equivalent to $153 per ton were used on the basis that the material would be used on the farm. The large on-farm and small community, wet stills may or may not have on-site stillage feeding potentials. Consequently, a lower price—$135 per ton—was used. DDGS from the two stills drying stillage was priced at $130 per ton. This is equivalent to the long-term wholesale price and was based on the assumption that these units would be competing in the national market for DDGS sales.

Because wet stillage is sometimes priced on a per gallon basis, the prices and conversion factors are summarized in Table 6-1.

OPERATING AND MAINTENANCE COSTS

The operating and maintenance costs include labor, energy, electricity, other supplies (enzymes, yeasts, etc.), repairs and maintenance, taxes, insurance, and general and administrative expense.

Table 6-1. Stillage Prices and Conversions

Type of still	DDGS price			Stillage per bushel		Stillage price ($/gal)
	10 percent moisture ($/ton)	Dry ($/ton)	Dry ($/lb)	Solids (lb)	Material	
Pot still	67	74	0.037	16	21.4[a]	0.028[d]
Small on-farm	153	170	0.085	16	15.7[b]	0.087[e]
Large on-farm	135	150	0.075	16	15.7[b]	0.076[f]
Small community, wet	135	150	0.075	16	15.7[b]	0.076[f]
Small community, DDGS	130	144	0.072	16	17.8[c]	–
Large community, DDGS	130	144	0.072	16	17.8[c]	–

[a] $\text{gallons} = \dfrac{16 \text{ lb solids}}{(9 \text{ percent solids})} \Big/ 8.3 \text{ lb per gal.}$

[b] $\text{gallons} = \dfrac{16 \text{ lb solids}}{12.3 \text{ percent solids}} \Big/ 8.3 \text{ lbs per gal.}$

[c] $\text{pounds} = \dfrac{16 \text{ lb solids}}{90 \text{ percent solids}}$

[d] $\$/\text{gal} = \dfrac{16 \text{ lb solids} \times \$0.037/\text{lb}}{21.4 \text{ gal}}$

[e] $\$/\text{gal} = \dfrac{16 \text{ lb solids} \times \$0.085/\text{lb}}{15.7 \text{ gal}}$

[f] $\$/\text{gal} = \dfrac{16 \text{ lb solids} \times \$0.075/\text{lb}}{15.7 \text{ gal}}$

Source: USDA, *Small-Scale Fuel Alcohol Production,* U.S. Government Printing Office, Washington, D.C., 1980.

CAPITAL COSTS

The estimate of annual ethanol capital costs is not as straightforward as feedstocks, coproduct credits, and operating and maintenance costs. These latter costs are expended annually, whereas capital costs must reflect investment costs that are capitalized (amortized) and recovered over the life of the plant.

A common approach is to annualize capital costs as the sum of straight line depreciation and an interest charge on average investment (investment divided by two). This method to convert capital expenditures into an equivalent annual cost, as well as other approaches, fails to reflect key financial parameters including the depreciation method, capital structure, cost of debt, cost of equity, investment tax credits, federal and state income taxes, and inflation rates. The inclusion of all of these factors in a cost estimate is not straightforward and involves considerable computational effort.

A computer model was selected which analyzed discounted cash flow (sometimes called a life-cycle cost), reflected all of these factors, and provided a capital cost recovery rate in terms of an average real cost in 1979 dollars.

The factors and resulting annual capital cost per gallon for the base case estimate are reported in Table 6-2. A multiplication of the appropriate factor times the total investment, including working capital, gives the equivalent annual capital cost. Dividing this annual cost by the gallons of ethanol production gives the per-gallon cost comparable to the other cost elements. It should be noted that a different capital recovery factor would be associated with any set of these factors that differs from those shown in Table 6-3, that is, with different feedstock costs, coproduct credits, and operating and maintenance costs.

As a point of reference, the coefficients used in estimating small-scale ethanol production are summarized in Table 6-3.

Estimated Costs of Production

The preceding cost data indicate the estimated costs of producing ethanol. As shown in Table 6-4, the estimated cost of producing 190-proof ethanol ranges from $1.13 to $1.63 per gallon for the on-farm units. It should be noted that these represent farm-gate costs. Costs of producing 200-proof ethanol in the community units range from $1.21 to $1.38 per gallon, depending on the presence of drying.

Care must be exercised in comparing these costs. The first three costs shown in Table 6-4 are for 190-proof ethanol and the last three are for 200-proof alcohol. In addition, the first four costs assume the direct use of wet stillage, a condition which eliminates drying costs. The effect of drying on costs is demonstrated by a comparison of the $1.21 per gallon for the small community, wet unit and $1.38 per gallon for the small community, DDGS unit. These additional costs for drying amount to $0.17 per gallon of ethanol

Table 6-2. Factors Used to Estimate Capital Recovery Factors for Base Case Costs

Item	Unit	Model plant					
		Pot still	Small on-farm	Large on-farm	Small community, wet	Small community, DDGS	Large community, DDGS
Investment value							
Facilities	$1,000	25.0	140.0	365.0	1,200.0	1,575.0	2,750.0
Working capital	$1,000	3.0	7.2	46.8	140.0	162.0	325.0
Total	$1,000	28.0	147.2	411.8	1,340.0	1,737.0	3,075.0
Facilities investment composition							
Land	Percent	0	0	0	1	1	3
Site	Percent	0	7	5	1	1	2
Buildings	Percent	0	0	5	4	3	5
Equipment	Percent	100	93	90	94	95	90
Plant life	Years	5	10	10	20	20	20
Capital structure							
Equity	Percent	50	50	50	50	50	50
Debt	Percent	50	50	50	50	50	50
Cost of capital							
Equity	Percent	14	14	14	14	14	14
Debt	Percent	12	12	12	12	12	12
Depreciation method	Straight line	SL	SL	SL	SL	SL	SL
Income tax rate	Percent	25	25	25	38	38	38
Investment tax credit rate	Percent	20	20	20	20	20	20
Inflation rate							
General	Percent	7	7	7	7	7	7
Energy	Percent	10	10	10	10	10	10
Capital cost factor							
Debt	Percent of capital	10.3	6.1	5.6	3.9	3.3	3.8
Equity	Percent of capital	9.7	5.6	8.4	7.8	8.0	8.4
Income taxes	Percent of capital	0.6	0.5	0.9	1.0	1.4	1.5
Total	Percent of capital	20.6	12.2	14.9	12.7	12.7	13.7
Capital cost	$/gal ethanol	0.36	0.30	0.17	0.17	0.22	0.19

Source: USDA, *Small-Scale Fuel Alcohol Production,* U.S. Government Printing Office, Washington, D.C., 1980.

Table 6-3. Summary of Estimating Factors for Base Case Costs

Item	Unit	Model plant					
		Pot still	Small on-farm	Large on-farm	Small community, wet	Small community, DDGS	Large community, DDGS
Annual ethanol production							
192 proof	gal/yr	16,000	60,000	360,000	—	—	—
200 proof	gal/yr	—	—	—	1,000,000	1,000,000	2,000,000
Feedstock cost							
Ethanol yield, 192 proof	gal/bu	2.4	2.5	2.5	—	—	—
Ethanol yield, 200 proof	gal/bu	—	—	—	25	2.5	2.5
Price	$/bu	2.50	2.50	2.75	2.75	2.75	2.75
Coproduct credit							
Wet stillage yield, 9 pct solids	gal/bu	21.4			—	—	—
Wet stillage yield, 12.3 pct solids	gal/bu	—	15.7	15.7	15.7	—	—
DDGS	lb/bu	—	—	—	—	17.8	17.8
Wet stillage price, 9 pct solids	$/gal	0.028			—	—	—
Wet stillage price, 12.3 pct solids	$/gal	—	0.087	0.076	0.076	—	—
DDGS price	$/T	67	153	135	135	130	130
Labor							
Requirement	hrs/gal	0.05	0.04	0.02	0.010	0.014	0.007
Price	$/hr	3.00	5.00	6.00	6.00	6.00	6.00
Energy							
Requirement	MBtu/gal	43	43	41	61	82	82
Price	$/MMBtu	2.315	2.315	2.315	2.315	2.315	2.315
Electricity							
Requirement	kWhr/gal	0.5	0.5	0.5	0.5	0.5	0.5
Price	$/kWhr	0.05	0.05	0.05	0.05	0.05	0.05
Supplies	$/gal	0.09	0.09	0.09	0.09	0.09	0.09
Repairs and maintenance							
Through 5 years	Percent of equipment cost	3	3	3	3	3	3
After 5 years	Percent of equipment cost	5	5	5	5	5	5
Taxes and insurance	$/yr	1,000	5,400	15,800	41,100	52,700	88,300
General and administrative	$/yr	—	—	15,000	30,000	40,000	50,000
Capital cost	$/gal	0.36	0.30	0.17	0.17	0.22	0.19

Source: USDA, *Small-Scale Fuel Alcohol Production,* U.S. Government Printing Office, Washington, D.C., 1980.

Table 6-4. Estimated Costs of Ethanol Production (1979 Dollars)

Cost item	Pot still	Small on-farm[a]	Large on-farm[a]	Small community, wet[b]	Small community, DDGS[b]	Large community, DDGS[b]
Direct						
Feedstock	1.04	1.00	1.00	1.10	1.10	1.10
Labor	0.15	0.20	0.12	0.06	0.08	0.04
Energy	0.10	0.10	0.09	0.14	0.19	0.19
Electricity	0.03	0.03	0.03	0.03	0.03	0.03
Supplies	0.09	0.09	0.09	0.09	0.09	0.09
Subtotal	1.41	1.42	1.33	1.42	1.49	1.45
Indirect						
Repairs and maintenance	0.05	0.07	0.03	0.03	0.04	0.04
Taxes and insurance	0.06	0.09	0.04	0.04	0.05	0.04
General and administrative	0.00	0.00	0.04	0.03	0.04	0.03
Subtotal	0.11	0.16	0.11	0.10	0.13	0.11
Coproduct credit	(0.25)	(0.54)	(0.48)	(0.48)	(0.46)	(0.46)
Total operating	1.27	1.04	0.96	1.04	1.16	1.10
Capital costs	0.36	0.30	0.17	0.17	0.22	0.19
Total cost	1.63	1.34	1.13	1.21	1.38	1.29

[a] $/190-proof gal.
[b] $/200-proof gal.

Source: USDA, *Small-Scale Fuel Alcohol Production,* U.S. Government Printing Office, Washington, D.C., 1980.

and consist of higher labor, energy, and capital costs. The estimated co-product credit for wet stillage would be greater by the eliminated costs for drying or $0.02 per gallon of ethanol. Capital costs increase significantly ($0.05 per gallon of ethanol) with drying, owing to the rather high investment costs and energy inputs of dryers.

The cost estimates shown also provide some indication of economies of size for the two major situations shown. Costs per gallon decline from $1.63 for the small pot still (16,000 gallons per year) to $1.13 for the large on-farm unit (360,000 gallons per year). Comparison of the small and large community stills (1.0 million and 2.0 million gallons per year, respectively) with drying facilities also indicates economies of size for producing 200-proof ethanol.

CENTRALIZED DEHYDRATION

While 190 or lower proof may be used directly as a liquid fuel, this quality of alcohol would have to be dehydrated to 200 proof to enter the gasohol market. A central dehydration unit serving satellite stills producing 190-proof ethanol could perform this process. The topping-cycle central plant would operate a collection service similar to a milk collection route. One man and one truck would make two pick ups per day on a five-day schedule collecting 2,500 gallons per trip of 190-proof ethanol from on-farm ethanol plants and transport the low-proof ethanol to the central plant. The collections would be in an average 10-mile radius over a 42-week year and would transport 1,053,000 gallons of 190-proof alcohol.

Such a topping-cycle central plant with a cooking and fermentation capacity of 1.0 million gallons and an additional capacity to process 1.0 million gallons from on-farm units would incur costs of $0.17 per gallon of ethanol. (See Table 6-5.) This cost added to those of producing 190-proof ethanol results in a 200-proof cost of $1.30-$1.80 per gallon of ethanol depending on the source of ethanol. This results in 200-proof costs of close to those of the 1.0-million-gallon community still without stillage drying and the 2.0-million-gallon still with stillage drying.

LOWER-PROOF ETHANOL

As indicated in Chapter 10, lower proofs can technically be used as a straight-ethanol fuel for spark ignition engines or for aspirating in diesel engines. The estimated costs of production for a small on-farm unit producing 160-proof ethanol compared to those for the small on-farm unit producing 190 proof are shown in Table 6-6. The cost of production of $1.00 per gallon of 160 proof reflects the reduced investment required due to the elimination of one distillation column. Other direct and indirect costs were estimated analogously to those for the plants described in Chapter 3. Adjusting the cost

Table 6-5. Estimated Incremental Costs for Centralized Dehydration of 190-Proof Ethanol

Cost item	Central dehydration ($/200-proof gal)
Direct	
Feedstock	–
Labor	0.01
Energy	0.05
Electricity	0.01
Transportation	0.01
Subtotal	0.08
Indirect	
Repairs and maintenance	0.01
Taxes and insurance	0.01
General and administrative	0.02
Subtotal	0.04
Coproduct credit	–
Total operating	0.12
Capital costs	0.05
Total cost	$0.17

Source: USDA, *Small-Scale Fuel Alcohol Production*, U.S. Government Printing Office, Washington, D.C., 1980.

Table 6-6. Estimated Costs of Production of Ethanol In a Small On-Farm Still by Proof Level

Cost item	Small on-farm still	
	160 proof ($/gal)	190 proof ($/gal)
Direct		
Feedstock	0.84	1.00
Labor	0.17	0.20
Energy	0.06	0.10
Electricity	0.03	0.03
Supplies	0.08	0.09
Subtotal	1.18	1.42
Indirect		
Repairs and maintenance	0.04	0.07
Taxes and insurance	0.06	0.09
General and administrative	–	–
Subtotal	0.10	0.16
Coproduct credit	(0.45)	(0.54)
Total operating	0.83	1.04
Capital costs	0.17	0.30
Total cost	1.00	1.34

Source: USDA, *Small-Scale Fuel Alcohol Production*, U.S. Government Printing Office, Washington, D.C., 1980.

of $1.34 per gallon of 190 proof to 160 proof on an ethanol content basis gives a cost of $1.13 per gallon. Production of 160 proof then is $0.13 per gallon less than the cost if 190-proof ethanol is diluted.

Estimates of the cost of producing 100-proof ethanol indicate that a production cost of about $0.62 per gallon is equal to the cost of 160-proof ethanol adjusted to 100 proof.

Sensitivity Analyses

To provide additional perspective to these cost estimates and also to provide information to users to estimate costs for specific situations, sensitivity analyses were done for the major cost elements. These analyses involve reestimating the ethanol-production cost by varying one variable at a time while holding all other variables constant.

The results of those analyses are presented graphically for each of the model plants in Figs. 6-1 through 6-12. To facilitate the presentation, annual operating variables are shown in one figure and the equivalent capital costs are shown in the following figure for each model plant. The figure is designed with the per-gallon cost of ethanol production on the vertical axis and the percent change in base case estimating factors shown in Table 6-4 indicated on the horizontal axis.

The slopes of the lines indicate the relative impact of the variable. The steeper the slope, the greater the impact of cost changes by the indicated variable. Conversely, those variables with relatively gentle slope produce little impact on production costs.

It is also noted that these graphs can be used to estimate the cost of production with combined changes of several variables. This is done by multiplying the total base case cost times the ratio of the total cost (from interpolation) associated with the percentage change of the variable to the total base case cost. More specifically, the formula is as follows:

$$\text{New combined cost} = \text{Total base case cost} \times \frac{\text{Revised total cost for variable } j}{\text{Base case cost}} \times \frac{\text{Revised total cost } j}{\text{Base case cost}} \times \cdots \times \frac{\text{Revised total costs } j}{\text{Base case cost}}$$

The resulting estimate should be within 1 or 2 percent, depending on the rounding error and the accuracy of interpolation.

RESULTS OF SENSITIVITY ANALYSES

Eleven variables were analyzed, including six operating and five capital variables. As can be seen in the graphs, operating variables were found to be more sensitive, that is, they affect production costs more than do capital costs. The specific list of variables and the relative effect of each are as follows:

Operating	*Effect*
Annual ethanol production	Large
Ethanol yield per bushel	Large
Feedstock price	Large
Wage rate	Small
Energy price	Small
Coproduct price	Large
Capital	
Plant investment	Large
Debt-equity proportions	Small
Interest rate	Small
Investment tax credit	Medium
Cost of equity	Small

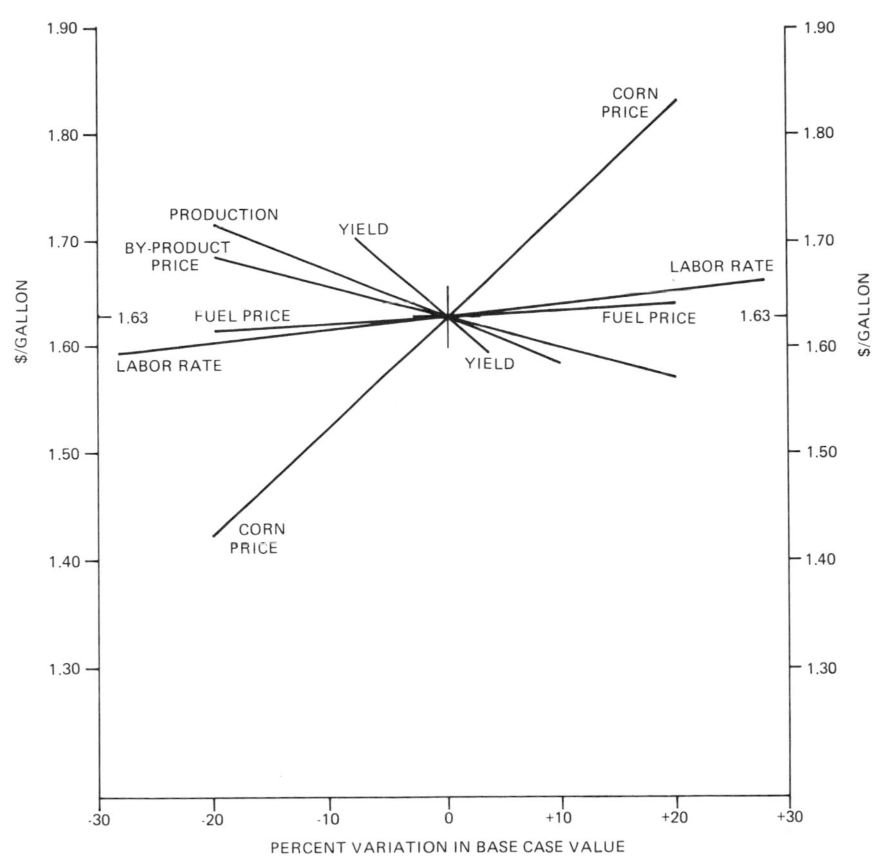

Source: U.S. Department of Agriculture. *Small-Scale Fuel Alcohol Production*. 1980. Washington, D.C. GPO No. 001-000-04124-0.

Figure 6-1. Annual operating variables—pot still.

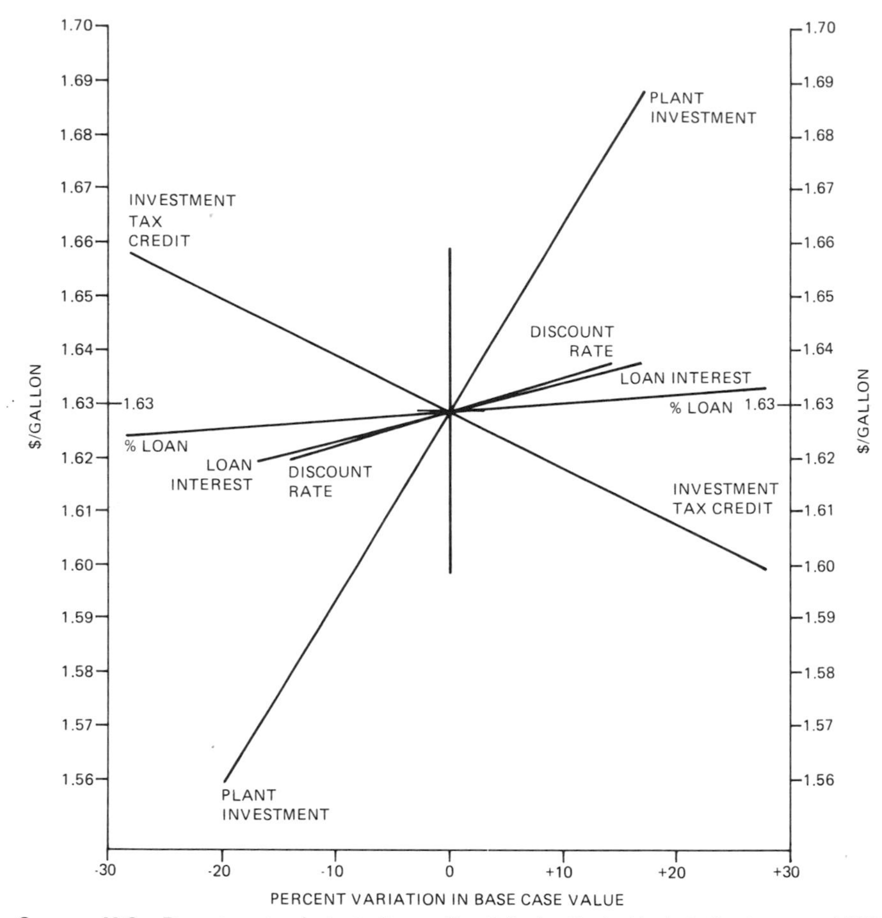

Source: U.S. Department of Agriculture. *Small-Scale Fuel Alcohol Production*. 1980. Washington, D.C. GPO No. 001-000-04124-0.

Figure 6-2. Equivalent capital cost—pot still.

The relatively large effect of production level, ethanol yield, feedstock price, and coproduct price would be expected. While all of these factors are important, feedstock prices dominate. A 10 percent increase, i.e., $0.25 per bushel, raises total cost nearly $0.10 per gallon of ethanol or, conversely, a 10 percent reduction reduces total ethanol cost about $0.10 per gallon. This is in contrast to a 10 percent change in interest rates which affect total cost by about $0.01 per gallon of ethanol.

It is noted that of the capital cost factors, the investment tax credit, and plant investment costs have the largest affect on total cost. However, relative to feedstock price, ethanol yields, and production levels, the effect of tax credits and plant investment are relatively small.

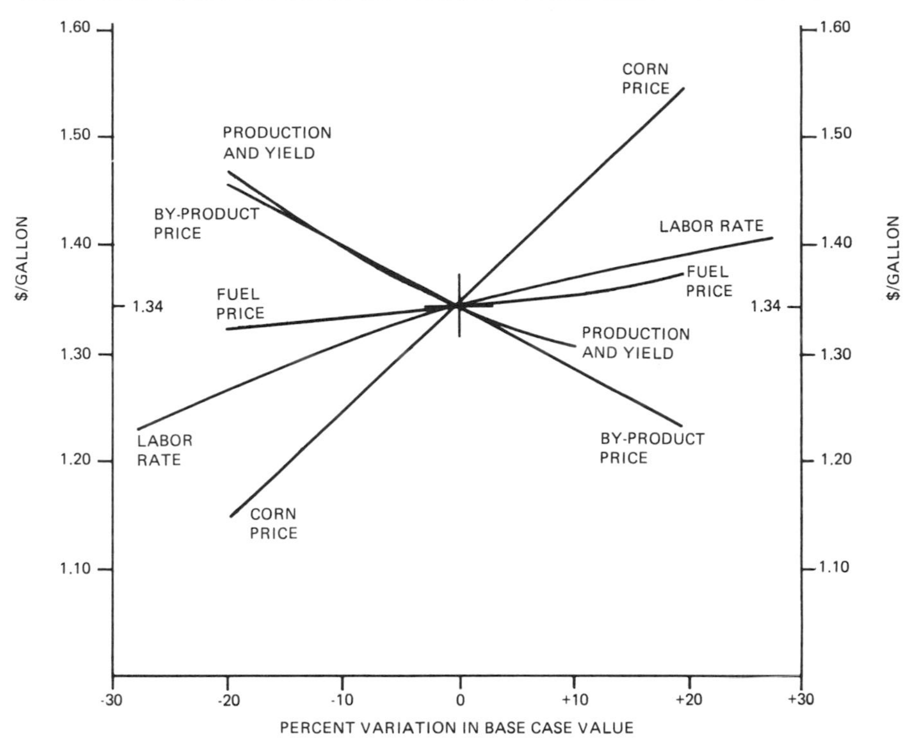

Source: U.S. Department of Agriculture. *Small-Scale Fuel Alcohol Production,* 1980. Washington, D.C. GPO No. 001-000-04124-0.

Figure 6-3. Annual operating variables—small on-farm.

Relationship of Production Costs to Fuel Value

To place the estimated costs of ethanol production in perspective, they should be compared with the fuel value of ethanol vis-à-vis the fuel for which ethanol potentially could substitute. Fuel values were estimated by applying the volumetric values developed in Chapter 10 to gasoline, diesel, and LPG prices as appropriate. Prices used reflect the prices of late 1979 and early 1980 paid by farmers for farm applications and consumers for consumer applications, i.e., gasohol. Four potential applications were examined including gasohol, straight alcohol, carbureting in diesel engines, and crop drying.

GASOHOL

As indicated in Chapter 10, ethanol in 10 percent mixture with gasoline would have a value of 75 percent of gasoline. In addition, there is the potential for an octane enhancement credit. If a base gasoline stock with a

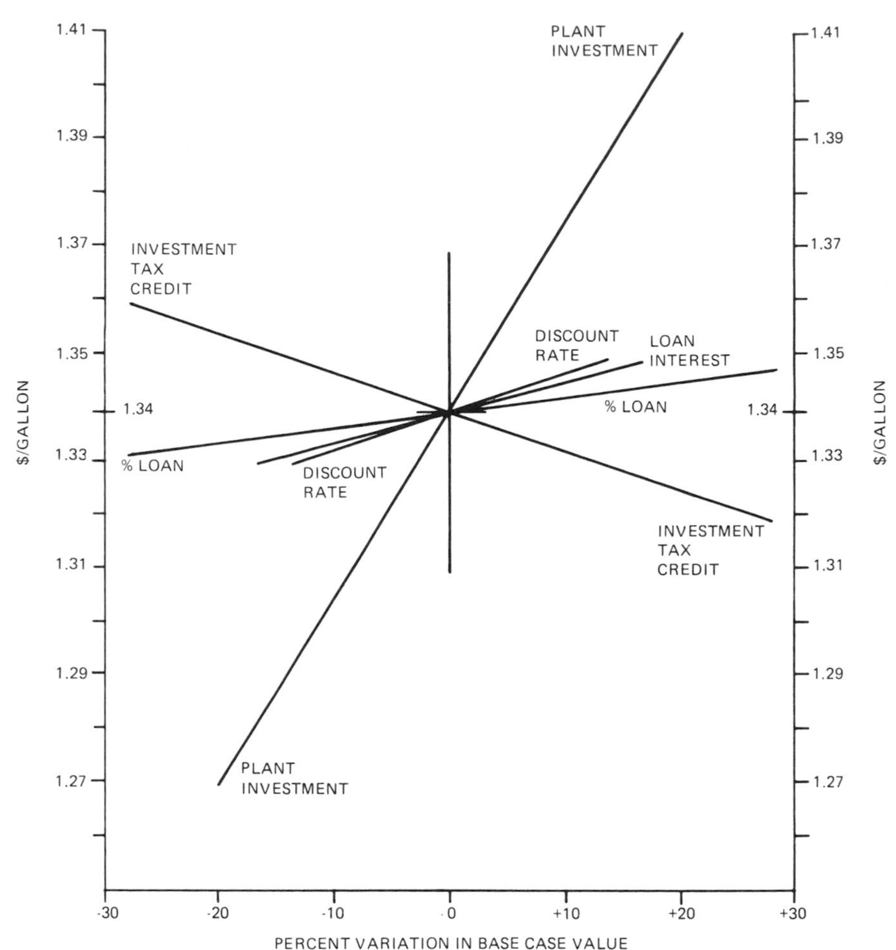

Source: U.S. Department of Agriculture. *Small-Scale Fuel Alcohol Production*. 1980. Washington, D.C. GPO No. 001-000-04124-0.

Figure 6-4. Equivalent capital cost—small on-farm.

lower octane is available for gasohol blending or if there is a need for a higher octane fuel, a credit of about $0.10 per gallon of ethanol would appear to be appropriate. However, a unique gasohol base petroleum stock is not likely to be available unless a large gasohol program emerges. With small local or regional programs, it is unlikely that a special gasohol base stock will be available. The need for higher octane is difficult to assess, for engine adjustments may produce the same result as the use of higher octane gasohol. If no octane enhancement credit was taken, the fuel value of 200-proof ethanol in gasohol would be $0.64 per gallon, against a refinery-gate unleaded regular gasoline price of $0.85 per gallon. Assuming that dealer margins, transportation and terminal costs, and fuel taxes for ethanol and gasoline

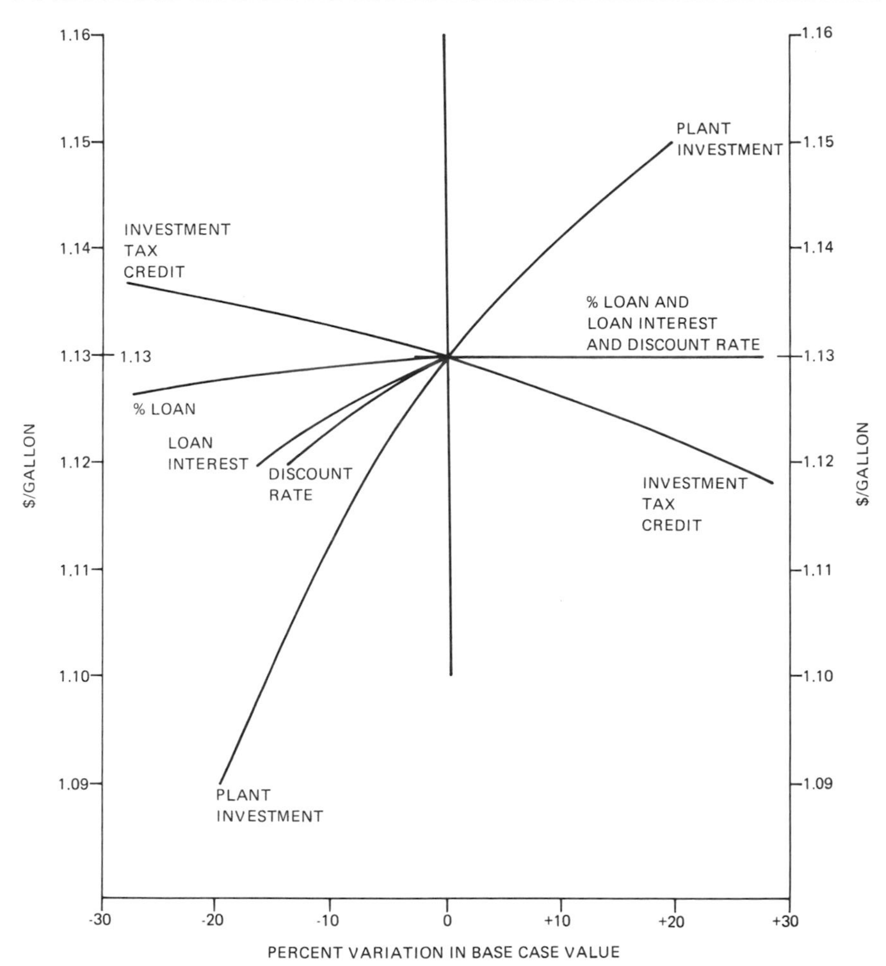

Source: U.S. Department of Agriculture. *Small-Scale Fuel Alcohol Production.* 1980. Washington, D.C. GPO No. 001-000-04124-0.

Figure 6-5. Annual operating variables—large on-farm.

were equal, the difference between production costs and fuel value would be \$0.54 to \$0.74 per gallon of ethanol depending on distillery type. Where the octane enhancement credit was applicable, this difference would be decreased to \$0.44-\$0.64 per gallon.

STRAIGHT ETHANOL

As suggested in Chapter 10, various approaches are available for using straight ethanol and include dual ethanol and gasoline systems and ethanol-only fuel systems. To illustrate the fuel value, the ethanol-only system was

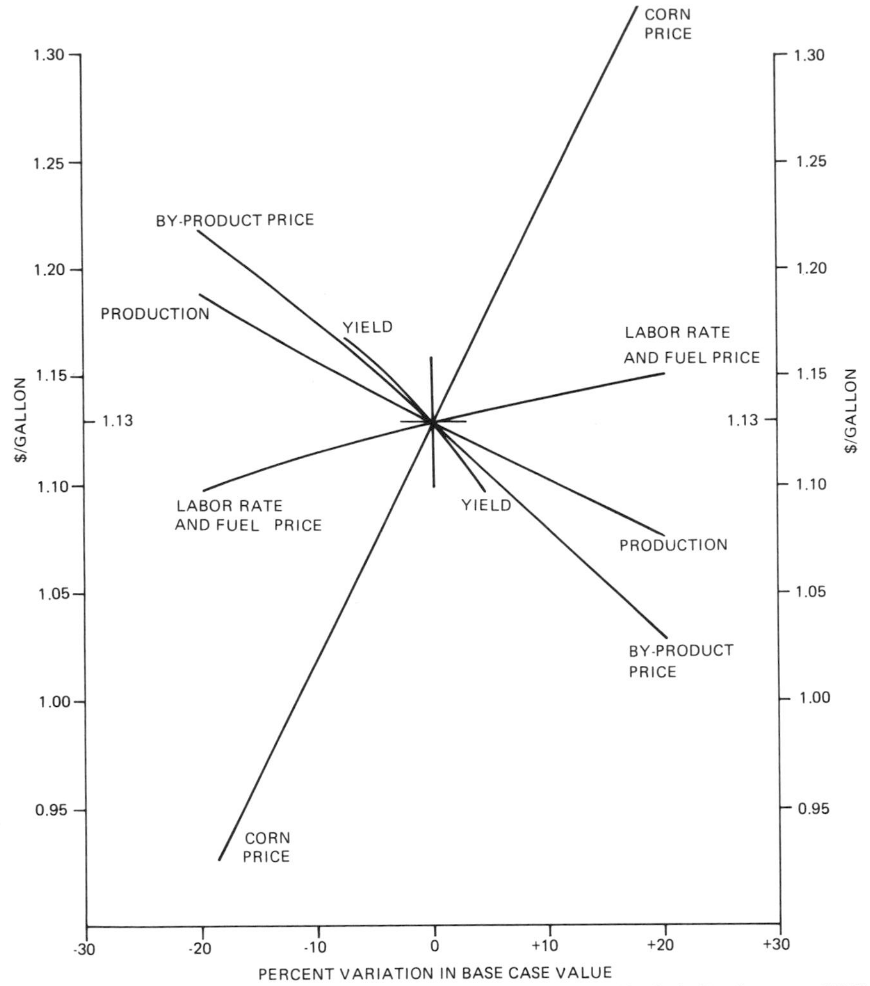

Source: U.S. Department of Agriculture. *Small-Scale Fuel Alcohol Production.* 1980. Washington, D.C. GPO No. 001-000-04124-0.

Figure 6-6. Equivalent capital cost—large on-farm.

considered. With a weighted gasoline price of $0.97 per gallon[1] to the farmer on a bulk basis and net of taxes, the ethanol-fuel value for various proofs would be as follows:

Proof	*Ethanol fuel value ($/gal)*
200	0.74
190	0.64
160	0.49

[1]Assumes $0.85 per gallon refinery-gate price for unleaded regular, $0.80 per gallon for leaded regular, $0.07 per gallon bulk dealer markup, $0.08 per gallon transportation and terminal costs, 40 percent unleaded and 60 percent leaded gasoline.

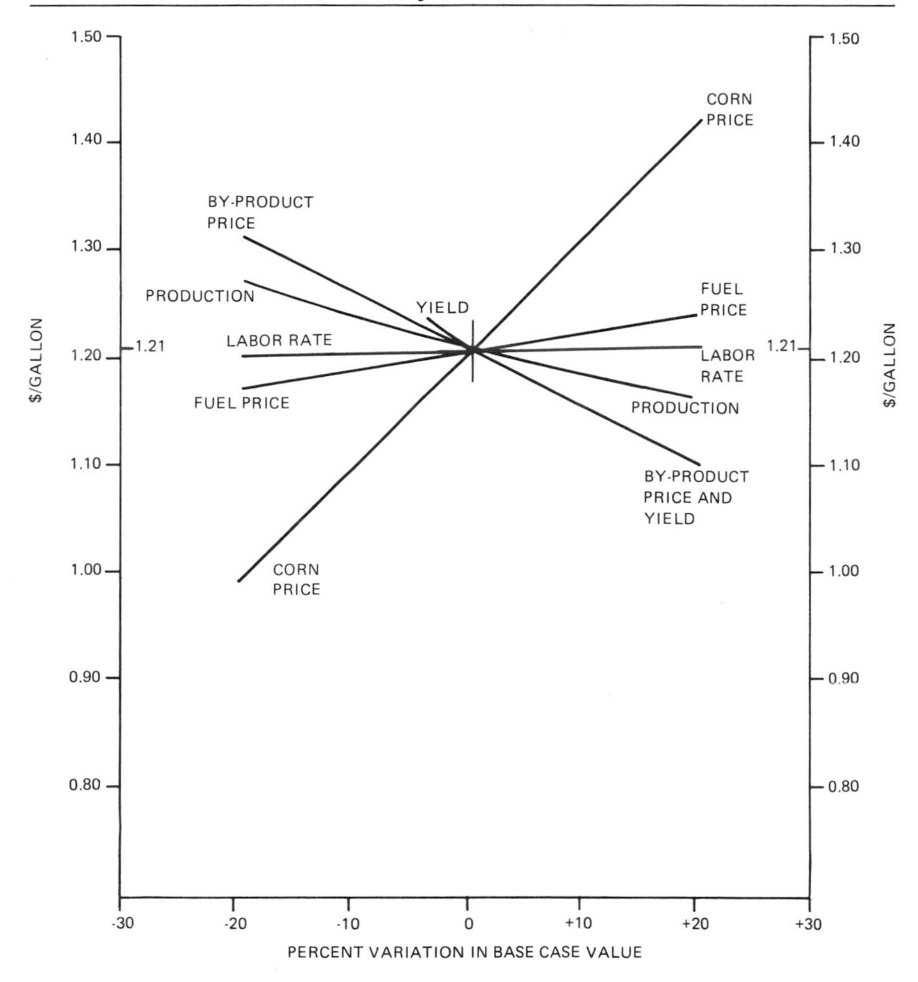

Source: U.S. Department of Agriculture. *Small Scale Fuel Alcohol Production*. 1980. Washington, D.C. GPO No. 001-000-04124-0.

Figure 6-7. Annual operating variables—small community, wet.

Use of 190-proof ethanol, with production costs of $1.13 to $1.63 per gallon in this manner, would result in costs exceeding fuel value by $0.49 to $0.99 per gallon.

CARBURETING ETHANOL INTO DIESEL ENGINES

Of the potential diesel applications, aspiration appears to be the most feasible in the near term. Assuming an on-farm diesel price of $0.90 per gallon, the fuel value of ethanol would be as follows:

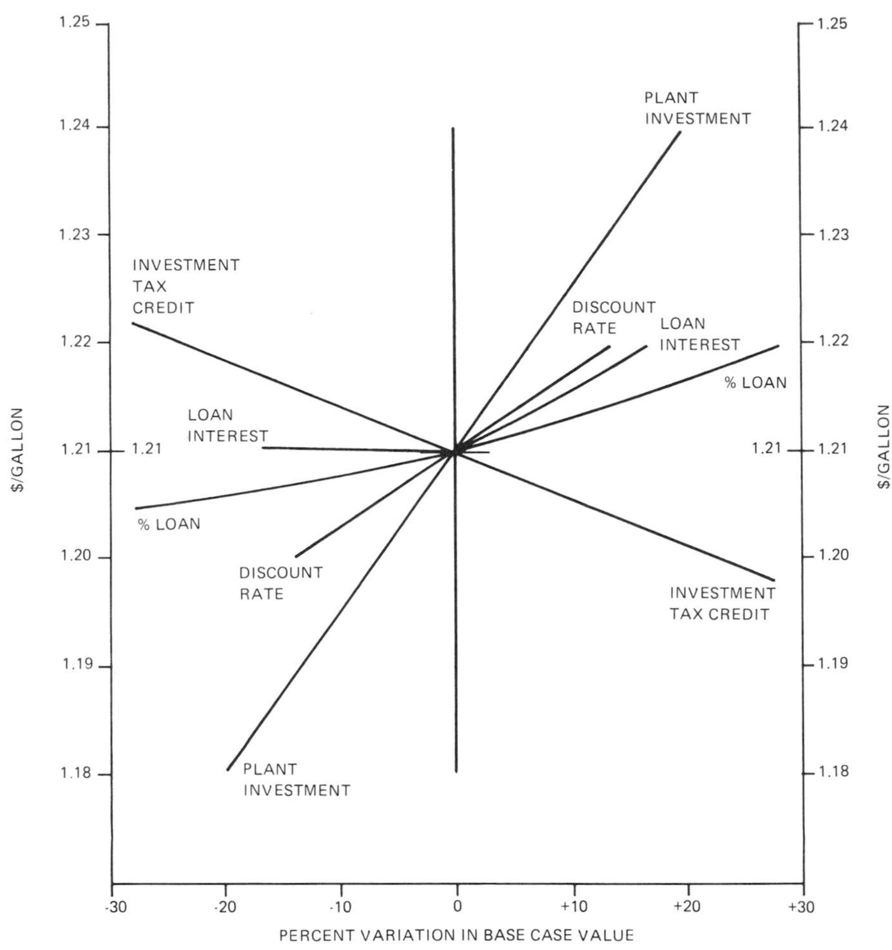

Source: U.S. Department of Agriculture. *Small-Scale Fuel Alcohol Production*. 1980. Washington, D.C. GPO No. 001-000-04124-0.

Figure 6-8. Equivalent capital cost—small community, wet.

Proof	*Ethanol fuel value ($/gal)*
200	0.50
190	0.46
160	0.40
120	0.30
100	0.25

The volumetric value estimates shown in Chapter 10 indicate that, on this basis, the costs of 190-proof ethanol would exceed its fuel value by $0.66-$1.17 per gallon, depending on the type and size of still. The 100-

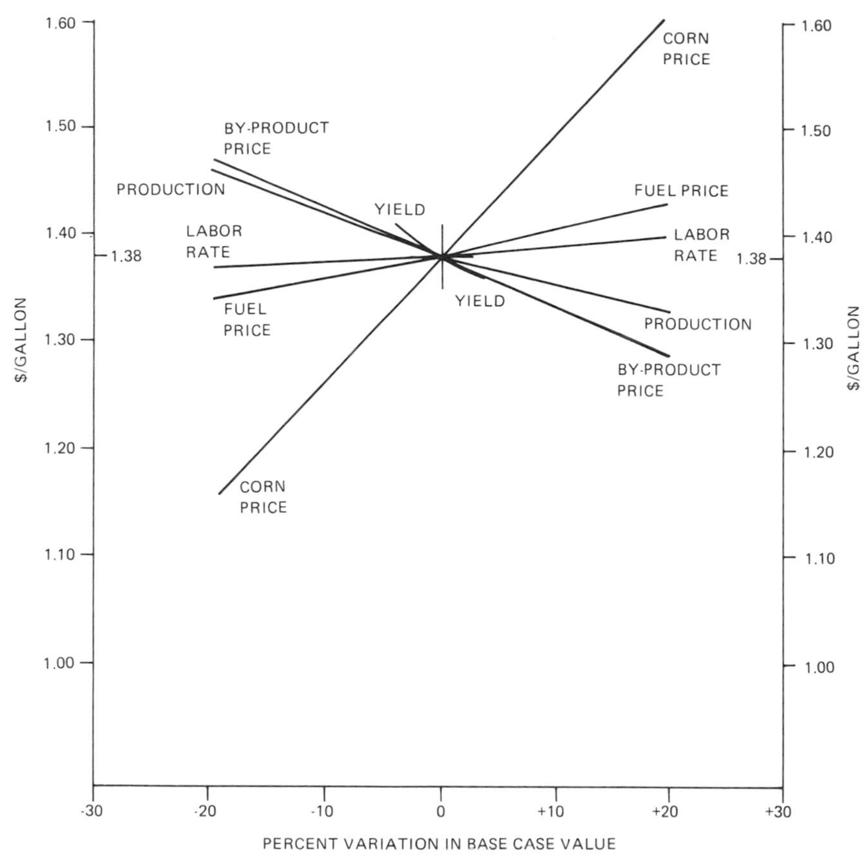

Source: U.S. Department of Agriculture. *Small-Scale Fuel Alcohol Production*, 1980. Washington, D.C. GPO No. 001-000-04124-0.

Figure 6-9. Annual operating variables—small community, DDGS.

proof ethanol cost would exceed its fuel by $0.37 per gallon at a base cost of $1.30. Larger stills with lower costs might reduce this difference to about $0.20 per gallon.

GRAIN DRYING

Of the other possible fuel applications for ethanol, grain drying appears to be the most promising. Although there is little research on this application, it appears technically feasible. Assuming LP gas prices of $0.55 per gallon delivered to the farm, 200-proof ethanol would exceed the fuel value of $0.51 per gallon. This would mean that production costs would have a fuel value of $0.70-$0.87 per gallon of ethanol in this application, depending on the type of production facility.

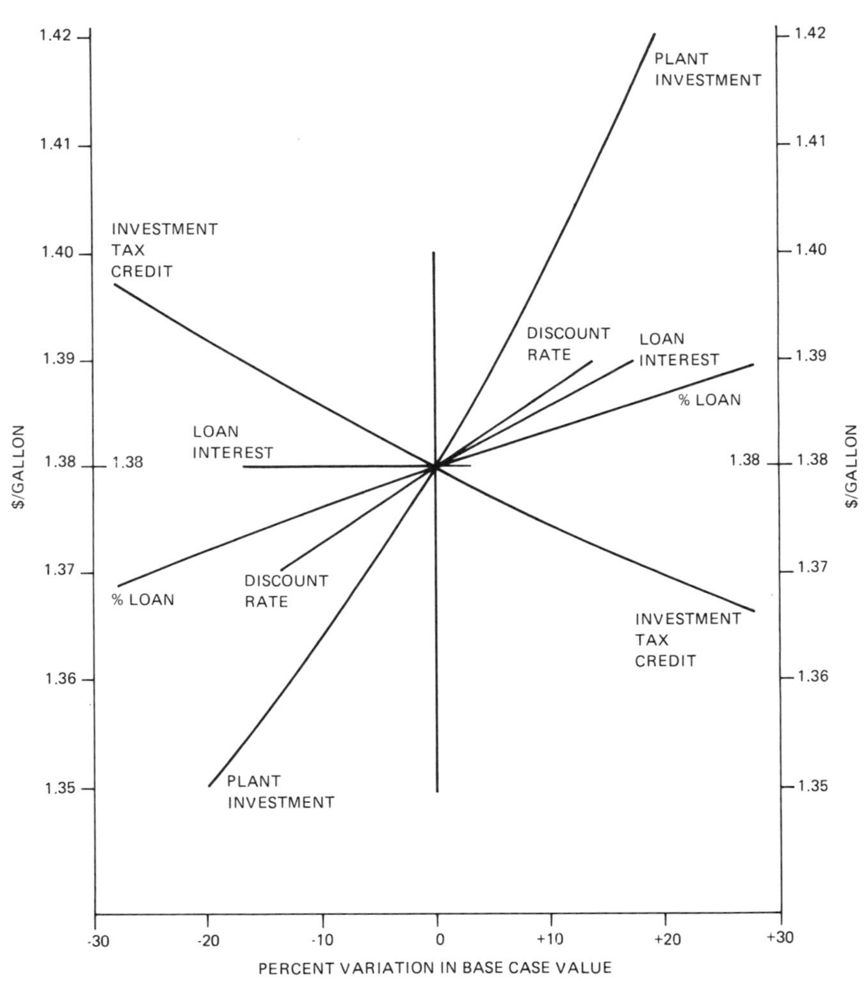

Source: U.S. Department of Agriculture. *Small-Scale Fuel Alcohol Production*. 1980. Washington, D.C. GPO No. 001-000-04124-0.

Figure 6-10. Equivalent capital cost—small community, DDGS.

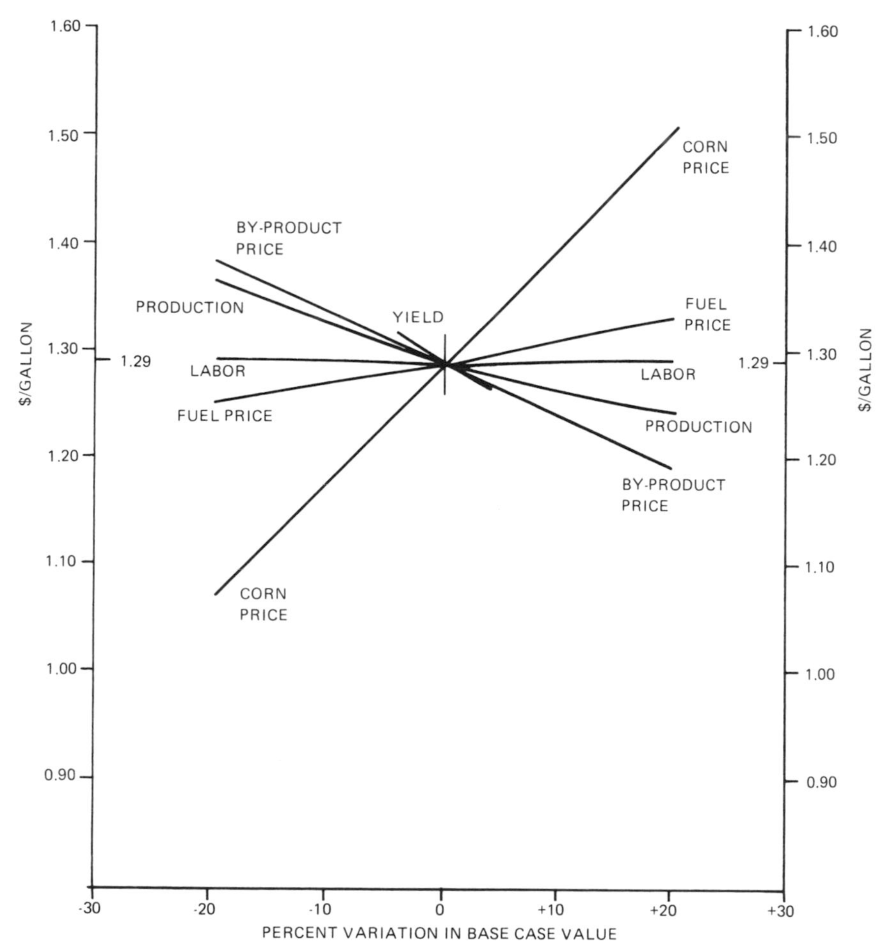

Source: U.S. Department of Agriculture. *Small-Scale Fuel Alcohol Production*. 1980. Washington, D.C. GPO No. 001-000-04124-0.

Figure 6-11. Annual operating variables—large community, DDGS.

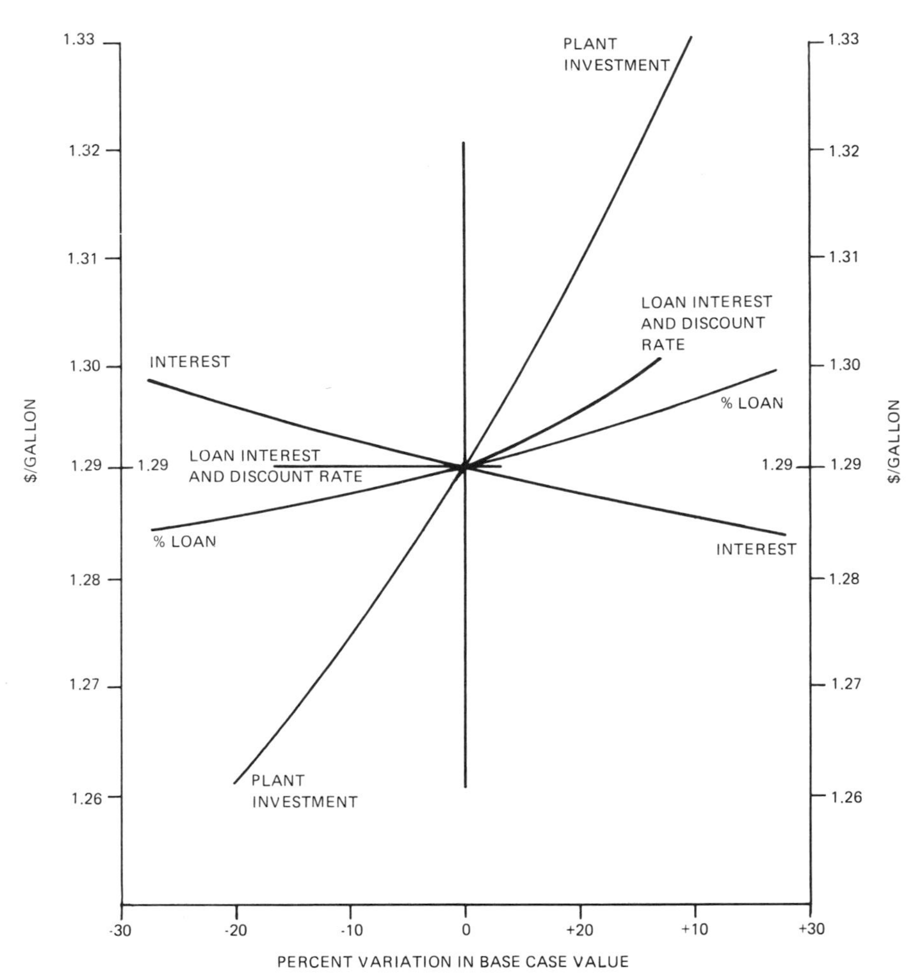

Source: U.S. Department of Agriculture. *Small-Scale Fuel Alcohol Production*. 1980. Washington, D.C. GPO No. 001-000-04124-0.

Figure 6-12. Equivalent capital cost—large community, DDGS.

Chapter 7

Overview of Process Technology[1]

Typical Conventional Technology

The current level of alcohol fermentation technology is represented by three existing U. S. plants that convert three distinctly different biomass resources: (1) Archer-Daniels-Midland Company (ADM) in Decatur, Illinois, uses corn; (2) Milbrew, Incorporated, in Juneau, Wisconsin, processes cheese whey; and (3) Georgia-Pacific Corporation in Bellingham, Washington, converts wood sugars in a sulfite pulp mill waste stream to alcohol. In-depth descriptions of these facilities are not available inasmuch as details of operating plants are generally considered proprietary for competitive reasons. In addition, recent emphasis on reduced energy usage for economic and conservation reasons has led to largely unrevealed process improvements in some existing plants.

ARCHER-DANIELS-MIDLAND (ADM) PLANT

The ADM plant began operation in mid-1978 at a production level of 5 million gallons per year, furnishing ethanol at 192 and 200 proof to beverage and industrial users. The raw materials included distressed corn, corn syrup, and wheat starch. Because some of the starting materials are acidic, 316-stainless steel was used throughout the facility. Benzene was used, at least initially, as the extractive agent for producing the anhydrous alcohol. The fusel oil was used as boiler fuel, and the carbon dioxide formed during fermentation was released to the atmosphere.

During 1979, ADM tripled its output of ethanol for gasohol by converting its entire beverage alcohol capacity to the production of anhydrous alcohol. Concurrently, the feed stream was tied in more closely to the adjacent ADM corn-processing facility. The objective was to first extract the major high-value nutrients from the corn to produce marketable food and feed products. Food products include corn oil, corn syrup, dextrose, proteins, gluten, and fiber; feed products encompass proteins, fiber, and waste yeast. An excess of starch is available for conversion by enzyme action to dextrose for diversion to the alcohol plant.

[1]Material abstracted from National Alcohol Fuel Commission Report, *Alcohol Fuels From Biomass: Production Technology Overview*, Aerospace Corporation, July 1980.

Because the ADM facility is relatively new, it incorporates several efficiency features in addition to those inherent in its large size. Fermentation is completed in 12 hours by injecting large batches of recycled yeast into the sugar solution. The yeast, it should be noted, is easy to separate from the beer when there are no suspended grain solids. Because fusel oil and aldehydes need not be removed from the fuel-grade ethanol, only one distillation column is used to concentrate the fermentation beer to about 192 proof for the final dehydration step.

Some of the plant's energy requirements are supplied by waste heat from the corn-processing facility. Additional steam is produced in a natural gas-fired boiler.

In early 1980, ADM announced plans to build a new distillery adjacent to its Cedar Rapids, Iowa, corn-processing plant to produce 78 million gallons per year of anhydrous ethanol. The boilers will probably be fired with natural gas, although coal has been mentioned as a possibility at both ADM plants. Design details of the new plant have not been disclosed. As of May, 1981, ADM had plans for additional production increases at Decatur and Cedar Rapids, as well as for the retrofit of an old Hiram Walker-Gooderham and Worts distillery recently purchased in Peoria, Illinois. By 1981, ADM expects to manufacture ethanol at the rate of 260 million gallons per year, over one-half of the Administration's national target capacity.

MILBREW PLANT

The Milbrew plant, located about 50 miles northwest of Milwaukee in prime dairy land, is the only alcohol-fuel source producing the product primarily from cheese whey. Very little information about the plant has been released. As of 1978, when it was furnishing 193-proof ethanol to Illinois for gasohol, the plant capacity was set at 5 million gallons per year, but facility expansion was reported in late 1979. As far as it is known, the plant does not produce an anhydrous product. High-protein yeasts and condensed fermentation solubles are coproducts of the whey distillation process. Process-energy sources include natural gas, propane, and fuel oil.

Milbrew is reported to be interested in building two new facilities capable of producing 10 million gallons per year from cheese whey.

GEORGIA-PACIFIC PLANT

The Georgia-Pacific plant was built during World War II to produce ethanol for synthetic rubber. It is an integral part of a 200,000-ton-per-year sulfite pulp mill. The waste liquor from the sulfite process contains 2-3.5 percent sugar, 65 percent of which is fermentable to alcohol after the acidic components of the waste are stripped from the liquor or neutralized with lime. Yields of over 18-22 gallons of alcohol per ton of pulp produced can be achieved. Continuous fermentation and reuse of the yeast, which is removed from the fermentation broth by a centrifugal separator, are important features of the process. Production in 1978 amounted to 5.5 million gallons per year and probably can be increased to 6 million per year. Of this quantity, about 2 million gallons per year of anhydrous product can be produced. Inasmuch as the alcohol facility processes the entire lignosulfonate waste stream from the mill, alcohol output cannot readily be increased.

The plant is constructed of stainless steel. The carbon dioxide formed during fermentation is released to the atmosphere. Very little fusel oil is formed, and it is used as plant fuel. The aldehydes are also removed and burned. Appreciable quantities of methanol present in the incoming sulfite liquor are separated and sold periodically in tank truck loads. The lignosulfonates left in the spent fermentation liquor are converted to a variety of products, from artificial vanilla to insecticides, herbicides, and the drug L-Dopa. The plant runs on a combination of wood waste, fuel oil, and natural gas.

An interesting feature of the plant is its use of diethyl ether as the entraining agent to produce anhydrous alcohol. The ether forms a binary azeotrope (with water) rather than the ternary benzene azeotrope usually employed for dehydrating ethanol. As a consequence, less steam and water are required, and the equipment is simpler. The patent[2] on this process has expired, and ether dehydration is being considered for several new ethanol-production facilities.

Near-Term New Technologies

In the near term, advances in engineering design and savings in energy for processes using starch and sacchariferous (sugar) feedstocks are expected to predominate over the commercialization of cellulose conversion processes. For the former (and more expensive) feedstocks, steps can be taken to maximize ethanol yield and to reduce operating costs, particularly in the area of energy costs. Several engineering companies offer variations of ethanol-manufacturing facilities. The following discussion is not intended to endorse any of the processes but rather to indicate the range of variations currently possible for new alcohol plants. In certain cases, process details are not included due to the proprietary nature of the information.

RAPHAEL KATZEN ASSOCIATES INTERNATIONAL, INC.

The Katzen process, shown in Fig. 7-1, for producing fuel alcohol incorporates a number of energy-saving measures into the standard ethanol-production sequence. A pressure-cascading technique permits substantial heat recovery and reuse. Two separate distillations are run at different pressures so that waste heat from the higher-pressure step can be used again

[2]U.S. Patent 2,152,164 (March 28, 1939).

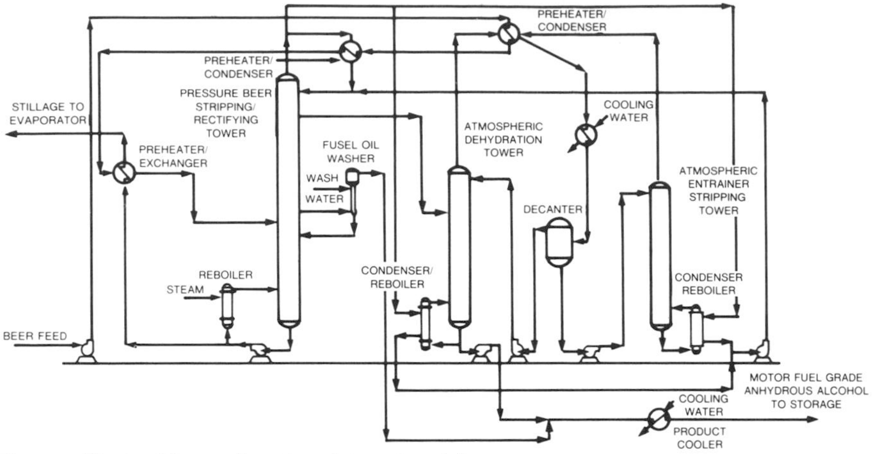

Source: Raphael Katzen Associates International, Inc.

Figure 7-1. Distillation of motor-fuel-grade anhydrous ethanol—Raphael Katzen Associates design.

in the lower-pressure step. Capital equipment requirements are also reduced in comparison to a beverage ethanol distillery by bypassing one of the distilling towers and by such innovations as using a common condenser and decanter for two separate towers.

In the alcohol-concentration sequence, the beer is preheated sequentially with azeotrope vapors; pressure vapors from the rectifying column; and, finally, hot bottoms from the beer stripping tower. (At the same time, bottoms are cooled to provide a feed to the stillage recovery operations for vapor recompression evaporation at an ideal temperature, requiring neither preheating nor flashing.) The overhead from the rectifying column is used to boil up the dehydration tower. Fusel oil is obtained from side draws, washed, decanted, and combined with the ethanol product.

The entraining agent in the dehydration step may be benzene, heptane, or cyclohexane. Consumption of the entraining agent is low in this process. The total energy consumption claimed by Katzen for the production of fuel-grade ethanol from a typical grain feedstock is 40,000-50,000 Btu per gallon of product.

A detailed discussion of the Katzen process is included in Chapter 8.

VULCAN-CINCINNATI, INC.

One feature of the design employed by Vulcan-Cincinnati is the use of diethyl ether as an extractive solvent for the dehydration step, along with a two-tower distillation system (including the dehydration tower). Cascading of distillation towers is employed, making one tower become the reboiler for the other; this technique requires balancing distillation temperatures by adjusting pressure. According to Vulcan-Cincinnati engineers, with ether vapor reuse, ethanol can be concentrated from the fermented mash at an energy expenditure of 19,000-20,000 Btu per gallon. The makeup volume of ether is quite small, approximately 1 gallon for every 3600 gallons of product. The energy requirement cited is only for the distillation/dehydration; overall energy consumption for the process will, of course, depend on the feedstock chosen. The construction of several distilleries is currently under study by Vulcan-Cincinnati; the distilleries range in size from 3 million to 100 million gallons per year of ethanol product.

VOGELBUSCH DIVISION—BOHLER BROTHERS OF AMERICA, INC.

Vogelbusch offers a thermally integrated plant based on a continuous fermentation, distillation, and rectification technology. Yeast recycle is combined with continuous fermentation to minimize the downtime generally associated with batch processes. An energy-saving design features a combined distillation, rectification, dehydration, and evaporation setup. For example, stillage is heated by alcohol vapors from the distillery and thickened for conversion into high-protein feed. Although the front end of the plant varies according to feedstock and will affect overall energy consumption, the energy requirement quoted by Vogelbusch from fermentation through anhydrous ethanol is 30,500 Btu per gallon. Cyclohexane is usually employed as the dehydrating agent (makeup volume is only 1 gallon per 1000 gallons of product), but gasoline is recommended as the most economically attractive dehydrating agent if it can also be used as a component of the denaturant and, therefore, not have to be entirely removed from the ethanol. The wastewater treatment operation is somewhat unusual, in that the distiller's wash is recycled, and condensate is treated to a level that permits discharge to a sewage system without additional, extensive treatment. Depending on the capacity of the plant, operation costs are projected by Vogelbusch to be between $0.36 and $0.44 per gallon of fuel ethanol product.

CHEMAPEC

The emphasis in the Chemapec process (Fig. 7-2; Process Engineering Company) is on total energy recuperation through maximum use of all portions of the feedstock. For cereal grains such as corn, the process involves enzymatic conversion of the grains, continuous alcohol fermentation, vacuum distillation, energy recovery, and coproduct utilization. The process employs heat at only 85°C (185°F). The major differences between the Chemapec process and other schemes are as outlined below:

Protein and fibers in the grain are separated prior to fermentation.

Fermentation is continuous, and a vacuum process is used to remove alcohol from the mash.

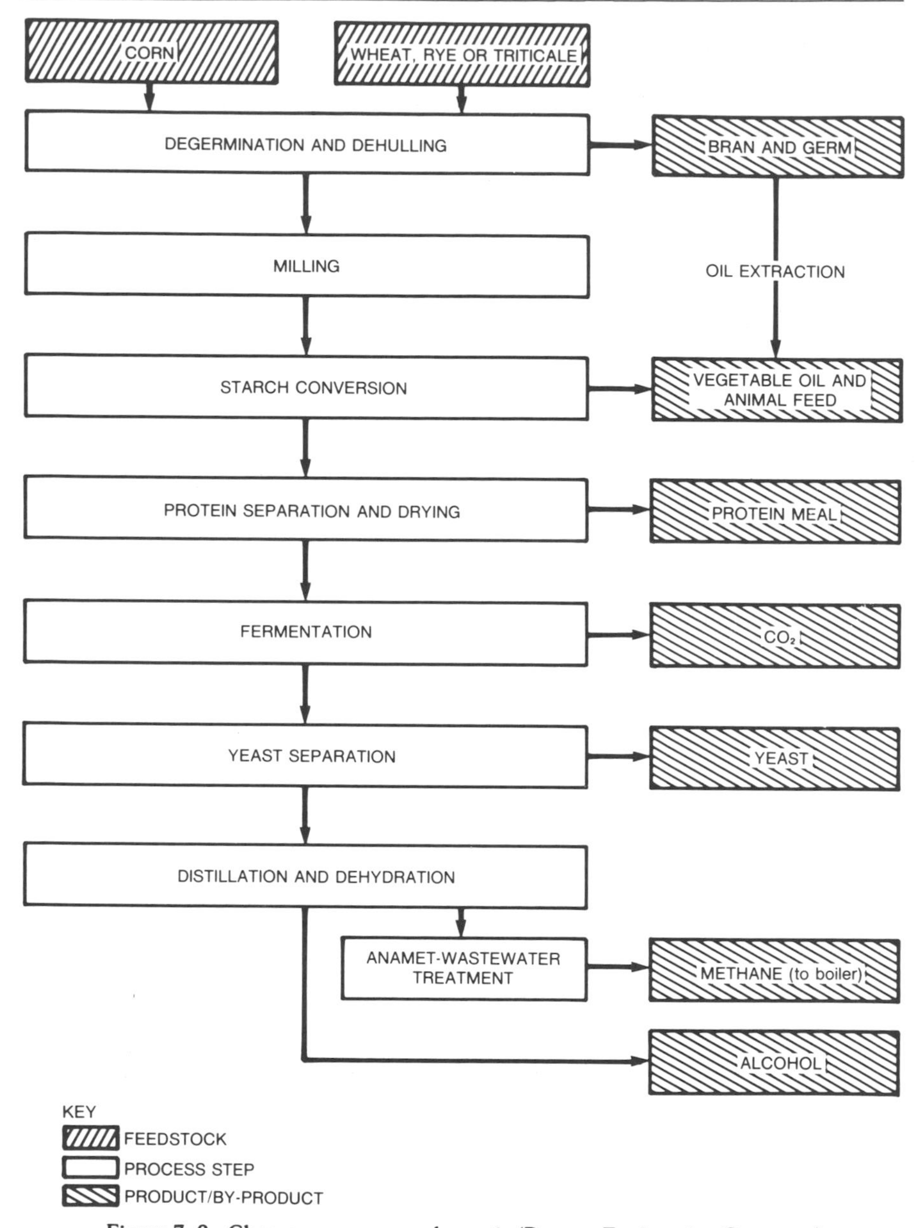

Figure 7-2. Chemapec process schematic (Process Engineering Company).

The leftover mash is converted to methane, which provides approximately 60 percent of the plant's energy needs.

All coproducts with substantial value (corn oil, edible meal, fodder yeast, fusel oil, carbon dioxide) are recovered.

The process can be adapted for use with grains other than corn, as well as with cassava, potatoes, molasses, and juices.

Although details of the actual process remain proprietary, the general approach is stated to be a continuous recovery of ethanol from a bypass system, eliminating the (ethanol) concentration-inhibition effect on the yeast. A technique called "thermocompression" is used in the distillation stage. The claimed products that can be obtained from one bushel of corn feedstock are as follows:

Ethanol	2.28 - 2.35 gal (100.6 proof)
Germs	5.00 - 6.2 lb
Bran	4.50 - 5.00 lb
Protein meal	4.8 - 6.2 lb
Yeast	0.85 - 1.00 lb
Carbon dioxide	10.00 - 11.80 lb
Fusel oil	0.07 - 0.08 lb
Methane	16.00 - 19.00 scf

By contrast, the typical yield of ethanol from other processes when corn is the feedstock is 2.5 to 2.6 gallons; in addition, approximately 17 pounds of distiller's dark grains are obtained. The most feasible scale for this process is 20 million gallons per year. For this size facility, the producer cost of ethanol, according to Chemapec, is approximately $0.60 per gallon.

ACR PROCESS CORPORATION

A very limited amount of information about the ACR process is available. Basically, less emphasis is placed on coproducts, which are considered to be capital intensive and energy consuming. An ACR spokesman says that a novel heat-exchanger design and new enzymes for converting starch into fermentable sugars are part of the process. A unique distillation setup uses any kind of commercially available gasoline as the azeotrope for the dehydration of ethanol. The use of gasoline at this stage has two advantages: it saves energy in the distillation and it avoids the need for denaturing the product. The process is touted as using only 45,000 Btu of process steam per gallon of alcohol, including evaporation to produce distiller's dark grain coproduct. The cost of producing ethanol is reportedly $1.00-$1.10 per gallon.

In February, 1980, the first entirely computerized, continuous grain-based facility in the United States was brought on line using the ACR process. The plant capacity is 3 million gallons per year.

Long-Term New Technologies

Several technologies undergoing research and development and not expected to be ready for commercialization before the mid-1980s are presented in this section. Two cellulose pretreatment processes are discussed, followed by brief descriptions of six cellulose hydrolysis processes. Lastly, advances in individual process steps are reported, with attention given to biological research, vacuum techniques, membrane technology, and dehydration techniques.

CELLULOSE PRETREATMENT PROCESSES

Iotech Process

The Iotech process is a pretreatment process in which wood is explosively decompressed after it is charged with steam, resulting in increased accessibility of the cellulose and lignin components to chemicals. The feedstock generally employed for the developmental work is commercial wood chips (1 $\frac{1}{2}$-in. diameter, $\frac{1}{8}$-in. thick). The raw material is heated in a pressure vessel to approximately 500°F (260°C) and maintained at that temperature for a short period before the pressure vessel is opened to explode the wood. Pressure in the vessel reaches 600 psi before release. Treatment times range from 20 to 90 seconds; the shorter treatment periods are used to maximize cellulose production, the longer times to maximize lignin production. As the powdery product mixture flows freely, the "gun" is easily unloaded, and turnaround time for the pretreatment is about 3 minutes.

The pretreated wood is susceptible to either acid or enzyme hydrolysis, and the material has been tested and found compatible with the organisms generally employed in downstream biological processes such as conversion to ethanol. Conversion of up to 80 percent of the available cellulose to ethanol has been achieved.

Separation of the lignin and cellulosic material after pretreatment can be accomplished by either of the following two general approaches:

- The lignin can be extracted in high yields with simple solvents such as methanol. Ninety-percent extraction of the lignin, for example, can be effected by 0.4 percent caustic soda treatment.
- The cellulose can be converted enzymatically to glucose, which can then be washed away, leaving the lignin as a residue.

The method chosen will depend on a variety of factors, but special attention is paid to the intended major product (i.e., cellulose or lignin), with less emphasis on the coproduct.

The incremental cost of the pretreatment is reported to be approximately $7.50 per ton of wood chips. This figure represents the cost of the steam explosion process only, as the raw-material-handling equipment would be necessary for any process using wood chips as feedstock. The cost of steam is approximately 12 percent of the total pretreatment expense. (It should be noted that although 1000 Btu of steam is required per pound to heat the wood, 80 percent of the steam is recovered and reused; thus, the net steam requirement is 200 Btu per pound of wood.) The economics for producing ethanol from an Iotech-treated feedstock are highly dependent on the selling price of the lignin coproduct. The minimum value of the lignin is as a fuel, with its maximum value as a feedstock for the manufacture of formaldehyde resins. If the larger credit is applicable for the lignin, the net cost of alcohol obtained through the Iotech process would be reduced correspondingly. Several advantages of the Iotech pretreatment process are enumerated below:

- The process provides for comminution of the starting material.
- The pretreated material is compatible with organisms employed for downstream reactions.
- The reactions that normally occur after the wood is heated are quenched; for example, repolymerization of the lignin and the cellulose is prevented.
- The resulting lignin is soluble and very reactive, and of relatively low molecular weight. Separation of the lignin and the cellulose is complete.
- The lignin obtained has the potential of being processed to other chemicals rather than merely serving as a fuel.
- The process achieves sterilization of the feedstock during the pretreatment, thus circumventing possible contamination problems downstream in the biological processes.

A $\frac{1}{2}$-ton-per-day pilot plant is in operation, and the Iotech process will be incorporated into pilot projects expected to come on line in the near future at Georgia Tech and in Kansas. Scale-up is envisioned as a number of parallel guns rather than one or two vastly enlarged units. To process 250 tons of wood per day, four 30-cubic-foot guns would operate in parallel.

Research is currently focused on improving the combined yields of cellulose and hemicellulose. Preliminary findings indicate that the yields may be increased by higher temperatures and shorter residence times.

U. S. Army Natick Research and Development Command

Over 100 different cellulosic materials have been evaluated under a variety of pretreatment conditions at Natick Laboratories. Numerous chemical and physical pretreatment methods have been explored, including milling, steaming, and solvent treating. The most effective pretreatment thus far for a feedstock with 20- to 25-percent moisture content (e.g., garbage) has been two-roll milling. After fewer than 20 passes—and in many cases fewer than 10 passes—through the two-roll mill, newspaper, other cellulosic wastes, and several virgin wood feedstocks are reduced to an easily powdered material. The mills used are of the type found in the rubber and plastics industry. The treatment is rapid (5 minutes or less) and effective in rendering the cellulosic material vulnerable to subsequent enzymatic hydrolysis. The energy required to run the pretreatment series is approximately 0.225 kilowatt-hour per pound of newspaper or other cellulosic waste. This figure is translated to 11.03 kilowatt-hours per gallon of ethanol, or approximately $0.45 to $0.65 per gallon, depending on local electrical utility rates.

A significant feature of the two-roll-mill-treated newspaper is that slurries of approximately 20 percent cellulose can be prepared from the pretreated material. In theory, then, a sugar solution of sufficient concentration (10 percent) can be attained, such that a concentration step would not be necessary before fermentation to ethanol.

CELLULOSE HYDROLYSIS PROCESSES

Arkansas/Gulf Process

The Arkansas/Gulf process (Fig. 7-3) is currently the only integrated (that is, feedstock-through-ethanol) process available employing enzymatic hydrolysis of a cellulosic feedstock. A 1-ton-per-day pilot plant has operated for a period of 1 year in a converted railroad car paint barn in Kansas. During this time the process has been refined, and detailed material balance data (not available to the general public) have been accumulated for each stream involved. No exotic construction materials are necessary—only carbon steel or stainless steel—because no particularly corrosive substances are used. Reactions are conducted at normal atmospheric pressure; thus, the need for expensive pressure vessels is precluded. Although a variety of cellulosic feedstocks have been used successfully, most economic data have been developed for the case of a mixed feedstock of municipal solid waste and pulp mill waste. The unique feature of the overall process is that saccharification and fermentation are carried out simultaneously.

Many pretreatment methods were tested, and a moderate mechanical pretreatment was selected for the pilot plant. A horizontal attrition mill—a cylindrical chamber filled with small steel balls and containing an internal agitator—is used to grind the feedstock. After physical pretreatment, a portion of the feed stream is sterilized and diverted to enzyme production; the remainder is pasteurized and fed directly to the simultaneous saccharification and fermentation (SSF) reactor. The enzyme slurry is produced by a mutant strain of the organism *Trichoderma reesei* in a continuous process involving two trains of three reactors connected in series. The total retention time in the enzyme-production units is 48 hours. Enzyme production in the pilot plant is at the 300-gallon scale.

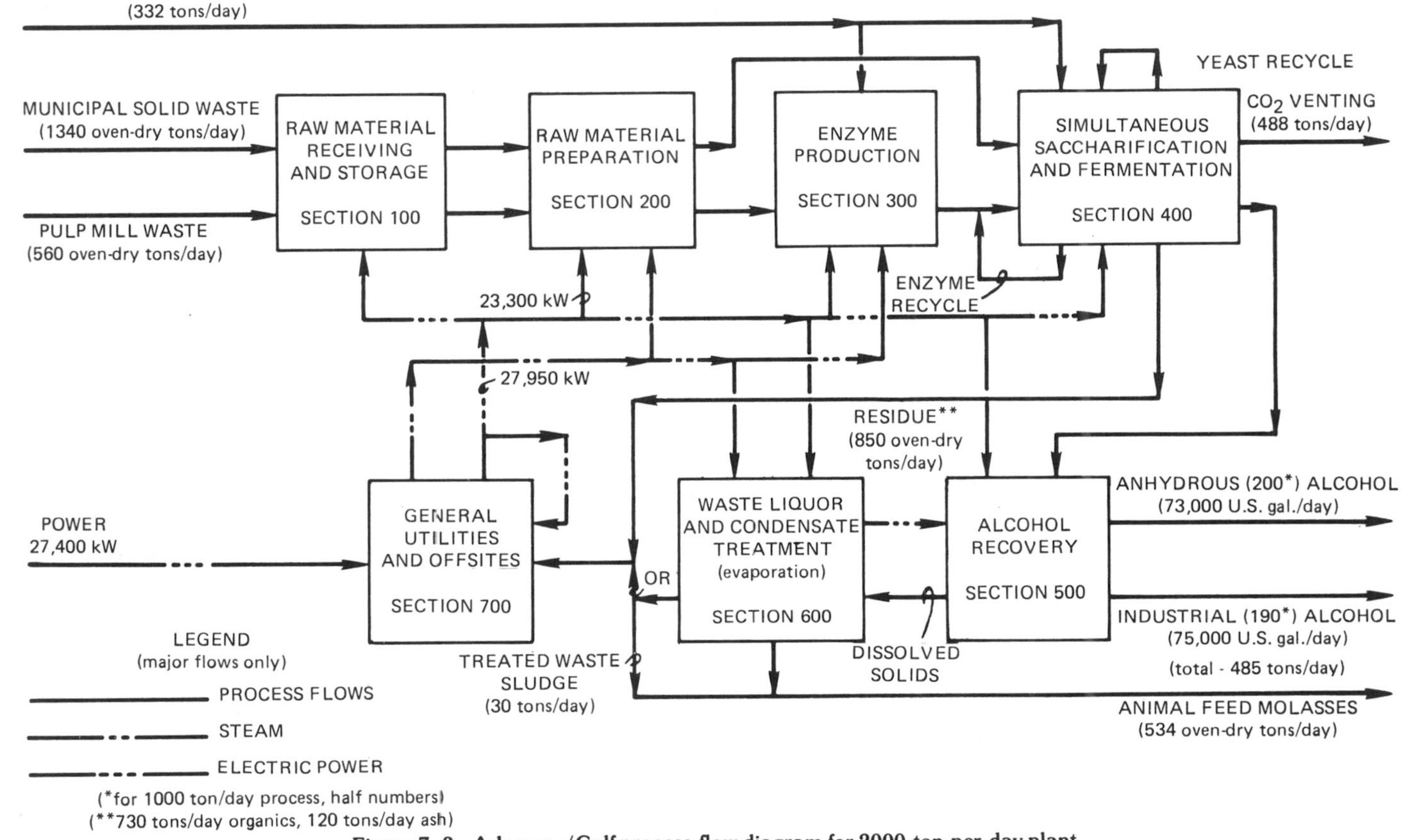

Figure 7-3. Arkansas/Gulf process-flow diagram for 2000-ton-per-day plant.

The simultaneous saccharification and fermentation takes place in the presence of lignin. Yeast and the enzyme slurry (containing approximately 8 percent cellulose) are mixed with the slurried feedstock. Four trains of three fermenters in series are used. After a retention time of 24 hours at 40°C, (104°F), 3.6 percent ethanol is obtained. Several types of yeast, including *Saccharomyces cerevisiae* and *Candida brassicae*, have been tried in the SSF process. Because products (e.g., glucose) formed during the saccharification that serve to inhibit that reaction are removed during the simultaneous fermentation, a 25- to 40-percent increase in ethanol yield is claimed using the SSF process.

The enzyme is recovered, and the cellular solids are removed. The beer slurry is neutralized before distillation. The mash is pumped into a "slurry stripper," a column in which ethanol is stripped using an upward flow of steam that concentrates the ethanol to 25 weight percent in the overhead stream. Waste solids appear in slurry form at the column bottoms. The solids are dewatered and used as fuel, for they consist primarily of lignin and unconverted cellulose. The protein-rich solubles are concentrated to a 60-percent solids syrup and used as animal feed. The remainder of the process consists of a commercially available energy-efficient distillation and dehydration of ethanol.

Economics for the Gulf process appear to be favorable. For the optimum-sized 54-million-gallon-per-year facility (2000 tons per day of municipal solid waste and pulp mill waste feedstock), the selling price of 95 percent ethanol in 1983 is projected to be $1.44 per gallon (a 1979 equivalent of $1.10 per gallon). If 80 percent of the plant financing can be accomplished through the issuance of municipal bonds—a reasonable possibility if municipal solid waste is used as the feedstock—the 1983 selling price is estimated at $0.95 per gallon (equivalent of $0.73 in 1979). The cost of enzyme production is approximately $0.21 per gallon of ethanol. These figures, supplied by the project participants, are not normalized. Although the year 1983 was used as a basis for the economic evaluation, late 1984 is not considered the earliest possible commercialization date. It should be remembered that the economic analysis is based on a plant across the fence from a municipal solid waste treatment facility. The number of locations that could support the optimal-sized plant without unduly large transportation and handling charges for the feedstock should be examined.

The net energy efficiency of the Gulf process when conducted on a large scale is estimated to be 59 percent, and the energy requirement for producing a gallon of ethanol is in the neighborhood of 22,000 to 24,000 Btu. Because a lignocellulosic residue is a coproduct of the process, the ethanol facility will be nearly self-sufficient in terms of energy; that is, only small amounts of fossil fuel and purchased electric power will be necessary to supplement the energy derived from feedstock residues. The developers of the process consider it ready for commercialization.

U. S. Army Natick Research and Development Command

The most promising work at Natick Laboratories involves the MCG 77 mutant of *T. reesei*, an organism that produces a mixture of cellulases. The enzyme is produced at 28-29°C (82-84°F), but hydrolysis is performed at 50°C (122°F). The organism has been "fine-tuned" to grow on actual wastes, such as newsprint, rather than on idealized (and costly) cellulosic substrates. After fermentation, the organism can be filtered out and the enzyme left in the broth, or the intact mixture can be transferred to the hydrolysis vessel. In the former case, the dead organism serves as feed for the growing organism, but separation costs must be balanced against the savings in nutrient costs.

Digestion of cellulosic feedstocks can be improved by removing cellobiose (an inhibitor) by the addition of B-flucosidase as a supplement. In practice, the combination of one unit of cellulase plus one unit of B-flucosidase is as effective as two units of cellulase, but the combination is substantially less expensive. Conditions have been developed whereby 10 percent sugar syrups are obtained in 14 hours from two-roll-milled waste cellulose slurries. The initial charge of substrate solids (20 percent) is increased to an effective 30-percent level by adding substrate as slurry viscosity drops during the early hours of hydrolysis. Maximum enzyme productivity obtained thus far is 125 IU/liter/hr (IU = international units); the ratio of enzyme to substrate employed in the hydrolysis is 10 IU/g.

The broth obtained after enzymatic hydrolysis of the cellulosic feedstock is compatible with *S. cerevisiae*, the yeast used to ferment sugar to ethanol. On a small scale, hydrolysis and fermentation have been carried out at the same time, with higher overall yields due to the removal of sugar from the system. Experiments are underway in immobilization of the yeast.

Process scale is at the pilot plant stage—250-liter (65-gal) hydrolyzer, 300-liter (78-gal) inoculum tank, 400-liter (104-gal) fermenter. The pilot plant is highly instrumented for testing variations in several process variables. Scale-up problems are anticipated to be minimal because the biological efficiency, according to researchers, should improve with increasing plant size.

Preliminary economic evaluation has been made of alcohol production using the pretreatment and enzymatic hydrolysis processes developed at Natick Laboratories. The cost analysis is based on a plant with a manufacturing capacity of 25 million gallons per year of 95 percent ethanol from municipal waste feedstock as shown in Fig. 7-4. The substrate is 45 percent hydrolyzable to fermentable sugars, which are converted to ethanol in 40-percent yields. The overall material balance is 0.135 pound of ethanol per pound of feedstock, or the equivalent of 1 gallon of ethanol from 49 pounds of waste newspaper. The cost of enzyme production is $0.456 per gallon of ethanol (which is 37.5 percent of the total cost), and pretreatment costs are $0.224 per gallon of ethanol (which is 18.4 percent of the total cost). After

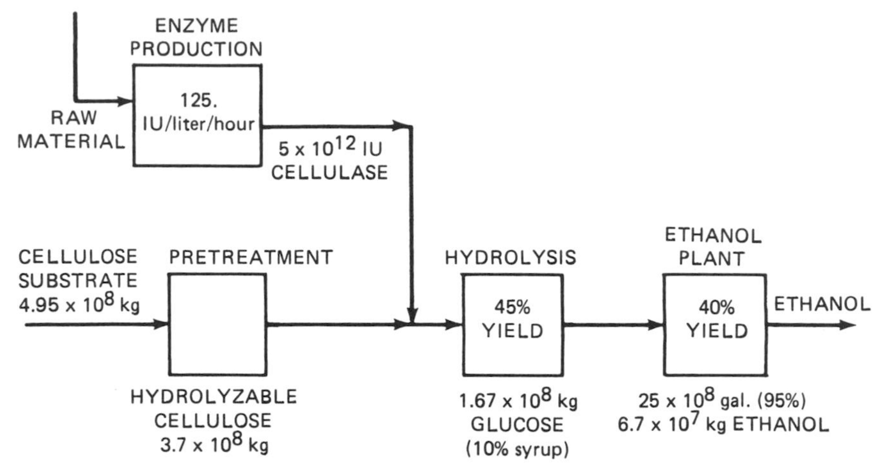

Figure 7-4. Natick process-flow diagram for 25-million-gallon-per-year plant.

process optimization, enzyme costs are projected to fall to approximately $0.23 per gallon by the mid-1980s.

Capital investment for the plant is projected to be $75 million, and the factory cost of ethanol would be $1.22 per gallon. Although no credits are taken in the economic analysis, those anticipated for process steam (from lignocellulosic material obtained by evaporating or centrifuging hydrolyzer residues to 50 percent solids) and for cellular biomass could reduce the factory cost substantially. Wastes could fulfill process-heat requirements, and the cellular biomass would be suitable as animal feed or fertilizer. The anticipated level of credits is $0.28 per gallon, reducing ethanol cost to $0.94 per gallon at the factory gate.

The process, the developers feel, would be compatible with a multifeedstock operation. Mr. Spano of the Natick Laboratories also suggests that enzyme production be centralized, with shipments made to locations with ample feedstock supplies. This might be more efficient overall than either installing relatively small enzyme production facilities at several locations or shipping the feedstock to a complete (that is, feedstock-through-ethanol) alcohol-production unit.

Massachusetts Institute of Technology Process

The aim of the research at the Massachusetts Institute of Technology (MIT) is a simple, one-step bioconversion (i.e., homofermentation) of cellulosic feedstocks to useful chemicals, including ethanol. A mixed culture of thermophilic bacteria is used for the simultaneous hydrolysis and fermentation of cellulosic wastes to a mixture of ethanol, acetic acid, and lactic acid. Work is underway to suppress formation of the acid coproducts and maximize ethanol production. The major recent advances in the work involving ethanol-production techniques are summarized in the following paragraphs.

The anaerobic, thermophilic organism *Clostridium thermocellum* is capable of degrading biomass by producing cellulolytic enzymes. The reaction could use improvement in that the production rate of ethanol is slow, the organism cannot tolerate high concentrations of ethanol, and organic coproducts (especially acetic and lactic acid) are formed. In addition to glucose, *C. thermocellum* produces five-carbon sugars, but does not metabolize them further. Relatively high concentrations of both five- and six-carbon reducing sugars accumulate during fermentation by the organism. Separately, steps were taken to increase the ethanol tolerance of the organism and to eliminate, or at least to minimize, the formation of lactic and acetic acid coproducts.

To increase the organism's ethanol tolerance, serial transfer methods were employed. This procedure entails gradually changing the environment of the organism with an adaptive technique. In this case, ethanol concentrations were increased by approximately 0.2 percent at each stage, arriving finally at a strain that can tolerate 8-percent concentrations of ethanol. After each adaptation, checks were routinely performed to ascertain that the organism could still degrade cellulose and that it still had the capacity to produce ethanol.

Coproduct formation has been decreased by conventional mutation techniques, yielding a strain, labeled S-7, that is unable to produce lactic acid and that produces relatively small amounts of acetic acid. The reason for the large amount of attention given to reducing the amount of acetic acid formed is that, if ethanol is produced by the unaltered organisms, tremendous quantities of acetic acid would also be produced. As an example, 1 quad of ethanol from biomass would yield approximately 50 times the current annual U.S. consumption of acetic acid.

One of the remaining difficulties is the conversion of the accumulated reducing sugars to ethanol. An additional problem area is finding suitable ways to use the xylose resulting from the degradation of hemicellulosic material in the reaction medium. The organism *C. thermosaccharolyticum* is able to use both five- and six-carbon sugars. During its fermentation, however, the same set of difficulties result as in the fermentation with *C. thermocellum*, that is, the organism does not tolerate high ethanol concentrations, and organic acid coproducts are formed along with ethanol. A series of experiments were performed involving mutation induced by nitrosoguanidine and selection for ethanol tolerance and low acid production. The result has been a strain, labeled HG-4, that possesses the desired characteristics.

A mixed-culture (the S-7 strain of *C. thermocellum* and the HG-4 strain of *C. thermosaccharolyticum*) fermentation has been carried out on an ideal cellulosic substrate (Solka floc) and on corn stover that had not been pretreated. Both organisms were engineered to produce relatively small quantities of undesired coproducts. The former organism degrades cellulose, produces a small amount of ethanol, and leaves reducing sugars in the medium. The latter organism, inoculated after the fermentation with *C. thermocellum* is well underway, uses the five- and six-carbon sugars to produce ethanol and small quantities of coproducts. Results of mixed-culture fermentations are listed in Table 7-1. The ethanol yield from Solka floc is 45 percent and from waste corn stover, 26 percent.

Conceptual process designs have been developed for the large-scale (27 or 41 × 10^6 gallons per year) manufacture of ethanol by the MIT process. Manufacturing costs of producing ethanol from corn stover are estimated to be comparable to those of ethanol from corn.

University of Pennsylvania/General Electric Reentry Systems Division

The thrust of the work by the University of Pennsylvania/General Electric team is the total utilization of cellulosic raw material. Using wood chips (poplar) as a feedstock, a preliminary separation yields three phases, each of which is used in a separate process as shown in Fig. 7-5. The cellulosic feedstock undergoes a preliminary solvent delignification with alkaline aqueous butanol for 15 to 30 minutes at 150-175°C (302-347°F). Lignin is solubilized in butanol; hemicellulose, in water. Pretreatment is being optimized for susceptibility of the cellulosic fraction to enzymatic hydrolysis. The cellulose is removed as solids, and the hemicellulose in the aqueous phase is hydrolyzed. Acidification precipitates the xylans, which yield xylose upon treatment with the xylanases derived from *Thermomonospora*. The remaining hemicellulose degradation products are fed as a dilute stream for use as a carbon source in an extractive fermentation process for the production of

Table 7-1. Summary of Results for the Mixed Culture (*C. thermocellum*, S-7, and *C. thermosaccharolyticum*, HG-4) Fermentation of Biomass

Substrate	Solka floc (g/liter)	Corn stover (g/liter)
Total fed	70	100
Residue	5	62.7
Amount consumed	65	37.3
Ethanol	29	9.7
Acetic acid	6.8	6.4
Lactic acid	3.2	0.65
Reducing sugars	0.5	0.8
Cells	3.6	1.3
Ethanol yield (g/g substrate consumed)	0.45	0.26
Total product yield (g/g substrate consumed)[a]	0.61	0.47

[a]Cells excluded.

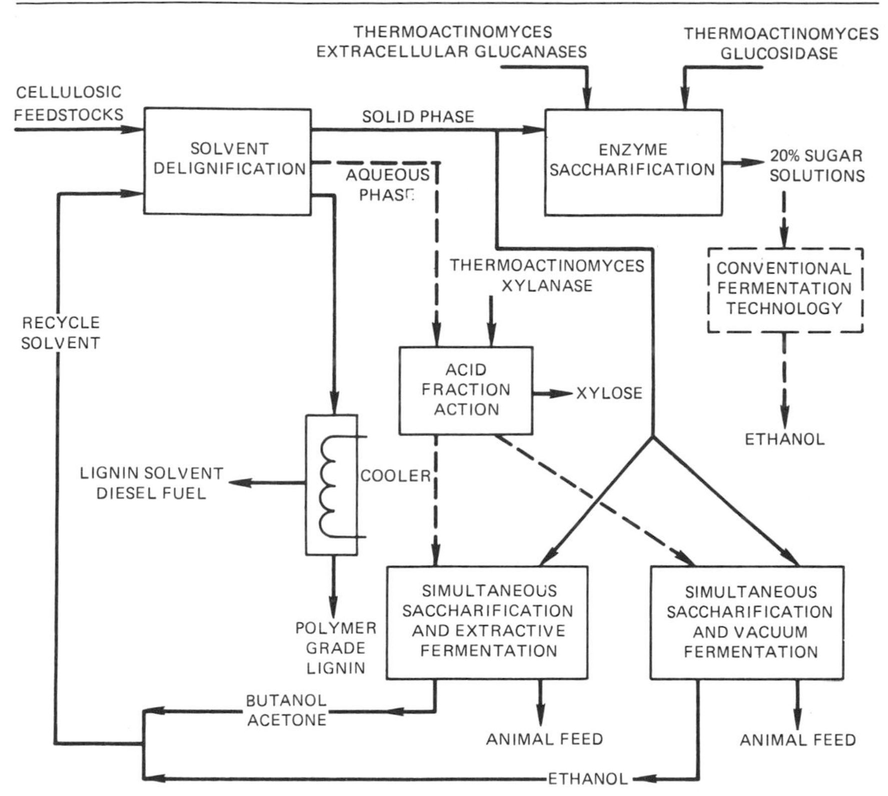

Figure 7-5. University of Pennsylvania/General Electric process for total biomass use.

butanol. The lignin in the butanol phase can be purified to polymer-grade lignin or concentrated for use as a Bunker C type fuel or diesel fuel in a butanol slurry. Use of the noncellulosic fractions is less advanced in development than the use of the cellulosic solids; a commercialization time scale for the former process is estimated by the researchers as no sooner than 6 or 7 years. Use of the cellulose, the portion of the process closest to commercial readiness, is outlined in the following paragraph.

The enzymatic saccharification of cellulose has been carried out at a 40-liter (10.4-gal) scale at 55°C (131°F) and a pH of 6.6 with a mixture of cellulases produced by *Thermomonospora*. Through the use of immobilized cells or cellobiose, the production of glucose syrups of greater than 20 percent solids is expected. (The goal of 20 percent sugar derives from the economic optimum with regard to steam usage in distillation, which is 8 percent ethanol in the broth, along with the practical yield of 44 percent ethanol from sugar.) The resulting sugar solution would be used for a typical yeast fermentation to produce ethanol. The cell immobilization should permit a high hydrolysis rate in the presence of high concentrations of glucose and product. Work is also being done in carrying out the cellulose saccharification simultaneously with cellobiose fermentation by *C. thermocellum* at 60°C (140°F). At this temperature, the broth could be vacuum flashed at pressures between 140 and 173 mm of mercury to yield a vapor phase approximately nine times enriched in ethanol compared with the broth.

The cellulose solids could, alternatively, be used in a simultaneous saccharification and fermentation to produce butanol. Because the organism that effects fermentation to butanol (*C. acetobutylicum*) exhibits low tolerance to the butanol product, work is underway to develop a continuous extraction process to remove butanol from the medium as it is formed. One promising solvent for such an extractive fermentation is dibutyl phthalate.

Preliminary economic analysis of the Penn/GE process is based on enzymatic hydrolysis costs being comparable to those reported by Gulf for the Gulf-Arkansas process and those reported by Katzen for the wood-handling and -treatment portions of the process. The materials balance assumptions are as follows: 56 percent cellulose in the feedstock; 50-percent yield to glucose; 50-percent conversion to alcohol. Credits are taken for animal feed and spent solids only. For a 25-million-gallon-per-year ethanol plant based on the Penn/GE process and the assumptions outlined above, the cost per gallon of ethanol product is projected as $1.07 at 31.7-percent ethanol yield, and $0.74 at 95-percent yield. Although the overall costs are very sensitive to yield, coproduct credits have a greater incremental effect on manufacturing cost when yields are low. If feedstock can be obtained at less than $20 per oven-dry ton, fermentation of cellulosic feedstock is competitive with ethanol from hydration of ethylene; the fermentation is purported to be marginally competitive even at $30 per ton of feedstock.

In addition to research on using the three major fractions obtained from wood feedstock in the manner described above, work is underway in several other areas, as follows:

Co-immobilization of yeast and glucosidase and other means of maintaining enzyme activity

Upgrading the lignin to polymer-grade material

Development of a two-reactor scheme (with recycle feature) for producing concentrated glucose solution

Adsorptive fermentation

Use of municipal solid waste as feedstock

Most of the work is under experiment at the 1- to 10-liter (0.26- to 2.6-gal) scale, and process integration has not yet been accomplished.

Continuous Acid Hydrolysis—New York University Process

A twin-screw extruder used in the plastics industry has been adapted by New York University (NYU) for use in a continuous acid hydrolysis process employing cellulosic wastes as feedstocks as shown in Fig. 7-6. The extruder conveys, mixes, and extrudes the materials under controllable conditions of temperature and pressure. A volumetric feeder is used to introduce the cellulosic feedstock, which is sheared into a slurry. As it is transported, the slurry becomes dense and forms a plug, preventing backflow. Steam is injected into the reaction zone and is used externally around the barrel to heat the feedstock to 220-230°C (428-446°F). During conveyance, additional shearing action breaks up the fiber, rendering it more accessible to hydrolysis. Near the end of the machine, dilute sulfuric acid (approximately 1 percent by weight) is injected such that the total contact time between feedstock and acid is less than 20 seconds. Pressures of approximately 500 psi develop in the extruder. A ball-valve is employed for semicontinuous discharge of the hydrolyzate, which is expelled as a viscous mud containing approximately 30-percent glucose.

Preliminary economic studies (using sawdust as feedstock) indicate that glucose can be produced for \$0.03 to \$0.04 per pound. Assuming state-of-the-art energy-efficient distillation, 60-percent conversion of α-cellulose to glucose, and 45-percent conversion of glucose to ethanol, ethanol can be produced, according to the researchers, for \$0.80 per gallon when this pretreatment method is used. An economic analysis has been completed for a 2000-ton-per-day (27-million-gallon-per-year) plant without utilization of the hemicellulose fraction of the feedstock. If extraction is followed by fermentation of the xylose to ethanol, in addition to the separate conversion of the residual hexosan to glucose, then a 2000-ton-per-day plant is projected to produce 48 million gallons of ethanol per year. An economic analysis for the latter case is also available.

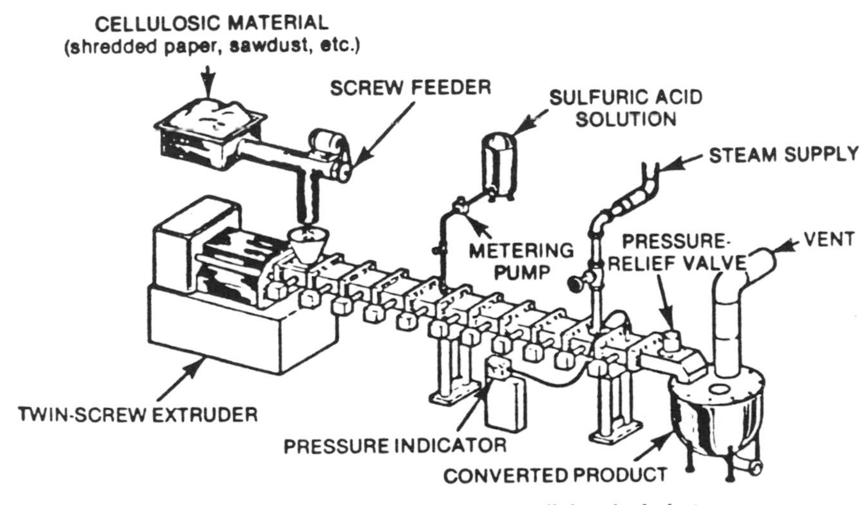

Figure 7-6. New York University continuous cellulose hydrolysis process.

The energy requirement for pretreatment is reported to be 1200 Btu per pound for sawdust. Seventy-five percent of the energy is in the form of steam, and 25 percent is electricity. This number does not include energy required to recover the sugars from hemicellulose, only from α-cellulose.

Corrosion of the extruder during the hydrolysis has not, according to the researchers, been a problem to date; rather, erosion/wear is of concern in isolated areas of the machine. Further research in the areas of materials and lubricants should improve the process. The problems remaining are the engineering problems associated with any scale-up, and general fine-tuning and process control (perhaps with special valves or computer control). Numerous improvements and refinements are expected, but the developers anticipate no barriers that would preclude early commercialization of the pretreatment process.

The extent of treatment of the hydrolyzate necessary before fermentation may be underestimated. Little or no work has been done to ensure that the "mud" discharged from the extruder after acid treatment is compatible with the biological reactions downstream. The assumption has been made that the success of the Madison process[3] provides a sufficient basis for optimism regarding the compatibility of the hydrolyzate with the remainder of the alcohol manufacture process.

Purdue/Tsao Process

The so-called Purdue process generally refers to an acid/methanol scheme. Cellulosic feedstock (corn stover) is treated with hot 2 percent sulfuric acid to remove pentosans, after which the solids are first dewatered and then dried by application of heat. Although the pentosans can be converted to butanediol, the market for this product may be somewhat limited. Furfural could also be produced from this fraction.

After the pentosans are removed, the cellulose is "dissolved" in 70-80 percent sulfuric acid; a modification of the cellulose/lignin structure is effected by the concentrated acid. Methanol is added to precipitate the cellulose (amorphous), and the solids, consisting of both cellulose and lignin, are separated. Methanol and sulfuric acid are recovered and recycled. The solids are hydrolyzed at 125°C (257°F) with 2.5 percent sulfuric acid, and the residual solids (lignin) are removed by filtration for use as fuel.

Several potential problems are associated with this process. Most importantly, it has not been demonstrated at any appreciable scale. Additionally, material problems should be expected under the acidic conditions employed: lime that is used for neutralization could cause scaling problems, side reactions between methanol and sulfuric acid have not been taken into

[3]The Madison process is an acidic hydrolysis process that was used in the United States during World War II. In this process alcohol is produced at the rate of 52 gallons per ton of dry, bark-free wood.

account, and solvent losses have been estimated as being very low but have not been demonstrated.

A preliminary economic evaluation of the process indicates a cost of $0.045 per pound of fermentable sugar (that is, hexose plus pentose). The economics were performed for the most favorable case and include neither storage costs for the huge amounts of stover that would be needed for the reference-sized plant (25 million gallons per year) nor waste treatment costs. One ton of dry corn stover is expected to yield 41.7 gallons of 95 percent ethanol. The energy requirement for the process is estimated to be 2800 Btu per pound of stover.

Recently, emphasis at Purdue has shifted to two other lines of research, dilute acid hydrolysis of grain and an alternate, three-stage cellulose hydrolysis process.

Dilute acid hydrolysis. If starchy feedstocks such as grain are subjected to weak acid hydrolysis, not only the starch but also a portion of the fiber can be converted to sugar. (The sugar derived from the fiber is, however, predominantly xylose rather than glucose.) The dilute acid hydrolysis can increase the yield of ethanol from 2.5 gallons per bushel to 3.0 gallons per bushel. Along with the 20-percent increase in ethanol yield as a reduction in the quantity of distiller's dried grains obtained, because some of the fibrous material is converted to product rather than coproduct. The fermentation of the sugars obtained upon dilute acid hydrolysis is accomplished with a modified yeast process, in which a mixed culture is employed to metabolize both the five- and six-carbon sugars.

Three-stage process. The three-stage cellulose process consists of three separate "cuts" at progressively higher cost per incremental gallon of ethanol. At any of the three stages, the residue can be treated for fuel use; after each of the first two steps, the residue can instead become the feedstock for the next phase of the operation. This process has been tried on a laboratory scale only; consequently, cost figures mentioned below, obtained from the researchers, should be regarded as preliminary estimates.

Stage 1—During stage 1, the hemicellulose fraction of the feedstock is used to manufacture ethanol. Fresh corn stover is treated with 0.5 percent acid at 95°C (203°F) for 6 hours. The resulting sugar can be converted to ethanol (45 gallons per ton of fresh dry stover; stover loses a substantial amount of its hemicellulosic fraction on storage) at a cost of $1.00-$1.10 per gallon. The energy requirement for the process is 55,000 Btu per gallon. Equipment required for the process is ordinary silage equipment. The cost of the sugar is $0.025 per pound. The lignocellulosic residue obtained can be burned to yield 9 million Btu per ton of stover.

Stage 2—The wet lignocellulosic residue from Stage 1 can be fed into a grinding refiner of the type used in pulp milling operations (one stationary and one rotary disk) to shear the residue into fine particles. Sulfuric acid (3-5 percent) is added to convert approximately 50 percent of the cellulose to glucose, which in turn yields 25 gallons of ethanol per ton of feedstock. The cost of alcohol in this stage is $1.20 per gallon. The residue is lignin and about one-half of the original cellulose, which can be used as a fuel or carried on to Stage 3.

Stage 3—The remaining cellulose is highly crystalline and requires severe reaction conditions to liberate the monomer units. Sulfuric acid (70-75 percent) is added to disrupt the crystalline structure, and water is added for hydrolysis. The mixture is neutralized with ammonia and lime, and the lignin is removed before fermentation. An additional 20-25 gallons of ethanol per ton of feedstock can be derived in this stage. If lime is used in the ethanol dehydration step of the overall distillation process, the calcium hydroxide thus obtained can double as a base to neutralize the acidic hydrolysis medium.

Summary and Comparison of Cellulose Hydrolysis Processes

Table 7-2 presents the key features of the cellulose hydrolysis processes discussed along with data on the commercialized Madison process for contrast. The entry under "feedstock" indicates the feedstock tested at the largest scale, which is also indicated. Unique features and major uncertainties in the development of each process are also given. Normalized ethanol cost figures are listed if available.

The estimates of time until commercial readiness—necessarily somewhat subjective—were made after consideration of published literature and discussions with project principal investigators. In the relatively near term, the Arkansas/Gulf, Natick, and New York University processes could be commercialized. Even these, however, have serious uncertainties associated with them, as noted.

OTHER ADVANCES IN PROCESS TECHNOLOGY

Biological Research

Much biological research in the field of ethanol production has focused on the following major areas:

Increasing the rate of ethanol production

Minimizing coproduct formation

Increasing the tolerance of organisms to end products

Increasing the rate and/or decreasing the cost of producing cellulase enzymes

Using sugars derived from hemicellulose

Table 7-2. Key Features of Cellulose Hydrolysis Processes

Process	Type of hydrolysis	Feedstock(s)	Scale	Unique features	Major uncertainties	Normalized ethanol cost ($/gal)[a]	Time to commercial readiness (years)
Madison	Acid	Wood chips	Commercial (220 tons/day in United States)	No pretreatment	—	—	0
Arkansas/Gulf	Enzyme	Municipal solid waste and pulp mill waste	1 ton/day pilot plant	Simultaneous saccharification and fermentation; integrated process demonstrated	Scale-up	0.68	0–3
Natick	Enzyme	Waste newspaper	Small pilot	Extensive pretreatment; concentrated sugar solutions obtained	Scale-up; feedstock adaptability	1.05	0–3
NYU	Acid	Sawdust	Small pilot	Feedstock through sugar in one continuous process	Scale-up; feedstock adaptability; materials wear; process integration	—	0–3
MIT	Enzyme	Corn stover	Bench	Homofermentation; minimizes organic coproducts; mixed culture uses five- and six-carbon sugars	Scale-up; feedstock adaptability	0.88/0.66[b]	3–5
Penn/GE	Enzyme	Wood chips	Bench	Total biomass use	Scale-up; process integration	—	>5
Purdue (three-stage)	Acid	Corn stover	Bench	Separate use of hemicellulose and cellulose	Scale-up; materials; process integration	—	>5

[a]Normalized.

[b]Higher figure for base case; lower figure for optimistic case.

Work in some of these areas has been summarized in the discussions of the Gulf, Natick, Penn, and MIT research programs. In all of them, mutation and selection techniques can be used to enhance desired traits in the organisms employed at various stages of the alcohol-production process. Less drastic measures such as changes in reaction conditions, nutrient composition, equipment, etc., can also be used to optimize the output of the organism for each type of bioconversion. In this second category (i.e., nonmutational) of research are projects such as those listed:

Continuous production of cellulase

Optimization of batch cultures and initiation of continuous culture studies

Use of high-cell-density systems involving cell recycle (to decrease the amount of carbohydrate consumed in yeast growth)

Recovery and reuse of enzymes—solid residues from enzymatic hydrolysis can be treated to release adsorbed cellulase from the unhydrolyzed cellulose (urea, for example, desorbs 45 percent of the adsorbed enzyme)

Improvement of hydrolysis reactor design—studies of kinetics of enzyme hydrolysis are expected to yield information that should facilitate the successful design of a reactor

Immobilization of enzymes

Development of tests to evaluate the effectiveness of new mutant strains

Work is in progress in all of these research areas. The major uncertainty involved in the isolated development of portions of the alcohol-production process is that it is not possible to predict the behavior of an integrated process involving one or more of the improved individual steps. For example, recovery of the enzyme complex—and individual step improvement—could lead to a deficiency in a specific component of the complex. Most of the other isolated research areas have similar potential limitations.

Membrane Technology

Membranes, both natural and synthetic, are able to discriminate between substances on the basis of several properties, including electric charge and size. If membrane technology advances to the point of the availability of large, inexpensive membranes, there are several potential applications for them in the production of ethanol by fermentation:

- Concentration of dilute feedstock (such as sorghum juice) before fermentation

- Treatment of fermentation broths or stillage to separate high-molecular-weight dissolved and suspended solids to produce high-quality distiller's dark grains
- Selective removal of alcohol from fermentation broth
- Removal of water from ethanol/water mixtures

A process suggested by Dr. Harry Gregor of Columbia University is still at the conceptual stage. It involves preliminary size reduction of a cellulosic feedstock, drying, treating with 40-42 percent hydrochloric acid at room temperature, filtration of colloidal lignin particles, removal of lignin fines through ultrafiltration (a membrane process), separation of sugars from the hydrochloric acid by electrodialysis (another membrane process), fermentation, and removal and concentration of ethanol from the fermentation broth (a third membrane process). Although the reaction scheme is quite attractive in that it involves minimal amounts of process heat, commercialization of the membrane steps is viewed as several years in the future.

Vacuum Techniques

Process development is underway using vacuum techniques in the fermentation step and/or in the distillation step of ethanol preparation.

Fermentation. Although the idea of removing ethanol from the fermenter as it is formed is attractive, vacuum fermentation ordinarily involves a significant energy and capital expense, in that sophisticated equipment is required to maintain a vacuum such that ethanol can be removed at the temperature tolerated by *S. cerevisiae*. Substituting a suitable thermotolerant organism would allow this technique to be used at higher temperatures and pressures, with lower associated costs. Removal of ethanol would serve to enrich ethanol in the vapor sent to the distillation tower (in effect, a preconcentration) and would lessen end-product inhibition of the yeast, thus increasing the rate of ethanol formation. However, vacuum fermentation has several disadvantages. The noncondensable gases carbon dioxide and oxygen are removed along with the ethanol, and pure oxygen has to be sparged into the fermenter during operation under vacuum. Moreover, the exit gas contains a substantial amount (approximately 11 percent) of the ethanol product that has to be scrubbed out and returned to the system.

A flash fermentation process has been developed by Wilke *et al.* at the University of California at Berkeley. In this procedure, fermentation is carried out at atmospheric pressure, thus maintaining two major advantages of conventional atmospheric fermentation: (1) oxygen required for yeast maintenance can be supplied by sparged air, rather than oxygen; and (2) carbon dioxide is vented from the fermenter, with no recompression required. The beer, containing about 3.5 weight percent ethanol, is rapidly cycled between the fermenter and a small vacuum flash vessel where ethanol is vaporized. Vapor recompression heating is employed. Ethanol (2.4 weight percent) is returned to the fermenter. A small amount of carbon dioxide in the cycling beer goes to the flash vessel, but very little ethanol (less than 1 percent of the product) is carried away with the vented carbon dioxide. The flash fermentation process requires 14 percent less energy than vacuum fermentation. Unfortunately, although the distillation column receives 13.2 weight percent ethanol, the energy required for distillation is only marginally less than that required for distillation of ethanol obtained by an ordinary batch process.

Distillation. At reduced pressures, the ethanol-water equilibrium is altered such that the azeotropic composition moves toward higher ethanol concentration; indeed, the azeotrope disappears below 90 mm of mercury. If vacuum distillation is combined with the flash fermentation process, the total energy required to concentrate the ethanol product can be reduced to 16,000 Btu per gallon of ethanol. This energy requirement contrasts with 20,400 Btu per gallon for flash fermentation plus atmospheric distillation. Overall, the rapid ethanol fermentation with ethanol removal and concentration in an auxiliary flash vessel, when coupled with vacuum distillation, results in a 42-percent energy savings and is estimated to produce a 54-percent reduction in manufacturing cost compared to conventional batch processes and atmospheric distillation. A side advantage of the former scheme is the very high productivities achieved by the yeast when the end product is prevented from accumulating in the fermenter; these productivities have been measured at 80 grams of ethanol per liter per hour.

Dehydration

Besides the conventional azeotropic distillation, several alternative dehydration methods are in varying stages of development and are summarized briefly below.

Extractive distillation. An extractive distillation differs somewhat from the azeotropic distillation described earlier. In the extractive process, a third component is added in the upper part of the distillation tower above the alcohol/water feed. The extractant is distributed in the liquid in all of the lower plates and is removed from the column with one of the components being separated at the column bottoms. As in the azeotropic distillation, the added component changes the distribution of the desired component among the phases present in the system.

Extractive fermentation. As the name implies, extractive fermentation consists of conducting the fermentation while removing the ethanol from the fermentation broth through the use of an extracting solvent. This procedure has the advantage of increasing the fermentation rate, because the ethanol is not permitted to accumulate in the broth and retard the bioconversion reaction. The problem associated with this technique is the toxicity of many potential solvents to the organisms used in the bioconversion reaction.

Solvent extraction. Several laboratories have been developing solvent extraction schemes. At the University of Pennsylvania, for example, work has been done on the use of water-immiscible solvents for alcohol, such as dibutyl phthalate. If the solvent has a much higher boiling point than ethanol, the alcohol can be driven off, with relatively little expenditure of energy, in a single distillation step. Other solvents being investigated include dodecanol, dioctyl phthalate, and diisobutyl phthalate.

Another extraction scheme under development uses a critical fluid to dissolve and extract ethanol from fermentation beer. Carbon dioxide at 50-80 atmospheres pressure is a critical fluid that can be used to extract ethanol from fermentation broth after the solids are removed by filtration. After the ethanol is dissolved in carbon dioxide, the water is easy to remove. Subsequent extraction of ethanol from carbon dioxide is much simpler than extracting it from water. The extraction and ethanol isolation process begins with the fermentation beer and ends with product ethanol and is said to require only 20,000 to 30,000 Btu per gallon of ethanol product. The testing thus far has been at the laboratory stage; a pilot plant to verify the numbers is, according to an Arthur D. Little spokesman, still about 2 years away. One advantage of the use of carbon dioxide is that it is a coproduct of the fermentation process, making the solvent costs minimal.

Adsorption/absorption. Another dehydrating technique, usually more expensive than the azeotropic distillations, is the use of an absorbent to take up the water. Quicklime was used in earlier days but has fallen into disuse. (The use of lime, which forms calcium hydroxide upon reaction with water, could be economical in a process involving acid hydrolysis of a feedstock because the coproduct caustic could be used for neutralization rather than being dried.) Modern absorbents are molecular sieves, which have openings large enough for water molecules to pass into and remain trapped, but too small for ethanol molecules. A selective absorption thus occurs, and the water is preferentially removed from the mixture. Absorbents used as dehydrating agents must be regenerated.

Recently, cornstarch and other readily available cellulosic materials have successfully been used as dehydrating agents. Although additional work must be done to perfect the process, the agents are inexpensive, nontoxic, and easily regenerable. Hot vapor is passed up through a column packed with the absorbent; later, desorption occurs with hot nitrogen. The largest dehydration tested to date was with a column containing 0.5 pound of cornmeal. The absorbent can be recycled at least 20 times, and the energy requirement for the dehydration is estimated as 1000 Btu per gallon of ethanol, compared with the typical 14,000-Btu value for the benzene azeotropic distillation. If cracked corn is used as the absorbent, it is possible to ferment it rather than dry it, further reducing the energy requirement for the process.

Textile fibers have shown some potential as absorbents, according to workers at the Textile Research Institute. Certain textile yarns retard the movement of water vapor but allow organic vapors such as ethanol to pass freely. Water can be removed by a loop of yard fibers that is pulled slowly through a tube containing alcohol/water vapors. Pure alcohol is recovered from the end of the tube. The energy requirements for this process, however, have not yet been determined and are probably substantial, as the yarn has to be dried before reuse.

Chapter 8

Ethanol-Plant Designs

This chapter describes typical and specific fuel-grade-ethanol plant designs for farm-scale, small-scale, and commercial-scale systems.

The U.S. National Alcohol Fuels Commission (NAFC), in a recent publication, described several farm-scale plants producing more than 15,000 gallons of ethanol per year. The system designed by Daryl and Gene Schroder has had wide publicity and was used as input for the SERI small-scale system concept.

The SERI representative plant described in *Fuel From Farms—A Guide to Small-Scale Ethanol Production* is described in this chapter along with the actual design developed by EG&G for the Department of Energy; this plant is currently operating at EG&G's facility in Idaho Falls, Idaho.

The third segment to this chapter describes a large commercial-scale ethanol plant. It is the Raphael Katzen Associates International Inc. design which was described in SERI's *A Guide to Commercial Scale Ethanol Production and Financing* released in December of 1980.

In terms of plant production, a joint Department of Energy (DOE)/U.S. Department of Agriculture (USDA) publication has defined the following ethanol plant capacity size ranges:

Farm scale Small scale	Less than 1 million gallons/year
Intermediate scale	1-5 million gallons/year
Large-commercial scale	15 million or more gallons/year

The information in this chapter is provided only as general information. If you decide to build a plant, contact the specific organizations or individuals for more information.

The production of fuel-grade ethanol can range from very small on-farm operations to very large chemical-plant-type operations. Production capacity costs range from $1.16 to 2.40 a gallon. Recent experience indicates that USDA and DOE loan guarantees find a typical requested funding cost of $2.24 per gallon, including all costs.

Farm-Scale Ethanol Plants

The U.S. National Alcohol Fuels Commission in their publication *Fuel Alcohol on the Farm—A Primer on Production and Use* described several farm-scale ethanol plants. Four of these were selected for inclusion in this chapter because they provide a representative cross section of actual on-farm fuel-alcohol plants. The system schematics are not drawn to scale and, therefore, should not be used as accurate design drawings of the alcohol units.

Some of the units have been constructed by individuals for their own use; others are prototypes of units intended for commercial scale. Each of these systems can be improved upon. They are featured here in order to give the reader an appreciation for the early stages of development of current small-farm-scale alcohol-fuel production. No endorsement of any particular process or still is implied by its inclusion in this handbook.

The reader should remember that there is presently no small-scale ethanol-production system that is mass produced, fully warranted, and independently tested to verify the manufacturer's claims. Current small-scale producers are pioneers in the truest sense of the word. The production of ethanol is a complex process requiring well-designed, -constructed, and -maintained equipment. Before investing in any equipment, a person should consult with a qualified engineering firm who can independently evaluate claims made by any manufacturer of ethanol fermentation and distillation equipment.

This handbook frequently warns of the risks involved in on-farm alcohol-fuel production, but the idea of such production does make sense.

Farmers have been hit hard by the rise in fuel costs; that rise has outstripped increases in farm income. Energy costs have become a significant portion of the delivered cost of agricultural products. Indeed, fuel and energy costs now threaten the existence of many small and middle-sized American farms.

Any interruption of fuel supply is potentially damaging to the United States, but it is particularly serious for a farmer if it occurs during planting or harvesting operations.

Many farmers are responding to the energy challenge by looking for new ways to become more self-sufficient. One facet of this response is increasing interest in the production of fuel ethanol on the farm. Farmers are interested in this fuel primarily for use in vehicles and equipment and as a new commodity for sale. Secondary reasons include the use of the fuel for crop drying and for heating.

Farmers enjoy several advantages over other potential producers of fuel ethanol:

- Feedstocks are readily available. Corn, wheat, milo, and barley, as well as spoiled crops, can be used to produce ethanol. For example, only 15-20 acres of corn are needed to produce enough ethanol to provide the liquid-fuel needs for the average size farm.

- The technology for the fermentation and distillation of crops into fuel ethanol is known. Manufactured, dependable small-farm-scale stills are not yet widely available, and the production of fuel ethanol is not a simple process. But hand-built ethanol stills are feasible in many instances. They produce alcohol fuel plus coproducts that include animal feed.
- Existing gasoline-powered farm equipment can be modified to run on high-proof ethanol. Diesel tractors present special but not insurmountable conversion problems. Tractors designed to run on pure alcohol are being produced abroad and, if domestic demand increased, could be produced here.

For these reasons, many farmers, viewing ethanol production as a valuable addition to their existing operations, are exploring the possibility of constructing on-farm ethanol distilleries. As in any other undertaking involving a substantial capital investment, the construction and operation of what is basically a small-scale industrial operation, risks as well as rewards are involved.

On-farm production and use of ethanol will require investment in not only the still itself, but also in modifications to equipment to permit the use of the alcohol. These costs should be considered prior to making a major financial commitment.

The stills featured in this section range in production capacity from 15,000 to 400,000 gallons per year.

The financial considerations involved in committing to on-farm fuel-ethanol production are fundamentally important. Both tangible and intangible costs and benefits must be considered. Assigning a value to tangible costs—capital investment, feedstock, energy, labor—requires a forthright analysis as described in Chapter 2. Assigning a value to intangible costs and benefits is more difficult and is eventually based on subjective judgments. An example is the value assigned to becoming energy self-sufficient. As long as alternative liquid fuels are available at a lower cost than ethanol, the benefits are largely personal satisfaction and the assurance of a fuel supply in the event that alternatives cease to be available. As the price of petroleum rises, this intangible benefit pays off in dollar savings.

The tangible cost of feedstock for the ethanol still is a critical one. Crops may have a direct market value higher than their value as feedstock for the still, even after the costs of processing (drying, etc.) and transporting them to the point of sale are calculated.

Similarly, a farmer considering ethanol production must look at available labor, possible slack time, and other factors in analyzing the financial aspects of constructing and operating a still.

The considerations that must enter into the decision to produce ethanol on the farm—even on a small scale—are complex and must take into account a number of interrelated factors affecting overall farming operations. A preliminary accounting of all these factors must be performed before any serious thought can be given to on-farm alcohol production.

APPLE AGRI-SALES

Phil Apple owns Apple Agri-Sales and has been operating his Tri-Star still since April, 1980. Figure 8-1 provides a general concept illustration of the Apple Agri-Sales Tri-Star still. The unit, a 500-gallon propane-fired batch still, is claimed to be capable of making 35-40 gallons of 180-proof alcohol per batch. A single-tank unit sells for $8000 and a three-tank unit for semicontinuous production is about $18,500.

For more information, contact Phil Apple, Apple Agri-Sales, Rural Route 6, Crawfordsville, Indiana 47933.

Feedstock/Preparation

Apple uses corn only. The corn, which comes directly from the local elevator across from Apple's shop, is ground through a $\frac{13}{16}$-inch screen (fine-grade, ground corn) for pulp feed.

Cooking

Approximately 200 gallons of water are heated to 150°F (66°C) and the proper amount of liquefaction enzyme is added. After a few minutes the ground corn (14-16 bushels) is added. Then the mixture is boiled and stirred constantly for 30-40 minutes. As the corn is cooked, the starch granules are broken down.

When the cook cycle is complete, the mash is cooled to 140°F (60°C) by adding water to the tank. The pH is adjusted to 4.3 and the saccharification enzyme is added. The mash, mixed constantly with a paddle-wheel stirrer, is held at 140°F for about 30 minutes. Finally, the balance of the water is added—a total of about 26-30 gallons per bushel of corn—and circulated through cooling coils to lower the temperature to 90°F (32°C).

Fermentation

When the mash has cooled to between 88 and 91°F (31-33°C) it is ready for fermentation. Apple uses GB-Red Star yeast, approximately a 1-pound package per batch. The dry yeast is added to the mash, stirred constantly for about 1 hour, and then left to ferment for approximately 60 hours. Automatic temperature sensors cycle water through the cooling coils to maintain the temperature between 88 and 93°F (34°C). When the cooling cycle is engaged, the mixing paddle starts up. During the winter when the cooling cycle is not required, Apple activates the mixer for about 5 minutes

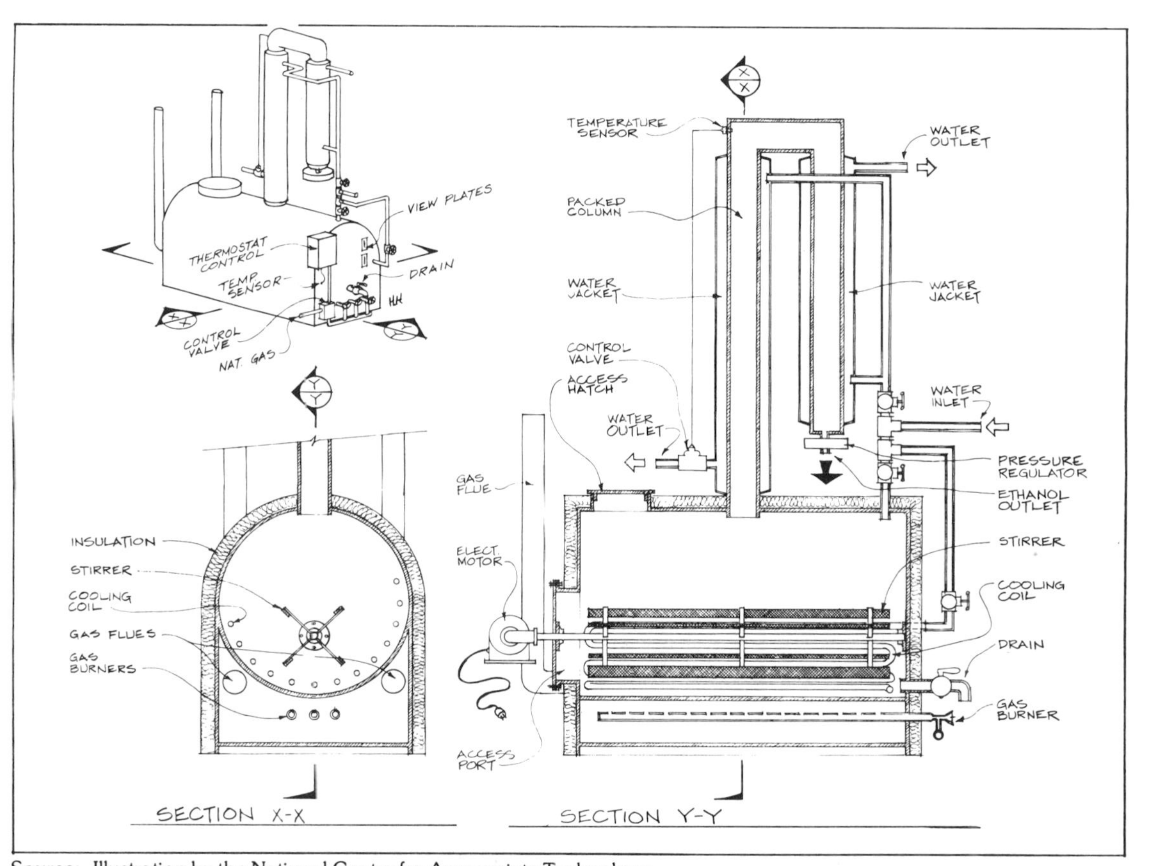

Source: Illustration by the National Center for Appropriate Technology.

Figure 8-1. Apple Agri-Sales Tri-Star farm-scale still.

each morning and night to stir up the mash. Alcohol concentrations in the beer have been estimated at 8-10 percent.

Distillation

The Tri-Star unit uses a triple-tube propane-fired burner. When the fermentation is complete, the system is turned on and the fermented mash is heated. The alcohol and water vapors are driven off the mash and travel up a distillation column packed with metal turnings. The temperature at the top of the column is maintained at 174°F (79°C) via a water jacket. The column and condenser are each 10-feet tall with a 5.5-inch inside diameter and a 6-inch outside diameter. The flow rate varies with the concentration of alcohol in the beer, but averages are claimed to be about 5-6 gallons of 180-proof alcohol per hour over the distillation cycle.

Distiller's Grains

Apple is using a screen box to separate the solids from the liquid in the spent beer. Upon completion of the distillation cycle, the tank is drained into a large, shallow box with a screen at the outfall end. The solids are dewatered for 24 hours and the liquid is discharged to a gravel drain field. The solids, at about 80 percent moisture, are fed to a neighbor's cows. The cattle raised on the wet distiller's grains were doing quite well.

To prolong the storage of this wet feed, it is sun-dried. According to Apple, an average of 16-18 pounds of distiller's grains is obtained per bushel of corn, and this feed averages about 28 percent protein.

Product Use

Apple uses the fuel alcohol in his 1977 Chevrolet truck. At first he experienced a problem with clogged fuel filters; but after enough alcohol was run through the fuel system to clean it up, the truck ran well, with no problems. Apple does use a fine-grade fuel filter now to strain out any sediment that might collect with the alcohol.

Energy Inputs

Apple estimates the energy input on his propane-fired single-tank unit at 64,000 Btu per gallon of ethanol. However, with proper insulation of the tanks and column, water recycling, and heat exchangers, Apple thinks energy inputs will drop significantly. Estimated electrical energy costs for the stirrer and process controls are 6-10 cents per gallon of ethanol at 180 proof.

Manpower Requirements

The Tri-Star unit is fairly automatic with temperature sensors and a timer to control most of the cycles. "Once you've loaded the system (made your mash), it pretty much runs by itself. You should check it periodically during the distillation cycle to make sure it's running OK."

Apple estimates 6-8 manhours are required per 30-40-gallon batch of alcohol. He plans to automate the system even more to make it simple enough to be part of the daily farm-chore routine.

Contamination Control

Apple recommends that the system be flushed out after every third or fourth batch. However, he has run more than 10 batches in succession without a cleanout. "Every time you boil the mash (for 30 minutes), you sterilize the system. We've never made a bad batch yet."

BECKMAN CONSTRUCTION COMPANY

Beckman Construction Company has developed a farm-sized fuel-alcohol plant as a prototype for a larger plant design. Their unit is a 1500-gallon batch plant, with a claimed capability of producing approximately 500 gallons of 189-proof alcohol in 6 hours (25 gph distillation columns). Larry Gosset, vice president of Beckman's alcohol engineering division, has operated the pilot plant since January, 1980. Including their experimental laboratory and the numerous design changes in the pilot plant, Beckman has invested over $175,000 to design and construct this prototype. Although Beckman does not intend to market small plants (the company plans to design and build larger commercial-sized units), Gossett says the prototype would sell for approximately $85,000. Figure 8-2 shows the system layout for the Beckman Construction Company project.

For more information, contact Larry D. Gossett, Vice President, Alcohol Engineering Division, Beckman Construction Company, 7201 West Vickery Street, Fort Worth, Texas 76116.

Feedstock/Preparation

A number of different feedstocks have been used including bakery waste, stale donuts, cookie dough, cookie waste, biscuit dough, sweet sorghum, sweet sorghum silage, sweet sorghum juice, and grains, such as milo.

Milo was used to standardize the plant. Using a hammer mill, the milo is ground to approximately 20-25 percent retention on a #20 mesh screen. Rolled milo was tried first, but it was found that hammer milled milo produced better results.

Cooking

Beckman cooks about 60 bushels of milo per 1500-gallon batch. The milo is added to about one-half tank of water. Using constant agitation, the liquefaction enzyme is added and steam is injected directly into the tank,

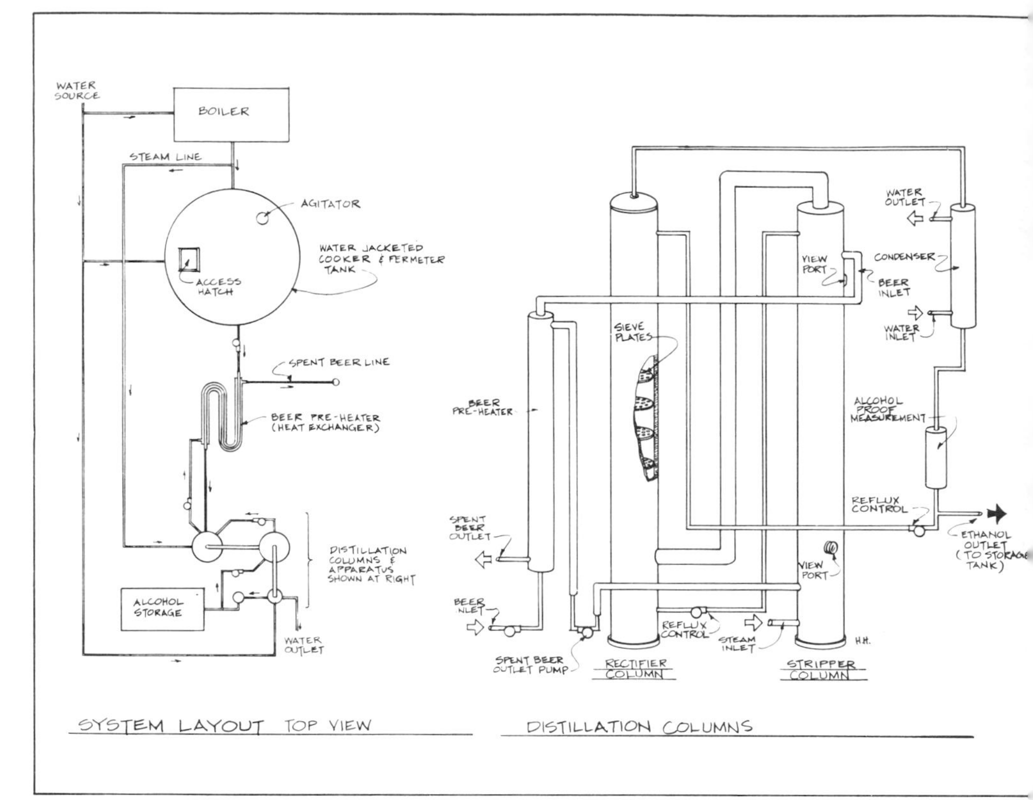

Source: Illustration by the National Center for Appropriate Technology.

Figure 8-2. Beckman Construction Company equipment.

raising the temperature to 195-200°F (91-93°C). The temperature is held at this point for 30 minutes.

When the cook cycle is complete, the mash is cooled to 140°F (60°C) by adding water directly into the tank and by circulating cooling water in the outside shell of the cooker. The saccharification enzyme is added and the mash is held at 140°F for approximately 30 minutes. The mash is then ready for the fermentation step. Beckman has evaluated many kinds of enzymes and has standardized its operation using Miles Laboratory Enzymes, i.e., Takatherm and Diazyme. Final volume of mash is approximately 25 gallons per bushel of milo.

Fermentation

The saccharified mash is cooled to between 85 and 105°F (29-41°C); the mean temperature of 95°F (35°C) is optimum. It is then innoculated with distiller's active dry yeast (from Miles Laboratory) at a concentration of 1 pound per thousand gallons (1.5 pounds per 1500-gallon batch). The yeast is premixed in 5 gallons of warm water at least 10 minutes before addition to the mash. The fermentation, according to Gossett, is running about 8.5 percent beer in 26 hours and 10 percent beer in 48-60 hours. There is no mechanical agitation during the fermentation process. When the fermentation cycle is complete (60 hours), less than 0.01 percent sugar and less than 3 percent starch are left in the mash.

Distillation

During distillation, the beer in the cooker/fermentation tank is agitated constantly for the first hour and last hour, and sporadically in between to keep the solids stirred up. Steam from the boiler is used to distill the beer being pumped into the stripper column. The stripper column is 12 inches in diameter, 16-feet tall, and made of plates spaced up the length. The beer is fed into the upper portion of the stripper column and steam is injected into the bottom.

A centrifugal pump moves the alcohol for reflux from the condenser back to the top of the stripper and rectifier columns. From the condenser, which is cooled by a stream of cold water, the alcohol flows into a storage tank. Beckman has set up a hydrometer in the line to monitor the proof of the alcohol going into the storage tank. The columns use a cartridge-type plate arrangement to facilitate cleaning, which is recommended once or twice a year to remove protein buildup on the plates.

The claimed distillation capacity is rated at 25 gph of 189-proof alcohol. At that rate, a 1500-gallon batch of mash is distilled in approximately 6 hours. The alcohol content remaining in the spent mash is less than 0.1 percent alcohol.

Distiller's Grains

Beckman has experimented with a centrifuge-type separation system to remove solids from stillage. Independent laboratory tests have shown that the solids from the spent mash contain 29.8-31.5 percent protein. Beckman sells the wet stillage to a local buyer who sells it to dairy farmers in the Fort Worth, Texas area.

Product Use

Beckman has tested the ethyl alcohol in a 1979 Ford truck and a Mercury station wagon. They have looked into alcohol conversion kits but, with their limited production, this does not seem feasible at this time.

Energy Inputs

Electricity is used to run the pumps and motors in the system. Oil or gas is used to fire the 47.5-hoursepower Kwani boiler. The boiler supplies steam for the cooking and distillation portions of the production cycle. Using milo, Gossett estimates the net energy costs are 5-6 cents per gallon of ethanol. Yields were claimed to be 2.54 gallons of ethanol per bushel of milo, with total energy inputs of 39,000 Btu per gallon.

Manpower Requirements

Beckman estimates approximately 8-8.5 hours of operator time per 1500-gallon batch of beer, including the time required for milling the grain and disposing of the stillage. For a full-sized operation with four fermentation tanks and with the plant running 7 days a week, 24 hours per day, Gossett projects five operators (i.e., one per shift, one for weekends, one as a backup).

Contamination Control

Beckman has not experienced any contamination problems as yet with the pilot plant. The tanks are washed down thoroughly, and disinfectants have not had to be used.

Designer's Comments

The Beckman plant uses milo because the grain is locally available, but the system should work equally well with any fermentable feedstock. Gossett would not discuss the yields obtained with other feedstocks because that work was paid for by private clients and results are confidential.

Gosset's personal comments are timely:

"The first thing I'd say to a farmer is to be honest with yourself. If you can get more for your grain down at the co-op than you can get for the alcohol, or an alcohol value to yourself, then sell your grain to the co-op. Don't get yourself into a corner where you have to be making alcohol. The farmer ought to get the best price he can for his crop."

"Another important point is to get good technical help if you need it, and get it from somebody who has made alcohol."

"Be sure to investigate the track record of those who intend to help you or sell you a plant. Have they actually helped anyone or built a successful, operating—and I mean still operating—plant? If you're buying equipment, make sure that it comes with a guarantee and some training."

RANDY BUTTERS

Randy Butters has been making alcohol fuel on his Homer, Michigan, farm since January, 1980. His 1500-gallon cooker tank will process 60 bushels of corn at once. He has invested about $14,000 in the alcohol-production system and an additional $8500 in a screw press he uses to dewater the mash. He claims that his twin columns produce 180-182-proof alcohol at a rate of 15-20 gallons per hours. The plant design is basically his own. See Fig. 8-3 for a schematic layout of Randy Butters's system.

For more information, contact Randy Butters, 4257 Two-and-one-half Mile Road, Homer, Michigan 48245.

Feedstock/Preparation

Randy Butters uses only corn from his farm as a feedstock. No special provisions are made to harvest or store this grain; it is the same grain used to feed the livestock (mostly hogs).

Butters grinds the corn through a mixer-grinder using either a $\frac{1}{18}$-inch or $\frac{5}{16}$-inch screen. He prefers to use ground corn because it is easier to salvage the grain coproduct and squeeze it down to a lower moisture content.

Cooking

Six hundred to 800 gallons of water are heated to approximately 160°F (71°C), and 60 bushels of ground corn are added over an hour-long period.

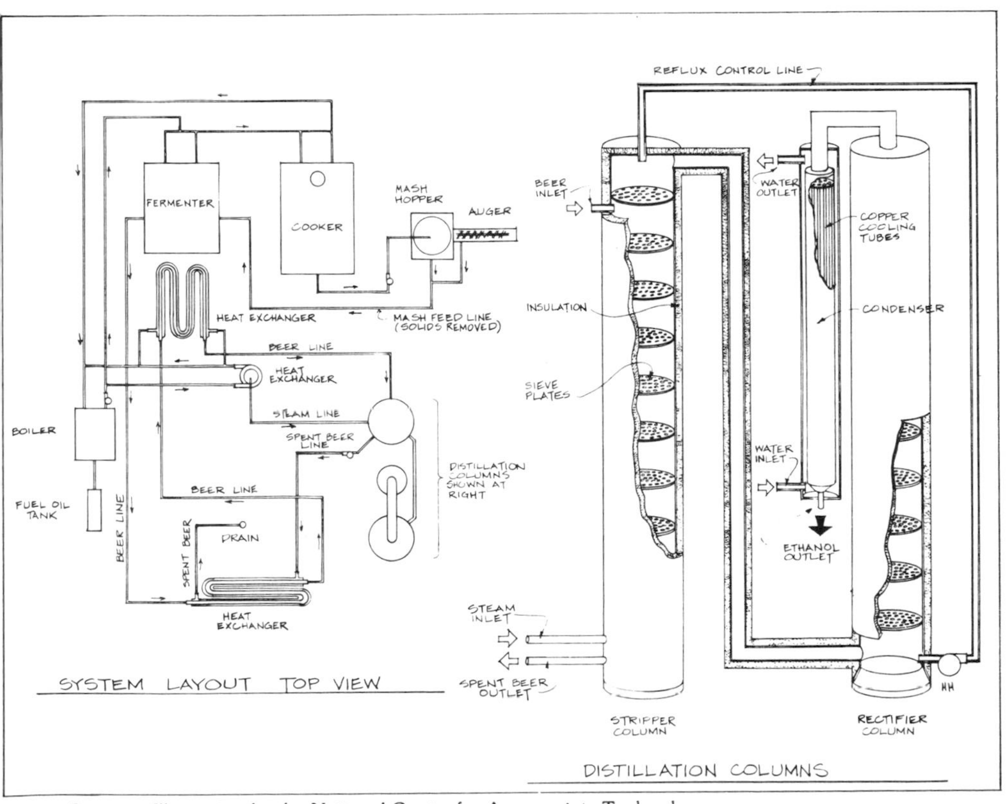

Source: Illustration by the National Center for Appropriate Technology.

Figure 8-3. Randy Butters's fuel-grade alcohol still schematic.

As the corn is mixed with the water, one-half of the liquefaction enzyme is added, and the cooking process is continued using steam injection throughout the cycle. When the temperature reaches approximately 210°F (99°C), the rest of the enzyme (total 2 ¼ quarts) is added, and the 210°F temperature is maintained for approximately 1 hour.

When the cook cycle is complete, the mash is cooled to about 140°F with heat exchangers and by adding cold water to the tank. The pH is adjusted to 4-4.5, and the saccharification enzyme is added. The 140°F temperature is maintained for another 30 minutes. At this point, the solids from the mash are removed with a shaker sieve and screw press. The remaining liquid mash is transferred to the fermentation tank and cooled to between 85 and 100°F, optimizing at about 90°F.

Fermentation

The 1500-gallon fermentation tank is innoculated with approximately 2 pounds of distiller's yeast. The fermentation cycle is completed in approximately 3 days without any outside agitation (unless the yeast does not start working right away, in which case compressed air is injected into the tank to get the yeast started). Temperatures in the fermentation tank vary from 85 to 100°F. Total volume in the fermentation tank is between 17 and 18 gallons per bushel of corn.

Distillation

A stripper and a rectifier column are used. The two sieve plate-type columns are both 12 inches in diameter and 17-feet tall. The stripper has ¼-inch holes in the plates, and the rectifier column has ⅛-inch holes.

Steam is injected directly into the stripper column to distill the beer. The alcohol coming out of the rectifier column is condensed in a shell and tube condenser that is 8 inches in diameter and 10-feet long. The condenser contains 400 feet of ½-inch copper tubing for the heat exchange.

A centrifugal pump supplies the alcohol for reflux to the top of the stripper and rectifier columns. Output is rated at 15-20 gallons per hour. The yield is said to be approximately 1.5 gallons of 180-proof alcohol per bushel of corn.

Distiller's Grains

Butters separates the solids before fermenting with a vibrating screen and a screw press. The solids are used on his farm for hog feed or sold to a neighbor for dairy-cattle feed. The dairyman said that the butter fat from his milking herd increases between 0.5 and 1 percent when the cows eat the high-protein distiller's grains (tests for protein content of the grains have averaged 28 percent).

Product Use

Butters runs his pickup truck on alcohol and also uses the alcohol fuel to run the boiler for the alcohol plant. He plans to convert other farm machinery and to sell the alcohol when the plant is in full production. Butters is also investigating various types of anhydrous units to make 200-proof alcohol for sale on the gasohol market.

Energy Inputs

The heat source is a multifueled hot-water boiler: diesel oil and, lately, fuel-grade alcohol. Six electric motors are used: on the grinder/mixer (for feedstock preparation); for fuel injection to the boiler; on pumps for beer transfer and to move mash; on the shaker sieve; and on the screw press to dewater the solids. The actual Btu's required to produce 1 gallon of alcohol is unknown at this time. Butters uses heat exchangers throughout the system, and all tanks, lines, and columns are spray insulated with cellulose. He is also looking into a hot-water storage tank to save even more energy.

Manpower Requirements

Butters says it takes 5 or 6 manhours for a 60-bushel batch: 1.5 hours for the cook cycle to add grain, make pH adjustments, and remove solids; 2-3 hours during the distillation cycle for start-up and shutdown; and about 2 hours for twice-a-day inspections during the self-operating fermentation cycle to control the temperature.

Contamination Control

As part of his general housekeeping routine, Butters uses water to clean out any tanks and lines in the system that will be idle for any period of time. To date, he has not had any problems.

Designer's Comments

Butters has invested approximately 6 months and more than $20,000 in his alcohol plant. He started off using a 400-gallon dairy bulk tank and 6-inch columns. But, after visiting the Schroder plant in Campo, Colorado, he designed his columns similarly. He says that the serious operator should use heat exchangers to conserve operating energy.

GENE SCHRODER

Gene Schroder and his father Daryll have been leaders in the fuel-alcohol movement for the past few years. They have invested $400,000 to develop an alcohol plant which has more than 47,000 gallons of fermentation capacity. They claim that their diesel-oil-fired boiler system, one of the most advanced farm plants in operation, produces 40 gallons per hour of 192-192.5-proof alcohol. In addition, an anhydrous system is used to produce 200-proof alcohol. The schematic for Gene Schroder's still design is presented in Fig. 8-4.

For more information, contact Gene Schroder, North Route, Campo, Colorado 81029.

Feedstock/Preparation

The Schroders have used corn in their operation, but have standardized the system using milo. The grain is harvested from their fields and stored in large elevators before being transferred to the alcohol facility. Once at the plant, the milo is cleaned and ground through a roller mill to at least a #60 mesh or finer. Before addition to the cooker tank, the grain is weighed.

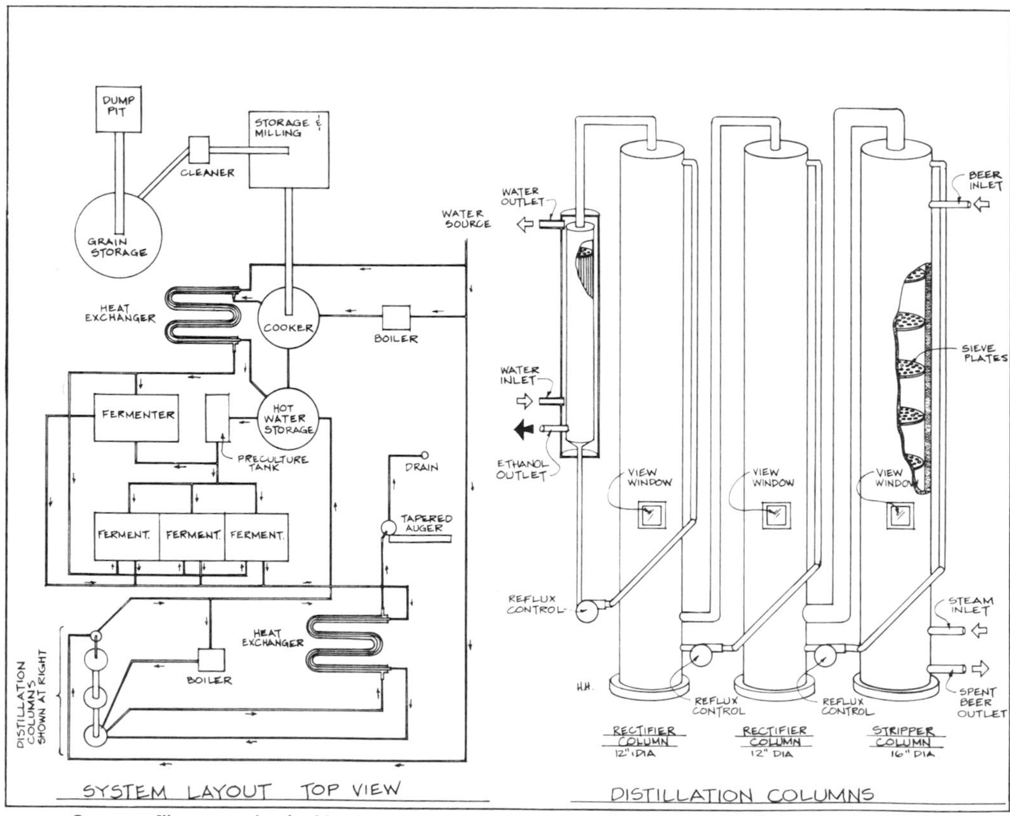

Source: Illustration by the National Center for Appropriate Technology

Figure 8-4. Gene Schroder's still schematic.

Cooking

The Schroders use steam injection in their cooking cycle. Water from a hot-water storage tank is transferred to the cook tank and heated to approximately 106°F (41°C). The ground milo and liquefaction enzyme are added to the 8900-gallon cook tank and the mixture is heated to 190-195°F (88-91°C). This process takes about 3 hours in the insulated cook tank, with a total volume of about 16 gallons per bushel of ground milo. Schroder noted that the pH of the cooker tank mash is generally 6.5.

Fermentation

When the cook cycle is complete, the mash is cooled through a heat exchanger and transferred to the fermentation tank. The hot water collected in the exchanger is piped to an insulated storage tank and used in the next batch of cooking mash.

The cooled mash [approximately 100°F (38°C)] is pH adjusted to 4.5 with sulfuric acid, and at this time the saccharification enzyme is added. After saccharification is complete, the sugar content has been estimated to be 20-22 percent.

The mash is allowed to cook during saccharification and precultured yeast is added. The yeast is precultured for about 48 hours in a 1000-gallon tank prior to introduction into the fermentation tank. The Schroders profess no preference in the brand of distiller's yeast. According to Gene, "We've used most of the commercially available brands, and if they are precultured, there is very little, if any, difference."

The fermentation tank is maintained at pH 4.5, and even though no cooling coils are used, the temperature remains at less than 100°F. The fermentation tanks are stirred intermittently throughout the process, and the alcohol content in the beer is approximately 12 percent by volume. The fermentation is allowed to run approximately 72 hours, depending on the amount and strength of yeast added from the preculture tank.

Distillation

The Schroders' distillation system consists of a 16-inch-diameter stripper column and two 12-inch-diameter rectifier columns. The three columns are insulated and stand 16-feet tall. Beer from the fermentation tank passes through a heat exchanger where it is heated by the hot spent beer solution exiting from the bottom of the stripper column. This cycle conserves energy by reusing waste heat. The beer that is introduced into the top of the stripper column is distilled with steam from a second boiler. The columns are of a sieve-plate (tray) design.

The condenser is a shell and tube design and cools the product with cold well water. The heated water from the condenser is recycled to the hot-water storage tank. The proof of the alcohol is measured at a consistent 192-192.5. The alcohol is then transferred by gravity to a storage tank prior to processing through the anhydrous system. The anhydrous system is set up using Union Carbide's molecular sieves. These sieves must be regenerated three times during a 24-hour period. Regeneration uses hot air from a propane heater.

Distiller's Grains

The spent beer is processed with a vibrating screen system to remove the bulk of the water from the solids. A screw press dewaters the solids further to about 60 percent moisture. These solids are stockpiled on the ground and picked up by a local cattle feeder. The protein content has been measured at about 28-30 percent.

Product Use

Most of the fuel alcohol is sold locally to individuals or used in the Schroders' farm trucks, pickups, and other vehicles. They claim that the alcohol-powered vehicles are performing well, although with a little less power.

Energy Inputs

Electricity powers the roller mill, pumps, and motors for the system. The diesel-oil-fired boiler generates steam to cook and distill the mash. The Schroders estimate yields of 2.5 gallons of 190+-proof alcohol per bushel of milo processed and energy inputs of 24,000 Btu per gallon of ethanol. The anhydrous operation is estimated to be an additional 4000 Btu per gallon in a column packed with molecular sieves to produce the 200-proof alcohol. The total estimated energy inputs equal 28,000 Btu per gallon of 200-proof alcohol.

Manpower Requirements

Gene Schroder estimates that, when the alcohol plant is in full operation, one person will be on duty at all times, and possibly two people during the day.

Contamination Control

The Schroders have had no contamination problems with their system. They do, however, wash out the tanks and line them periodically with caustic soda and steam as a precautionary measure.

Designer's Comments

The Schroders are using the alcohol plant as a means of gaining a fair return on the investment of growing crops on their farm. By selling the alcohol and the distiller's grains, they say that they are able to earn more per bushel on their milo than by selling it as a whole grain through the local co-op.

Small-Scale Ethanol Plant

In *Fuel from Farms—A Guide to Small-Scale Ethanol Production,* the Solar Energy Research Institute (SERI) provides a description of a typical 25-gallon-per-hour ethanol plant. The DOE has provided funds for EG&G of Idaho Falls, Idaho, to construct and test a 25-gallon-per-hour plant that reflects the concept provided by SERI. This section describes the SERI typical plant and includes a more-detailed description of the actual hardware built by EG&G.

SERI REPRESENTATIVE PLANT

Following is a description of the SERI representative ethanol plant producing ethanol and wet stillage. This representative plant normally produces 25 gallons of anhydrous ethanol per hour. The distillation section can be operated continuously with shutdown as required to remove protein buildup in the beer column. Heat is provided by a boiler that uses agricultural residue as fuel. The plant is designed for maximum flexibility, but its principal feedstocks are cereal grains, with specific emphasis on corn.

This representative plant should not be construed as a best design or the recommended approach. Its primary purpose is to illustrate ethanol-production technology.

Overview of the Plant

As shown in Fig. 8-5, the representative plant has seven main systems: (1) feed preparation and storage, (2) cooker/fermenter, (3) distillation, (4) stillage storage, (5) dehydration, (6) product storage, and (7) boiler. Grain from storage is milled once a week to fill the meal bin. Meal from the bin is mixed to make mash in one of three cooker/fermenters. The three cooker/fermenters operate on a staggered schedule—one starting, one fermenting, and one pumping out—to maintain a full beer well so that the distillation section can be run continuously. The beer well provides surge capacity so that the fermenters can be emptied, cleaned, and restarted without having to wait until the still can drain them down. Beer is fed from the beer well to the beer still through a heat exchanger that passes the cool beer countercurrent to the hot stillage from the bottom of the beer still. This heats up the beer and recovers some of the heat from the stillage.

The beer still is a sieve-plate column. The feed is introduced at the top of the stripping section. Vapors from the beer column flow into the bottom of the rectifying column where the ethanol fraction is enriched to 95 percent. The product is condensed and part of it is recycled (refluxed) to the top of the column; if ethanol is being dried at the time, part of it is pumped to the

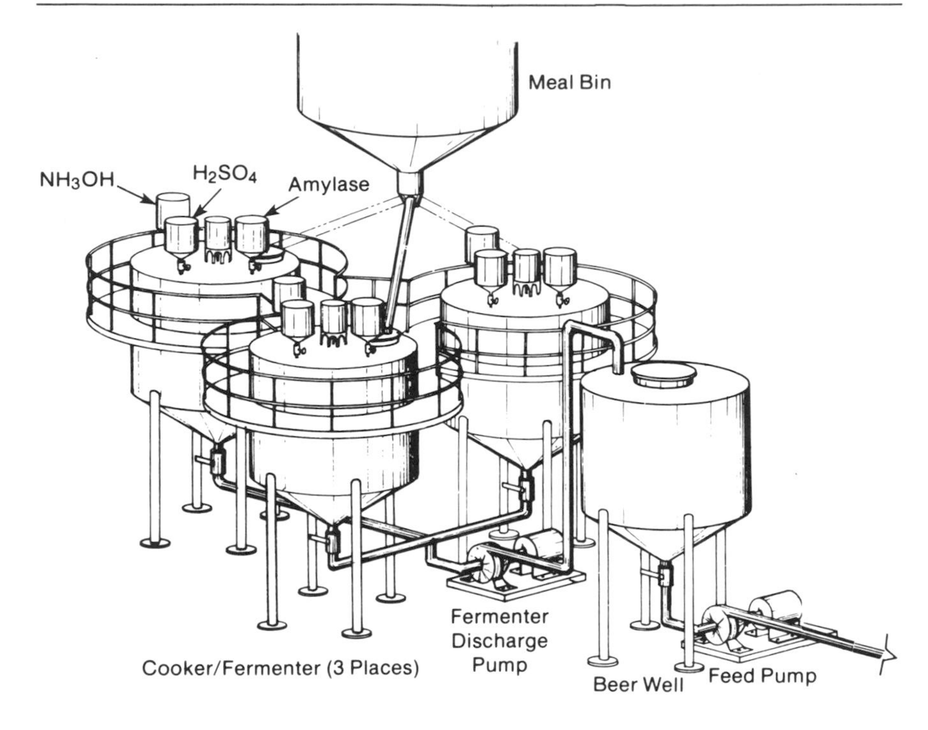

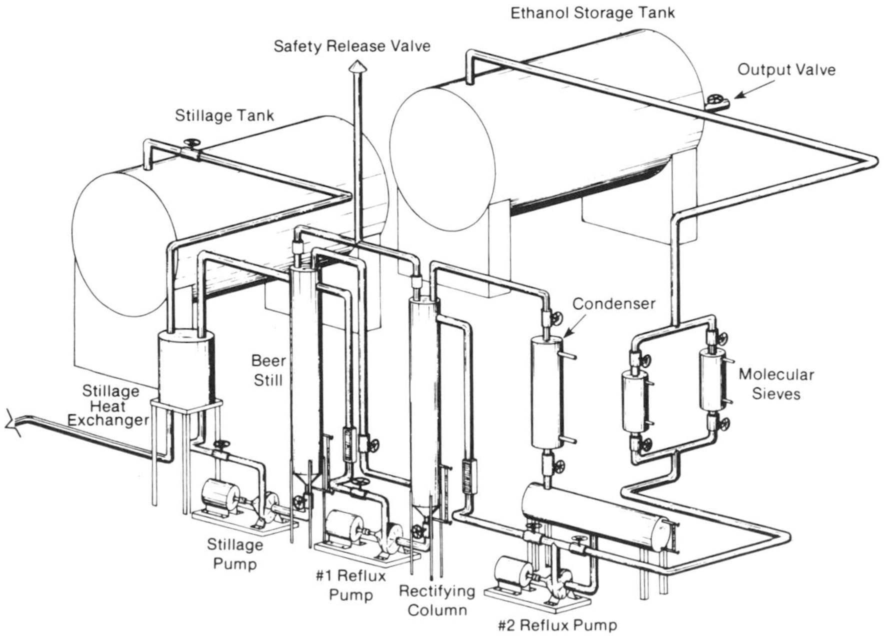

Source: *Fuel From Farms—A Guide to Small-Scale Ethanol Production*, SERI, Denver, 1980.

Figure 8-5. Generic anhydrous ethanol plant.

dehydration section. If the ethanol is not being dried, it flows directly to a storage tank for 190-proof ethanol (a separate tank must be used for the anhydrous ethanol).

The stillage that is removed from the bottom of the beer column is pumped through the previously mentioned heat exchanger and is stored in a "whole stillage" tank. This tank provides surge capacity when a truck is unavailable to haul the stillage to the feeder operation.

The distillation columns are designed for inherent stability once flow conditions are established, so a minimum of automatic feedback control and instrumentation is required. This not only saves money for this equipment, but also reduces instrument and/or controller-related malfunctions. Material flows for cooling and fermentation are initiated manually but proceed automatically. A sequencer microprocessor (a miniature computer) controls temperature and pH in the cooker/fermenters. It also activates addition of enzymes and yeast in the proper amounts at the proper times. At any point, the automatic sequence can be manually overridden.

The period of operation is quite flexible and allows for interruptions of operation during planting or harvest time. The 25-gallon-per-hour production is a nominal capacity, not a maximum. All support equipment is similarly sized so that slightly higher production rates can be achieved if desired.

The control and operating logics for the plant are based on minimal requirements for operator attention. Critical activities are performed on a routine periodic basis so that other farming operations can be handled during the bulk of the day. All routines are timed to integrate with normal chore activities without significant disruption.

A complete equipment list is given in Table 8-1. The major components are described in Table 8-2.

Start-Up and Shutdown

The following is a sequence for starting up or shutting down the plant.

Preliminaries. For the initial start-up, a yeast culture must be prepared or purchased. The initial yeast culture can use a material such as molasses; later cultures can be grown on recycled stillage. Yeast, molasses, and some water should be added to the yeast culture tank to make the culture. Although yeasts function anaerobically, they propagate aerobically, so some oxygen should be introduced by bubbling a small amount of air through the culture tank. The initial yeast culture will take about 24 hours to mature.

At this time, the boiler can be started. Instructions packaged with the specific boiler will detail necessary steps to bring the unit on-line (essentially the boiler is filled with water and the heat source started). These instructions should be carefully followed, otherwise there is the possibility of explosion.

The next step is the milling of grain for the cooker/fermenter. Enough grain should be milled for two fermentation batches (about 160 bushels).

Table 8-1. Equipment for Representative Plant

Equipment	Description
Grain Bin	• ground carbon steel • 360 bu with auger for measuring and loading cooker/fermenter
Back-Pressure Regulators	• 0–50 in. of water
Back-Pressure Regulator	• 100–200 psig
Beer Storage Tank	• 6,000-gal • carbon steel
Condenser, Distiller	• 225 ft^2, tube and shell • copper coil (single tube, 1½-ft diameter) • steel shell cooled
Condenser	• 50 ft^2 • copper coil (single tube) • steel shell
Cooker, Fermenter	• 4,500-gal • hydraulic agitator • carbon steel
Microprocessor	• to control heat for cooking, cooling water during fermentation, and addition of enzymes
Beer Still	• 18-ft height • 1-ft diameter • sieve trays • carbon steel
Alcohol Still	• 24-ft height • 1-ft diameter • sieve trays • carbon steel
CO_2 Compressor	• 1,500 ft^3/hr, 200 psig
Frangibles	• 4–5 psig burst • alarm system • high and low pressure
Heat Exchanger	• 150 ft^2, tube and shell • copper coil (single tube, 2-in. diameter) • steel shell
Heat Exchanger	• 100 ft^2 • stack gas • carbon steel
Hydraulic System for Pumps	• with shut-off valves tied to microprocessor monitoring pump pressures and frangible vent temperature
Grain Mill	• 300 bu/hr • roller type
Beer Pump	• positive displacement • hydraulic drive • variable speed • carbon steel, 50 gal/min
Yeast Pump	• positive displacement • hydraulic drive • variable speed • carbon steel, 10 gal/min

Table 8-1. Equipment for Representative Plant (Continued)

Equipment	Description
Feed Pump	• 300 gal/hr • variable speed • positive displacement • hydraulic drive • carbon steel
Stillage Pump	• 300 gal/hr • variable speed • positive displacement • hydraulic drive • carbon steel
Column 2 Bottoms Pump	• 250 gal/hr • open impeller • centrifugal hydraulic drive • carbon steel
Column 2 Product and Reflux Pump	• 200 gal/hr • open impeller • centrifugal hydraulic drive • carbon steel
Ethanol Transfer Pump	• centrifugal • explosion-proof motor • 50 gal/min
Water Pump	• electric • open impeller • centrifugal • 300 gal/min
Rotameter	• water fluid • glass • 25 gal/min
Rotameter	• glass • 0–250 gal/hr
Rotameter	• glass • 0–150 gal/hr
Rotameter	• glass • 0–50 gal/hr
Rotameter–CO_2	• glass • 200 psig • 100 actual ft^3/hr
Boiler	• 500 hp, with sillage burning system
Stillage Pump	• electric motor • positive displacement • 600 gal/hr
Pressure Gauges	• 6, 0–100 psig • 1, 0–200 psig
Pressure Transducers	• 4, 0–100 psig
Ethanol Drying Columns	• includes molecular sieve packing 3-angstrom synthetic zeolite
Condensate Receiver	• 30-gal, horizontal • carbon steel
Ethanol Storage Tank	• carbon steel • 9,000-gal

Table 8-1. Equipment for Representative Plant (Continued)

Equipment	Description
CO_2 Storage	• 100-gal, 200 psig
Stillage Storage Tank	• 4,500-gal • carbon steel
Thermocouples	• type K, stainless sheath
Multichannel Digital Temperature Readout	• 15 channels
Ball Valve-65	
Metering Valve-6	
Three-Way Valve-4	
Snap Valve	
Water Softener	• 300 gal/hr
Yeast Culture Tank	• carbon steel • 200-gal

Source: *Fuel From Farms – A Guide to Small-Scale Ethanol Production,* SERI, Denver, 1980.

Prior to loading the fermenter, it should be cleaned well with a strong detergent, rinsed, decontaminated with a strong disinfectant, and then rinsed with cold water to flush out the disinfectant.

Mash preparation. The amount of meal put in the cooker/fermenter depends on the size of the batch desired. For the first batch, it is advisable to be conservative and start small. If the batch is ruined, not as much material is wasted. A 2000-gallon batch would be a good size for this representative plant. This will require mixing 80 bushels of ground meal with about 500 gallons of water to form a slurry that is about 40 percent starch.

Cooking. The water and meal are blended together as they are added to the cooker/fermenter. It is crucial to use rates that promote mixing and produce no lumps (the agitator should be running). The alpha-amylase enzyme can be blended in during the mixing (the enzyme must be present and well mixed before the temperature is raised because it is very difficult to disperse the enzyme after gelatinization occurs).

Since cooking in this representative plant is initiated by steam injection during slurry-mixing, the enzyme must be blended in simultaneously. (Dry enzymes should be dispersed in a solution of warm water before mixing is started. This only takes a small amount of water, and the directions come on the package. Liquid enzymes can be added directly.) If the pH is lower than 5.5, it should be adjusted by addition of a calculated amount of sodium hydroxide. If the pH is higher than 7.0, a calculated amount of sulfuric acid should be added. Steam is added at a constant rate to achieve uniform heating. When the temperature reaches 140°F (60°C), the physical characteristics of the mash change noticeable as the slurry of starch becomes a

Table 8-2. Features of Major Plant Components

Components	Features
Feedstock Storage and Preparation	
Grain Mill	• roller mill that grinds product to pass a 20-mesh screen
Meal Bin	• corrugated, rolled galvanized steel with 360-bu capacity
Auger	• used for feeding meal to cooker/fermenter
Trip buckets	• used to automatically measure meal in proper quantity; as buckets fill, they become unbalanced and tip over into the cooker/fermenter; each time a bucket tips over, it trips a counter; after the desired number of buckets are dumped, the counter automatically shuts off the auger and resets itself to zero
Cooker/Fermenter	
3 Cookers	• 4,500-gal right cylinder made of cold-rolled, welded carbon steel

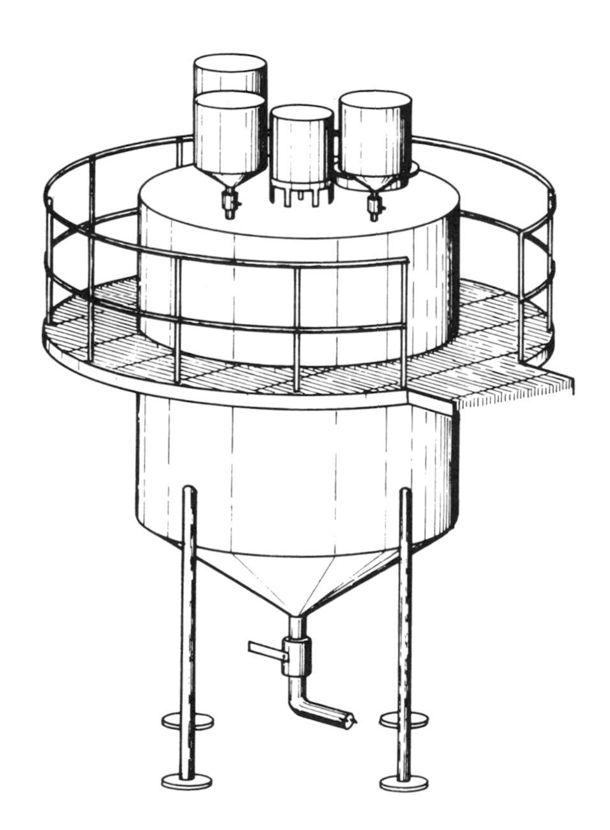

Cooker/Fermenter

Table 8-2. Features of Major Plant Components (Continued)

Components	Features
	• flat top
	• conical bottom
	• ball-valve drain port
	• top-mounted feed port
	• hydraulic agitator
	• cooling coils
	• pH meter
	• sodium hydroxide tank
	• dilute sulfuric acid tank
	• temperature-sensing control, preset by sequences
	• steam injection
Glucoamylase Enzyme Tanks	• 5-gal capacity
	• fitted with stirrer
	• ball-valve port to cooker/fermenter triggered by sequencer
Sequencer	• controls cooking fermentation sequences
	• actuates ball-valve to add glucoamylase enzyme after temperature drops from liquefaction step
	• sequences temperature controller for cooker/fermenter
	• sets pH reading for pH controller according to step
	• 6,000-gal capacity
	• cold-rolled, welded carbon steel
	• flat top
	• conical bottom
	• ball-valve port at bottom
	• man-way on top, normally kept closed (used for cleaning access only)
Beer/Stillage Heat Exchanger	• 2-ft diameter, 3-ft tall—beer flows through coil, stillage flows through tank

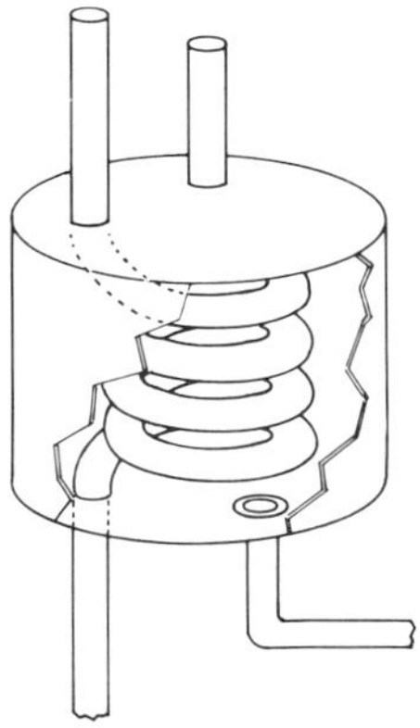

Beer/Stillage Heat Exchanger

Table 8-2. Features of Major Plant Components (Continued)

Components	Features
Beer Pump	• pump from any of the three cooker/fermenters to beer well, hydraulic motor on pump

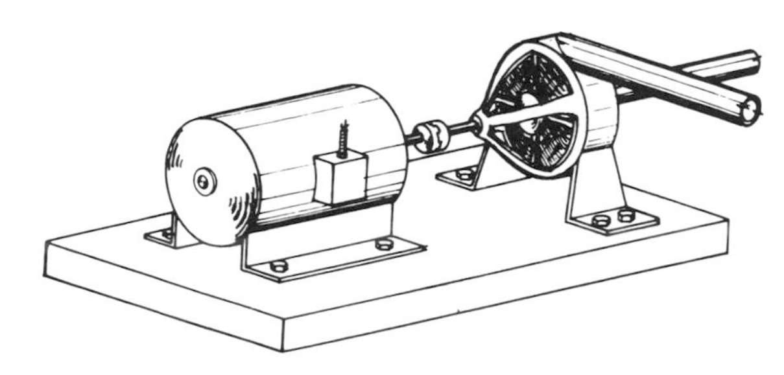

Beer Pump

Components	Features
Feed Pump	• pump beer to distillation system, hydraulic motor on pump
Distillation	
Beer Still	• 1-ft diameter • 20-ft tall coated carbon steel pipe with flanged top and bottom • fitted with a rack of sieve trays that can be removed either through the top or bottom • steam introduced at bottom through a throttle valve • pump at the bottom to pump stillage out, hydraulic motor on pump • input and output flows are controlled through manually adjusted throttle valves • safety relief valves prevent excess pressure in column • instrumentation includes temperature indication on feed line and at the bottom of the still, sight-glass on bottom to maintain liquid level, pressure indicators on the outlet of the stillage pump

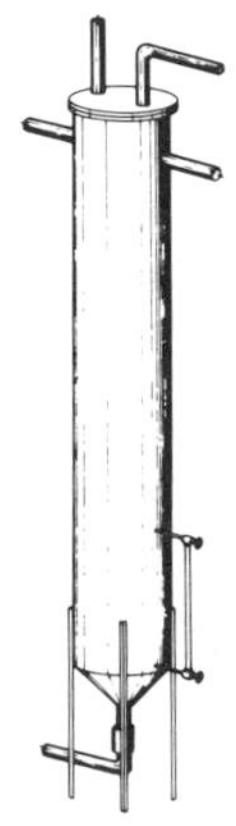

Beer Still

Table 8-2. Features of Major Plant Components (Continued)

Components	Features
Rectifying Column	• 20-ft tall • 1-ft diameter • coated carbon steel pipe with flanged top, welded bottom to prevent ethanol leaks • fitted with rack of sieve-plates which can be removed through the top • pump at bottom of column refluxes ethanol at set rate back to beer still, rate is set with throttle valve and rotameter, hydraulic motor on pump

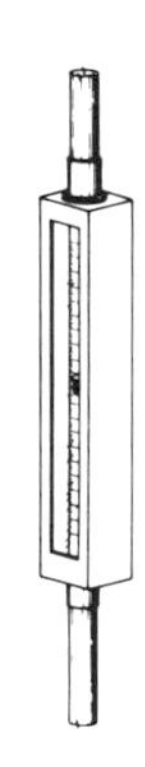

Rectifying Column Rotameter

• instrumentation consists of temperature indication at top and bottom of column and level indication at bottom by sightglass, pressure is indicated on the outlet of the recirculation pump

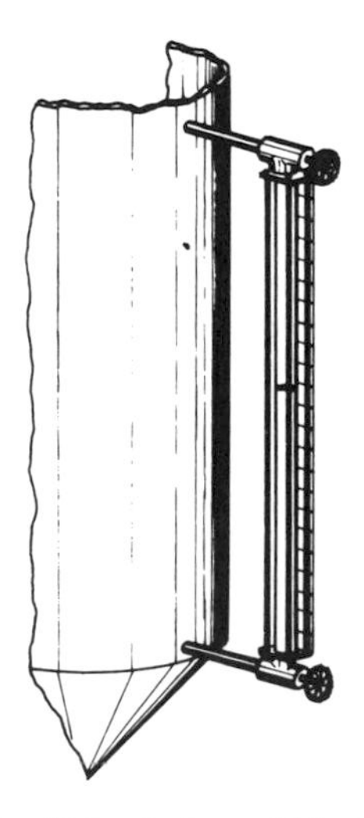

Rectifying Column Sight-Glass

Table 8-2. Features of Major Plant Components (Continued)

Components	Features
Condenser	• ethanol condenses (in copper coil), water flows through tank
	• cooling water flow-rate is manually adjusted
Dehydration Secton	
2 Molecular Sieves	• packed bed
	• synthetic zeolite, type 3A-molecular sieve material
	• automatic regeneration
	• automatic temperature control during regeneration
	• throttle flow control to sieves adjusted manually
	Molecular Sieves
CO_2 Compressor	• 2-stage air compressor with reservoir (conventional)
Denaturing Tank	• meets Bureau of Alcohol, Tobacco, and Firearms specifications
Ethanol Storage	
2 Ethanol Storage Tanks	• 3,000-gal capacity each
	• same as gasoline storage tanks
	• cold-rolled, welded carbon steel
Stillage Storage Tank	• 6,000-gal capacity
	• cold-rolled, welded carbon steel

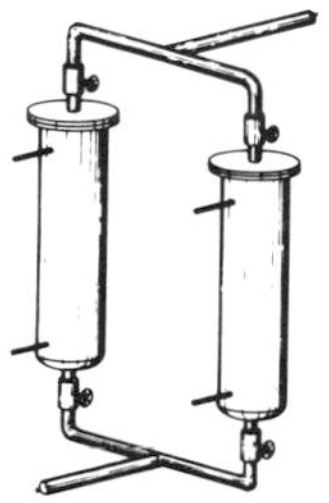

solution of sugar. If there is insufficient enzyme present or if heating is too rapid, a gel will result that is too thick to stir or add additional enzyme to. If a gel does form, more water and enzyme can be added (if there is room in the tank) and the cooking can start over.

Once liquefaction occurs, the temperature is uniformly raised to the range for optimum enzyme activity [about 200°F (93°C)] and held for about half an hour. At the end of this time, a check is made to determine if all of the starch has been converted to sugar. A visual inspection usually is sufficient; incomplete conversion will be indicated by white specks of starch or lumps; a thin, fluid mash indicates good conversion. The mash is held at this temperature until most of the starch is converted to dextrin.

Saccharification. Once the mash is converted to dextrin, the microprocessor is manually started and (1) reduces the temperature of the mash to about 135°F (57°C) by circulating cooling water through the coils; and (2) adds dilute sulfuric acid (H_2SO_4) until the pH drops to between 3.7 and 4.5 (H_2SO_4 addition is controlled by a pH meter and a valve on the H_2SO_4 tank). Once the pH and temperature are within specified ranges, the microprocessor triggers the release of liquid glucoamylase (which must be premixed if dry enzyme is used) from its storage tank. Either sodium hydroxide or sulfuric acid is added automatically as required to maintain proper pH during conversion. The microprocessor also holds the mash at a constant temperature by regulating steam and/or cooling water flow for a preset period of time. The sequencer can be overridden if the conversion is not complete.

Fermentation. After hydrolysis is complete, the sequencer lowers the temperature of the mash to about 85°F (29°C), by adding the remaining 1500 gallons of water (and by circulating cooling water thereafter as necessary). The water addition will raise the pH of the solution so the sequencer automatically adjusts the pH to between 4.5 and 5.0. Next, the sequencer adds a premeasured quantity of dispersed distiller's yeast from the yeast tank. (Note that the yeast tank is not on top of the cooker/fermenter as high temperatures during cooking would kill the culture.) Thereafter, the sequencer maintains the temperature between 80 and 85°F and the pH between 4.0 and 5.0. The agitator speed is reduced from that required during cooking to a rate which prevents solids from settling, but does not disturb the yeast. The batch is then allowed to ferment for 36-40 hours.

Pump-out and cleanup. After a batch is complete, it is pumped to the beer well and the fermenter is hosed out to remove any remaining solids.

Distillation. Once the beer well is full, the distillation system can be started up. This process involves the following steps:

1. Turn on the condenser cooling water.
2. Purge the still with steam. This removes oxygen from the system by venting at the top of the second column. When steam is seen coming out of the vent, the steam can be temporarily shut off and the vent closed. Purging the still with steam not only removes oxygen, but also helps to preheat the still.
3. Pump beer into the still. The beer is pumped in until it is visible at the top of the sight-glass.
4. Turn steam on and add beer. This process of adding beer and watching the liquid level movement to adjust the steam level will be repeated several times as the columns are loaded. Initially,

steam flows should be set at a low level to prevent overloading the trays which might require shutdown and restart. During this period, the valves in the reflux line are fully opened, but the reflux pump is left off until enough liquid has built up in the condensate receiver. This prevents excessive wear on the pump. The reflux line between the two columns should also be opened and the reflux pump should be left off. The liquid level in the bottom of the beer still should be monitored and, when it drops to half way, beer should be fed back into the column to refill the bottom of the still. The liquid level should continue to drop; if it does not, additional steam should be fed into the still bottom.

5. Start reflux pump between the beer still and rectifying column. When liquid starts to accumulate in the bottom of the rectifying column, the reflux pump between this still and the beer still is started. Flow in this line should be slow at first and then increase as more and more material reaches the rectifying column. When reflux is started to the beer still, the steam feed-rate will have to be slightly increased, because reflux tends to cool down a column.
6. Start pump for reflux from the condensate receiver to rectifying column. Eventually, enough vapor will have been condensed to fill the condensate receiver. Then, the pump for the reflux to the rectifying column can be started. Flow for this reflux line should be slow at first and then increased as more and more material distills. It should be noted that temperatures in the columns will be increasing as this process takes place. When the top temperature of the rectifying column is no longer changing, the liquid levels in the bottom of the two columns are changing, and the condensate receiver level is no longer changing. Then, the reflux flow rates are at their designed flow and the column has reached equilibrium.
7. Set beer feed pump, stillage pump, and product takeoff at their designed flow. Initially, the beer feed entering the beer still will be cooler than normal; the heat exchanger has not heated up yet. For this reason the steam to the beer still will need to be slightly increased. The thermocouple at the feed point will indicate when the feed is being heated to its designed temperature. At this time, the steam rate can be slightly lowered. Some minor adjustments will probably be needed. It must be kept in mind that this is a large system, and it takes some time for all points to react to a change in still conditions. All adjustments should be made, and then a period of time should be allowed before any additional adjustments are made.

8. Check product quality. Product quality at this time should be checked to ensure that ethanol concentration is at the designed level. If it is lower than anticipated, the reflux ratio should be increased slightly. An increase in reflux cools the columns and additional heat must be applied to compensate for this. Also, the product flow-rate will be slightly decreased; therefore, flow-rate to the still should also be varied. The ethanol concentration in the stillage should be checked to ensure that it does not exceed design concentration significantly.
9. Drying ethanol. After the ethanol leaves the distillation column, it must be further dried by passing through the molecular sieve drying columns and then stored in the ethanol storage tank. Literature from the vendor of the molecular sieve material will indicate at what temperature that flow must be switched to the other unit.
10. Regenerate spent sieve material. Carbon dioxide (CO_2) is used to regenerate the molecular sieve material. The CO_2 is collected from the fermentation system and compressed CO_2 storage tank. To regenerate the molecular sieve material, the lines for regeneration are opened. Next, the CO_2 line is opened to allow flow to the stack heat exchanger and then on to the sieve columns. A rotameter in the CO_2 line is set to control the CO_2 flow-rate to the desired level. The molecular sieve columns are heated to about 450°F (232°C) during regeneration. After regeneration is complete, the column is cooled down by CO_2 which bypasses the stack heat exchanger.

This essentially covers all the steps involved in the start-up of the plant. It should again be emphasized that caution must be exercised when operating any system of this complexity. If proper care is taken, and changes to the system operation are thought out sufficiently, successful plant operation will be achieved.

Shutdown. The second period of operation which differs significantly from normal operation is that period when the plant is being shutdown. Proper care must be taken during shutdown to ensure both minimal losses of product and ease of restarting the process.

As the fermenters are individually shutdown, they should be cleaned well to inhibit any unwanted microbial growth. The initial rinse from the fermenters can be pumped to the beer storage tank. Subsequent rinses should be discarded. The processing of this rinse material through the stills can continue until the top temperature of the beer column reaches 200°F. At this time, the unit should be put on total reflux.

During this shutdown period, the product quality will have degraded slightly, but the molecular sieve column will remove any additional water in the ethanol product. The stillage from the distillation system can be sent to the stillage storage system until the stillage is essentially clean. At this point, the steam to the column should be shut off and the column should be allowed to cool. During cooling, the column should be vented to prevent system damage. The pressure inside the column will be reduced as it cools. The air which enters the column at this time can be purged with steam prior to the next period of operation. The molecular sieve drying columns can be regenerated if necessary. The boiler should be shut down. If the shutdown period is of any significant duration, the boiler should be drained. If the plant is to be shut down for a short term, the fermenters should not require any additional cleaning. After an extended shutdown period, it is advisable to clean the fermenters in a manner similar to that performed at the initial start-up.

Shutdown periods are the best time to perform preventive maintenance. The column trays can be cleaned, pump seals can be replaced, etc. The important thing to remember is that safety must not be overlooked at this time. Process lines should be opened carefully because, even after extended periods of shutdown, lines can still be pressurized. If it is necessary to enter tanks, they must be well vented. It is suggested that an air line be placed in the tanks and that they be purged with air for several hours before they are entered. Also, a tank should never be entered without another person stationed outside the tank in case an emergency situation arises.

Daily Operation

The day-to-day operation of the representative on-farm plant required the attention of the operator for two periods of about 2 hours each day.

Each morning, the operator begins by checking the condition of the plant. All systems are operative because the operator should have been alerted by the alarm if there had been a shutdown during the night. A quick check will confirm that the beer flow and reflux flows are near desired values. The temperature of the top plate of the rectifying column and the proof of the product before drying should be checked. Even if the proof is low, the final product should be dry because the dryer removes essentially all of the water, regardless of input proof. However, excessively low entering proof could eventually overload the regeneration system. If the proof before drying is low, reflux flow is adjusted to correct it.

Next, the fermenter that has completed fermentation is checked. The concentration of ethanol is checked and compared to the value indicated by the sugar content at the beginning of fermentation. If the concentration is suitable, the contents of the fermenter are dumped into the beer well. The inside of the fermenter is washed briefly with a high-pressure water stream. Then the fermenter is filled with preheated water from the holding tank.

The operator next checks the condition of the boiler and bale burner. The bale burner is reloaded with two of the large, round bales of corn stover from the row outside of the building. A frontend loader is used for this. The operator returns to the fermenter that is being filled. It is probably half filled at this time, and the flow of meal into the tank is begun from the overhead meal bin. The flow-rate is continuously measured and indicated and will cut off when the desired amount is reached. The agitator in the tank is started. The liquefying enzyme is added at this time. The operator checks the temperature. When the tank is nearly full, steam is admitted to bring the temperature up to cooking value. The operator checks the viscosity until it is clear that liquefaction is taking place.

The operator now prepares for the automatically controlled sequence of the remaining steps of cooling and fermenting. The microprocessor controls these steps, and it will be activated at this time. However, the operator must load the saccharifying enzyme into its container. The enzyme is dumped into the fermentation tank on signal from the microprocessor. After cooking is complete, the microprocessor initiates the flow of cold water into coils in the vessel which cools the mash to the temperature corresponding to saccharification. When the appropriate temperature is reached, the enzyme is introduced. After a predetermined time, the converted mash is cooled to fermentation temperature, again by circulating cold water through the coils. When fermentation temperature is reached, the yeast is pumped into the fermenter. All of these operations are controlled by the microprocessor and do not require the operator's presence.

Once the fermentation is initiated, the operator can check the condition of the distillation columns and turn his attention to the products. The driver of the truck which delivers the whole stillage to the dairies and feeding operations will have finished filling the tank truck. If it is time for the pick-up of ethanol, the distributor's driver would then load the truck and start back to the bulk station.

In the evening, the operator repeats the same operation with the exception of grinding meal and delivering the product.

Maintenance Checklist

Table 8-3 provides a general timetable for proper maintenance of a representative ethanol plant.

EG&G SMALL-SCALE ETHANOL PLANT

This section describes a design for a 25-gallon-per-hour small-scale ethanol plant as shown in Fig. 8-6. This design is based on the SERI 25-gallon-per-hour concept, but is amenable to scale-up or scale-down from this production rate. This design effort is primarily aimed at the rural production sector, but could be used by small industries where fermentable coproducts are available. The design presented is a preliminary design, and should be viewed from that perspective.

Table 8-3. Maintenance Checklist

Bale Burner	
Remove ash	daily
Lubricate fans	monthly
Check fan belts	monthly
Water Softener	
Regenerate and backwash	weekly
Check effectiveness	yearly
Boiler	
Blow flues and CO_2 heater	monthly
Check tubes and remove scale	monthly
Roller Mill	
Check for roller damage	weekly
Check driver belts	monthly
Elevator Leg to Meal Bin	
Lubricate	monthly
Yeast Tubs	
Change air filter	monthly
Fermenters	
Sterilize	every 3rd week
Wash down outside	weekly
Back Pressure Bubblers	
Clean out	weekly
Beer Well	
Sterilize and wash down	weekly
Steam Lines	
Blow condensate	daily
Beer Preheater	
Clean both sides	weekly
Beer Column	
Clean out	weekly
Sight Glasses	
Clean out	weekly
Flow Meters	
Clean out	as needed
Condenser	
Descale water side	monthly

Table 8-3. Maintenance Checklist (Continued)

Stillage Tank	
Clean and sterilize	monthly
Pumps	
Check seals and end play	weekly
Lubricate	per manufacturer
Hydraulic System and Motors	
Check for leaks	daily
Change filter	per manufacturer
Top-up	as necessary

Sources:

1. Ladish, Michael R.; Dyck, Karen. "Dehydration of Ethanol: New Approach Gives Positive Energy Balance." *Science.* Vol. 205 (no. 4409): August 31, 1979; pp. 898–900.
2. SERI. *Fuel From Farms – A Guide to Small-Scale Ethanol Production.* May 1980. Denver, CO.

Figure 8-6. Photograph of EG&G ethanol plant.

The design presented is modular in concept, including modules for feedstock preparation, cooking, saccharification and fermentation, distillation, alcohol drying, coproduct dewatering, instrumentation and control, and the boiler. The majority of the information presented herein is considered to be standard technology and felt to have a high level of engineering confidence.

Two concepts presented here are not typically practiced in rural-based ethanol production and, therefore, have a somewhat lower level of engineering confidence associated with them. The use of continuous cooking, although routinely practiced in the food industry, is not typically used in rural-based ethanol production and may require some developmental research after completion of the prototype system. The use of molecular sieves for the drying of alcohol is a new concept, and minimal information exists as to the performance of a system of this type.

Those systems felt to have a high level of engineering confidence are the feedstock preparation, fermentation, distillation, boiler, instrumentation, and control elements, and are anticipated to present minimal problems during initial operation of the prototype system.

The initial concept of the plant was to incorporate a coproduct-drying step into the design. Considerable effort has been made to minimize costs (both capital and operational) of equipment for this effort. For plants of this scale, it is suggested that the coproducts be used in wet form, with the use of a dewatering step to reduce shipping costs if the stillage is not used in close proximity to the plant.

General Functional/Operational Requirements

The objective of this design is to provide a basic fuel-grade ethanol-production system design to be used in a rural setting. The system design has been aimed at meeting the general requirements of safety, minimal environmental impact, high product quality, ease of operation, ease of maintainability, high energy efficiency, and minimal manpower input. When feasible, components specified in the system are off-the-shelf equipment, readily available throughout the country.

Product Quality

The products from the system are fuel-grade ethanol and animal feed-grade coproducts. The distillation section is designed to produce a 190-proof ethanol/water blend. If the desired product is to be a lower concentration ethanol/water blend, minor adjustments in the operating conditions of the distillation can be easily made (primarily this is a reduction in the rate of reflux). The animal feed coproducts will be in a wet or dry form. The system is so designed that no known foreign material will be purposely introduced into the coproduct material to deter from the coproducts' feeding properties. The distinction is made that distressed crop material may be used in the production of ethanol, but the coproducts generated may not be acceptable for animal feeding, and if so, must be properly disposed.

Operation and Maintenance

Equipment is designed and located in such a manner as to ensure ease of operation and maintenance in both routine preventive and emergency modes. The system is automatically controlled and monitored where economically possible. The manpower requirements for the plant operation do not exceed 4 hours per 24-hour operating day; this includes both normal operation as well as the performance of routine preventive maintenance. The system is designed and equipment specified for a predicted 10-year lifetime of major components, e.g., distillation columns and pumps. The plant will be operable a minimum of 80 percent of the time.

Energy Efficiency

Every consideration has been made to ensure minimal energy input to the system, with energy cascading employed where economically possible. The primary area of energy cascading is with the distillation column bottoms that are used to preheat the feed to the distillation section and to preheat the water used in the cooking step.

Plant Sizing

The process is designed to produce 25 gallons per hour of 190-proof ethanol on a continuous basis. The distillation section is operated continuously, the saccharification step and fermentation step are batch operations, and the cooking and the milling steps are semicontinuous.

The system is primarily designed to process corn. Wheat can be processed with the same equipment, with minor changes in the operating parameters of the cooking section. Potatoes can be processed with modification to the feed-handling system before the milling/grinding step and again with minor changes required in the operating parameters of the cooking section. Sugar beets can be processed by removing the cooking stage and replacing it with a sugar extraction stage.

Specific Component Functional/Operational Requirements

In addition to the above generally applicable requirements, the system has been designed to adhere to the following requirements for specific unit operational steps.

Distillation. The distillation unit has been designed with the distillation column built in two segments to decrease overall height requirements. It is heated by direct steam injection. The condenser is cooled with tap, well, or other water assumed to be at 40-60°F (4-16°C). The internals to the still columns are trays and standard distillation packing material, with easy removal of the internals for cleaning. The feed into the distillation section is 8-10 percent ethanol/water solution with slurried solid coproducts. The distillation section can use molecular sieves to remove water from the 190-proof solution of ethanol water to produce 25 gallons per hour of 99.9 percent ethanol. The distillation system is continuously operated and fully automated. The system consists of two distillation columns, piped in series. The first column has metal Intalox packing above the feed, and trays below the feed point. The second column has metal Intalox packing. The separation has 6 theoretical stages (12 physical trays) below the feed, and a total of 16 stages of separation above the feed.

The columns will be constructed of carbon steel pipe, with stainless-steel internals.

The feed will be preheated against the overhead vapors from the second column, then against the bottoms from the beer (first) column. Specific design considerations are listed below.

Process design. A product purity very near the azeotropic concentration was selected to provide ethanol as dry as possible to the molecular sieve dehydration system. It was felt that in terms of capital cost and energy, the distillation is a more efficient separation, up to 92 percent, than molecular sieves, and should provide as much separation as is practical.

A maximum column height of 20-25 feet was selected to avoid the requirement for extremely high buildings and to avoid the problems of construction and safety associated with taller structures. Given this height limitation, the separation stage requirements dictated the two-column design.

The energy costs of producing ethanol increase with increasing reflux ratio. To obtain minimum possible reflux, the maximum stages that could be provided in two columns are used. The theoretical reflux ratio required is 3.0. A design reflux ratio of 4.5 is used for conservatism. This is roughly equivalent to the conservatism provided by 30 percent extra stages of separation.

Column design. Sieve trays are used below the feed point (as opposed to packing) because it could not be verified that packing would handle the potential foaming. The advisability of using silicon-based foam control agents has not been fully investigated. Downcomer sieve trays have been selected based on the ease of cleaning sieve trays, the difficulty of limited operational flexibility with downcomerless trays, and Archer-Daniels-Midland's good experience when feeding whole mash to trays with downcomers.

A packed bed was selected for use above the feed point because of its good operational flexibility, ease of column control, and high throughput for a given column diameter. The experience with the packed column at DOE's experimental geothermal site near Raft River, Idaho, has supported this decision.

Metal Intalox packing was selected because of its high throughput and high efficiency. Although the cost per cubic foot is higher than for pall rings, the cost per stage is lower.

Stainless-steel column internals have been selected because the major portion of their cost is fabrication, not material. The columns themselves will be carbon steel pipe, which is about 20 percent the cost of stainless steel. (A stainless-steel column would cost about $6000 for the pipe alone.)

Front end processing. This section includes necessary storage, cleaning, size reduction, cooking, saccharification, and fermenting operations of the system for grains (corn and wheat), potatoes, and sugar beets. The feedstock storage, cleaning, and size reduction requirements are based on the following rationale.

Farmers or cooperatives are most likely to have their own storage facilities. The plants most likely would be built next to these large storage areas; however, some minimum should be maintained, i.e., a 1-day supply. A normal plant supply would be about 15 days, necessitating more costly storage facilities.

The daily routine for plant startup would consist of manually filling all process elements, i.e., enzymes, yeast, chemicals, and feedstock storage tank. Input by bucket elevator is manual and has no tie-in to the master controller. The input auger is sized to fill the storage bin within 1 hour or less, with a 6 × 4-inch cup and an approximate height of 25 feet. A manually controlled three-way valve is set to allow for the bucket elevator output to be directed to the storage bin.

The storage bin holds 300 bushels and is oversized by 25 percent to allow for variances in size. The capacity is based on the plant producing 600 gallons a day, using 240 bushels of grain at a rate of 2.5 gallons a bushel. A visual indication of the fill level in the bin is made by the operator.

The three-way valve is set to allow for the bucket elevator output to be directed to the feedstock processing equipment. The master controller begins operation of the bucket elevator, vibrating feeder, vibrating separator, and size reduction mill.

Cyclones, aspirators, and gravity separators were eliminated in favor of a simple vibrating screen separator. The intended purpose is to eliminate all large material, i.e., stones, vines, or corn ears. Particles similar in size are allowed to pass through. Some light particles will be carried off by an air blower. A magnetic bar provides adequate removal of metallic particles before the size-reduction step.

Trade-off study. The assumed replacement of hammers, caused by foreign material entering the hammer mill, is less than the cost of a cyclone and gravity separator. Grinding size is large enough not to cause problems associated with fine flour cleaning.

A 20-mesh (0.040-inch) final product size is required. The type of grinder to be selected is dependent on cost, reliability, energy requirements, and operation characteristics under the no-load conditions.

The feedstock is to be processed (cleaned and milled) within a 12-hour period. Longer periods of time result in throughput rates for which equipment availability is reduced. Shorter periods of operation result in equipment items being more costly than required. The hammer mill throughput rate was determined to be about 20 bushels per hour or 1120 pounds per hour. A master time relay allows for the processing equipment to run for 12 hours and then automatically shutdown.

Cooking. A 12-hour continuous cooking cycle was chosen to allow a small steam usage for cooking, compared with the overall steam load. This allows use of a smaller capacity boiler than if the whole batch had to be cooked in 1 or 2 hours. The 12-hour cooking cycle also allows half the batch to be loaded in 6 hours, permitting maximum time for saccharification and fermentation.

A plug flow reactor was chosen to allow continuous cooking, gelatinization, and conversion to dextrins because the overall expense and complexity is much less than a continuously stirred tank reactor of the same reacting capacity. The reactor remains filled with dextrins during the no-flow portion of the cycle (12-24 hours) and is insulated to maintain temperature. At the start of the next cycle, the dextrin solution begins filling the next fermentation tank.

Hot water is purged through the fermenter loading system (except for the plug flow reactor) during the time when the mash feed auger is shutdown. This purge serves two purposes: (1) the saccharification temperature is maintained at 140°F, and (2) the feed line is cleaned.

The corn mash slurry in the mixing tank is heated just below the gelatinization temperature (140°F) by direct injection of hot water [about 174°F (79°C)] from the still bottoms heat exchanger. Gelatinization occurs downstream of the steam injector, which raises mash temperature to 200°F, located just above the plug flow reactor. The intent is to delay gelatinization until the mash enters the large open flow area of the plug flow reactor, so that the system pumping power required to move the mash will be minimized.

Saccharification. To produce fermentation sugars, the mash temperature is reduced to 140°F by an in-line heat exchanger. After saccharification is completed, the batch is cooled to the fermentation temperature by adding about 800 gallons of water to the fermenter and circulating it through the heat exchanger. Use of this method allows a much smaller heat exchange area and thus less expensive heat exchangers than direct cooling in the fermentation tank.

Saccharification takes place in the fermenter tank, reducing the number of vessels and the capital expense required for the system.

Agitation will be accomplished by jet-type nozzles located in the saccharification/fermentation vessels, powered by a 400-gallon-per-minute circulating pump.

Commercial microbial alpha-amylase and glucoamylase are used for saccharification rather than barley malt because commercial enzymes are cheaper than on-farm malted barley and are more reliable.

Fermentation. One fermentation batch processed per day was chosen so that the fermentation tank would be small enough (8000 gallons) to clean without getting inside the fermenter. In addition, one batch per day allows the operator to establish a daily routine of emptying a completed fermentation batch into the beer well and refilling the empty tank with mash.

Although continuous fermentations have been run successfully by some, the technology is considered too new, and subject to too many problems in control. For on-farm usage, batch fermentation is still recommended. Additional fermentation research and development is required to use continuous fermentation in a rural ethanol plant.

Instrumentation and control. A small computer will control the rural ethanol plant most of the time. The computer, called a microprocessor, will remotely measure various temperatures, pressures, tank levels, etc., throughout the plant and automatically control valves, pumps, and other equipment necessary for operating the plant. It will also monitor unusual and emergency conditions within the various processes and shutdown appropriate parts of the plant if necessary.

If the operator wishes, he can take over control from the computer and operate most of the plant directly from a control panel. There are, however, certain functions that the operator must perform personally every day, even when the plant is being run by the microprocessor. These include filling grain hoppers, chemical tanks, and the like. In addition, the operator must tell the computer when to begin processing another batch of brew by pressing the appropriate switch on the control panel.

Automatic control operating requirements. From a control viewpoint, it is helpful to divide the plant into four general parts for purposes of explanation. These are:

Input feedstock processing

Gelatinization, saccharification, and fermentation

Distillation

Product dehydration

The control of each of these sections is unique and is discussed below.

The input feedstock processing (cleaning and milling) will not be controlled by the microprocessor. Instead, each day the operator will fill the input feed hopper using a motorized bucket elevator, and then start the cleaning and milling machinery. A timer will automatically shut equipment down when the process is complete.

The gelatinization, saccharification, and fermentation process will be run by the computer automatically. Once the operator has filled the various enzyme, yeast, and chemical tanks, and cleaned out and set up the valves for a particular vat, he will start the brewing process by pressing the appropriate switch on the master control panel (Example: BEGIN FERMENTATION). The computer will then automatically control the entire fermentation process for the next 72 hours. Although no operator control should be necessary during this period, the operator can adjust saccharification inlet temperature, steam injection inlet temperature, fermentation vat temperature, and the amount of water injected into the vat. The operator can also turn the vat agitator on and off. If anything serious goes wrong during this sequence (loss of steam, jammed slurry auger, etc.), the computer will automatically stop the process and alert the operator. At the end of 72 hours, the operator will manually open the drain valve and pump the fluid out of the fermentation vat and into the beer well, using a centrifugal pump.

The computer also controls the fractional distillation tower. Since fractional distillation is a continuous process, the operator will "set up" the column by adjusting temperature, reflux accumulator pressure, input feedflow, and reflux flow from the control panel until the column is operating properly. The computer will then keep the column running in the same condition until either the operator readjusts it, it runs out of input beer, or there is a failure somewhere in the plant. Again, a plant failure (such as loss of steam pressure) will cause an alarm to alert the operator.

Site Description

The small-scale fuel-alcohol plant will be located in an area about 122 feet by 75 feet as shown in Fig. 8-7. It is divided into two main sections: the feed process and fermentation area where fire and explosion hazards are negligible, and distillation and production-holding areas, which must be well protected against these hazards.

Feed process and fermentation area. The feed bin, saccharification/fermentation tanks, beer well, and supporting equipment are contained in an area 40 by 40 feet square as shown in Fig. 8-8. This area is completely enclosed by a berm, 46 by 46 feet square and 3 feet high, which would hold the total spill from all tanks and vessels in this area, should such a spill occur. Other than the berm, the area is fully open to the atmosphere so that any spill which might occur will quickly dissipate. In areas where climatic conditions are such that subfreezing temperatures are reached, all lines susceptible to freezing are steam traced to prevent line freezing.

The water-treatment and boiler equipment are located adjacent to the feed process and fermentation area. They are contained inside a heated building, 16 feet by 40 feet, to protect against freezing. This building will also house the control system for the plant.

Distillation and product-holding areas. Owing to the high flammability of ethanol, much greater design precautions must be included in the distillation

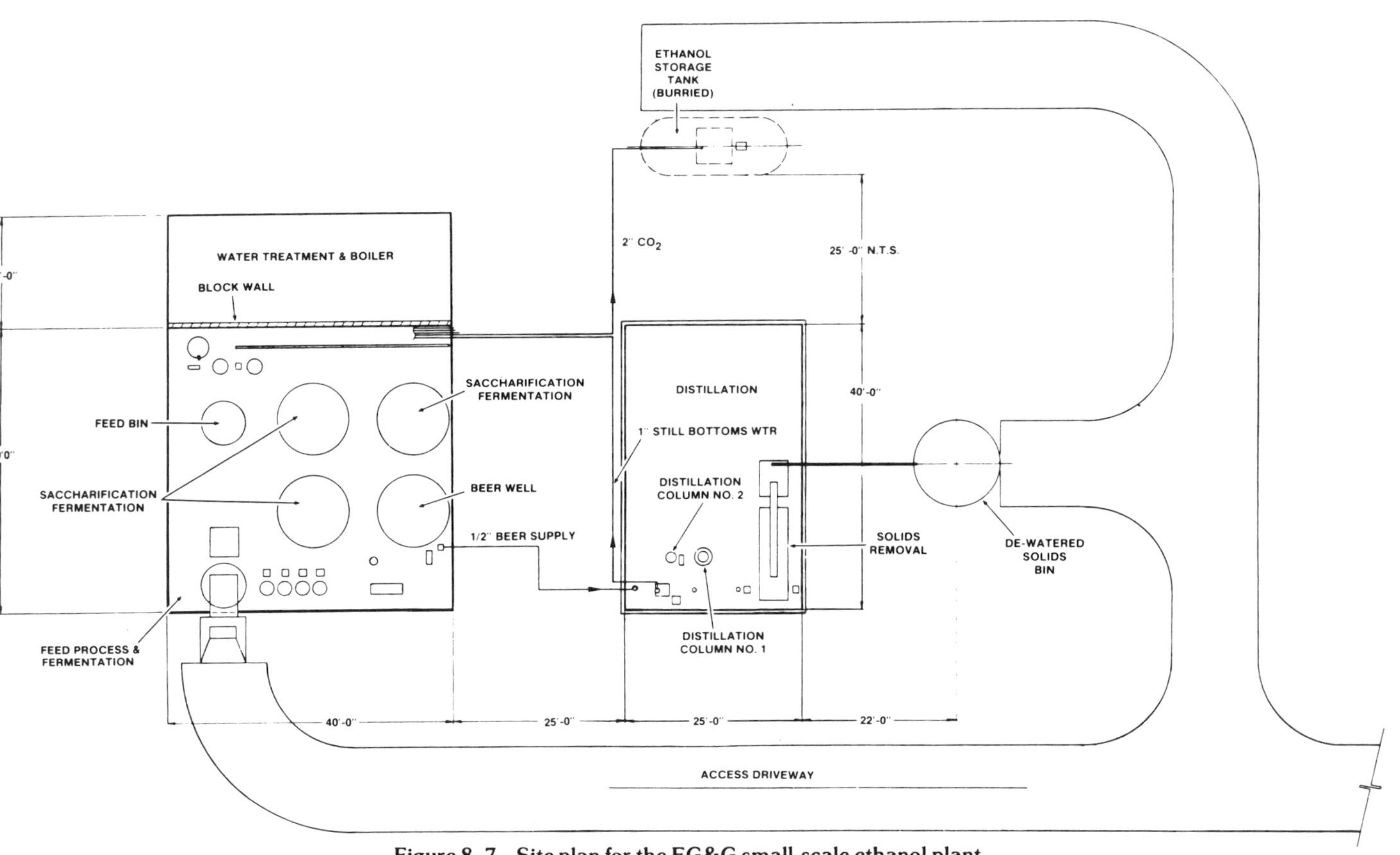

Figure 8-7. Site plan for the EG&G small-scale ethanol plant.

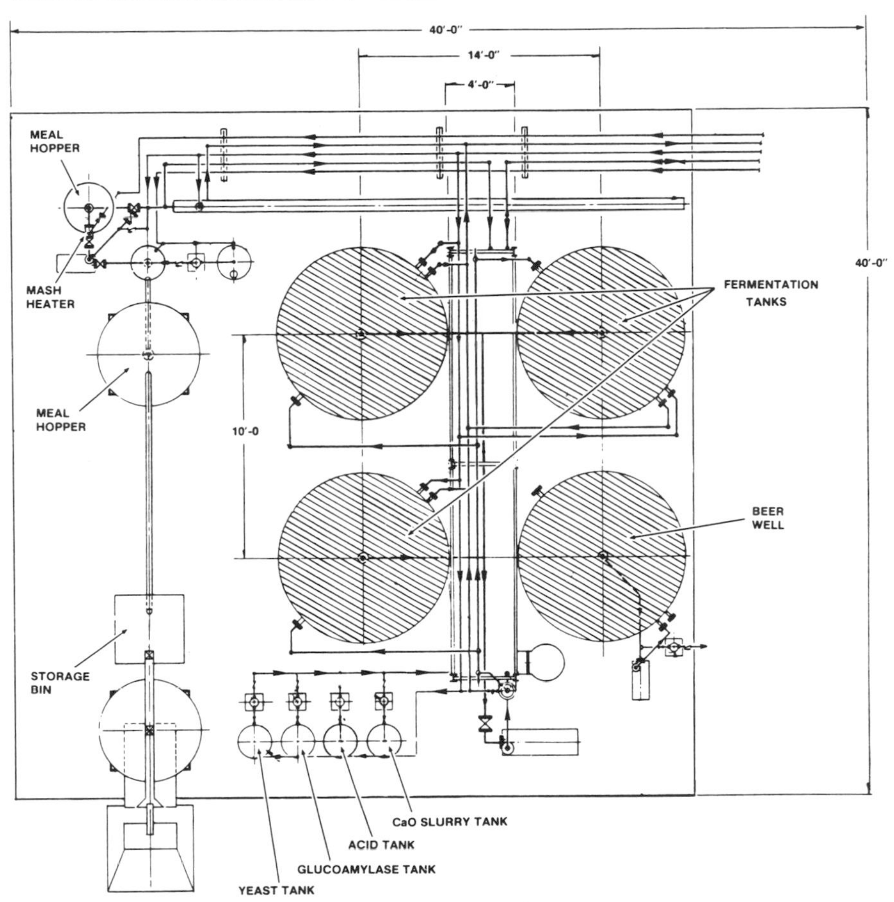

Figure 8-8. Fermentation tank arrangement.

and product-holding areas. The distillation equipment is located in an area 25 feet by 40 feet and must be separated from the feed process and fermentation area by a minimum distance of 20 feet as shown in Fig. 8-9. The area is also completely enclosed by a berm, 26 feet by 46 feet and 2 feet high, designed to hold a total spill from all equipment in this area, in case a spill should occur. Like the feed process and fermentation area, the distillation area is fully open to the atmosphere and all lines in this area that are susceptible to freezing are steam traced. Electrical equipment (i.e., pump motors) in the distillation area is specified explosion-proof.

Parallel to the 25-foot side of the distillation area and at a distance of 25 feet, an ethanol storage tank is located, buried in the ground below the frost line. A buried tank was chosen in preference to one above ground because it can be constructed at less expense while still maintaining a sufficient measure of safety in storing the flammable ethanol.

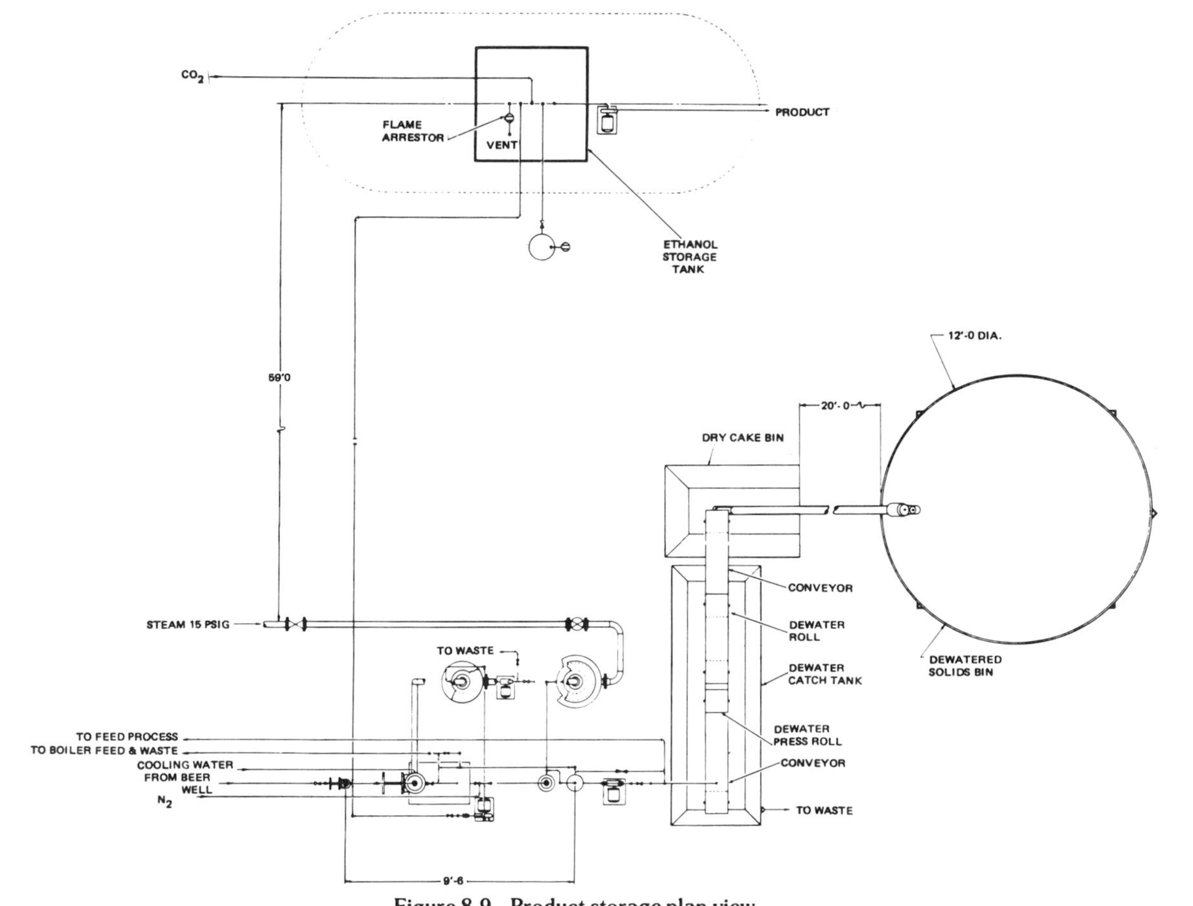

Figure 8-9. Product storage plan view.

A dewatered solids bin is located outside the distillation area, on the side opposite from the feed process and fermentation area as shown in Fig. 8-9. The bin is centered at a distance of 22 feet from the distillation area.

Another berm, approximately 3 feet high and 100 feet long, is located midway between the feed process and fermentation area and the distillation and product-holding areas. Since ethanol vapor is more dense than air, any vapor escaping from the distillation area will be blocked by this berm. This will allow the ethanol vapor to dissipate into the atmosphere sufficiently.

PLANT DESIGN—DESCRIPTION AND OPERATION

The plant as designed is broken into the following main sections, listed in the order of process flow. *Instrumentation and control* includes necessary field instrumentation for automatic processing and a microprocessor for control and operator interface. *Feedstock preparation* includes storage bins, conveyors, cleaners, and milling equipment. *Cooking* includes vessels and equipment for automatic addition of pH buffers (HCl or CaO) and alpha-amylase, equipment for the mixing of ground carbohydrate material with water, equipment for the cooking of the carbohydrate material, and heat exchange equipment for the cooling of the carbohydrate material before its entrance into the saccharification step. *Saccharification and fermentation* uses batch reactors for the conversion of starch to glucose with subsequent conversion of the glucose to ethanol and carbon dioxide; necessary equipment for the addition of pH buffers, glucoamylase, and yeast; and storage for the beer produced. *Distillation* includes necessary distillation equipment for the concentration of beer (8-10 percent ethanol) to 92 percent (190-proof) ethanol and necessary storage vessels for the ethanol product. *Dewatering* includes coproducts remaining after the fermentation. *Ancillary equipment* includes a boiler and other support equipment for the process.

Feedstock Preparation

Feedstock preparation is an integral part of the alcohol-production activity. A multifeedstock capability has been included wherever possible. The feedstock preparation activity consists of three areas: storage, cleaning, and size reduction as shown in Fig. 8-10. Whole product processing is followed for selected feedstocks (corn, wheat, potatoes, and sugar beets). The following operating procedures will deal with corn and not be controlled by the microprocessor.

Feedstock storage. A daily production capacity of at least 240 bushels of corn is stored in a bin as shown in Fig. 8-11. This storage bin may be slightly oversized to allow for variances in feedstock densities. The correct operation is accomplished manually by the following steps:

1. The three-way valve is set to allow corn to be directed into the storage bin.

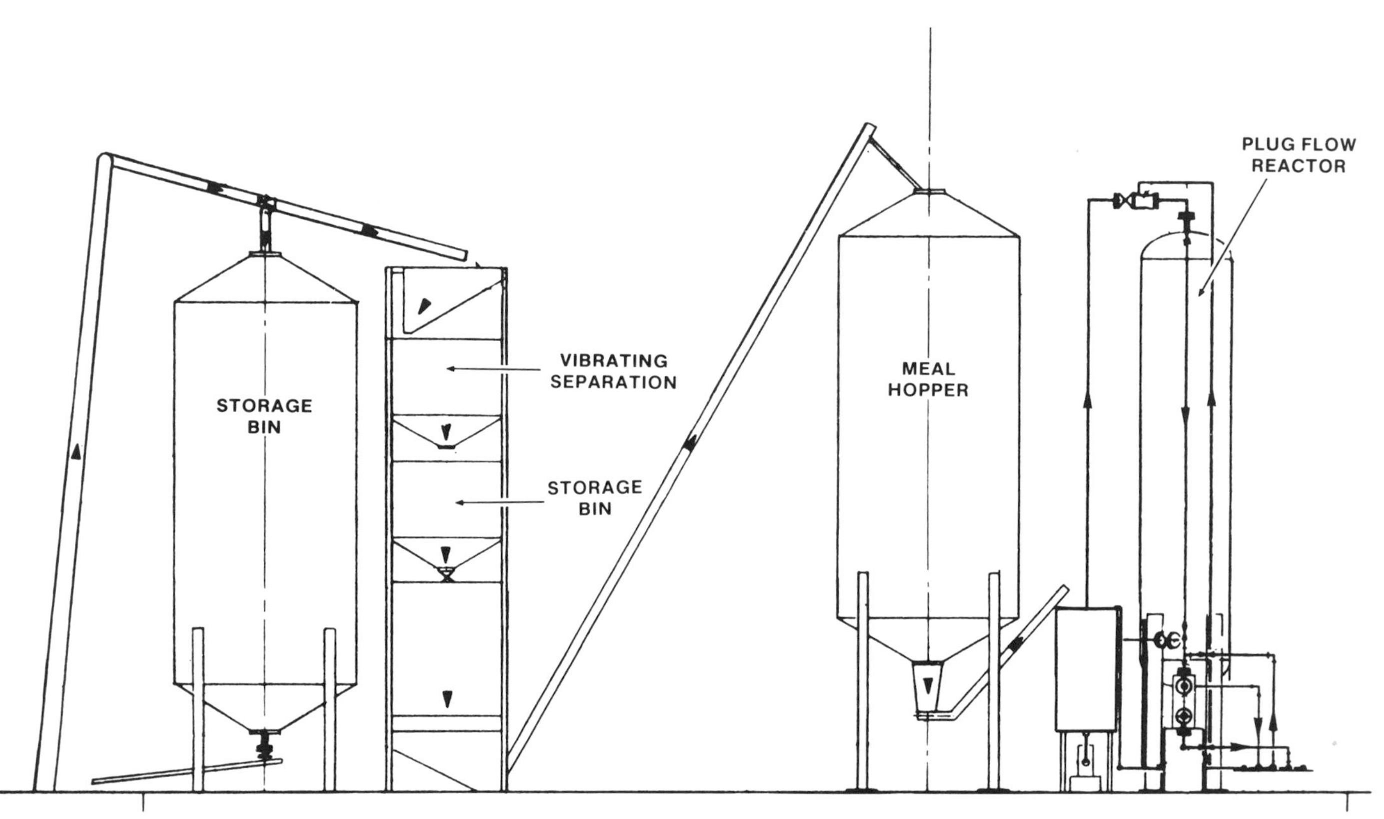

Figure 8-10. Feed process sections.

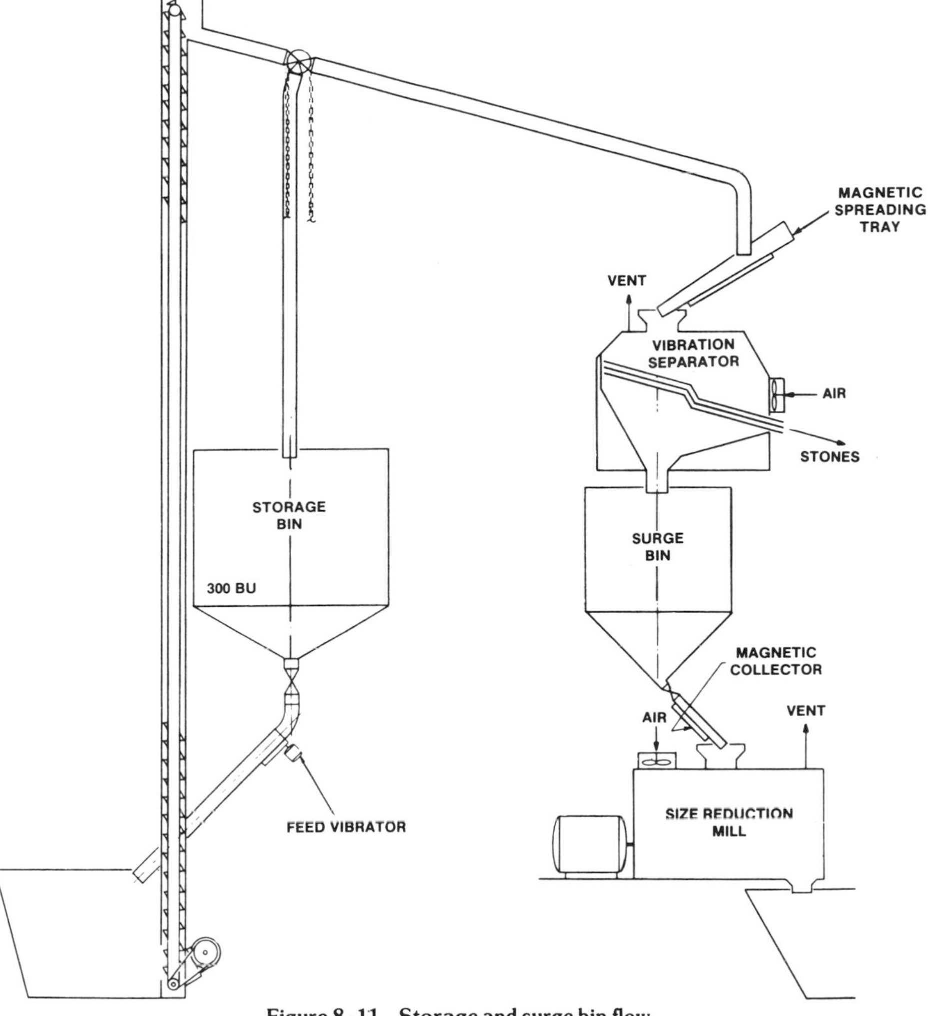

Figure 8-11. Storage and surge bin flow.

2. The bucket elevator is manually started by a local control switch.
3. The bucket elevator is allowed to run until the proper fill line, indicated on the storage bin, has been reached. Inspection of this fill line is visual. Operation of the input storage subsystem has been designed to be accomplished with a 1-hour period.

Cleaning and size reduction. This operation step begins by setting the three-way valve to allow corn to be directed to the processing equipment. Engage start switches for the bucket elevator, vibrating separator, and size reduction mill. These switches are the bucket elevator, vibrator feeder, separator motor, separator blower motor, size-reduction mill motor, and size-reduction mill air blower. This cleaning and milling activity is designed to be accomplished within a 12-hour duration, and local elapse timers set at 12 hours control the shutdown. A surge bin of 20-bushel capacity is located as an input hopper to the size-reduction mill. Any excesses not contained within this surge bin will be controlled through an independent relay control level switch.

Emergency Operations

Overload switches will be used to control problems associated with grinder mill plugs, elevator failures, etc. The 12-hour timers will disconnect power to all operating equipment after the 12-hour duration. Loss of power will result in no adverse effects on product quality or the restart operation.

Normal Shutdown

The normal operating procedure used for shutdown is to allow the timer to disconnect power to the elevators, cleaner, and mill.

Routine Maintenance

Normal maintenance of the feedstock preparation equipment can be accomplished through random inspection checks on the elevators, vibrator screens, vent line, and grinder blades. The grinding mill will require closer inspection on blade conditions to ensure proper size reduction. The magnets will need to be checked and cleaned periodically. Lubrication should follow manufacturer's specifications.

Saccharification and Fermentation

The fermentation process will be run automatically after the operator has filled the various tanks and selected valve positions. Pushing the start button on the control panel will begin the process. Refer to Figs. 8-12 and 8-13 for the flow through saccharification, fermentation, and beer well.

Start-up

The start-up process is as follows (refer to Fig. 8-14 for process flow).

1.0 Prerequisites
- 1.1 Cooling water pressure—30 psig.
- 1.2 Meal hopper being filled.
- 1.3 Electric power on.
- 1.4 Steam pressure—10 psig.
- 1.5 Fermentation tanks have been cleaned and steam sterilized for 1 hour.

2.0 Procedure

2.1 Fill alpha-amylase tank with 20 gallons of cold water and add 12 pounds of alpha-amylase enzyme. Start the agitator.

2.2 Add water to the fill mark on tank. Add five gallons of 1.0 *N* sulfuric acid. Open control valve.

2.3 Fill glucoamylase tank with 10 gallons of cold water and add $3\frac{1}{3}$ pounds of glucoamylase enzyme. Start the agitator.

2.4 Blend together 25 pounds of baker's dried yeast and 17 gallons of lukewarm water in yeast tank. Open control valve.

2.5 Open the following valves:
plug flow reactor inlet (see Fig. 8-15);
plug flow reactor vent;
inlet to fermentation tank;
mash cooler cooling water outlet.

2.6 Shut the following valves:
inlet to fermentation tank;
beer outlet.

2.7 Vent or prime all pumps in cooking and fermentation system if not self-priming. Caution—take precautions with sulfuric acid pumping; wear rubber gloves, raincoat, and face protection.

2.8 Start fermenter fill cycle at microprocessor control station.

2.9 Return to the plant to check the following:
valve for meal hopper discharge open;
feed auger motor operation;
alpha-amylase pump operation;
stream line hot downstream of steam control valve;
proper temperature.

2.10 Fifty minutes after start-up, go to the plug flow reactor vent. When mash starts to exit the vent, vent, shut valve, check temperature 203 ± 2°F.

2.11 In 10 more minutes, check temperature for 140 ± 5°F, and make sure pump is operating.

2.12 Two hours after start-up check that tank level is approximately 500 gallons, tank temperature is at 140 ± 2°F, and tank agitation started. Check tank pH at 3.8-4.5. Revent the plug flow reactor. Check that the glucoamylase tank has started emptying.

2.13 Sample tank contents from fermentation tank. The sample should be free of visible starch granules and have a slightly sweet-and-sour taste.

2.14 After 12 hours of operations, the steam flow, meal auger, alpha-amylase supply, and freshwater supplies should shut off.

2.15 At 15.5 hours after start-up, the fermenter tank should begin cooling to 90°F by cooling water. The cooling should be

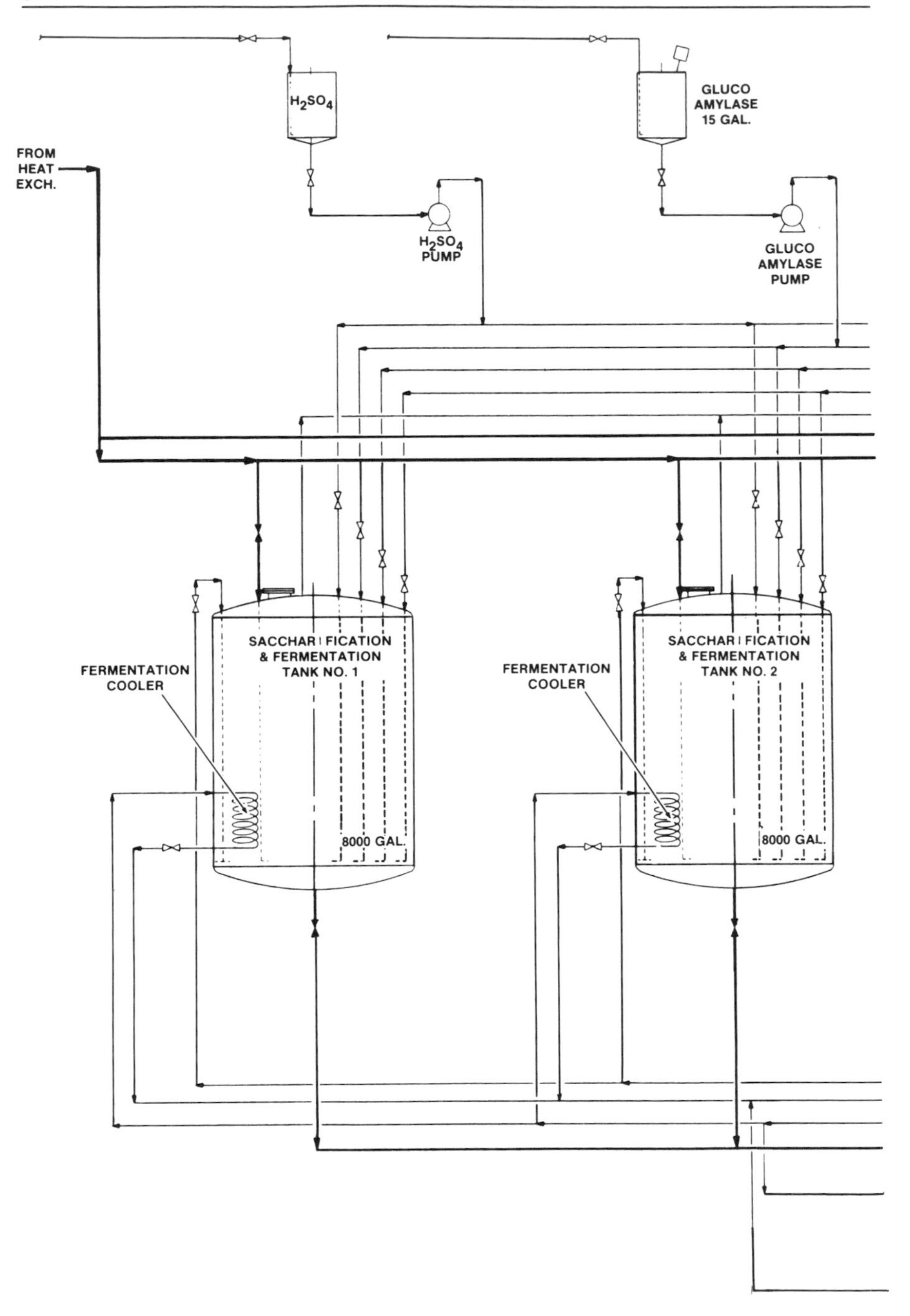

Figure 8-12. Saccharification and fermentation tanks.

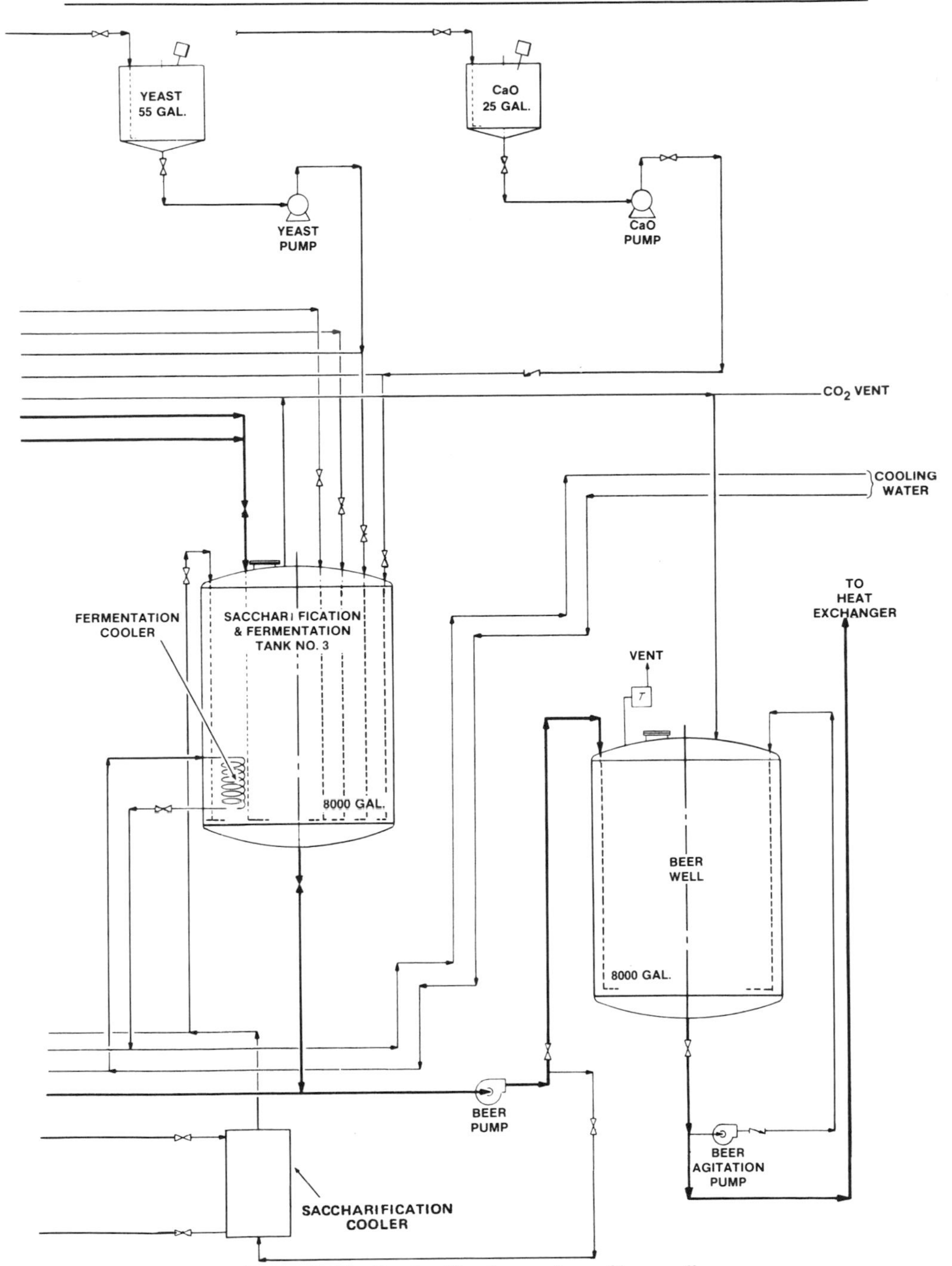

Figure 8-13. Saccharification cooler and beer well.

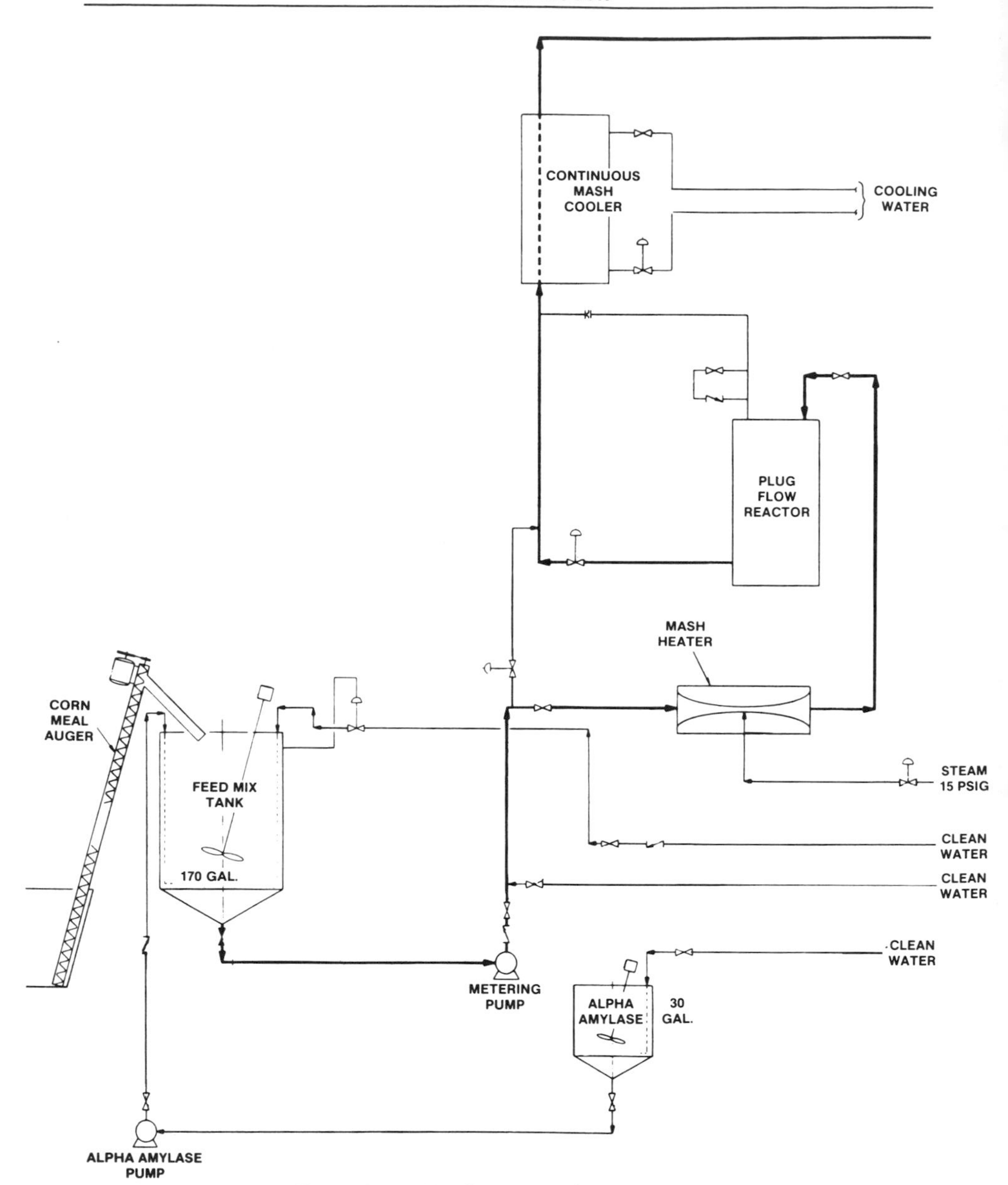

Figure 8-14. Feed mixing and mash handling.

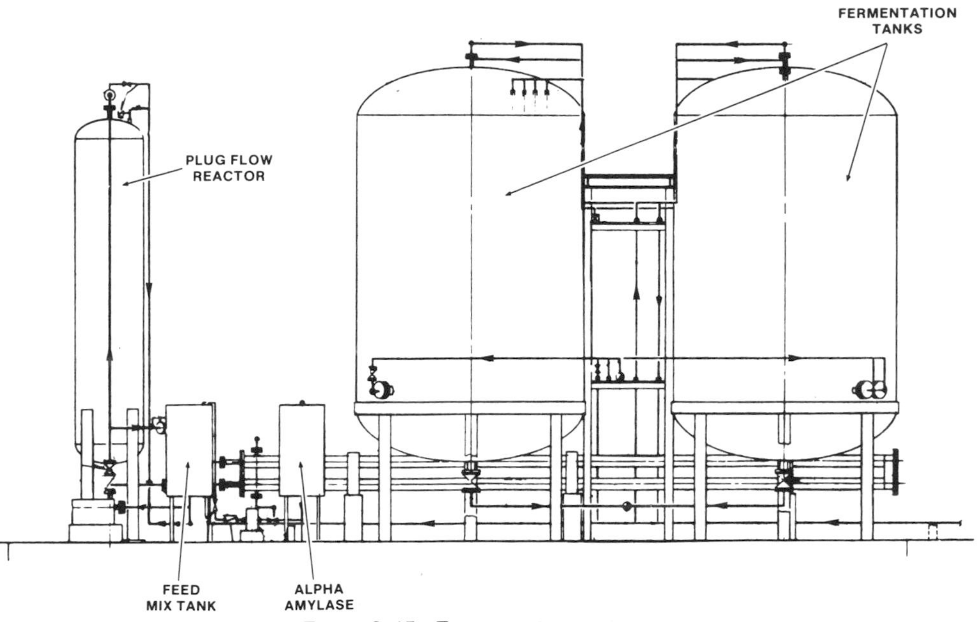

Figure 8-15. Fermentation sections.

completed by 20.5 hours after start-up to 90°F. Enough water will be added by metering pump at 18 hours after start-up to bring the fermenter to 5700 gallons.

2.16 After 30 minutes, check pH controlling at 4.8-5.0. Check every 8 hours that temperature is 80-90°F, pH 4.8-5.0.

2.17 Yeast should empty into the fermenter at 20 hours after start-up. Confirm emptying.

2.18 After 72 hours, pump out the fermentation tank to the beer well.

Normal Daily Routine

The normal daily routine is as follows:

1.0 Begin procedure.

1.1 Fill storage line.

1.2 Start feed mill cycle to begin filling feed hopper.

1.3 Pump out applicable fermenter to beer well if fermentation is complete and satisfactory. To perform pump-out, start agitator in tank to be pumped out, wait 5 minutes, open the beer discharge valve, and start beer pump. Make sure adequate space exists in the beer well.

1.4 Refill enzyme tanks, yeast hopper, yeast mix tanks, CaO tank, and acid tank.

1.5 After 30 minutes, the fermentation tank should be empty. Check that pump is stopped.

1.6 Perform the following on the fermentation tank after emptying:
hose out fermentation tank if necessary;
shut fermenter beer discharge valve;
shut CO_2 vent to product storage;
open auxiliary vent, shut fermenter mash filling valve, turn on sterilizing steam and purge for 1 hour.

1.7 Open the mash filling valve on the fermenter just emptied. Shut the mash filling valve on the fermenter that most recently completed filling.

1.8 Start the mash filling cycle at the microprocessor for the fermenter just emptied.

Shutdown

1.0 Shutdown

1.1 Purge the mash filling line with water (fully open valve) for 1 hour. Cycle open fermenter filling valves for 20 minutes each.

1.2 Check to be sure that all enzyme, acid, and yeast tanks empty. Hose out each and pump to each fermenter to purge lines with freshwater.

1.3 Open fermenter drains and hose out.

1.4 Drain the plug flow reactor.

1.5 Shut off cooling water at source.

1.6 Shut the following valves:
fermenter mash fill valves;
fermenter beer discharge valves;
manual discharge valves on each enzyme, yeast, and acid tank.

1.7 Open the following valves:
mash cooler cooling water discharge valve;
plug flow reactor isolation valve.

1.8 Open all high-point vents and low-point drains to drain the system.

1.9 Lock the fermenter access manholes shut.

1.10 Shut off electrical power to the system.

1.11 Cover all motors for protection from the elements.

Primary Set Points

Following are the primary set points:

Gelatinization temperature, 203°F
Saccharification temperature, 140°F

Saccharification pH, 4.15
Fermentation temperature, 85°F
Fermentation pH, 4.9
Batch size, 5700 gallons
Corn meal charge, 240 bushels

Batch Sequencing

Batch sequencing is as follows:

0 hours	Start feed
1.5 hours	Start glucoamylase addition
7.5 hours	Glucoamylase addition complete
12 hours	Stop feed
15.5 hours	Batch begins cooling to 90°F
18 hours	Dilution of batch to 85°F, 5700 gallons started
18 hours 5 minutes	Dilution complete
20 hours	Yeast added, fermentation starts
20.5 hours	Batch reaches 90°F
72 hours	Fermentation complete, pump to beer well
24 hours	Begin feed cycle on second fermenter
48 hours	Begin feed cycle on third fermenter

Safety Considerations

Safety considerations should include the following:

1. Beware of hot pipes.
2. Beware of pulleys, auger shafts, and other moving parts. No loose clothing should be worn around the plant.
3. Always wear safety glasses in plant.
4. When handling sulfuric acid or CaO, wear rubber gloves, rubber raincoat, and full faceshield.
5. Do not repair energized equipment.

Emergency Procedures

Procedures to be followed in case of emergencies are as follows:

1.0 Loss of electrical power.
  1.1 Control pH in each fermenter by manual additions of acid. CAUTION: Use protective equipment when handling acid.

Use a portable pH meter and periodically sample contents of each fermenter. Add acid sparingly.

1.2 Yeast addition to fermenters may be made manually if applicable.

1.3 Manually control temperature in fermentation tanks with cooling water as necessary using manual operators on valves. If all power has failed, perform all possible parts of the shutdown procedure before freezing of the lines and tank contents, using a portable pump and water supply if available.

2.0 Loss of steam.

2.1 Stop the fermentation tank filling cycle manually at the microprocessor if it has not already stopped automatically.

2.2 Perform shutdown procedure before freezing of lines and tanks.

2.3 Attempt to restore steam.

3.0 Loss of instrument air.

3.1 Stop the fermentation tank filling sequence manually at the microprocessor if not already stopped automatically.

3.2 Perform all possible parts of the shutdown procedure before freezing of lines and tanks.

3.3 Attempt to restore air pressure.

Troubleshooting Hints

If alcohol yields are low, check the following:

1. Temperature control at all points in the cycle.
2. pH control at all points in the cycle.
3. Activity of enzymes on small samples.
4. Yeast activity.
5. Possible microbial contamination of feedstock may be controlled by adding penicillin at the rate of 1 gram per 300 gallons of beer and by sterilizing the fermenters more thoroughly.

Routine Maintenance

Routine maintenance procedures are as follows:

1. Lubricate motors as necessary.
2. Check V-belts and pulleys for excess wear.
3. Repack leaking pumps, valves, and seals as necessary.
4. Repair steam leaks promptly.
5. Repair damaged electrical insulation promptly.

Distillation Section—Dry Start-up

A. Valve Line-up

open	feed isolation
open	column 1 bottoms isolation
open	bottoms isolation
open	feed pump isolation
open	bottoms pump isolation
closed	product takeoff
closed	column 1 vent
closed	column 2 vent
open	column 2 pump outlet
open	specific gravity inlet isolation
open	column 2 bottoms isolation
closed	condenser outlet
closed	reflux drum outlet
open	specific gravity outlet isolation
N/A	reflux check valve
closed	product pump recirculation
open	bottoms pump outlet
closed	bottoms/feed recirculation
closed	column 2 drain
closed	bottoms pump recirculation
closed	product storage outlet
closed	product denaturing control
open	product storage inlet isolation
N/A	product storage CO_2 check
open	condenser cooling water inlet
closed	condenser cooling water outlet
open	feedwater heater outlet
open	feedwater heater inlet
closed	feedwater heater bypass
open	condenser to boiler flow control
closed	pressure control, feed
closed	pressure control, bleed
closed	nitrogen supply isolation
closed	steam inlet isolation
N/A	column 1 pressure relief
closed	steam flow control

B. Establish Pressure Control

1. Open control valve.
2. Establish pressure of 4 psig.

3. Adjust controls to maintain 3.0 + 0/ − 0.5 psig.
4. Close control valve.

C. Column Heatup

1. Slightly open valve to establish cooling water flow to the condenser.
2. Put temperature control in manual.
3. Turn on columns 1 and 2 bottom level control systems.
4. Verify valve closed.
5. Open valve and verify a steam pressure of 10-15 psig.
6. Manually feed steam to column. Set steam column. Set steam flow to 482 pounds per hour.
7. When the level in the bottom of column 1 reaches the high set point, start pump.
8. When column 1 top temperature reaches 180°F, turn on pump and adjust feed flow to 3 gallons per minute.
9. Slightly open vent valve and vent column 1 until water vapors are being vented. Then close valve.
10. Place temperature control system on automatic to control steam flow.
11. When the level in the bottom of column 2 reaches the high set point, start pump.
12. When overhead product begins to accumulate in the reflux drum, start pump and begin refluxing at 0.5 gallon per minute.
13. Continue to adjust reflux to maintain a level in the reflux drum of 0-10 percent. Do not let the reflux pump run dry.
14. When the reflux rate has reached 1.0 gallon per minute, increase the feed to 4.0 gallons per minute or other normal operating set point.
15. Continue to increase reflux flow to maintain level of 0-10 percent in the reflux drum, until reflux flow has been increased to 2.1 gallons per minute or other normal operating point.
16. Turn on the reflux drum level control system.
17. Open valve.
18. Start pump.
19. Partially close valve until condenser outlet temperature reads about 143°F.

20. Partially close bottoms pump outlet valve so as to minimize cycling of valve. The flow should be adjusted so column 1 bottoms level barely drops with control valve closed.
21. Partially close product takeoff valve so as to minimize cycling.

NOTE: Following the initial system start-up, the optimum reflux flow and temperature set point should be determined as follows:

1. Set reflux flow at 1.0 gallon per minute.
2. Place reflux drum level control in manual and maintain drum level below 10 percent. This can be done by closing the control valve controlling product flow.
3. Increase the temperature control set point in 1°F increments until the product purity is not acceptable. Allow about 30 minutes to 1 hour between each set point change for the system to stabilize. When the product quality shows degradation, lower the set point by 1°F.
4. Increase reflux flow by 0.1 gallon per minute increments until product quality begins to degrade. Allow 30 minutes to 1 hour for stabilization between flow changes.
5. Record the temperature set point and reflux flow for reference. These are to be used as optimum operating points as long as feed flow rate and alcohol content do not change.
6. Place reflux drum level control in automatic, and open valve.

Distillation Section—Wet Start-up

The wet start-up procedure follows. Wet start-up occurs when liquid is in the bottom of both columns and the reflux drum.

A. Establish valve line-up per dry start-up procedure.
B. Establish pressure control as per dry start-up procedure.
C. Column heatup

1. Perform dry start-up column heatup Steps 1 through 11.
2. When column 2 top temperature reaches 165°F, start pump and begin refluxing at 0.5 gpm.
3. Continue to increase reflux flow to maintain level in the reflux drum at nominal setpoint as indicated on sight glass.

NOTE: If temperature exceeds 175°F, increase reflux as necessary to control temperature, allowing reflux drum level to drop.

4. When the reflux flow has reached 1.5 gallons per minute, increase the feed to 4.0 gallons per minute or other normal operating set point.
5. Continue to adjust reflux flow maintaining column top temperature at 174°F and reflux drum near the nominal level set point until a flow of 2.1 gallons per minute or other normal operating flow has been reached.
6. Turn on the reflux drum level control system.
7. Open valve.
8. Start pump.
9. Partially close valve until condenser outlet temperature stabilizes near 143°F.
10. Perform Steps C-20 and C-21 from dry start-up procedure.

Normal Operating Procedure

After the column has been manually brought to normal operating conditions, by either the dry or wet start-up procedure, the control system set points and control modes (manual versus automatic) should be checked for proper values or position.

A. Control mode positions

1. Temperature control, automatic.
2. Reflux drum level control, automatic.
3. Column 1 bottoms level control, automatic.
4. Column 2 bottoms level control, automatic.

B. Set points

1. Temperature, 180°F (or as determined during optimization).
2. Feed flow, 4.0 gallons per minute.
3. Reflux flow, 2.1 gallons per minute, (or as determined during optimization)
4. Pressure, 2-3 psig.

C. Check the product quality every 24 hours. If necessary, adjust set point by 1°F per hour until desired quality is achieved. Lower temperature to improve product quality, but keep as high as possible. The microprocessor automatically controls the distillation process after the column has been started up. The process continues automatically until controls are adjusted, the beer well becomes dry, or an emergency condition occurs.

D. The following nominal values and limits are listed for all displayed process parameters to allow complete verification of proper plant operation.

Shutdown

Shutdown procedures are as follows:

A. Shutdown to full reflux

1. Turn off feed by stopping pump.
2. Decrease cooling water flow to the condenser by partially closing control valve. Set flow to maintain water outlet temperature at 140-150°F.

NOTE 1: The distillation section of the plant is now in a total reflux or hot standby condition. It can be left in this condition indefinitely, if desired. Do not shut the distillation section down to cold shutdown unless for maintenance, or the shutdown is for an extended period.

NOTE 2: Start-up from total reflux is accomplished by slowly increasing the feed to the normal operating set point, and increasing condenser cooling water as required.

B. Cold shutdown

1. Turn off feed by stopping pump.
2. Turn off column 2 bottoms pump.
3. Turn off steam flow by setting temperature control set point to minimum.
4. Turn off reflux and product pumps.
5. Turn off column 1 bottoms pump.
6. Close steam supply shutoff valve.
7. Allow columns to cool to 160°F as indicated at the bottom of column 2.
8. Stop condensor cooling water flow by closing valves.
9. Purge column for 5 minutes by opening vent valve.
10. Close vent valve.
11. Close valve to complete column isolation.

NOTE: The distillation section is now shutdown and pressurized with nitrogen. Leave the system in this status to prevent air inleakage unless the system must be opened for repair.

C. System depressurization

1. Close nitrogen feed isolation valve.
2. Open valve to vent column to atmospheric pressure.

Emergency Conditions

A. Action in event of annunciator

1. High-pressure alarm.

NOTE: This alarm indicates excessive pressure in the distillation columns and can be caused by any of the following conditions.

1. Failure open of steam supply valve.
2. Failure open of pressure control feed valve.
3. Failure closed of pressure control bleed valve.
4. Loss of cooling water to the condenser.

This procedure will place the plant in a safe status for any of the above failures.

a. Shut off steam to the column by placing temperature control in manual and setting to minimum set point.
b. Close steam supply shutoff.
c. Close pressure control feed valve.
d. Open pressure control bleed valve.
e. Stop feed flow by stopping pump.
f. Stop reflux flow by stopping pump.
g. Stop column 2 bottoms pump.
h. If alarm condition has not yet cleared, operate vent valve.
i. Complete column shutdown per steps B-7 through B-11 of the shutdown procedure.
j. Determine failure and repair before start-up.

2. High-temperature alarm

NOTE: This alarm indicates that safety rupture sensor has ruptured probably due to high column pressure. Response is identical to that required above.

3. High-temperature alarm

NOTE: This alarm indicates off-specification product at the top of column 2, and can be caused by the following:

1. Low or no reflux flow
2. High steam flow
3. High pressure

a. Check reflux drum level. If the drum is empty, proceed as follows:

   i. Stop pumps.
   ii. Place reflux drum level in manual (assumed failure of automatic control) and close product take-off valve.
   iii. As reflux drum refills, start reflux pump and slowly increase flow back to the normal operating flow.
   iv. Investigate reflux drum level control system, and repair as necessary.

b. If the reflux drum is not empty (step a), increase reflux flow to lower temperature.
c. Check steam flow. If it is above normal operating, adjust it manually and determine if a controller failure has occurred.
d. After steam flow control is reestablished, decrease reflux flow to normal operating set point.
e. Check system pressure and verify both are within the normal operating range.

4. Low-temperature alarm

NOTE: This alarm indicates low temperature at the top of column 2 and could be caused by any of the following:

a. Excessive reflux flow
b. Extremely subcooled reflux
c. Loss of steam supply
d. Column full of liquid
e. Loss of vapor flow from column 1.

1. Check system pressures. If either pressure is above the high limit, proceed with system shutdown as required.
2. Check reflux flow, reflux drum temperature, etc., to isolate the problem. No safety hazard exists, so immediate action is not necessary.

B. Loss of power

NOTE: Loss of 120-volt power will stop all pumps in the system, and control valves will close. Steam flow will be stopped on loss of power by an upstream control valve. This procedure prepares the plant for an orderly start-up when power is restored.

1. Shutoff steam supply.
2. Turn off pumps.
3. Place temperature controller in manual and close control valve.
4. Place reflux drum level controller in manual and close valve.
5. Proceed with wet start-up procedure when power is restored.

C. Loss of steam

1. Stop reflux pump.
2. Stop feed pump.
3. Place temperature controller in manual and close valve.
4. Proceed with shutdown, per procedure, or wet start-up if steam is restored quickly.

D. Loss of instrument air—no effect

E. Loss of tank purge—no effect

Routine Maintenance

A. Tray cleanout

NOTE: The frequency of tray cleanout will vary with type of feedstock, water purity, and other factors. Initially, the cleanout should be performed every 2 weeks.

1. Remove feed, steam, and bottoms takeoff connections from the lower section of column 1.
2. Remove all wiring and instrument lines from the bottom section of column 1.
3. Unbolt the flange connection holding the lower section. Loosen the bolts uniformly to lower the section to the ground.
4. Remove the column section and remove the trays from the column through the top.
5. Clean the trays and reassemble.

NOTE: If the specific location allows 2-foot clearance below the column, the internal tray support and trays can be removed from the bottom of the column without disconnecting the process lines.

Troubleshooting hints. Troubleshooting hints are as follows:

A. Foaming

Excessive foaming of the feed could cause loss of efficiency of the stripping section, resulting in low recovery of alcohol from the beer and less product.

Foaming could also cause flooding of the stripping section, causing high pressure drop and reduced column throughput.

The column has a window at the feed tray to observe foaming behavior. The foam should not extend above the top of the first tray, or above the downcomer. It may be necessary to add a foam-control agent to the beer if foaming is excessive.

B. Fusel oil buildup

Excessive fusel oil buildup in column 2 could cause flooding of the column and reduced product takeoff. The situation can be corrected by shutting down and pumping the bottoms of column 2.

If high-purity product is not required, fusel oils can be removed overhead by reducing the reflux flow.

C. Low product recovery (poor ethanol mass balance)

If the column appears to be working properly, but product recovery is not equal to at least 90 percent of the alcohol in the feed, the reflux flow is probably too low. Perform the reflux optimization steps following the dry start-up procedure.

Preliminary Operating Procedure for Solids Removal System

General. The solids removal system is not controlled by the microprocessor. All control is local.

Equipment. The solids removal system includes the following equipment:

1. Solids removal conveyor
2. Dewatering conveyor
3. Six-inch auger
4. Drip pan
5. Dewatered solids storage bin
6. Air line with nozzles

Prerequisites. The start-up of the solids removal system can begin before start-up of the distillation column, refer to Fig. 8-16. The following is a step-by-step procedure for start-up.

1. Open valve.
2. Close control valve.
3. Start solids removal conveyor.
4. Start dewatering conveyor.
5. Start auger. The auger will start only when level switch is activated.

This system will run continuously with no operator adjustment. The shutdown procedure is as follows:

1. Shut off all motor controllers.
2. Close valve.

Troubleshooting. If solids in the bin are too wet, adjust squeeze conveyor.

Distillation Section—Physical Description

The distillation section consists of two distillation columns as shown in Fig. 8-17, five heat exchangers, the reflux drum, and the necessary piping, pumps, and valves.

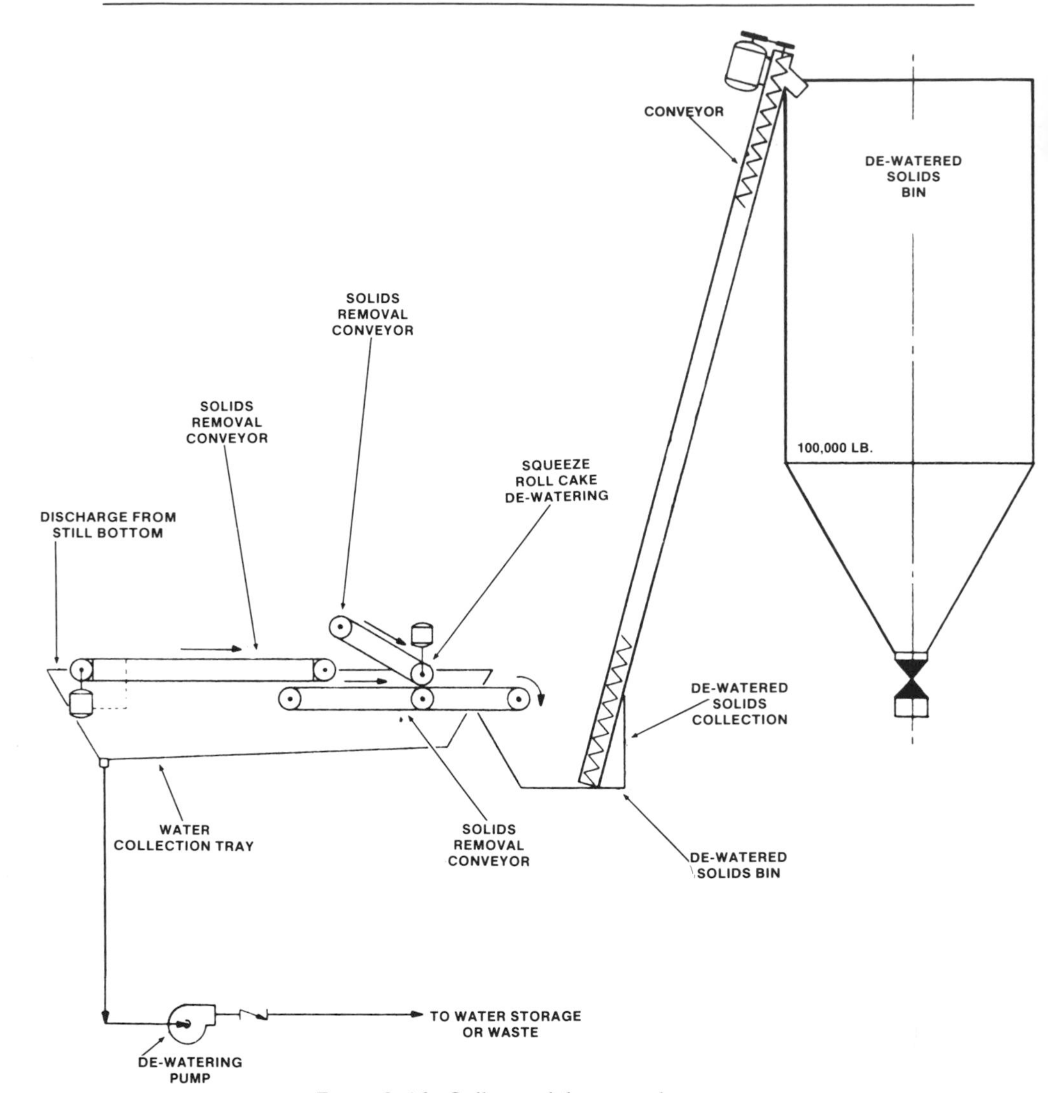

Figure 8-16. Stillage solids removal processing.

Each distillation column is about 21 ½ feet in overall height. The first column contains 12 sieve trays in the stripping section below the feed. This is equivalent to about six theoretical stages of separation. The first column also contains 5.2 feet of packed column above the feed point. This packed bed is equivalent to about 3.7 stages of separation. The column is made from 20-inch standard wall pipe below the feed point, and 12-inch standard wall pipe above the feed.

The sieve trays are built as complete units including the downcomer and side walls, and are constructed of 14-gauge 304 stainless steel. The trays

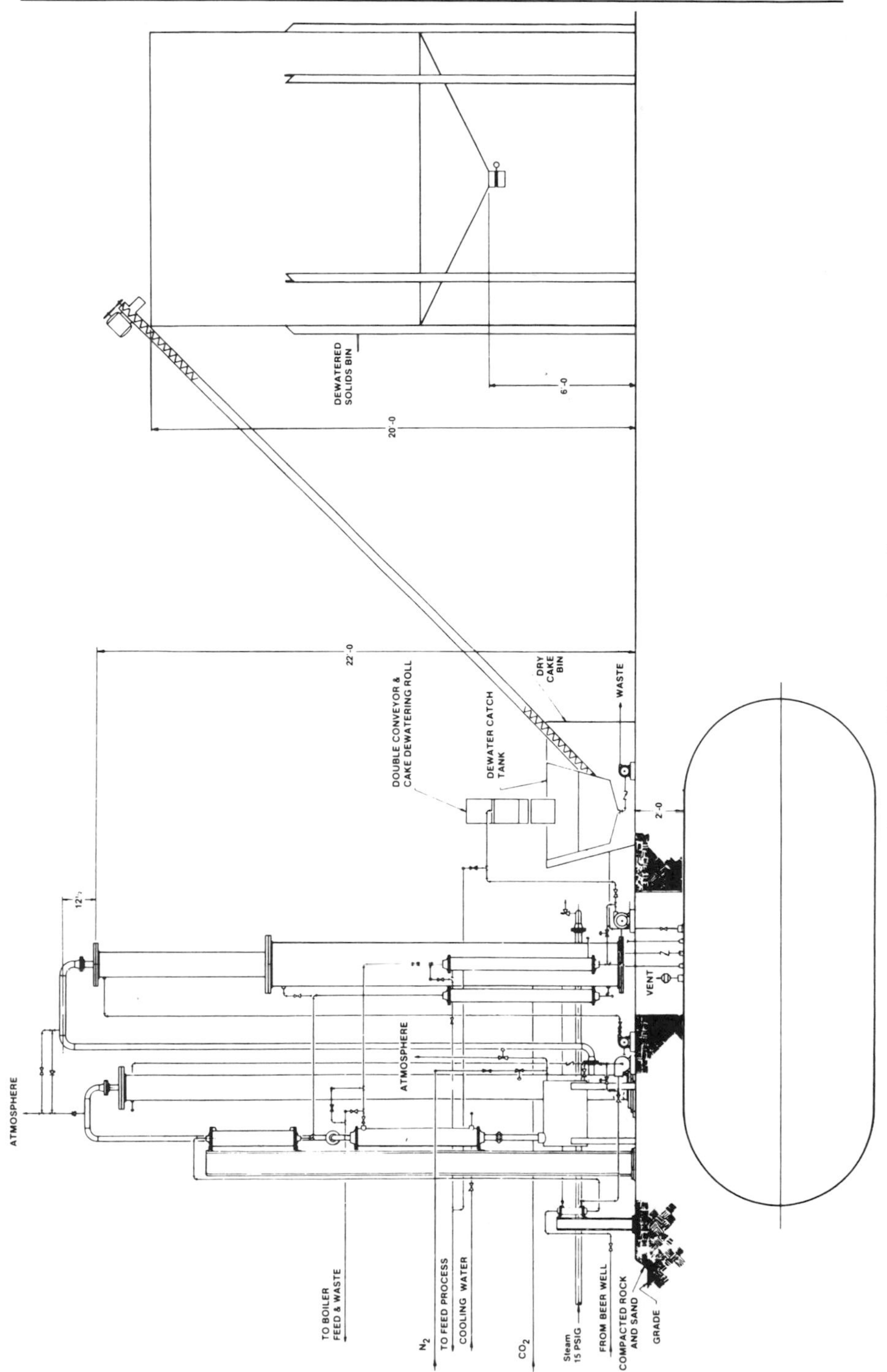

Figure 8-17. Elevation of distillation and product storage sections.

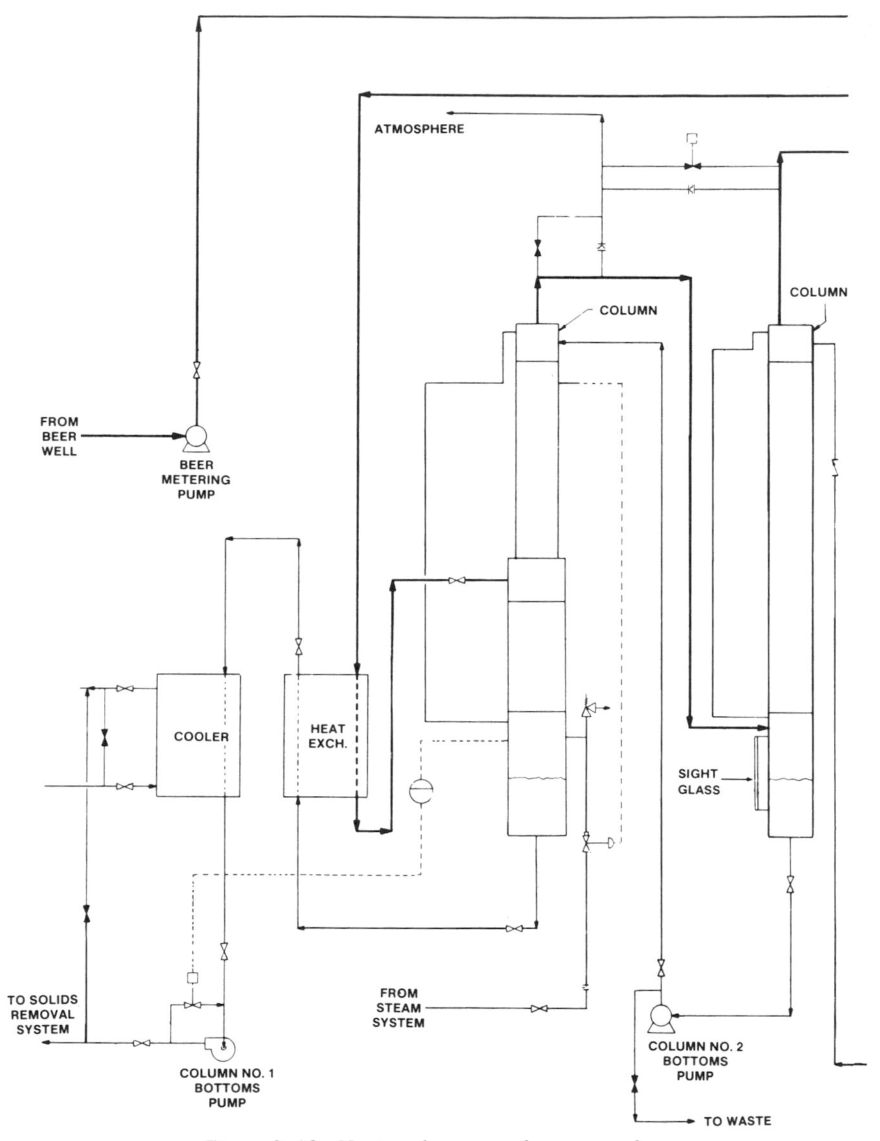

Figure 8-18. Heat exchanger and process columns.

stack on top of each other in the column, with the bottom tray resting on an internal support. The feed nozzle is positioned to feed into the downcomer of the top tray. The column will be supported above the feed point, with the stripping section suspended. The stripping section can be disconnected and the trays removed from cleaning without disassembly of the entire column. The internals for the packed section are 304 stainless steel.

The second column is piped in series with the first column to provide an extension of the rectifying section of the still. This column has a packed bed

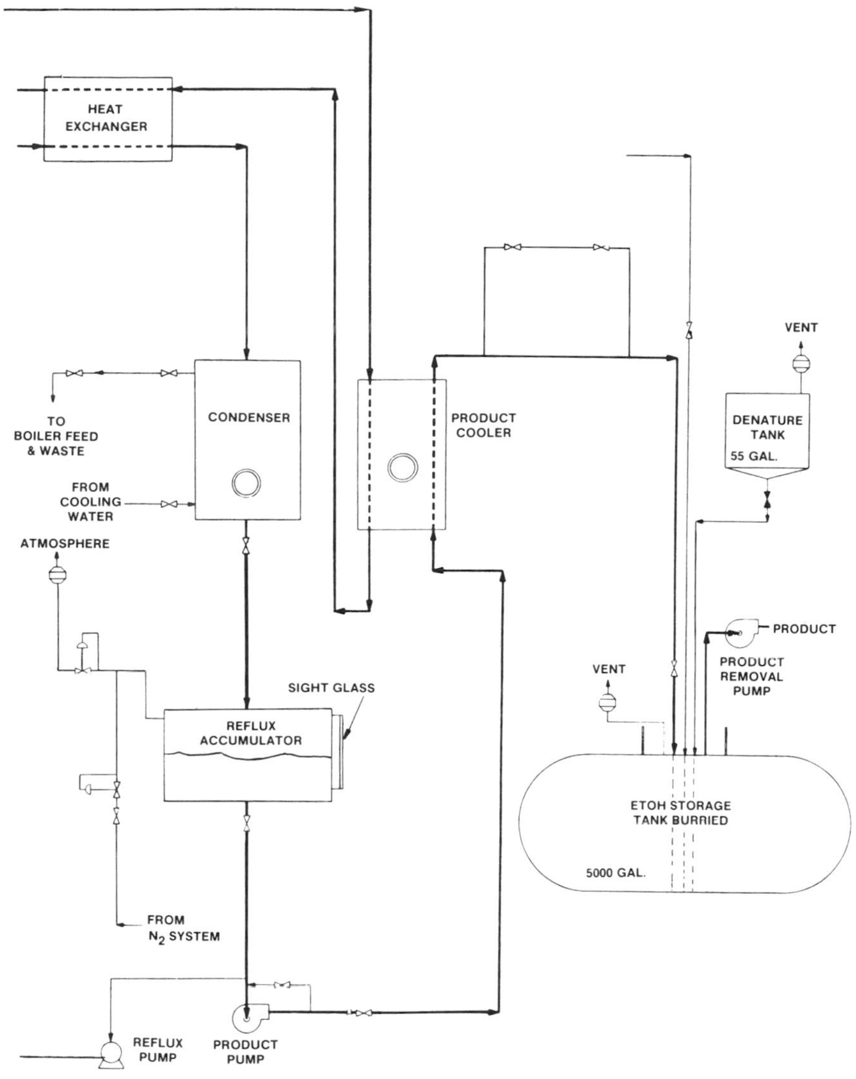

Figure 8-19. Ethanol cooler and storage.

17.2 feet long and has 12.3 theoretical stages of separation. The bed support, bed limiter, and liquid distributor are 304 stainless steel. The column is equipped with three "wall-wiper" liquid redistributors to reduce liquid channeling on the column wall. The heat exchangers are all shell and tube units.

The product cooler, shown in Fig. 8-18, cools the product going to the molecular sieve against the beer feed. The heat exchanger is to be mounted vertically, with the beer feed (whole mash) going down through the tubes, and the ethanol product flowing up on the shell side.

The vapor/feed heat exchanger heats the beer against the overhead vapor from the second column, while condensing about 25 percent of the vapors. This heat exchanger is to be mounted vertically with the beer flowing down through the tubes, and vapor and condensate flowing down on the shell side. This unit must be mounted above the condenser so that condensate can flow into and through the condenser due to gravity. (Refer to Fig. 8-19.)

The vapor and condensate mixture enters the condenser, where the remaining vapor is condensed. This heat exchanger is sized to condense all the vapors so the unit can still operate without feed flow and partial condensation in the vapor/feed heat exchanger. The condenser is a vertical shell and tube heat exchanger, with water flowing up through the tubes, and condensate down on the shell side. The condenser is mounted above the reflux drum to provide gravity flow of condensate into the reflux drum.

The bottoms/feed heat exchanger heats the feed against the still bottoms. The heat exchange is performed after the feed has been heated against the overhead vapors. This heat exchanger is a vertically mounted shell and tube, with feed flow down through the tubes, and bottoms flow down through the shell side. Both flows are down to minimize the potential problems with solids separation and settling.

After the bottoms have been used to heat the feed, they then flow through another heat exchanger to heat water for use in the cooking process. The heat flows down through the tube side of the bottoms/water heat exchanger.

Cooling water requirements are minimized by reusing it where possible. The water used to cool the fermenters will be mixed with additional cooling water, and used in the condenser. The warm water from the condenser will be used for boiler feedwater, and also as feedwater to the bottoms/water heat exchanger, and then to the cooking process. The reflux drum will be a 40-gallon domestic hot-water heater. This provides an insulated and glass-lined tank at a cost lower than a special tank could be obtained. The reflux drum is to be mounted at least 2 feet above the reflux pump.

Portions of the distillation section requiring thermal insulation include both distillation columns, the liquid and vapor lines between columns, the steam injection line, and the reflux return line.

System Control

The feed and reflux flow rates are controlled to predetermined values. The product takeoff is controlled by a level control on the reflux drum. Composition control is obtained by varying feed flow to maintain a predetermined temperature at a specific location in the packed section of the first column. Bottoms flow from both columns are controlled by the level in the bottom of the columns.

Major Equipment

Major equipment includes the following:

1. *Reflux drum*—40 gallom Rheem domestic hot-water tank, glass-lined and insulated.

Closet model, #65HS-40D	
Overall height	31 inches
Diameter	22 $\frac{1}{4}$ inches
Connections	four—two upper, two lower
Connection size	$\frac{3}{4}$ NPT
Height of upper connections	22 $\frac{3}{8}$ inches
Height of lower connections or equivalent	2 $\frac{5}{15}$ inches
Estimated cost	$183
Delivery	1 week

2. *Product cooler*—Young shell and tube heat exchanger.

Model	F-301-HR P
One pass	—
Fixed tube bundle	—
Overall length	13.88 inches
Diameter	3.62 inches
Heat transfer surface	2.66 square feet
Baffle spacing	1.13 inches
Tube size	$\frac{3}{8}$ -inch outside diameter
Shell pressure rating	150 psig
Shell material	Brass
Baffle material	Brass
Tube material	Admiralty or equivalent
Estimated cost	$185
Delivery	3-4 weeks

3. *Bottoms/feed heat exchanger*—Young shell and tube heat exchanger.

Model	F-606-ER-1P
One pass	—
Fixed tube bundle	—
Overall length	77.76 inches
Diameter	6.12 inches
Heat transfer surface	52.7 square feet
Baffle spacing	4.50 inches
Tube size	$\frac{3}{8}$ -inch outside diameter
Shell pressure rating	150 psig
Shell material	Brass
Baffle material	Brass

Tube material	Admiralty or equivalent
Estimated cost	$955
Delivery	3-4 weeks

4. *Bottoms/water heat exchanger*—identical to bottoms/feed heat exchanger.
5. *Vapor/feed heat exchanger*—Young shell and tube heat exchanger.

Model	HF-804-ER-1P
One pass	—
Fixed tube bundle	—
Overall length	45.24 inches
Diameter	8.25 inches
Heat transfer surface	67.1 square feet
Baffle spacing	4.5 inches
Tube size	$\frac{3}{8}$-inch outside diameter
Shell pressure rating	250 psig (150 psig not available)
Shell material	Brass
Baffle material	Brass
Tube material	Admiralty or equivalent
Estimated cost	$1572
Delivery	3-4 weeks

6. *Condenser*—Young shell and tube heat exchanger.

Model	HF-806-AR-1P
One Pass	—
Fixed tube bundle	—
Overall length	63.24 inches
Diameter	8.25 inches
Heat transfer surface	100 square feet
Baffle spacing	9.0 inches
Tube sizing	$\frac{3}{8}$-inch outside diameter
Shell pressure rating	250 psig (150 psig not available)
Shell material	Brass
Baffle material	Brass
Tube material	Admiralty or equivalent
Estimated cost	$1891
Delivery	3-4 weeks

7. *Boiler*—The following instructions can be considered typical procedures for operation of a firebox firetube boiler.

   To start steam generator. The following procedure should be followed to start the steam generator:

   1. Attach the flue pipes.
   2. Connect field water supply.
   3. Open make-up water valve.

4. Connect field steam or water piping to outlet on tip of steam generator.
5. Connect blowdown line.
6. Turn all switches on panel to "off" position.
7. Connect separately fused AC field electrical service to main switch supply terminals.
8. Close blowdown valves and open feedwater valve.
9. Turn main switch to "on" position.
10. If a manual-type combination water column is used, the reset pin must be pushed.

To stop generator. Turn the main switch to "off" position.

Operation.

A. Whenever going on duty, check the water level in the steam boiler at once.

B. Boiler should be "blown down" regularly to maintain chemical concentrations at the desired level and to remove precipitated sediments. This is normally required once daily. Boiler operation is not interrupted.

1. Blowdown procedure for boiler:
   a. Some pressure must be left on the boiler for proper blowdown.
      IMPORTANT
      NOTE: Boiler water must be settled for proper blowdown.
   b. Open boiler blowdown valve slowly for 30 seconds then close slowly. Use this same procedure with the low water cutoff blowdown valve. Then go back to the boiler blowdown valve, open and close the valve very fast two more times; repeat this procedure with the low water cutoff valve. This will rid the boiler of most precipitated sediments.

Items to be checked periodically. Items that should be periodically checked include the following:

A. Gauge glass must be kept clean and tight.
B. Combination low water control under operating conditions.
C. Any leaks must be repaired immediately.
D. Pressure-fired units must be kept gas tight.
E. Heating surfaces must be kept clean. Tubes in HRTs must be brushed periodically. Check tubes twice monthly. If the boiler is properly maintained, tube cleaning should be necessary only once or twice a year.
F. Internal inspections should be done at least twice a year. Hand hole

assemblies must be removed and their gaskets replaced. Tighten replaced hand hole gaskets after 2 days of operation.

Description of Boiler Operation

1. Level Control

Boiler water level control consists of a combination low water cutoff and water level controls. One switch operating stoker opens on fall. When the water level in the boiler drops to a predetermined level, the valve opens to feedwater to the boiler. As the level in the boiler rises, the float rises and closes the water valve.

If the water supply should fail, the water level in the boiler will continue to drop below the point where the valve opened. The cutoff switch opens, breaking contact to the burner.

2. Probe-Type Low Water Cutoff

The manual reset "Probe," when used, is a supplementary safety device for protecting the boiler against damage from low water if the float-operated low water cutoff should become clogged or fail to operate for any reason.

This unit consists of a relay which is energized by a circuit between the probe rod and the boiler drum through the water in the boiler. If the water level falls below the probe rod, the energized circuit is broken which causes the switch in the relay to open, thus breaking the circuit to the stoker. This control must be manually reset after power interruption.

3. Pressure or Temperature Control

The pressure or temperature control consists of a pressure- or temperature-actuated switch with adjustable range and adjustable differential. This switch opens on rise, breaking the circuit to the stoker. It should be set for a maximum operating pressure or temperature desired (not more than 6.8 kilograms or 100°C gauge). The differential is usually set at a minimum to cause the burner to operate as quickly as possible upon steam demand.

4. Induced-Draft Fan

An induced-draft fan should be installed when a constant back pressure or positive pressure is present at the burner. If this situation is not relieved, serious damage to stoker parts could occur.

This small-scale plant is currently being operated by EG&G, Idaho Falls, Idaho for the DOE. Extensive testing of the concept has been started to provide a test bed. Inquiries are welcomed from DOE and EG&G regarding this system. For more details, contact EG&G Idaho, Ind., P.O. Box 1625, Idaho Falls, ID 83415, (208) 526-0509. A complete set of blueprints, process information and operating procedures are available by contacting EG&G.

Alcohol Dehydration

This plant is designed to produce approximately 25 gallons per hour of 190-proof alcohol (92 percent by weight). It is felt that many producers of ethanol, operating at this plant scale, would desire the capability of producing anhydrous ethanol. Typically, ethanol has been dehydrated by the addition of a third solvent (e.g., benzene, cyclohexane, polyglycols, gasoline) and then performing a distillation or solvent-extraction step, depending on the solvent used. In most of these dehydration techniques, the added solvent is removed by distillation and recycled internal to the operation. One exception is the ACR process where gasoline is used as the azeotype breaker, and the gasoline is left in solution with the dehydrated alcohol. This solution of gasoline/ethanol is later mixed with additional gasoline to produce gasohol. Presently this technology is proprietary to ACR and awaiting the approval of patent applications. Additional technologies for the dehydration of ethanol have been and are continuing to be researched. These include semipermeable membrane techniques, absorption techniques, and vacuum distillation techniques.

One technique that has shown significant promise is the use of molecular sieves (using artificial zeolite) to remove water from hydrous alcohol solutions. In this technique, the alcohol/water solution is passed through a packed bed of molecular sieve material, and the water is preferentially absorbed by the bed. The bed is later regenerated by heating with a heat-carrying gas to drive off the water absorbed in the bed. At this time, this process is not employed in any sizable ethanol-production facility. A process is presented in this section for the dehydration of ethanol using molecular sieves.

In this design, *n*-hexane is used in the regeneration of the molecular sieve as the heat carrier. The regeneration is carried out at two temperature cycles. A 250°F cycle is used to remove ethanol remaining on the bed. The *n*-hexane is boiled using steam from the boiler and then superheated to 250°F. During this cycle the backpressure on the molecular sieve is maintained at 7 psig and most of the *n*-hexane is condensed at the molecular sieve; this increases the heat transfer rate to the bed when compared with regeneration with a noncondensable gas such as CO_2 or nitrogen. The next cycle is started after the bed reaches approximately 235°F. In this step the *n*-hexane is superheated to 450°F to remove water from the bed. After the *n*-hexane gas passes through the bed it is condensed. Water is essentially immiscible in *n*-hexane, resulting in a two-phase liquid in the *n*-hexane tank. Ethanol is preferentially soluble in water; so the majority of ethanol goes into the water phase. The bed is cooled by reducing the temperature of the *n*-hexane and continued cycling of the *n*-hexane through the bed. At the completion of regeneration, *n*-hexane remaining on the bed is blown with nitrogen back to the *n*-hexane holding tank. The water/ethanol (approximately 50 percent ethanol) phase in the *n*-hexane holding tank is pumped back to the beer tank.

Emergency Conditions

In the event of an emergency, certain procedures should be followed:

A. Loss of power
   1. If loss during adsorption step and pump cavitates, the pump is shut off and valves are closed. Standby until suction is available, then restart as normal.
   2. If loss occurs while transferring contents of tank 2 and pump 2 cavitates, pump is shut off. Standby unless less than 150 gallons remains in tank 2, then restart adsorption step if desired.

B. Loss of steam
   1. If loss occurs during regeneration warm or hot purge, close all valves and shut off all pumps. That step should be restarted at the beginning of that particular sequence.
   2. If loss occurs at other times, continue sequence until warm purge is needed. Then, close all valves and shut off all pumps until steam is regained.

C. Loss of instrument air
   1. Control valves fail closed (if air operated), shut off pumps. Treat as loss of power.

Safety Considerations

Ethanol, 99.9 percent pure, is stored at ambient temperature and slight positive pressure. The flash point is 55°F and flammable limits are 4.3 and 19.0 volume percent. The saturated vapor-air mixture above the liquid ethanol will be flammable between 50 and 110°F. Hence, a carbon dioxide blanket will be used in the ethanol storage tanks.

n-Hexane exists as a superheated vapor at 450°F and 10 psig. The flash point is 0°F. The flammable limits are 1.2 and 7.5 volume percent. The saturated vapor-air mixture above the liquid hexane will be flammable between -20 and 40°F.

Pressure-relief valves with flame arresters will be installed on the sieve vessel and hexane settling tank.

Routine Maintenance

Molecular sieve material will have an effective lifetime of 2000 regenerations, or at the rate of one regeneration per day for five days per week equaling 260 per year, about a 7.5-year lifetime without fouling from such compounds as sulfur. For a plant lifetime of 10 years, the sieve may have to be replaced once.

Typical maintenance on pumps and valves is required. If cooling water is clean with low total dissolved solids, fouling of condenser tubes should be minimal. Sulfur compounds in the flue gas may necessitate cleaning the flue gas exchanger tubes periodically.

Troubleshooting Hints

If the ethanol product exits the sieve less than 99.9 percent pure, several factors could be existing:

1. Water concentration in the feed above 6 weight percent.
2. Feed flow rate above 83 gallons per hour.
3. Adsorption step longer than 12 hours.
4. Regeneration of sieve incomplete:
   a. temperature, flow rates, or flow durations too low;
   b. too much water in hexane passing through sieve;
   c. sieve lifetime exceeded;
   d. sieve damaged by sudden thermal shocks from rapid temperature changes.

Commercial-Scale Ethanol Plant

PROCESS DESCRIPTION

The process described in this section is a recognized process, and information is available from a DOE report entitled *Grain Motor Fuel Alcohol Technical and Economic Assessment Study*, by Raphael Katzen Associates International.

The illustrative example used in this section is a 50-million-gallon-per-year, corn-based, coal-fired, fuel-ethanol plant. The processing elements of the plant are described in this section as a means of familiarizing the reader with the elements of commercial-scale fuel-ethanol production at a level of technical detail with which the owner/investor should be familiar in order to work closely with a consulting engineer or architectural and engineering firm. The Katzen process is based on a highly integrated, thermally efficient conceptual design. Energy efficiency and coal firing are very desirable aspects of this or any commercial-scale design.

The energy and feedstock efficiency claimed for the Katzen-engineered plant are illustrated in the following figures:

Alcohol yield = 2.57 gallons/bushel of corn
DDG yield = 18.2 pounds/bushel of corn
Coal required (Illinois #6) = 41,700 Btu/gallon of alcohol
Electrical energy required = 1.32 kWhr/gallon of alcohol

The physical plant layout is illustrated in Fig. 8-20.

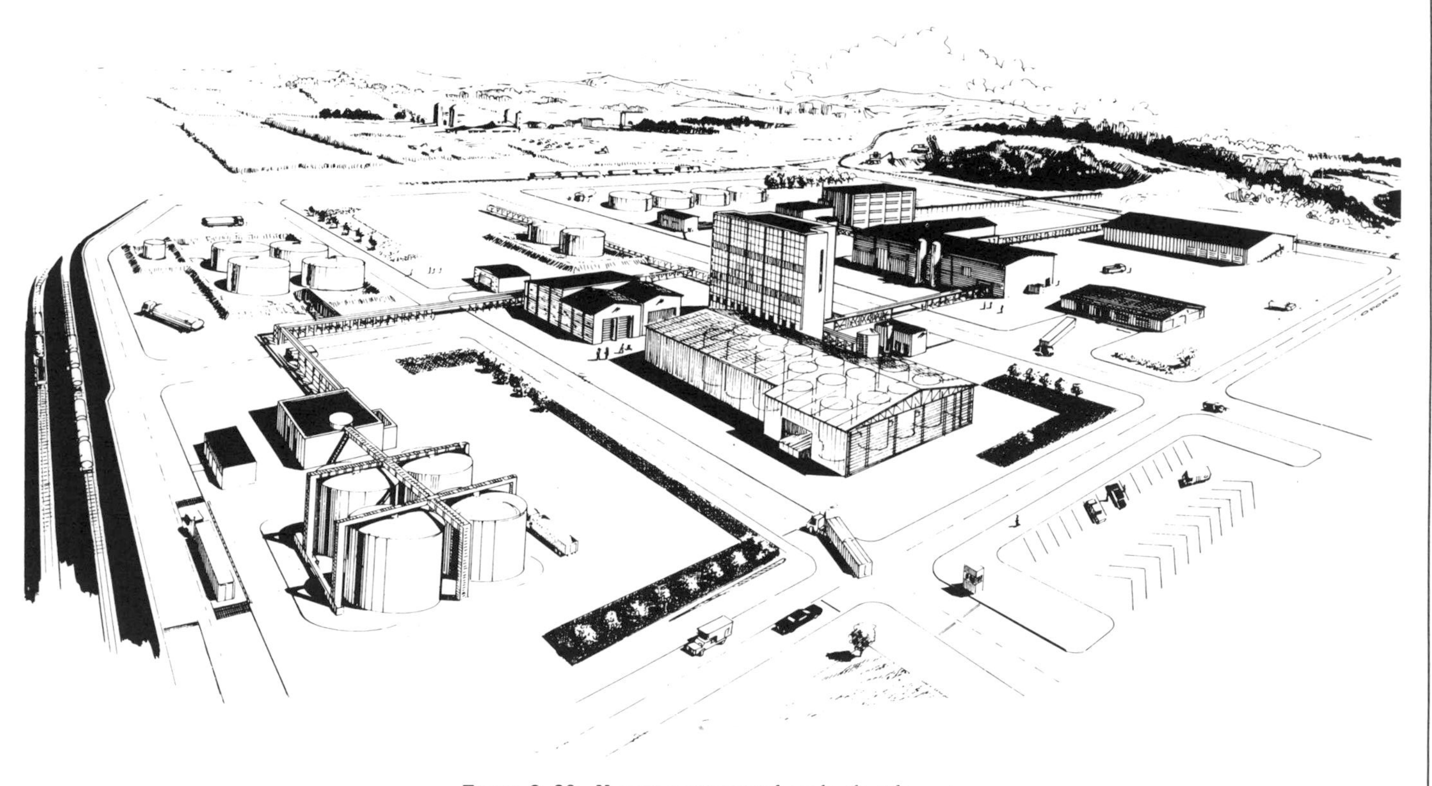

Figure 8-20. Katzen commercial-scale plant layout.

RECEIVING, STORAGE, AND MILLING

Shelled corn is delivered to the plant by railroad hopper cars or grain trucks. A single railroad unloading station and two truck unloading stations have been provided. The unloading arrangement has been planned so that a railroad hopper car and a truck can be unloaded simultaneously. (See Fig. 8-21.) Railroad hopper cars hold about 52.5 tons of grain (1875 bushels of corn). The fully loaded hopper car weighs about 67.4 tons. Grain trucks come in various sizes, but a typical truck will hold about 800 cubic feet or 643 bushels of corn. Unloading conveyors have been specified to accommodate a total unloading rate of 7500 bushels per hour, enough grain in a single 8-hour shift to operate the plant for a full day.

Trucks delivering grain to the plant are lifted by means of truck dump hoists and are weighed on one of two truck dump scales. The grain passes from the truck into a bin housed in a pit. Grain is discharged from these bins through star valves into either of two truck unloading conveyors.

In case of grain delivered to the plant by railroad hopper cars, the car is weighed on a rail car scale, and the grain is then dumped from the car through a bin and star valve to a rail car unloading conveyor. The grain passes into either of two rail car unloading cross conveyors. A truck unloading conveyor, coupled with a rail car unloading cross conveyor, then can be delivered into a bucket elevator. The grain is lifted to a position above the grain storage bins, and passes from the bucket elevator into one of two distributing conveyors. These conveyors are arranged to deliver to the storage bins or may convey their grain directly into storage by-pass conveyors which deliver directly into a surge hopper.

The surge hopper has been sized to hold 7500 bushels of grain. This provides a nominal holdup time of 3 hours. When the surge hopper is full, grain can be diverted into storage in any of the grain storage bins. The total grain storage capacity is equivalent to grain usage for 1 week. When grain is being received, the operation could have grain passing directly to the surge hopper. When it is filled, grain then would be diverted to the storage bins. When grain is not being received, it would pass from the storage bins to the surge hopper through the individual storage bin bottom conveyors, through a collecting conveyor, and into a bucket elevator. The grain is thus lifted and discharged into the surge hopper.

Grain discharges from the surge hopper at the rate of 2453 bushels per hour into the grain cleaner, which separates materials in the grain which are foreign to the process, including sand and tramp metal. Light materials in the grain are picked up from the screens and air transported through a blower, to the baghouse in the coproduct-recovery section, where they become part of DDG. Tramp metal and other oversize materials are rejected from the grain cleaner and periodically removed from a collecting bin.

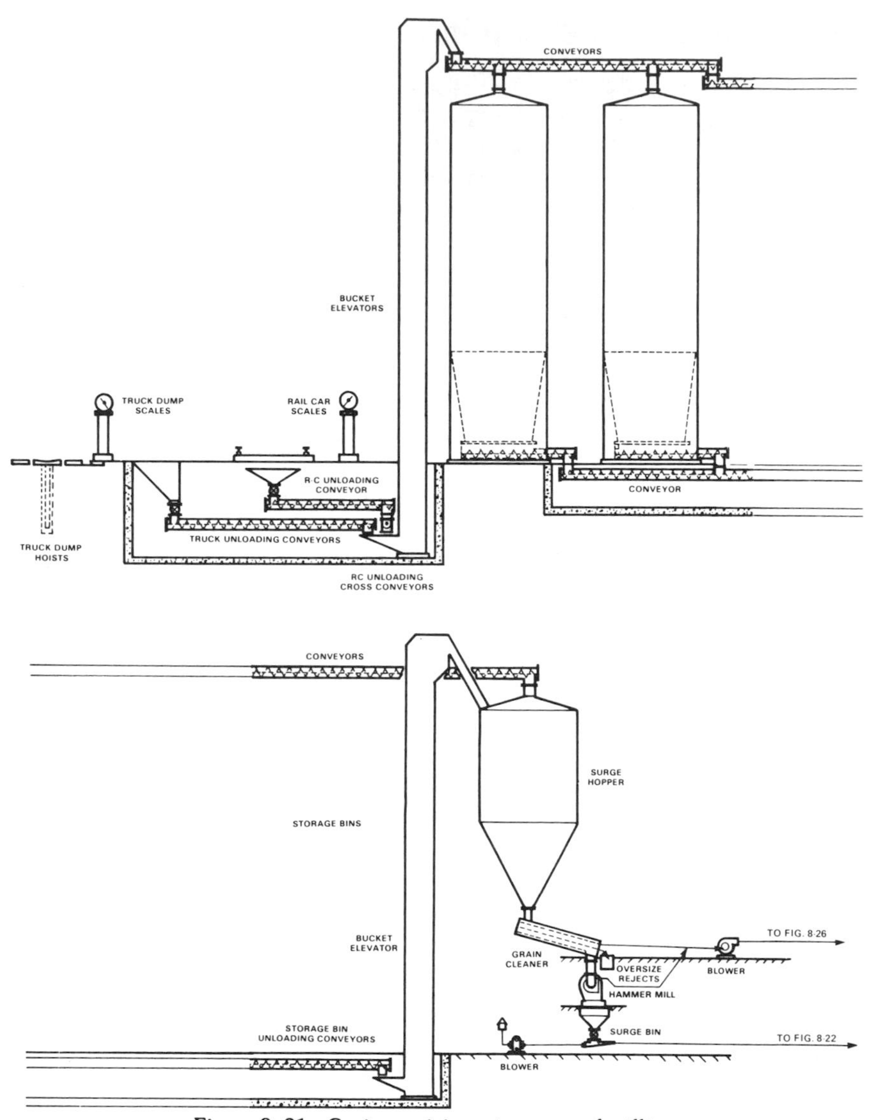

Figure 8-21. Grain receiving, storage, and milling.

Grain, suitable for processing, passes into the hammer mills which deliver into a surge bin. The ground grain then passes through a star valve at the base of the surge bin and is pneumatically conveyed to the process section for mash cooking and saccharification.

The grain receiving, storage, and milling area has been separated from other plant processing areas because of the dust problem associated with these front-end operations.

MASH COOKING AND SACCHARIFICATION

Corn meal is received from the milling area in the surge tank. (See Fig. 8-22.) This tank is sized to allow continuous meal input while the output to the batch weigh tank is shut off when the batch tank is being emptied into the

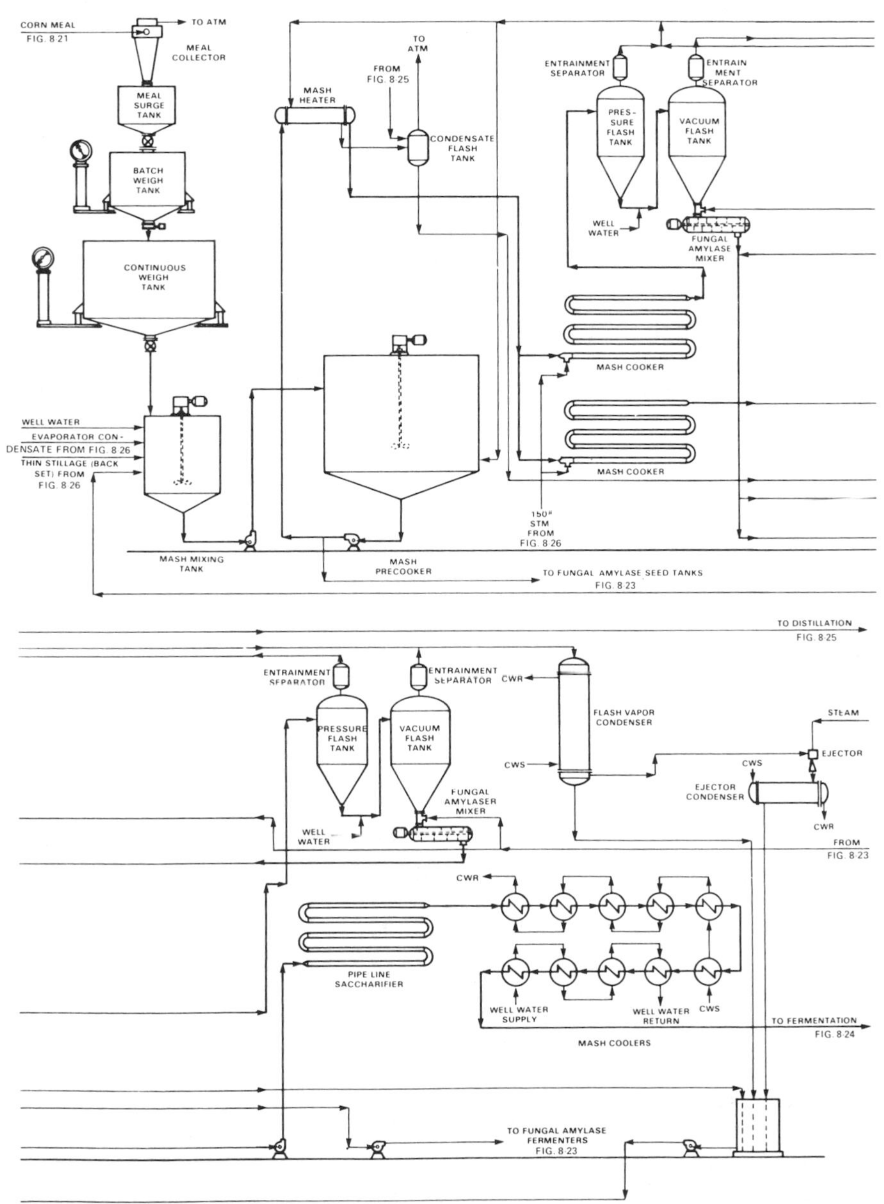

Figure 8-22. Cooking and saccharification.

continuous weigh tank. The batch weigh tank provides an accurate record of the total grain used, and the continuous weigh tank provides a reading of how much grain is used within any given period.

The continuous weigh tank feeds the mash mixing tank where the other mashing ingredients are added. This tank is sized for a 2.5-minute residence time and is fitted with an agitator to promote thorough mixing. The other main ingredients are recycled stillage (backset) and water. The water comes from recycled condensates and from makeup freshwater. The condensates are hot, and their use is regulated to maintain a tank temperature of 145°F. The total water input to this tank is controlled to produce about 22 gallons of mash per bushel of grain input (56 pounds per bushel basis). The thin stillage is added in an amount about 10 percent of the final mash volume going to the fermenters.

The mash is transferred from the mixing tank to the mash precooker. This tank has provision for adding live steam in case insufficient condensate is available to attain the 145°F precooking temperature. This tank is sized for a residence time of about 7 minutes.

The mash is further heated in the mash heater located downstream of the precooker. This heater uses 15 psig steam from the pressure flash to heat the mash to 229°F. Final cooking of the mash takes place in the mash cookers by injection of live steam to attain a temperature of 350°F. The cookers consist of several 20-foot lengths of 10-inch-diameter pipe conncected with 180° return bends, and they are sized to privide a cooking time of 1.5 minutes.

The cooked mash is flashed to 15 psig in the pressure flash tanks. Some of the steam from this flash is used to preheat the mash as discussed above, and the remaining flash steam is used for beer heating in the distillation section. Additional water is added at this point in an amount to provide a final mash volume of 30 gallons per bushel. The mash is then further flash-cooled in the vacuum flash tanks. The temperature when leaving these tanks is 145°F, which corresponds to a pressure of about 3.3 psia. This vacuum is maintained by the flash vapor condenser, its associated steam ejector, and the ejector condenser. The condensate from all of these heaters and condensers goes to a hot well, from where it is pumped back to the mash mixing tank as discussed above.

After the vacuum flash, the mash discharges directly into the fungal amylase mixers, where the fungal amylase is added. From the mixer the mash is pumped to the pipeline saccharifier, where the starch is converted to fermentable sugars. Part of this mash (approximately 1.5 percent) is delivered to the fungal amylase section.

The pipeline saccharifier is sized for a 2-minute residence time. From there, the converted mash is fed through the mash coolers to the fermenters. The first six mash coolers use cooling tower water at 85°F to reduce the mash temperature to 100°F. Then, in the remaining four coolers, well water at 60°F is used to complete the cooldown to 80°F before the mash enters the fermenters.

FUNGAL AMYLASE PRODUCTION

The system consists of seven seed tanks, seven batch fermenters, a system for delivering sterile compressed air, and a pump for delivering the product to the cooking and saccharification section. (See Fig. 8-23.)

Fungal amylase is prepared batchwise using a seed tank to grow the inoculum, which is initially started in the laboratory. One seed tank is used as a starter for a fermenter tank which is sized at 33,750 gallons to provide fungal amylase for 1 day of operation (27,000 gallons plus 25 percent

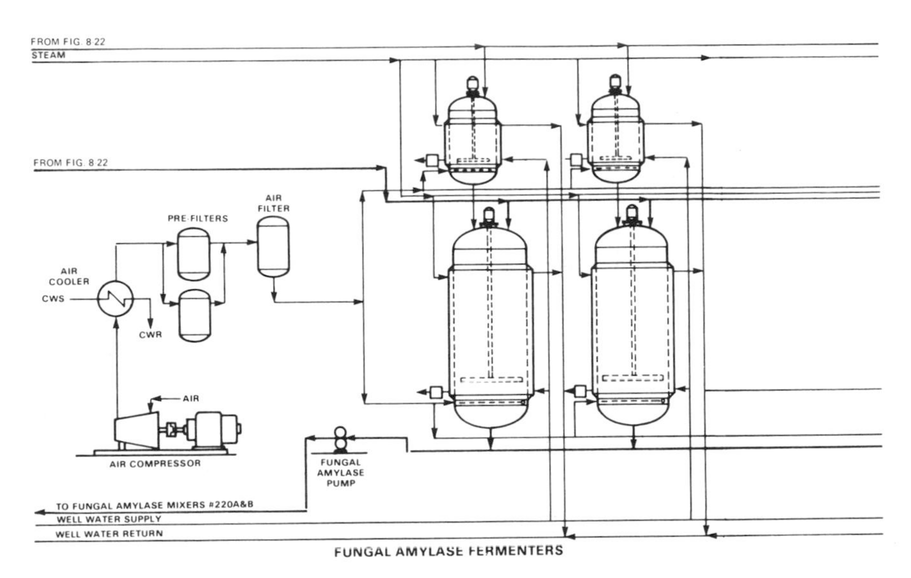

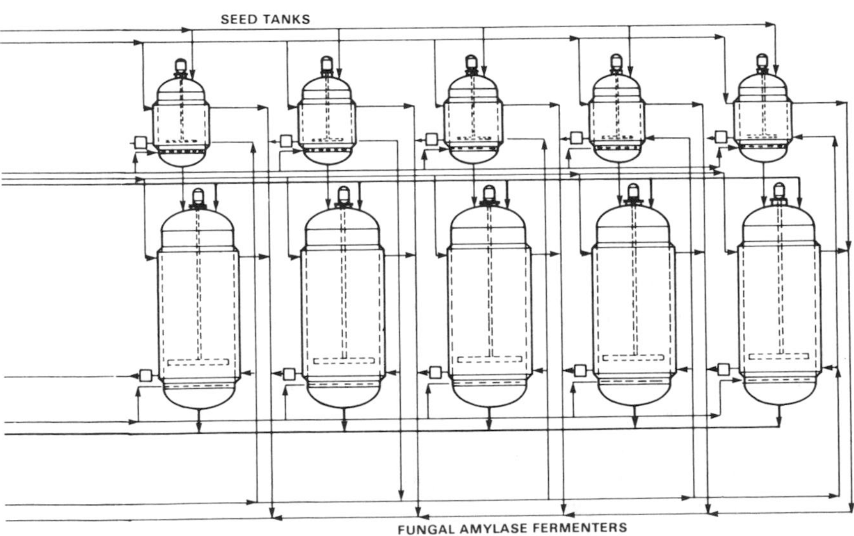

Figure 8-23. Fungal amylase production.

freeboard). The total batch cycle is 1 week, consisting of 1 day for tank cleaning, charging and sterilizing; 5 days fermentation; and 1 day for usage. Twelve thousand five hundred standard cubic feet per minute of compressed air is supplied at 25 psig, which corresponds to 0.5 standard cubic feet per minute per cubic foot of fermenter volume for five fermenters at a time.

Prior to filling the fermenters with mash, they are thoroughly cleaned, and after filling, the mash is heated to 250°F using steam in the tank jacket in order to sterilize the tank and its contents. After sterilization, the mash is cooled to 90°F using well water in the tank jacket. A small amount of additional cooling will be required throughout the fermentation period in order to remove the heat added by the agitators.

The contents of one seed tank are added to the fermenter after it has been cooled down. A period of about 5 days is then required for completion of the batch. During this time, the tank is agitated and aerated with compressed air.

Moyno pumps are used to transfer the finished fungal amylase to the saccharification section.

FERMENTATION (BATCH)

The fermenters receive mash continuously from the mash converter located in the mashing and saccharification section. The fermenters are batch-operated and consist of 16 250,000-gallon vessels which are arranged in sets of four with one heat exchanger and circulation pump for each fermenter set. (See Fig. 8-24.) The fermenters are designed for a liquid loading of 80 percent of maximum capacity. Since cooling is needed for only about 8 or 10 hours out of the 48-hour fermentation cycle, one exchanger will service the needs of four fermenters. The fermenters are filled on a 3-hour cycle.

The yeast, *Saccharomyces cerevisiae,* is manually added to a yeast mixtank and then transferred to the fermenter as it is being filled (about 300 pounds of yeast per batch). The yeast is purchased rather than manufactured on location. The inlet mash temperature is about 80°F, and the temperature gradually rises to a maximum of about 95°F during the fermentation period. Cooling is provided during peak period by recirculation of the mash through the fermenter cooler; well water at 60°F is the cooling medium. Each fermenter requires a flow of approximately 1200 gallons per minute during the peak period. The recirculation for cooling also serves to agitate the tank. At the end of the fermentation period, the fermenter contents are transferred to the beer well, from which they are pumped continuously to the distillation section.

The fermenters are cleaned and sterilized by means of automatic spraying machines installed in each fermenter tank. Each tank has two such spraying machines. After each fermentation cycle, the tank is washed with a cleaning solution, sterilized with an iodine solution, and rinsed with clean sterile water in preparation for the next cycle.

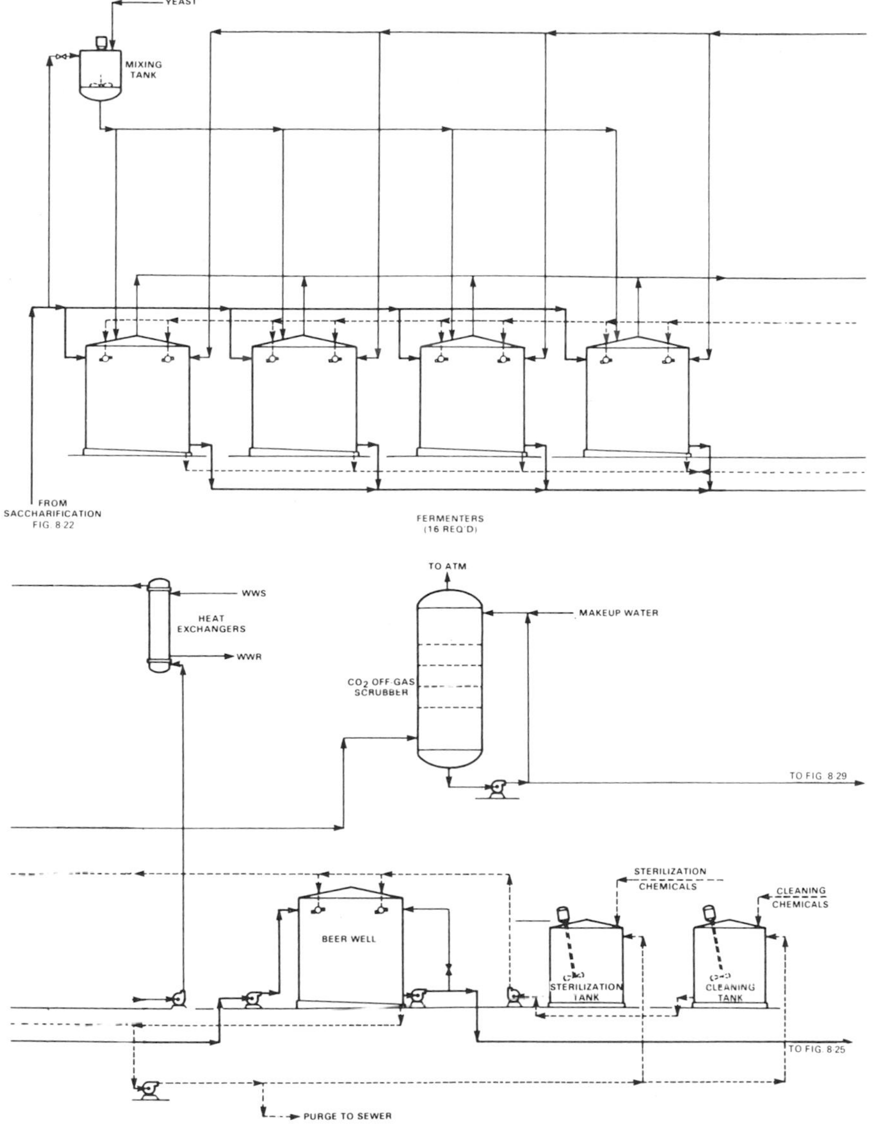

Figure 8-24. Fermentation.

DISTILLATION

Dilute beer feed from the fermentation section of the plant is collected in the beer well, from which it passes continuously to a heat exchanger in the distillation section of the plant. (See Fig. 8-25.)

The dilute beer feed, amounting to approximately 1150 gallons per minute, contains 7.1 weight percent alcohol and 6.92 percent solids. Solids

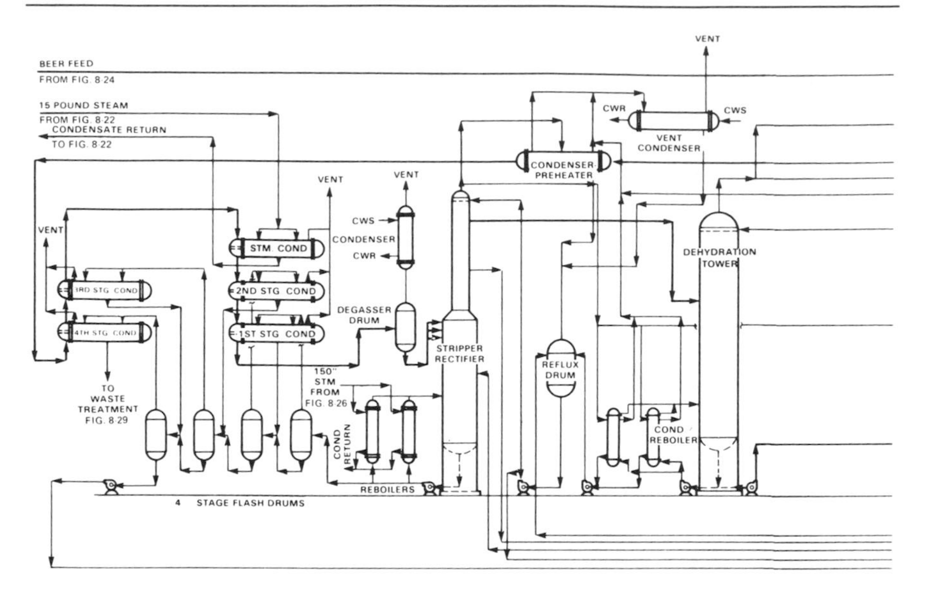

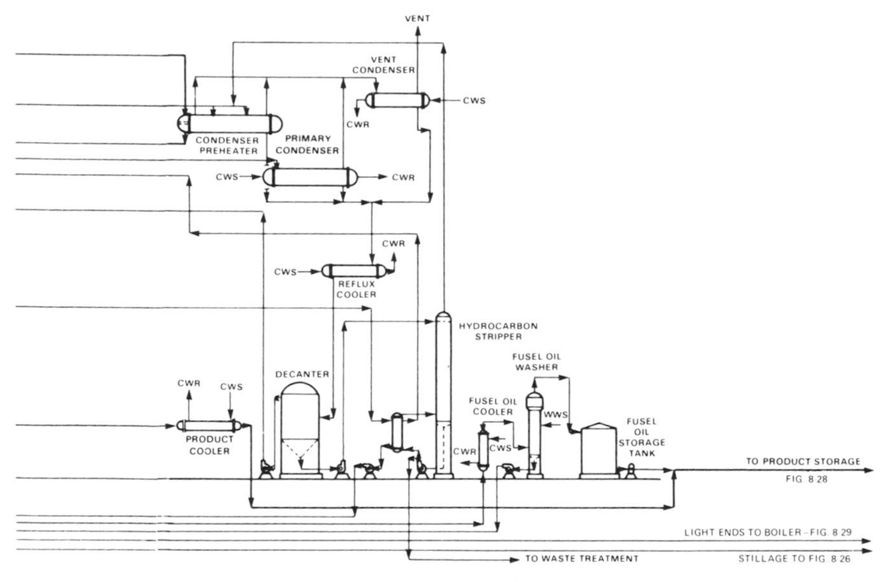

Figure 8-25. Distillation.

consist of both dissolved solids and suspended solids, in approximately equal amounts. The beer leaves the beer well at a temperature of 90°F and undergoes a series of preheating steps before it enters the first stage of distillation. The dilute beer first passes into the tube side of a condenser-preheater. In this unit, approximately 23 percent of the total preheating is accomplished. This first preheating step utilizes a portion of the vapors condensed from the dehydration tower. These vapors are condensed to supply this first-stage preheating. The warmed dilute beer feed next passes to a condenser-preheater, where additional preheat amounting to about 8.5 percent of the total is added. In this condenser-preheater, a portion of the overhead vapors from the pressure stripper-rectifier (PSR) is condensed to supply second-stage preheat. The dilute beer feed next passes through two stages of feed preheating, wherein a portion of the heat in the bottoms stream from the PSR tower is utilized in a two-stage flash operation. These stages add about 21.5 percent of the total feed preheating. The warm dilute beer feed next passes into a steam condenser where low-pressure steam is used to accomplish additional preheating. Approximately 23 percent of the total feed preheat is added in the steam condenser. The heating medium, in this case, consists of low-pressure steam taken from other parts of the plant. The feed is finally preheated, approximately to saturation temperature, in an additional two-stage heating step using flash heat taken from the bottoms stream out of the PSR. Approximately 24 percent of the total feed preheat is accomplished there.

The hot, saturated, dilute beer feed next passes into the degassing drum, where dissolved carbon dioxide is flashed off. This represents one of the products of the fermentation reaction. It is not recovered. Any alcohol or water vapor, accompanying the vented carbon dioxide, is condensed in a vent condenser from which it drains back to the flash drum.

The saturated dilute beer feed enters the midsection of the PSR tower. Because of the high suspended solids content of the beer feed, the lower section of the PSR has been designed as a disk-and-donut-type tower. This represents an effective contacting device which tends to be self-purging and does not allow the buildup of solids which would block ordinary distillation trays. The PSR operates with a head pressure of 50 psig. The nonvolatile dissolved solids and suspended solids in the dilute beer feed wash down through the stripping section of the PSR and a very dilute alcohol steam, containing less than 0.02 weight percent alcohol, is removed from the bottom of the tower. The dilute stillage containing the dissolved and suspended solids leaves the base of the PSR tower at about 304°F. In the bottom section of the PSR tower, alcohol is effectively stripped from the dilute beer. The aqueous bottoms stream then passes through a series of flash stages. These stillage bottoms are subjected to progressive reductions in pressure through four flash stages. The flash vapor that develops in these stages is utilized to accomplish a portion of the feed preheating as described previously. In these four flash stages, the temperature of the hot stillage is reduced from 304 to approximately 212°F. The whole stillage, containing about 7.5 percent total solids, is next pumped to the coproduct-recovery section of the plant where the solids in the stillage are recovered as an animal feed coproduct.

Heat is supplied to the base of the PSR by means of condensing 150 psig steam on the shell side of parallel forced-circulation reboilers. Total steam supplied to the base of the PSR tower through the shell sides of the reboilers is 110,000 pounds per hour.

The upper portion of the PSR contains perforated trays and has a reduced diameter compared to the stripping section. The lower section of the PSR tower is 138 inches in diameter while the top section of the tower, containing 28 perforated trays, has a diameter of 102 inches.

Alcohol-rich vapors generated in the PSR pass overhead from the tower at a temperature of 250°F and a pressure of 50 psig. These vapors may be utilized as a source of heat by condensing in the reboilers which are attached to the base of the dehydration tower and the hydrocarbon stripper. Sufficient vapor is generated in the PSR to allow a portion of the total overhead vapor to be utilized in a condenser-preheater to do some of the feed preheating which has been described previously. Of the total overhead vapor generated in the PSR, 10.9 percent is utilized for feed preheating, 81.6 percent is used to supply heat to the dehydration tower, and 7.5 percent is employed to supply heat to the hydrocarbon stripper. The overhead vapor from the PSR contains 95 volume percent alcohol (190 proof). The total condensate is returned to the top tray of PSR.

The upper five trays of the PSR operate in a total reflux condition. The liquid product from the PSR is removed as a liquid side draw stream about five trays from the top of the tower. From there, it passes to the midsection of the dehydration tower.

The dehydration tower is 138 inches in diameter and contains 50 perforated trays. The tower operates at essentially atmospheric pressure. The bottoms stream represents the anhydrous motor fuel-grade alcohol and has a concentration of 99.5 volume percent ethanol (199 proof); the balance is water. The bottoms stream from the dehydration tower is pumped through a product cooler which utilizes cooling water to reduce the temperature of the product alcohol to about 100°F. The cooled product next passes to product storage.

Heat is supplied to the base of the dehydration tower through parallel forced-circulation reboilers. The overhead product from the dehydration tower is a ternary minimum boiling azeotrope consisting of hydrocarbon, alcohol, and water. A portion of these overhead vapors is utilized for feed preheating and the balance condensed in the primary condenser, which utilizes cooling water to remove the heat of condensation in these vapors. The condensed vapors pass to a reflux cooler where they are further subcooled by cooling water prior to being fed to the decanter. The subcooled liquid entering the decanter separates into two layers. The upper layer is the larger in volume, and represents the hydrocarbon-rich layer. The lower, which separates, is a water layer containing some alcohol and hydrocarbon. The upper layer from the decanter is pumped back to the top tray of the dehydration tower. The lower layer from the decanter is pumped to the top tray of the hydrocarbon stripper which removes the remnants of hydrocarbon and alcohol contained in the feed to the top tray. The bottoms stream of the hydrocarbon stripper is essentially aqueous, and is removed and sent to the waste treatment plant. Thermal energy is supplied to the base of hydrocarbon stripper via alcohol-rich vapor condensing on the shellside of a reboiler. The condensed alcohol-rich vapor is passed by to a reflux drum where it joins other vapor condensate before being returned to the top tray of the PSR. Overhead vapors containing alcohol, hydrocarbon, and water from the atmospheric pressure hydrocarbon stripper pass to a condenser-preheater, where they are condensed. The condensate is returned through the reflux cooler to the decanter. The aqueous stream passing from the bottom of the hydrocarbon stripper contains less than 0.02 weight percent alcohol.

This distillation system is covered by a patent issued to Raphael Katzen Associates, International, Inc.

FUSEL OIL AND HEAD REMOVAL

In the yeast fermentation process, certain extraneous products, in addition to ethyl alcohol, are formed. These are generally higher alcohols, i.e., higher-molecular-weight alcohols known as fusel oils, and light ends which include such materials as aldehydes.

The distillation system provides for the removal of these extraneous components in the following manner. (See Fig. 8-25.) The fusel oils have the property of being more volatile than alcohol in dilute aqueous solution, but are less so than alcohol in concentrated alcohol solution. For this reason, they tend to concentrate on some tray in the rectifying section of the PSR. These fusel oils, thus having concentrated, can be removed as a liquid side draw stream from the PSR. They are removed and passed through a fusel oil cooler and to a fusel oil washer, a water washing extraction column in which the alcohol content of the fusel oils is washed from them, under reduced temperature, by countercurrently contacting the cooled fusel oil side stream with a stream of cold water. The heavy aqueous stream, containing the extracted ethyl alcohol, is removed from the base of the fusel oil washer. This stream is returned to the lower section of the PSR for alcohol recovery. The light fusel oil stream is decanted from the top of the fusel oil washer and passes into the fusel oil storage tank.

In general, the fermentation process, when utilizing corn, will produce about 4-5 gallons of fusel oil for every 1000 gallons of anhydrous alcohol product. These fusel oils do have a heating value and can be reblended into the product. If this reblending operation is not desired, then the fusel oils may be passed to the plant boiler where they are used as fuel. Fusel oils should have no harmful effect on motor fuel-grade alcohol.

Light extraneous fermentation products such as aldehydes are effectively removed in this distillation system by withdrawing a very small purge from the total reflux stream passing back to the top tray of the PSR. This light component purge, in general, cannot be reblended into the alcohol to be used for motor fuel blending, because these light products would tend to cause vapor lock when the ethanol is blended with gasoline to produce gasohol.

Therefore, these materials are removed and sent to the plant boiler where their fuel value is recovered.

The fusel oil and light ends must be removed because their presence would upset the equilibrium associated with the dehydration step, and could cause problems in the decantation step.

The distillation scheme[1] for producing motor fuel-grade alcohol, as described here, utilizes only 17.5 pounds of process steam per gallon of anhydrous motor fuel-grade alcohol product. This great reduction in energy use is accomplished by optimizing the feed preheating scheme and by utilizing the heat content of high-pressure vapors produced in the PSR to supply the reboil heat for both the dehydration step and the hydrocarbon-alcohol stripping.

EVAPORATION AND DRYING OF STILLAGE RESIDUE

Stillage from the distillation area is delivered to the whole stillage tank, where it is pumped to the solid bowl centrifugals that operate on a continuous basis. (See Fig. 8-26.) These centrifugals separate the whole stillage into two fractions: thin stillage containing 6.5-10 percent total solids and thick stillage containing about 35 percent total solids. Part of the thin stillage referred to as back set (corresponding to 10 percent of the total mash) is recycled to the mash mixing tank in the cooking and saccharification section.

The remaining thin stillage is evaporated in a vapor recompression evaporator to about 55 percent solids. Because of a cooling effect in the centrifugal separators (caused by evaporation of the stillage in contact with air), the thin stillage must be reheated from about 165 to 208°F before it enters the evaporator. Heating is accomplished by using evaporator condensate cooled from 230 to 185°F. Power for driving the evaporator's vapor compressor (approximately 6200 horsepower), is provided by a steam turbine which uses 580 psig steam and exhausts at about 160 psig. This exhaust steam is used for distillation, mash cooking, and heating supplemental air for spent grains' drying. The vapor compressor operates at approximately atmospheric pressure at the inlet and compresses the vapor to about 21 psia. The compressor outlet steam is superheated; however, before it enters the evaporator bodies, it is desuperheated by injection of condensate. This is done to get the best heat transfer possible and to prevent "baking" of solids on the evaporator surfaces. The water vapor from the evaporator bodies passes through entrainment separators and then to the vapor compressor. However, before entering the compressor, it must be superheated by mixing with a recycle flow of superheated vapor from the compressor outlet, to maintain dry, noncorrosive conditions in the vapor compressor.

[1]Patents allowed but not yet issued.

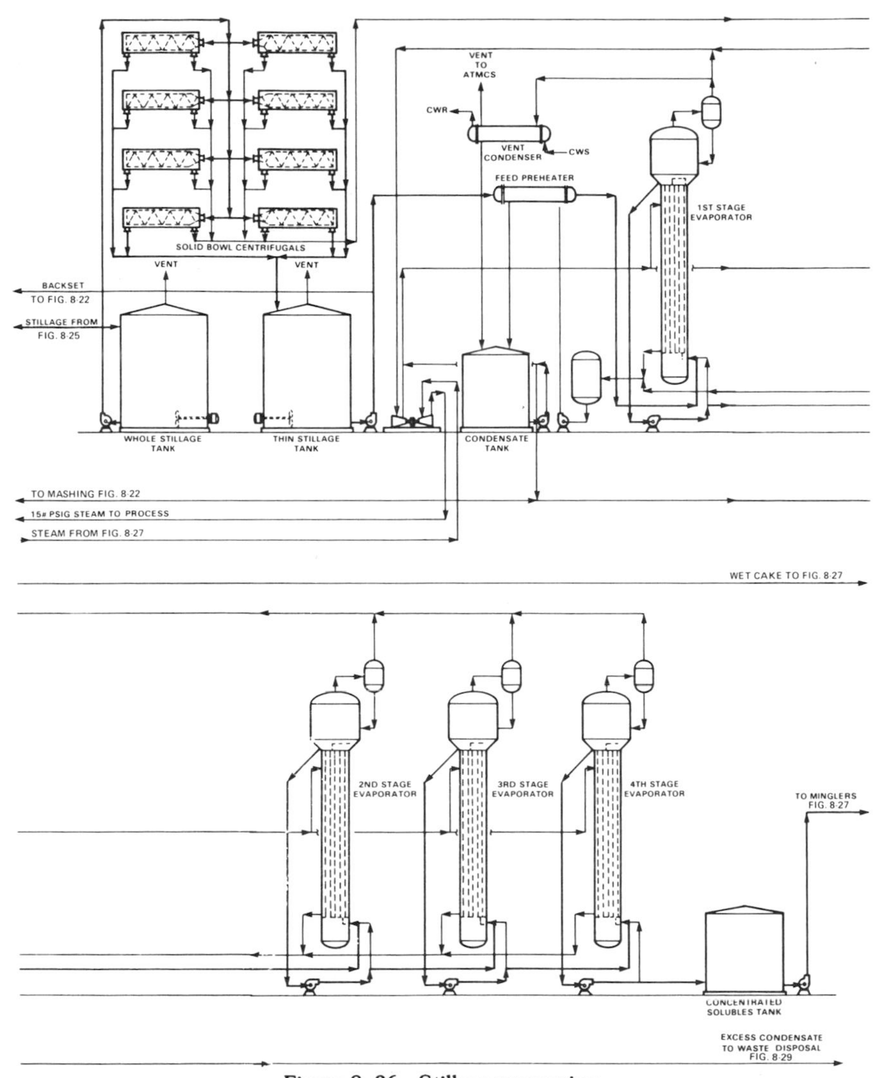

Figure 8-26. Stillage processing.

The thick stillage is mixed with the concentrated thin stillage and recycled dry grains in the wet grains' minglers. (See Fig. 8-27.) The amount of dry grains' recycle is regulated to maintain a wet grains' moisture content of 30 percent to minimize stickiness.

The wet grains are then fed to rotary dryers where they are tumbled in contact with hot flue gas from the power boiler. Supplemental hot air for drying is provided by an air heater using 150 psig steam for heating. The hot gas enters the dryers at about 600°F with a wet bulb temperature of about 145°F. The gas and dry grains leave the dryers at about 190°F. The flue gas goes through cyclone collectors to remove most entrained solids before it is delivered to the flue gas scrubber.

The dry grains, at about 10 percent moisture, are transferred to the dry grains' hopper. About 75 percent of the dry grains is recycled in order to regulate the moisture content of the wet grains. The remaining dry grains is ground in a hammer mill and then cooled in the product cooler, an auger-type heat exchanger that uses well water to cool the grains to about 100°F. The cooled coproduct is then transferred pneumatically to storage and shipping.

ALCOHOL, AMMONIUM SULFATE, AND DRY GRAINS' STORAGE AND SHIPPING

The alcohol is received from the distillation section and stored in receiver tanks, each of which is sized for 1 day of production. (See Fig. 8-28.) While one tank is being filled, the other is checked for quality and quantity by the government inspector. After inspection, its contents are sent to long-term storage. The four storage tanks hold a total of about 4.2 million gallons—about 28 days of production. Upon transferring the product from storage to a tank car or truck, denaturant (gasoline) is added at a rate of 1 gallon per 100 gallons of alcohol. The denaturant tank holds about 50,000 gallons. Accurate metering is provided by positive displacement meters.

A water solution, containing 40 percent ammonium sulfate, is produced by the flue gas scrubbing system. This solution is to be sold as field fertilizer and is stored in four tanks having a capacity of 1 million gallons each. This corresponds to storage of the total ammonium sulfate production for about 9 months, which allows for storage during the fertilizer off-season.

Dry grains are stored in an A-frame-type building with a storage capacity of about 295,000 cubic feet, equivalent to about 1 week's production. (See Fig. 8-28.) Shipping of the dry grains from storage is done on a first in/first out basis and utilizes a front-end loader to load the pneumatic conveyor system which transfers the grains to the live bottom surge bin at a rate of 88,000 pounds per hour. This rate is based on shipping out the dry grains for an average of 12 hours each day. Shipment may be made by either truck or rail, and shipping scales are provided for weighing the shipments.

COAL-FIRED BOILER

Steam for the plant is provided by a coal-fired boiler rated at 250,000 pounds per hour of steam at 600 psig and 600°F. (See Fig. 8-29.) Calculated plant usage is about 200,000 pounds per hour of steam. The firing rate is 12.6 tons per hour of coal, having a gross heating value of 10,630 Btu per pound (Illinois No. 6 coal). The coal contains approximately 3.8 percent sulfur (moisture-free basis). Small fuel inputs are also provided by light ends from distillation of the alcohol and by dewatered sludge from waste treatment.

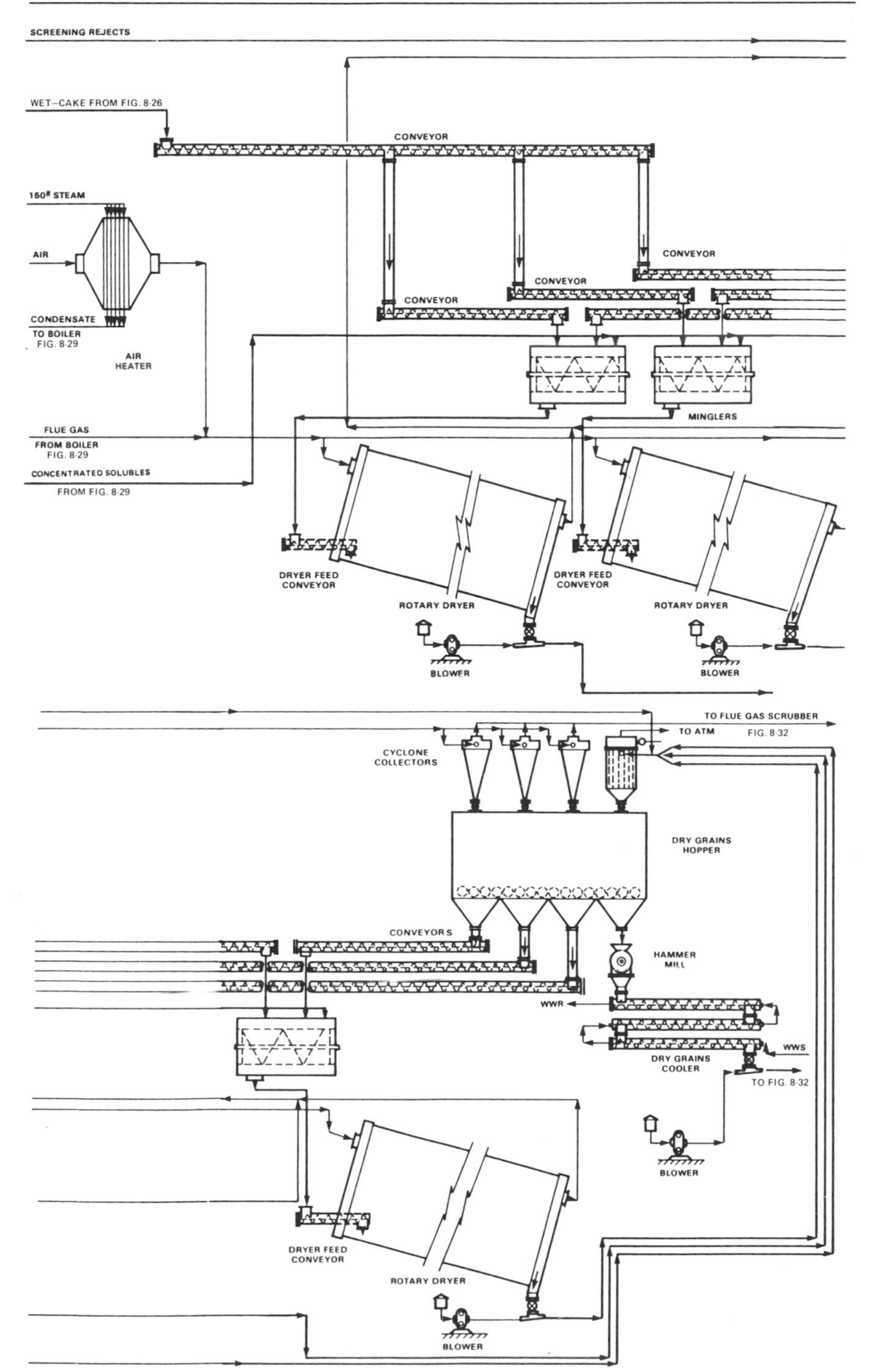

Figure 8-27. Residue feed processing.

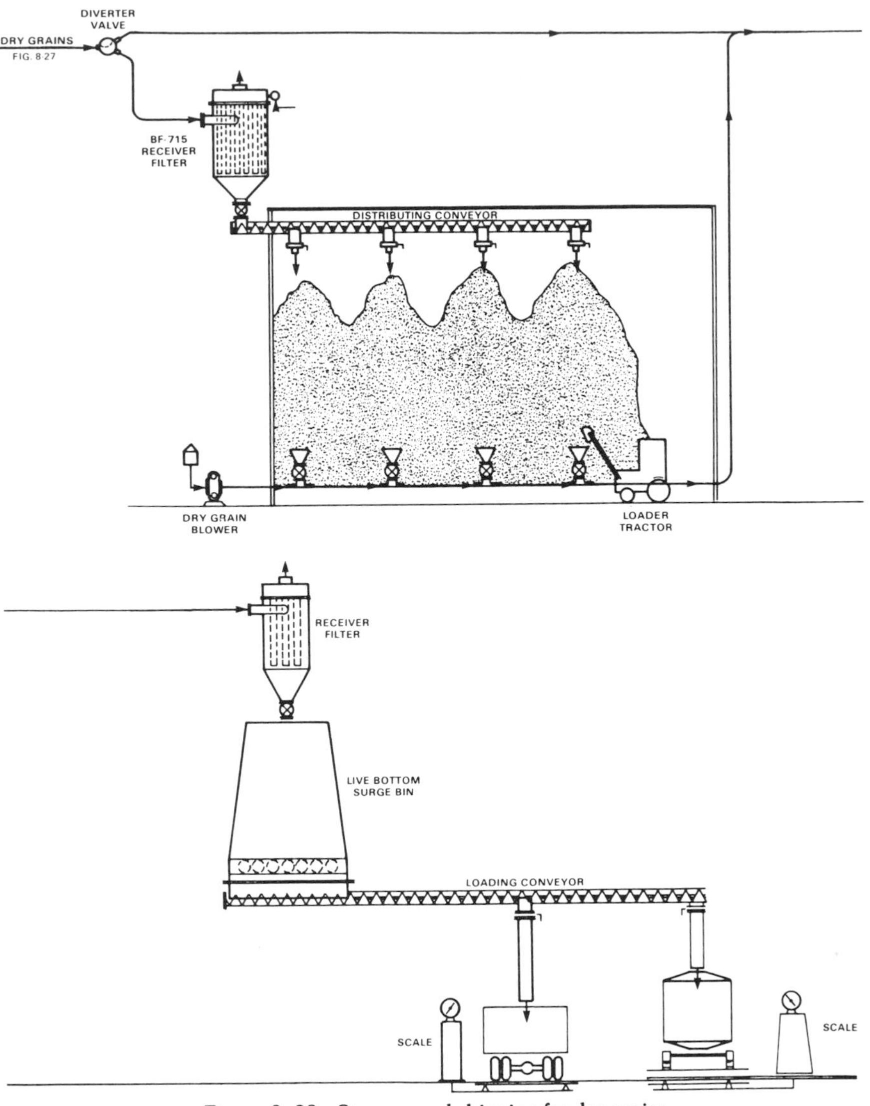

Figure 8-28. Storage and shipping for dry grains.

The coal unloading facility provides for direct transfer to the coal bunker or to a storage pile. A front-end loader is used to transfer coal from the pile back to the unloading area where it can be transferred to the coal bunker.

Boiler feedwater is provided by condensate return from the process, where possible, and by makeup water that has been filtered and conditioned in a conventional boiler feedwater treatment system. About two-thirds of the boiler feedwater is condensate return.

The flue gas passes through the cyclone collector for particulate removal. The collector consists of numerous small cyclones (multiclones) housed in a

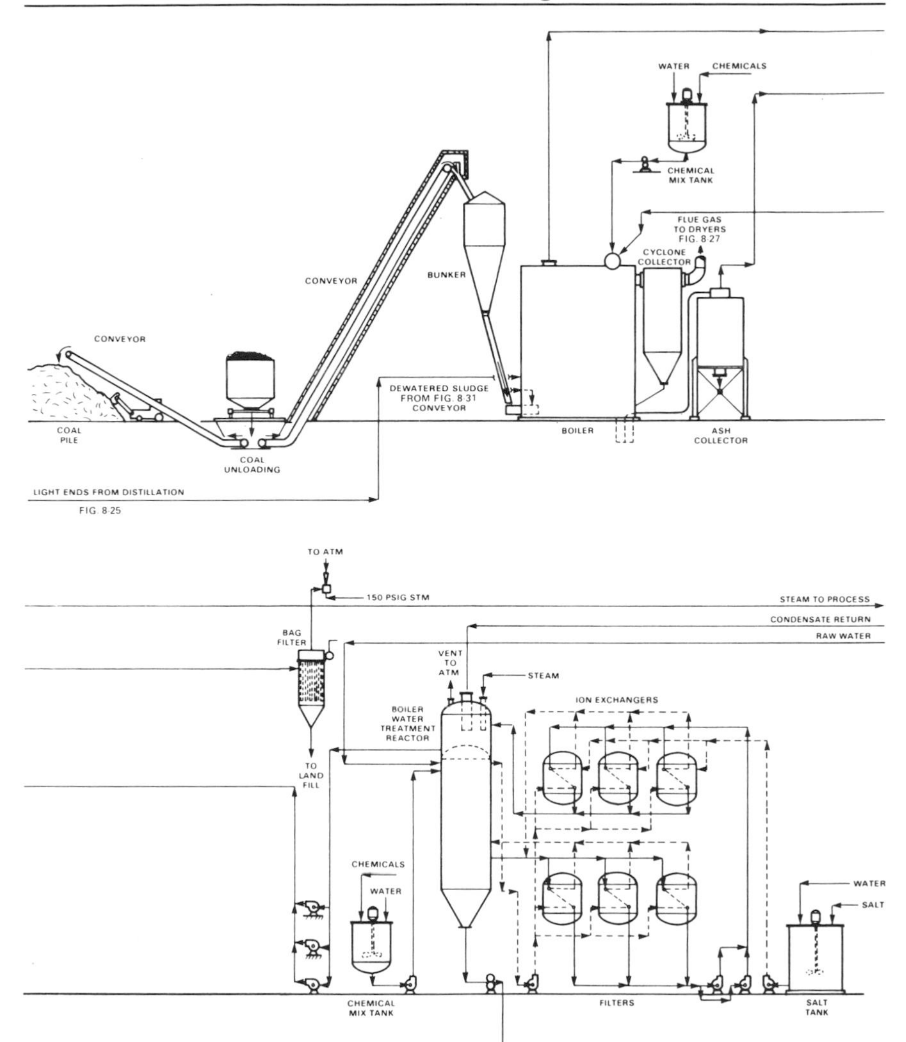

Figure 8-29. Utilities, boiler.

single chamber. The recovered particulate goes to the boiler at a temperature of about 725°F. This is a high flue gas temperature by normal standards, but in this case, it does not adversely affect the overall plant thermal efficiency since the hot flue gas, tempered with air to 600°F, goes to the stillage drying section where its heat is used to dry the distiller's grains. For this reason, no flue gas heat economizer or tack is required with the boiler.

The coal is fed to the boiler by four stoker-spreader units. The spreader feeding system was selected (rather than pulverized blown coal) because of the small boiler size and because the boiler inefficiency, due to excess air, does not affect the process thermal efficiency.

WATER SUPPLY

The water supply system provides well water for meeting process makeup requirements, for process cooling where cooling tower water is not cool enough, and for maintaining a supply of fire protection water. (See Fig. 8-30.) Well water is also used to provide makeup to the cooling tower. Three wells are provided with a capacity of 1800 gallons per minute each. The well water storage tank has a capacity of 1,000,000 gallons and provides for surges in demand and allows short-term shutdowns of the wells, as required for maintenance or repairs, without interrupting the supply of water to the plant.

The fire protection system consists of the fire protection tank which has a capacity of 300,000 gallons, four fire water pumps (two electric and two diesel), and an underground fire water distribution system, along with the appropriate fire hydrants and spray headers. Each pump has a capacity of 2000 gallons per minute. The diesel-powered pumps are used only in the event of an electric power failure.

The cooling tower is designed to provide 16,000 gallons per minute of water at 85°F from a warm water return temperature of 115°F, and an ambient design wet bulb temperature of 75°F. The tower consists of two cells

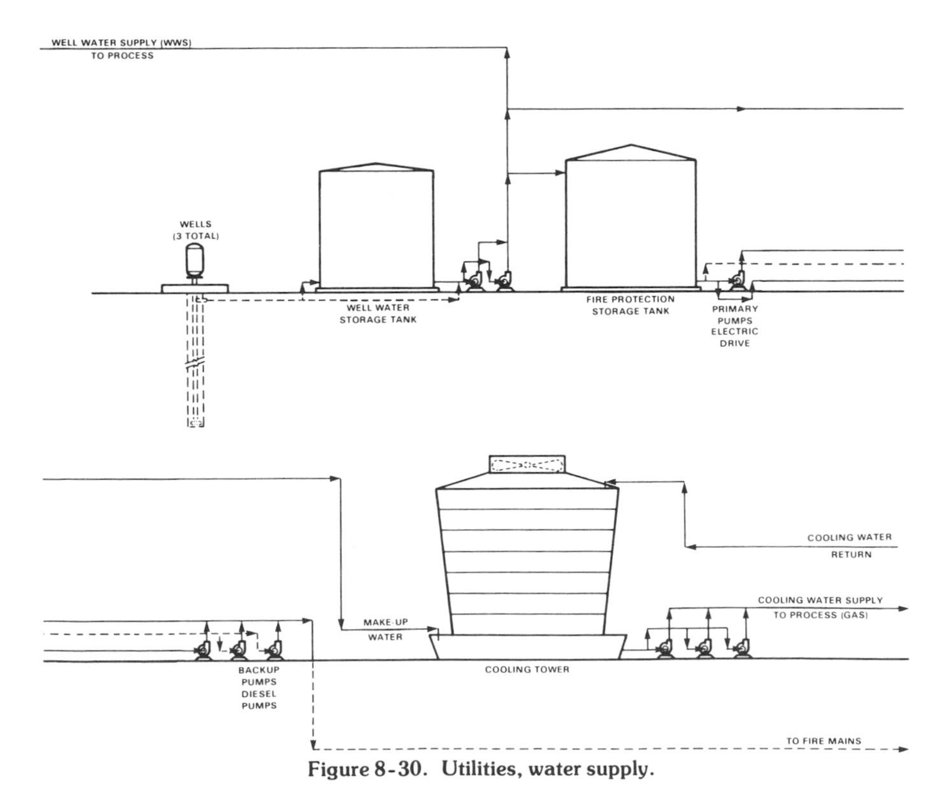

Figure 8-30. Utilities, water supply.

with a two-speed fan for each cell. The three cooling tower pumps are rated at 7500 gallons per minute each.

WASTEWATER TREATMENT

The wastewater flows include process wastewater and sanitary wastewater. (See Fig. 8-31.) The wastewater is collected at a lift station and pumped to the treatment plant.

The wastewater treatment plant is designed for secondary treatment with two extended aeration tanks and two settling tanks. The aeration tanks are 95 feet in diameter. The sludge thickening tank is 20 feet in diameter.

The influent stream passes through a bar screen and grinder prior to entering the first aeration tank. Water from the first stage is split; the major portion is recycled to the first aeration stage, while the remainder is sent to the thickening tank. Clarified water from the first settling stage overflows to the second aeration tank, whereby additional biochemical oxygen demand (BOD) is removed. Nutrients may be added to either aeration tank but, due to the nature of the wastewater flows, should not be needed. The water from the second aeration tank overflows to the second-stage settling tank.

The sludge from the second-stage settling tank is recycled to either the first- or second-stage aeration tank. A stream is sent to the thickener. The

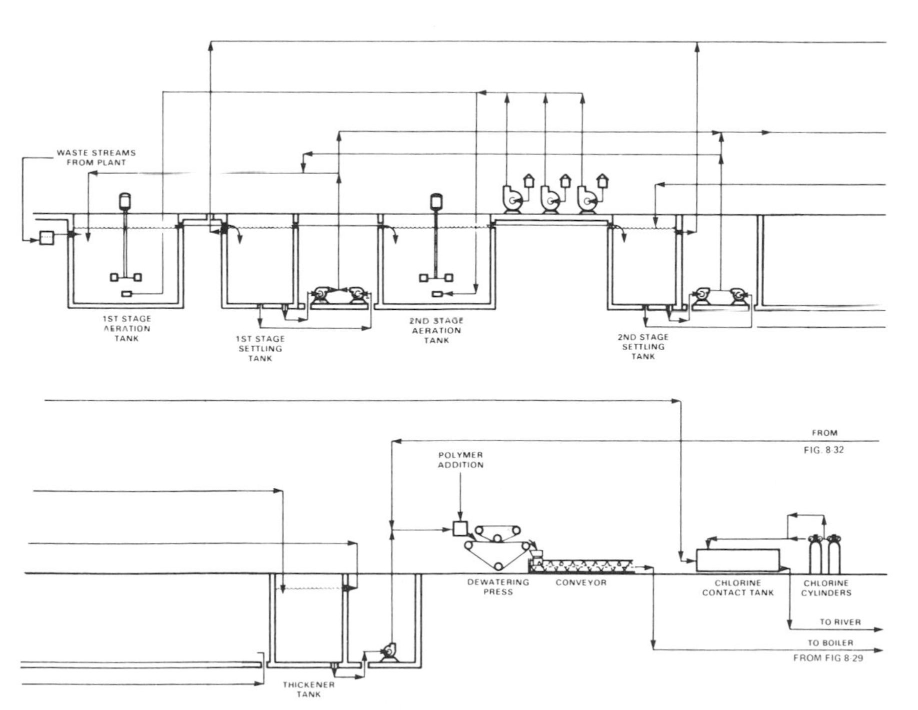

Figure 8-31. Utilities, wastewater treatment.

two-stage aeration system, coupled with the flexibility to recycle sludge from either stage, allows good control of effluent BOD.

The clarified water from the second-stage settling tank flows by gravity to the chlorine contact tank. Chlorine is added to the contact tank for destruction of final traces of impurities. The effluent from the chlorine tank flows to a sewer connection.

Sludge from both the first- and second-stage aeration tanks is collected in the sludge thickener and pumped to the dewatering press. The sludge is mixed with primary sludge collected from the flue gas desulfurization system and dewatered to about 25 percent solids. The solids from the dewatering press are chopped up in a flaker and conveyed to the boiler for burning.

The treatment plant is designed to remove 95 percent of the effluent BOD.

FLUE GAS SCRUBBER

The hot flue gas from the coal-fired boiler is sent to dry grains' recovery where it mixes with dilution air for grain drying prior to being sent on to the flue gas scrubbing system. (See Fig. 8-32.) Because of the high sulfur content of Illinois No. 6 coal (and also the high purchase price of low- versus high-sulfur coal), a flue gas desulfurization system is required. The desulfurization system recovers the SO_2 as ammonium sulfate and differs from the more conventional limestone scrubbing systems in that no calcareous sludge is produced in the system. The only coproduct is the ammonium sulfate.

In the flue gas scrubbing system,[2] water sprays cool the gas and remove particulates in three stages. The first stage removes particulates only. The next two stages cool the gas and remove additional particulates. Ammonia is used in a two-stage absorption section for removal of sulfur dioxide from the flue gas. The ammonium sulfite/bisulfite solution from the scrubbers is neutralized to ammonium sulfite and oxidized to ammonium sulfate. The ammonium sulfate is suitable for sale as agricultural fertilizer. A more-detailed description of the system is given below.

The hot gas enters the spray quench section where it is quenched to within 5 degrees of its wet bulb temperature, 143°F.

The saturated gas enters the first section of the flue gas scrubber where it is washed with a very high rate of water flow from spray nozzles. The gas is then cooled in two successive spray cooling sections to 110°F. Part of the heat removed in the first cooling section is used to reheat the exit flue gas in the heat exchanger, to improve the flue gas buoyancy. The remaining heat from the cooling sections is removed by cooling water using plate-and-frame-type heat exchangers. Particulates, removed in the quench and cooling sections, are removed by the cyclone cleaner. This material is concentrated

[2] U.S. Patent 3,957,951 and patents pending, licensed by Raphael Katzen Associates International.

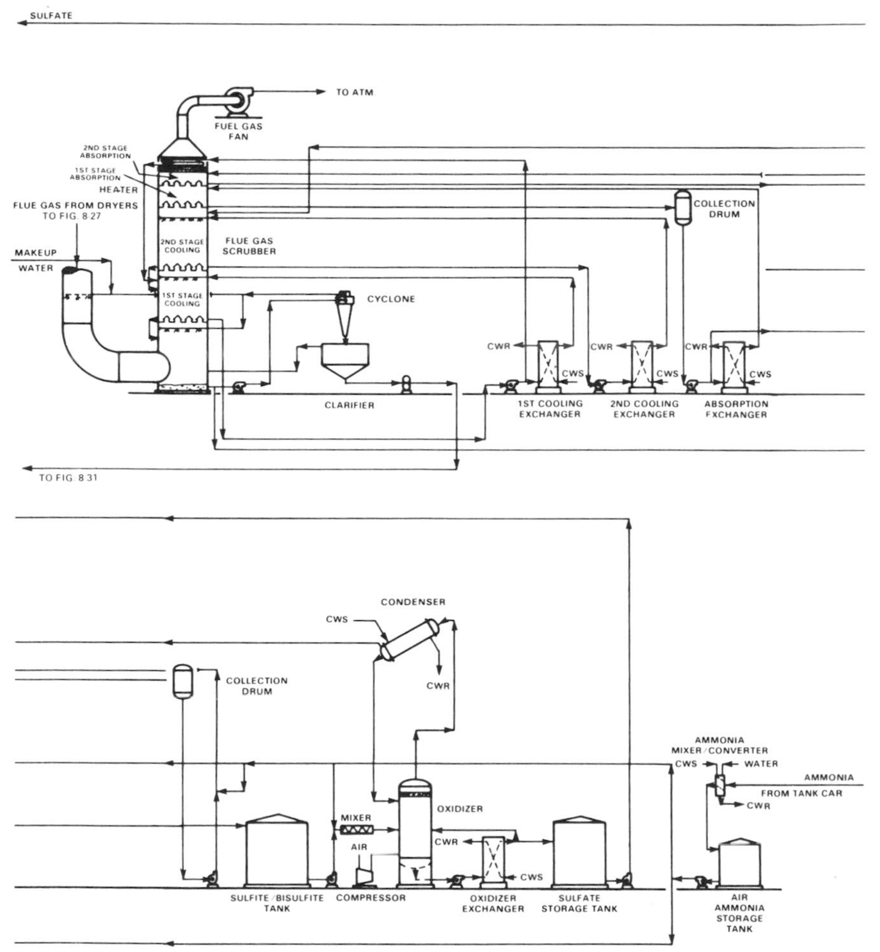

Figure 8-32. Utilities, flue gas scrubber.

further in a clarifier and then combined with sludge from the waste treatment system. The mixture is then dewatered in the dewatering press and sent to the coal-fired boiler. The boiler is equipped with special feeder/spreaders to handle the sludge.

In the absorption stages, the pH of the liquid is carefully controlled in order to minimize the gas phase relation between ammonia vapor and sulfur dioxide gas. The pH is controlled by adjusting the absorption solution circulation rate and the rate of ammonia addition to each stage. The heat of reaction from SO_2 absorption is removed by another plate-and-frame exchanger.

The flue gas draft for the scrubbing system, as well as for the spent grains' driers and interconnecting ducting, is supplied by the flue gas fan located at the scrubber outlet. The fan is driven with a 1000-horsepower motor and will handle 150,000 standard cubic feet per minute with a 30-inch-water guage pressure increase to the flue gas.

The product ammonium sulfite/bisulfite solution from the scrubber is taken from the lower absorption loop downstream of the first-stage absorption loop pump. This solution is neutralized with aqueous ammonia to ammonium sulfite and then oxidized to ammonium sulfate in the oxidizer reactor. This solution is then cooled to 100°F in the oxidizer exchanger. The water makeup to the scrubber will be controlled to produce a final oxidized solution strength of 40 percent ammonium sulfate, which can be used for direct application to fields for fertilizer or for a blending material by fertilizer manufacturers.

MATERIALS FLOW

Table 8-4 summarizes the daily materials flow through the system. Most noticeable is the large amount of water required by the process, i.e., about 7.25 gallons of water per gallon of ethanol produced.

ENERGY FLOWS

Table 8-5 summarizes the energy flows through the system by major operations in the process. Distillation requires over one-half the steam input to the process, followed by cooking and saccharification, which require about 30 percent of the total. Enzyme production, DDG recovery, and utilities are the major components of electricity demand.

The efficiency of the system (ratio of energy out to energy in) may be expressed in several ways:

> If only the fossil fuel inputs are considered (electricity being based on a 10,000 Btu/kWhr equivalent): alcohol/(coal + electricity) = 154.4 percent or about 55,000 Btu per gallon of ethanol.
>
> If the energy used to produce the corn and a credit for DDG production are included: (alcohol + DDG)/(coal + corn + electricity) = 105.1 percent (corn production requires about 95,000 Btu/bushel).
>
> If the total thermal energy in the corn and DDG is considered: (alcohol + DDG)/(coal + corn + electricity) = 68.9%.

Efficiency is not really a concern to the investor except as it affects economics. It is more a matter of national policy to determine what cost in fossil fuel (coal) can be tolerated to displace a barrel of imported oil.

PLANT INVESTMENT

The total plant investment is estimated to be $58 million (end of 1978 dollars). Table 8-6 shows a breakdown of the investment by plant section.

Table 8-4. Materials Flow for a 50-Million-Gallon-Per-Year Ethanol Plant

ITEM	UNITS	DAILY QUANTITY
INPUTS		
Corn	Bushel	58,900
Coal	Tons	296.7
Yeast	Tons	1.2
Denaturant	Gallons	1,500
Ammonia	Tons	9.2
Water	Gallons	1,250,000
OUTPUTS		
Ethanol	Gallons	151,515
Distillers Dried Grains and Solubles	Tons	536.7
Ammonium Sulfate	Tons	31.6
Wastewater	Gallons	1,100,000

Source: Katzen, R., "Grain Motor Fuel Alcohol Technical and Economic Assessment Study". Raphael Katzen Associates for the U.S. DOE, Dec., 1978.

Table 8-5. Energy Flow for a 50-Million-Gallon-Per-Year Ethanol Plant

ITEMS	PROCESS STEAM		ELECTRICITY	
	LB/HR	%	kW	%
Receiving, Storage, Milling			507	6.1
Mash Cooking & Saccharification	61,000	30.5	216	2.6
Enzyme Production	1,400	0.7	1,696	20.4
Fermentation	400	0.2	333	4.0
Distillation	117,000	58.5	133	1.6
DDG Recovery	12,800	6.4	2,253	27.1
Storage & Denaturing			58	0.7
Utilities	5,400	2.7	3,076	37.0
Buildings	2,000	1.0	42	0.5
TOTAL	200,000	100.0	8,314	100.0

Source: Katzen, R., "Grain Motor Fuel Alcohol Technical and Economic Assessment Study". Raphael Katzen Associates for the U.S. DOE, Dec., 1978.

The two major items are the utilities and the DDG-recovery sections, which together account for over one-half the total investment. For a plant in the range of 50 million gallons per year, the investment is on the order of $1.16 per gallon capacity (1978 dollars).

For plants of 10 million gallons capacity, the investment was estimated at about $25.2 million (1978 dollars) or about $2.50 per gallon capacity. The investment for a 100-million-gallon plant was estimated at $100 million or about $1.00 per gallon capacity. Figure 8-33 summarizes the plant investment data generated by Katzen and others. It is apparent that large investment economies are achieved when the capacity increases from 10 to about 50 million gallons per year. Above this range, the economies of scale are much less significant, since many pieces of equipment in the larger plants

Table 8-6. 50-Million-Gallon-Per-Year Plant Investment—Base Case

IDENTIFICATION	DECEMBER 1978 COST (DOLLARS)	PERCENT OF INVESTMENT
Receiving, Storage & Milling	$ 2,086,800	3.96
Cooking and Saccharification	2,824,300	5.36
Fungal Amylase Production	3,485,900	6.61
Fermentation	4,195,600	7.96
Distillation	5,123,800	9.72
Dried Grain Recovery	13,018,400	24.69
Alcohol Storage, Denaturing & Coproduct Storage	4,399,900	8.34
Utilities	15,090,000	28.62
Building, General Services, and Land	2,494,000	4.73
Subtotal	52,719,000	100.00
+ 10% Contingency	5,272,000	
TOTAL PLANT COST	$57,991,000	

Source: Katzen, R., "Grain Motor Fuel Alcohol Technical and Economic Assessment Study". Raphael Katzen Associates for the U.S. DOE, Dec., 1978.

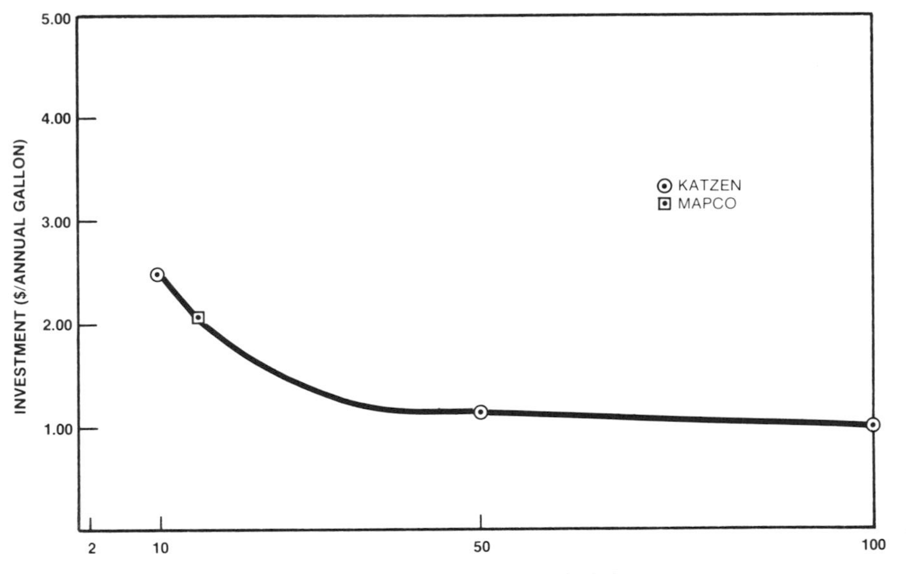

Source: Katzen, R., "Grain Motor Fuel Alcohol Technical and Economic Assessment Study." Raphael Katzen Associates for the U.S. DOE, Dec., 1978.

Figure 8-33. Ethyl alcohol plant costs in dollars per annual gallon.

are multiples of those used at the 50-million-gallon capacity and some of the large pieces of equipment such as the distillation towers must be field- rather than factory-assembled.

OPERATING COSTS

The annual operating cost for the 50-million-gallon plant is estimated at $44.51 million or $0.89 gallon (1978 dollars). The cost includes straight line depreciation over 20 years. Table 8-7 summarizes the cost items. The first column gives the 1978 equivalent cost; the second column gives the operating cost for the first year. The major cost item is corn and the second is the credit for the DDG coproduct. The net cost of corn (corn cost minus credit) amounts to about 58 percent of the annual operating cost. Therefore, ethanol production will be very sensitive to fluctuations in feedstock and coproduct market prices.

Under the same assumptions as those used in Table 8-7 (20 years straight line depreciation), the operating costs for 10- and 100-million-gallon plants are $1.22 and $0.84 per gallon, respectively.

Table 8-7. Plant Operating Costs for 50 Million Gallons per Year—Base Case

		EQUIVALENT 1978 COST		1st YEAR OPERATION* 1983 COST	
		ANNUAL, $ MILLION	$/GAL	ANNUAL, $ MILLION	$/GAL
FIXED CHARGES					
Depreciation	20 Years	2.900	0.058	3.200	0.064
	10 Years	5.8	0.116	6.4	0.128
Licenses Fees		0.029	0.001	0.040	0.001
Maintenance		1.829	0.036	2.560	0.051
Tax & Insurance		0.914	0.019	1.280	0.026
SUBTOTAL	20 Years	5.672	0.114	7.080	0.142
	10 Years	8.6	0.172	10.3	0.206
RAW MATERIALS					
Yeast		0.320	0.006	0.449	0.009
NH_3		0.373	0.007	0.522	0.010
Corn		44.770	0.896	62.679	1.254
Coal		2.410	0.048	2.273	0.067
Miscellaneous Chemicals		0.180	0.004	0.252	0.005
SUBTOTAL		48.053	0.961	67.276	1.346
UTILITIES					
Electric Power		1.646	0.033	2.305	0.046
Diesel Fuel		0.012	0.000	0.017	0.000
Steam (from plant)		0.000	0.000	0.000	0.000
C.W. (from plant)		0.000	0.000	0.000	0.000
SUBTOTAL		1.658	0.033	2.322	0.046
LABOR					
Management		0.240	0.005	0.337	0.007
Supervisors/Operators		2.194	0.044	3.072	0.061
Office & Laborers		1.202	0.024	1.683	0.034
SUBTOTAL		3.636	0.073	5.091	0.548
TOTAL PRODUCTION COST, TPC	20 Years	**59.019**	**1.181**	**81.769**	**1.640**
	10 Years	**61.9**	**1.24**	**85.0**	**1.70**
COPRODUCTS					
Dried Grains		19.175	0.384	26.845	0.537
Ammonium Sulfate		0.413	0.009	0.578	0.012
SUBTOTAL		19.588	0.393	27.423	0.548
MISCELLANEOUS EXPENSES					
Freight		2.504	0.050	3.506	0.070
Sales		1.930	0.039	2.705	0.054
G&AO		0.644	0.013	0.901	0.018
SUBTOTAL		5.078	0.102	7.102	0.142
TOTAL OPERATING COST	20 Years	**44.509**	**0.890**	**61.456**	**1.230**
	10 Years	**47.4**	**0.948**	**64.7**	**1.29**

*Assumed 7—percent inflation rate

Source: Katzen, R., "Grain Motor Fuel Alcohol Technical and Economic Assessment Study". Raphael Katzen Associates for the U.S. DOE, Dec., 1978.

ECONOMIC ANALYSES

Economic analyses were performed for the 50-million-gallon plant described above to determine the price of ethanol which will cover the production cost and provide a suitable return on equity. The return on equity is based on a discounted cash flow—interest rate of return (DCF-IROR) analysis.

The schedule of plant life and financial assumptions used in the analysis are summarized in Table 8-8.

Figure 8-34 summarizes the results of the analysis for three plant sizes. Ethanol price is expected to range from about $1.40 to about $2.20 per gallon in 1983.

SENSITIVITY OF ETHANOL PRICE

The sensitivity of ethanol price to various parameters is discussed below.

Sensitivity to Feedstock Price

The base case discussed above assumed corn price at $2.30/bushel. Figure 8-35 shows the dependence of ethanol price on that of corn for two levels of DCF-IROR. As is expected from the previous analysis, ethanol price is very sensitive to corn prices: an increase of 10 cents per bushel in corn price results in an increase of about 5 cents per gallon in ethanol price.

Sensitivity to Coproduct (DDG) Price

Figure 8-36 shows the impact of variations in DDG prices on the price of ethanol. An increase of about 10 percent in DDG price results in a decrease of about 8 percent in ethanol price.

Table 8-8. Financial Assumptions

Engineering Period	1979-80
Plant Construction	1980-83
Start-up	1982-83
Operationg Period	10 or 20 Years
DCF-IROR	15%
Working Capital	10% of operating costs
Inflation	7%
Depreciation	10 or 20 Years
Taxes (federal/local)	50% of profits after expenses
Investment Tax Credit	10%
Equity	100%

Source: Katzen, R., "Grain Motor Fuel Alcohol Technical and Economic Assessment Study". Raphael Katzen Associates for the U.S. DOE, Dec., 1978.

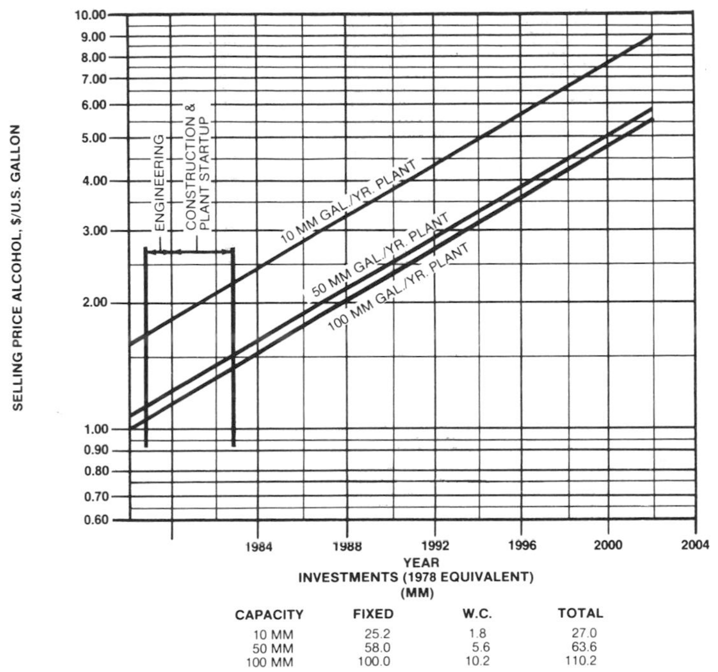

CAPACITY	FIXED	W.C.	TOTAL
10 MM	25.2	1.8	27.0
50 MM	58.0	5.6	63.6
100 MM	100.0	10.2	110.2

Source: Katzen, R., "Grain Motor Fuel Alcohol Technical and Economic Assessment Study." Raphael Katzen Associates for the U.S. DOE, Dec., 1978.

Figure 8-34. Plant capacities other than base case—selling price versus production year.

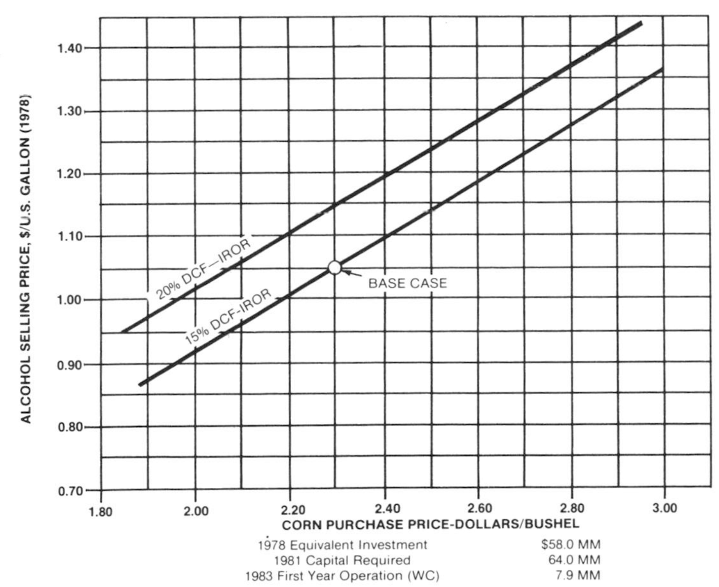

1978 Equivalent Investment	$58.0 MM
1981 Capital Required	64.0 MM
1983 First Year Operation (WC)	7.9 MM

Source: Katzen, R., "Grain Motor Fuel Alcohol Technical and Economic Assessment Study." Raphael Katzen Associates for the U.S. DOE, Dec., 1978.

Figure 8-35. Corn purchase price sensitivity analysis.

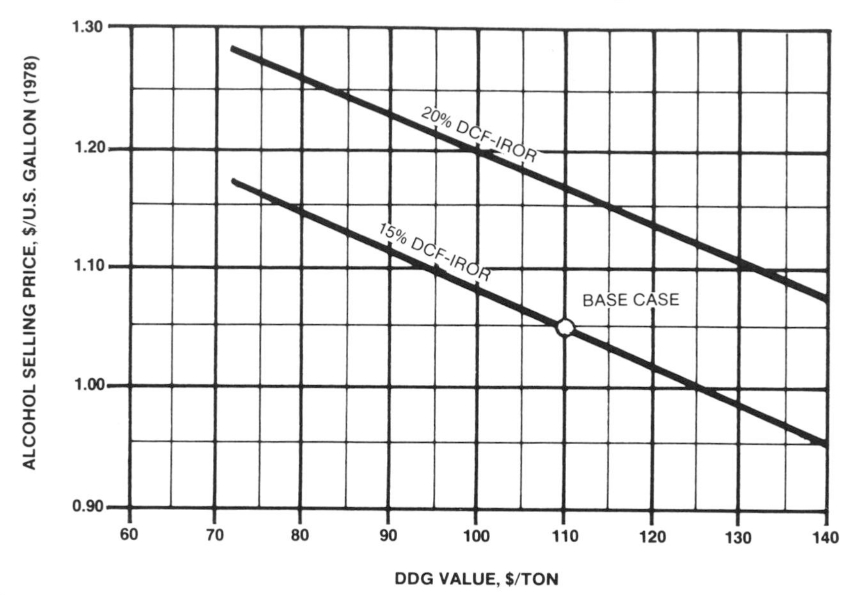

Source: Katzen, R., "Grain Motor Fuel Alcohol Technical and Economic Assessment Study." Raphael Katzen Assoicates for the U.S. DOE, Dec., 1978.

Figure 8-36. Alcohol selling price (1978) versus DCF-IROR.

Sensitivity to Changes in Feedstock and Fuel

An analysis similar to that performed for corn was performed for wheat, milo, and sweet sorghum feedstocks. In the latter case, there is some uncertainty in the results because a complete plant design was not available. Table 8-9 summarizes the results of the analysis for the base case of a 50-million-gallon plant.

The similarity in design of the corn, wheat, and milo plants is reflected by equivalent investments for the plants. Sweet sorghum is a sugar crop which requires a front-end processing unit different from corn (juice extraction and clarification). Sweet sorghum is only available for about 6 months of the year. To avoid extensive storage facilities, it was assumed that the plant operates one-half of the year on sweet sorghum and one-half of the year on corn. As was pointed out in an earlier section, the double feedstock capability results in higher investment costs because of the need for two separate front-end feedstock processing trains.

Of the feedstocks analyzed, grain sorghum shows a slight advantage over corn under the conditions of the analysis.

Switching from coal to corn stover for fuel results in a slight increase in ethanol price. This option could result in environmental problems as the removal of stover from the fields could increase the risk of erosion.

Table 8-9. Ethanol Prices for Various Feedstocks and Fuels (50-Million-Gallon Plant — 1978 Dollars)

FEEDSTOCK	FEEDSTOCK PRICE ($/bu)	INVESTMENT ($ millions)	OPERATING COSTS ($ millions/yr)	ETHANOL PRICE ($/gallon)
Corn	2.30	58.0	44.5	1.05
Wheat	3.15	58.0	57.1	1.31
Milo (grain sorghum)	2.20	58.0	42.6	1.02
Sweet				
Sorghum*	14.42/ton	91.6	58.8	1.40
Fuel				
Corn Stover	25.0/ton**	57.0	46.3	1.09

*Half corn, half sweet sorghum feedstock.
**Dry basis, 8000 Btu/lb.
Source: Katzen, R., "Grain Motor Fuel Alcohol Technical and Economic Assessment Study". Rephael Katzen Associates for the U.S. DOE, Dec., 1978.

Sensitivity to Financial Parameters

The sensitivity to parameters such as debt-to-equity ratio, DCF-IROR, and investment tax credit is shown in Table 8-10. The data were derived from the Katzen report, to which the reader is referred for details. In most cases, the impact of variations in financial parameters is smaller than that resulting from variations in feedstock or DDG prices.

Table 8-10. Sensitivity of Ethanol Price to Financial Parameters

PARAMETERS	VARIATION OF PARAMETER (%)	IMPACT ON ETHANOL PRICE (%)
DCF-IROR	+33[1]	+10
Investment Tax Credit	+300[2]	-2
Plant Depreciation Period	+100[3]	+2
Debt/Equity Ratio	+80[4]	-10

[1] DCF-IROR increased from 15 to 20%

[2] Investment Tax Credit increased from 10 to 30%

[3] Depreciation Period increased from 10 to 20 years

[4] Debt increased from 0 to 80 percent of capital

Source: Katzen, R., "Grain Motor Fuel Alcohol Technical and Economic Assessment Study". Raphael Katzen Associates for the U.S. DOE, Dec., 1978.

Chapter 9

Environmental and Safety Impacts

The environmental problems associated with ethanol production are minor in comparison to most other energy technologies. However, it is important to understand the impacts upon the ethanol plant due to environmental issues. This chapter provides a thorough overview of the environmental impacts on small- and large-scale ethanol production.

There is great concern that insufficient attention is paid to safety in ethanol plants. Many small-scale plants which are self-designed fail to pay attention to safety considerations. Ethanol is a flammable product produced with high temperatures easily accessible to plant personnel. Injury has occurred due to suffocation in the fermentation tanks in small-scale plants. Insufficient safety training and operating manuals are available. More attention must be paid to safety. This chapter also summarizes key safety concerns.

Environmental Considerations

The discussion of environmental considerations will focus on the conversion of feedstock to ethanol, but the major issues associated with feedstock production and end use of the ethanol in blends with gasoline will be summarized briefly. The role of environmental assessment in business decisions will also be addressed.

The major environmental concerns associated with ethanol production from grain are: emissions from the process heat source; and two multisource environmental problems, distillery wastewaters and occupational exposure to process and coproduct chemicals.

Distillery wastewaters are acidic and high in suspended solids, biological oxygen demand (BOD), and chemical oxygen demand (COD), and may present a potential health hazard if not properly treated or disposed. Stillage is particularly high in BOD and mineral salts (especially alkali salts), and both surface and groundwaters would be affected by improper or inadequate treatment or disposal practices. Standard wastewater treatment systems are commercially available.

Toxic and corrosive process chemicals employed in the conversion of starch feedstocks to ethanol or arising as coproducts in the conversion series include acids, dehydrating agents, denaturants, fusel oil, and aldehyde coproducts, and chemicals used for equipment maintenance. Environmental concerns attendant with the use of such substances include spills and other accidental emissions and potential occupational health hazards resulting from long-term exposure. The practice of good industrial hygiene should minimize danger to workers.

Impacts due to air emissions and solid wastes generated during combustion of fossil fuel, especially coal, to meet process heat requirements need to be considered, with special attention to commercial-scale alcohol conversion facilities with on-site industrial-size boilers. Both the environmental problems and the control measures are generic to coal-fired boilers, and thus represent no new problems peculiar to the fuel-alcohol industry.

Environmental issues arising in the production of feedstock grains are erosion and use of scarce resources such as water and land. Erosion serves not only to reduce soil fertility and productivity, but also as a transport mechanism whereby entry of pollutants into waterways and the food chain as a whole is facilitated.

The utilization of ethanol/gasoline blends as a motor fuel could have environmental consequences for both users of the fuel and the general population. Health and safety risks occur for individual users, and combustion of the fuel entails potential air-quality degradation and ecosystem effects.

Bioconversion of Grain Feedstocks to Ethanol

OVERVIEW OF EMISSION AND EFFLUENT SOURCES

For the discussion of environmental impacts during fermentation of biomass to ethanol and recovery/purification of the product, the reference plant will be a 50-million-gallon-per-year facility, employing the energy-efficient Katzen design. Dry-milled corn is the biomass feedstock and coal is the fuel for generation of process steam requirements. For such a reference plant, Table 9-1 indicates the major sources of emissions and effluents during the conversion process.

In the feedstock storage and preliminary processing sections of the operation, the major atmospheric emission is particulates from the physical preparation of the feedstock. Liquid effluents include various wash waters and flash cooling condensate. The only solid of concern is grain dust, which may be contaminated by pesticides or fungi. Aflatoxin contamination of midwestern corn is rare.

During hydrolysis and fermentation, the major air emissions arise from fermenter vents; CO_2 and accompanying volatile organic fermentation products and coproducts escape to the atmosphere. Wash waters are high in BOD, dissolved solids, and suspended solids; in addition, they may contain

Table 9-1. Bioconversion of Grain to Ethanol – Emission and Effluent Sources

OPERATION	EMISSIONS TO ATMOSPHERE	LIQUID EFFLUENTS	SOLID WASTES
Feedstock Storage	Transfer operations (fine dust)		Grain, dirt
Milling and Cooking	Mechanical collectors for milling operations (particulates)	Wash water (dissolved and suspended solids, organics, pesticides, alkali) Flash cooling condensate (dissolved and suspended solids, organics)	Grain dust (from mechanical collectors)
Hydrolysis and Fermentation	Fermentation vents (CO_2, hydrocarbons)	Wash water (dissolved and suspended solids, organics, alkali)	
Distillation and Dehydration	Condenser vents on columns (volatile organics)	Rectifier bottoms (organics), dehydration bottoms (organics)	
Storage	Storage tanks (hydrocarbons)		
Coproduct Recovery	Dryer flue gases (NO_x, SO_x, CO, particulates, hydrocarbons). Evaporator condenser vents (hydrocarbons)	Evaporator condensate (dissolved and suspended solids, organics)	Grain dust (from direct-contact dryers)
Steam Production (coal-fired)	Flue gases (NO_x, SO_x, CO, particulates)	Boiler blowdown (inorganics), cooling water blowdown (dissolved and suspended solids, organics)	Coal dust, ash (bottom ash and fly ash)
Environmental Control Systems	Evaporation from biological treatment ponds (organics)	Scrubber blowdown (dissolved and suspended solids, organics)	Sludge (from flue gas treatment), biological sludge (from wastewater treatment)

Source: R.M. Scarberry and M.P. Papai, "Source Test and Evaluation Report: Alcohol Synthesis Facility for Gasohol Production," Radian Corporation, January, 1980.

traces of chemicals used as nutrients or used to control the growth of undesired organisms during the fermentation. For starch feedstocks, the associated pesticide residues are thought to be destroyed during the cooking process.

In the alcohol purification (distillation and dehydration) and denaturation steps, the major sources of atmospheric emissions are condenser vents on columns. Evaporator condensates may contain materials that require treatment before disposal. No solid wastes are obtained.

Evolution of criteria pollutants occurs in coproduct processing through the exhaust of flue gas used in drying. In addition, evaporator condensate contains dissolved and suspended solids and various organic materials. The direct-contact dryers generate grain dust.

Steam production using coal-fired boilers causes emissions of criteria pollutants. Aqueous streams of environmental concern include boiler blowdown and cooling water blowdown. Solid wastes include coal dust, fly ash, and bottom ash.

Environmental control systems may produce secondary wastes. For example, biological treatment ponds yield evaporative emissions. Scrubber blowdown contains dissolved and suspended solids, as well as various organics. Biological sludge from wastewater treatment may contain pesticides, benzene, ammonia, and various metals.

IMPACTS ASSOCIATED WITH SPECIFIC PROCESS STEPS

Specific impacts in the three major categories of air quality, water quality, and solid wastes are discussed in more detail for each stage of the conversion process. A separate discussion of multisource environmental problems, specifically wastewater treatment and occupational safety and health, is also presented.

Feedstock Storage

Feedstock storage facilities may represent a potential health hazard by serving as breeding grounds for rodents and other pests. Spoilage of grain due to natural fermentation and other decay processes gives rise to a variety of substances which, depending on the degree of containment of the storage area, could be discharged to waterways or to the atmosphere. The use of fungicides to retard spoilage could affect worker health, as well as groundwater and surface-water quality. Finally, the generation of large amounts of fine dust in the transfer of grain into and out of the storage area results in potential respiratory health hazards to workers, as well as an explosion hazard.

Experience in controlling dust in the grain industry should be transferable to similar operations associated with fuel-alcohol plants.

Milling and Cooking

If corn is processed by hammer mills, particulates in the form of grain dust can be a problem, along with high levels of noise (85 to 88 decibels, absolute) generated by the processing equipment. Controls for dust from grain are similar to those used for fly ash particulates. Grain dust, chaff, and impurities collected in cyclones are not considered solid wastes, as they are cycled to the dryer for inclusion in the DDG coproduct. During the cooking process, a beneficial side effect is the apparent destruction of pesticides, many of which are vulnerable to heat and decompose to simpler compounds during the heat treatment. In a recent sampling program conducted at the Midwest Solvents plant in Atchison, Kansas, pesticides were detected in the feedstock grains, but no traces were found in the solid wastes or wastewater effluents downstream of the gelatinization step.

Flash cooling of the mash in multieffect evaporators yields a condensate (from a surface condenser employed at the last stage) which is high in dissolved solids and suspended solids, and volatile organic materials.

Hydrolysis and Fermentation

Nearly equal weights of ethanol and CO_2 are produced in the fermentation reaction. Other coproducts include a myriad of oxygenated organic compounds, primarily acids, alcohols, and aldehydes. The CO_2 can be trapped or merely vented to the atmosphere. If the CO_2 is recovered, the gaseous stream is passed through a water scrubber to condense water and volatile organic materials (including ethanol) before the cryogenic recovery of CO_2. Alternatively, if CO_2 is vented, the water scrubber may still be employed. The gaseous stream is accompanied by volatile organic coproducts, as well as some of the product ethanol. Temperature is carefully controlled during the fermentation reaction to minimize evaporative losses of ethanol, which increase by a factor of 1.5 for every 5°C increase in temperature. Closed fermenters with off-gas condensation and scrubbing may be used to minimize ethanol losses to the atmosphere.

A variety of chemicals added to the fermenting mash include acids for pH control, yeast nutrients (ammonium salts, urea, phosphates), and chemicals or antibiotics to control the growth of undesired organisms without affecting the yeast. Because of the variety of coproducts and nutrients/chemicals used in the fermentation process, it is not surprising that the fermentation wash water contains high levels of dissolved solids, suspended solids, organics, and alkali.

Care must be taken, when personnel enter the fermenters during cleaning, to avoid the possibility of asphyxiation in an atmosphere with a high concentration of CO_2.

Distillation and Dehydration

During alcohol concentration and purification, the major atmospheric emissions are from the vent condensers employed at distillation column openings to retard (but not entirely eliminate) the escape of organic materials to the environment. As the vent streams contain volatile organic materials too dilute for economic recovery, flares are often used to control emissions. Alternatively, the vent streams could be routed to the process burners as air feed. The evaporator condensates, rich in dissolved solids, suspended solids, and organic materials, are sent to wastewater treatment. Distillation and dehydration bottoms contain benzene (if used as the dehydrating agent) and other organics. Because benzene is a known leukemogen (i.e., can cause leukemia), strict precautions must be taken in work areas during its use. The federal standard (29 CFR 1910.1000) for exposure to benzene is 10 ppm (8-hour, time-weighted average), with a ceiling concentration of 25 ppm and a peak of 50 ppm allowable for 10 minutes. A recent sampling program conducted at the Midwest Solvents plant by the Radian Corporation revealed benzene concentrations in the low range of 2.7-59.4 ppm for the liquid streams. As these values are near the detection limit for benzene, they are assumed to have limited accuracy; nevertheless, it is known that 27 gallons of benzene (approximately 1 gallon for every 5500 gallons of product ethanol) would be lost daily somewhere in the process for a 50-million-gallon-per-year plant.

Hydrocarbon Storage

Evaporative emissions during storage can be controlled through the use of tanks with floating roofs or internal floating covers to reduce the air space above the stored liquids. Another option is the inclusion of a vapor recovery system.

Coproduct Recovery

If stillage is not dried or otherwise treated before disposal, water quality could be degraded by this waste stream. Stillage is relatively high in protein content, as well as unconverted starches and sugars, various fermentation products, and yeast. Stillage drying operations may affect air quality. Particulates, primarily in the form of grain dust generated during the drying process, are emitted, and dryer flue gases, which use boiler flue gases from the plant's heat source, contain criteria air pollutants, including SO_2, CO, and NO_x. Grain dust which is collected may be added to the DDG coproduct. Aqueous wastes, such as evaporator condensate from coproduct-recovery operations, are sent to the wastewater treatment system.

Aflatoxin contamination of midwestern corn, the primary source of corn for fuel-ethanol production, is rare. Although more than 50 percent of southeastern corn is contaminated, it represents only 8 percent of the national crop, and most of it is confined to intrastate use. In the rural Southeast, contaminated corn is directly ingested, and epidemiological information indicates no higher human liver cancer incidence than in other areas of the United States. As aflatoxins are not distillable, any which survive the fermentation process will be concentrated in the DDG coproduct. It has, however, been demonstrated that mature animals rapidly metabolize and dispose of aflatoxins. The only potential sensitivity could arise from feeding contaminated DDG to dairy cattle, as one type of aflatoxin remains in the milk.

Steam Production

Combustion of coal in industrial-size boilers results in atmospheric emissions, aqueous effluents, and solid wastes. To fulfill process steam requirements for the reference-size ethanol plant, 296.7 tons/day of coal is consumed, and 27.4 tons/day of ash is generated. Ash from a typical plant would be hauled to a nearby landfill for disposal. Coal dust collected during storage and transfer of the fuel can be routed to the boiler.

The atmospheric emissions from a coal-fired boiler at a fuel-alcohol plant that are most likely to require controls are sulfur dioxide and particulates. Control of nitrogen oxides is required in some states, but emission levels are expected to be within legislated limits in most cases.

Several standard options are available for control of particulates, including fly ash and coal dust entrained in the flue gas. For systems of the size required for a commercial-scale fuel-alcohol plant, fabric filters are usually the best alternative for particulate control.

For sulfur dioxide emissions, the simplest control in many cases is the use of low-sulfur coal, which could obviate the need for sophisticated control systems downstream. Several flue gas desulfurization systems are commercially available, but are expensive in terms of both capital costs and operating expenses. To remove both sulfur dioxide and particulates, a flue gas scrubbing system may be employed (at an initial cost of $4.2 million in 1978). Ammonia is used to remove sulfur dioxide, followed by neutralization of the ammonium sulfite/bisulfite solution and its oxidation to ammonium sulfate, which can be sold as a fertilizer. This type of chemical recovery desulfurization system has the advantage of producing no calcareous sludge, the secondary waste generated in conventional limestone scrubbing systems.

Water quality could be affected by leachate from coal or ash storage or ash disposal piles (although coal ash is generally disposed of in off-site landfills), as well as by cooling tower blowdown. Depending on the soil and groundwater characteristics in the area, leachate collection and treatment may become necessary. Boiler and cooling water blowdown are routed to the wastewater treatment section of the plant.

Because a fossil-fuel-fired steam generator represents one of the major emission sources at an ethanol-production facility, an obvious all-around environmental control measure is the reduction of the amount of process steam required. Any improvement in plant efficiency will result in a corresponding decrease in fossil fuel combustion and its associated atmospheric emissions, and aqueous and solid waste disposal problems.

Besides improvements in engineering design, substantial gains in this area can be obtained by using waste heat from other industries or by siting new plants imaginatively.

Environmental Control Systems

Environmental control systems may generate secondary pollutants, such as wastewater from various scrubbers. As mentioned above, certain types of flue gas treating systems give rise to sludge, as does the wastewater treatment system. The only secondary atmospheric pollutant is the evaporative loss from biological treatment ponds; this emission consists of an assortment of volatile organic compounds.

MULTISOURCE ENVIRONMENTAL CONCERNS

Wastewater streams arise from numerous separate sources in a fuel-ethanol plant and potential health and safety hazards pervade the occupational environment. These two multisource problems are discussed briefly.

Wastewater

Wastewaters from distilleries (approximately 1.1 million gallons per day for a 50-million-gallon-per-year plant, if extensive efforts are made to recycle water) are acidic and contain high levels of total solids (25,000 pounds per day for the reference-size plant); suspended solids (3000 pounds per day); and BOD (7300 pounds per day). Experience in the beverage alcohol industry indicates that levels of dissolved organics are below concentrations that require treatment before discharge to surface waters. Moreover, no significant problems are expected with regard to acidity levels in wastewaters, as the pH can be adjusted before discharge. High BOD loadings, however, will necessitate treatment. Suspended solids can be removed by preliminary screening and sedimentation in a holding tank before the biological oxidation step. Several biological oxidation systems are commercially available; spare aeration or equalization basins are recommended for added safety. Treatment can occur at the alcohol plant site or at publicly owned treatment works (POTW), if the latter have sufficient spare capacity to treat the large volumes of wastewaters associated with ethanol production. At many potential farm belt locations for ethanol plants, the local POTW may have insufficient capacity to receive distillery effluents, and treatment systems will thus become necessary parts of fuel-alcohol facilities.

Occupational Safety and Health

In a fuel-alcohol plant, the major potential threats to worker health and safety are explosion, fire hazards, and various modes of exposure to toxic and corrosive chemicals. Adequate controls or mitigating measures are currently available to cope with the problems.

Chemical exposure. Ethanol itself can cause mild irritation of the eye and nose and can defat the skin, causing dermatitis. Prolonged inhalation produces irritation of the eyes and upper respiratory tract, headache, drowsiness, tremors, and fatigue. It may increase the toxicity of other inhaled, absorbed, or ingested chemical agents. Moreover, since ethanol and some common prescription drugs interact unfavorably when ingested, the possibility of synergism between these drugs and inhaled or absorbed ethanol should be considered. The Federal standard for workplace exposure to ethanol is 1000 ppm (1900 milligrams per cubic meter). Work areas should be well ventilated, and normal safety precautions should be taken in handling the liquid.

The same general considerations apply to exposures to other organic compounds employed in or generated during the ethanol-production process. The primary entry routes to the body are inhalation and dermal absorption. Workers should be educated regarding the proper handling of chemicals; protective gloves and aprons should be worn when appropriate; and emergency spill containment procedures should be well established in the occupational environment.

Fire/explosion/burn hazards. Whenever any volatile organic compounds are in use, standard precautions must be taken to prevent ignition of leaks or fumes. Explosion-proof motors, for example, along with other specially protected electrical equipment, should be used routinely. Equipment should be available and emergency procedures in place to deal with chemical fires.

In any industry that employs large quantities of process heat, particularly when it is in the form of steam, precautions need to be taken against burns. Certain routine preparations and minor equipment modifications, such as using baffles to direct steam gasket leaks away from the work area, can be made. Prevention of contact burns from steam lines can be accomplished by making the lines conspicuous or by insulation.

ENVIRONMENTAL CONTROL COSTS

As indicated throughout the preceding discussion, hardware for adequate control of various pollutants generated in the production of fuel alcohol from biomass is commercially available and is similar to equipment used in other industries (e.g., food processing, drug manufacture, chemical manufacture, and beverage alcohol production). The major capital items include a flue gas scrubbing system, $4.2 million; wastewater treatment system, $2.0 million for the reference 50-million-gallon-per-year fuel-ethanol plant; fire protection system, $0.60 million; ash collection package, $0.17 million; vapor-controlled storage tanks, $0.16 million (incremental cost over cone roof tanks); and assorted vent condensers, $0.13 million. The cost of environmental controls, $7.5 million, accounts for approximately 13 percent of the total fixed investment for a fuel-ethanol plant. This estimate is for direct controls only (i.e., physical hardware) and does not include installation, maintenance and operating costs, or the cost of implementing procedures (such as conducting seminars for employees on appropriate actions during spills and other potential workplace emergencies). It should be mentioned that an alternative to the capital-intensive, on-site wastewater treatment system is the discharge of distillery wastewater to POTW, provided that the latter has the excess capacity to treat the industrial wastes. In a comparison of self-treatment costs to costs of treating at a POTW (including industrial cost recovery charges), it has been shown that for a POTW with a 15-million-gallon-per-day or greater capacity, it would generally be less expensive for distilleries of the reference size to discharge to the POTW. Actual charges for treating distillery effluent at a specific POTW, however, vary widely, because they depend not only on financial parameters such as interest rate, but also on the assimilative capacity of the stream into which the treated water is discharged.

REGULATORY CONSTRAINTS ON DEVELOPMENT OF THE FUEL-ETHANOL INDUSTRY

There are no current major federal environmental regulatory obstacles associated with biomass conversion to ethanol; moreover, no roadblocks are anticipated. The major effects of environmental considerations on commercial-scale fuel-alcohol development are in the interrelated areas of cost (discussed earlier), siting limitations, and uncertainties for investors.

The major legislation influencing fuel-alcohol development are: the Clean Air Act and amendments of 1977, the Federal Water Pollution Control Act of 1972 (Clean Water Act), and the Resource Conservation and Recovery Act (RCRA) of 1976. The provisions of these acts which pertain to the alcohol industry are summarized briefly below. It should be borne in mind that more stringent state environmental regulations could be superimposed on any of the existing federal regulations. The investment climate thus could be affected adversely by the uncertainties in environmental legislation at the state and federal levels and the attendant uncertainties in investment costs.

Clean Air Act

All of the criteria pollutants, except lead, subject to the National Ambient Air Quality Standards (NAAQS) are emitted during the preparation and conversion of feedstock to ethanol. Under the Clean Air Act, major existing and new sources of air pollutants within an area presently attaining the primary and secondary NAAQS may be subject to emission limitations and permitting requirements to ensure the prevention of significant deterioration (PSD) of air quality in the region. Sources thus controlled include coal-fired steam boilers of more than 250 million Btu per hour heat input. (For comparison, the Katzen-designed 50-million-gallon-per-year plant requires heat input of 263.3 million Btu per hour.) PSD limitations regulate criteria pollutant emissions so that the secondary NAAQS are not violated within the air quality control regions (AQCR) concerned. Although all of the criteria pollutants, as well as certain other pollutants not associated with alcohol production, are subject to PSD limitations, to date the permissible emission increments have been announced only for particulates and sulfur dioxide. It is possible, therefore, that standards yet to be proposed by the EPA could have an impact on the siting of energy facilities such as ethanol-from-biomass plants.

Certain new, modified, and reconstructed stationary sources, including large fossil-fuel-fired steam generators such as those used in distilleries, are subject to EPA's New Source Performance Standards (NSPS). Limitations are placed on particulates, opacity, nitrogen oxides, and sulfur dioxide. Smaller industrial boilers are expected to be covered by revised NSPS standards in late 1981.

Siting of new fuel-alcohol plants could be barred in so-called nonattainment areas. In order to begin construction or major modification of a major source in a nonattainment area, a construction permit must be obtained. Before the permit is granted, the source must obtain an emissions offset from existing sources in the region. In addition, the source must limit its air pollutant emissions to the lowest achievable emission rate or the lowest obtainable in practice, regardless of energy and economic considerations. The extent to which the fuel-alcohol industry could be affected is demonstrated by noting the number of counties in each of the 48 contiguous states which have been designated as nonattainment areas, as shown in Table 9-2. Although the number of counties affected may not be proportional to their land area, the figures convey at least a general indication of the extent of the problem in a given state. In 31 states, 30 percent or less of the state's counties had been classified as nonattainment areas for one or more pollutants as of December, 1979, and in 20 of these states, 15 percent or less of the counties were affected. At the other extreme, in 13 states, 71 percent or more of the counties had been designated as nonattainment areas (see Fig. 9-1), and all counties in Connecticut, Massachusetts, New Hampshire, New Jersey, New York, Pennsylvania, Rhode Island, and Vermont were in nonattainment status.

Because of proximity to feedstock supplies, the states included in the USDA's "cornbelt" growing region (Iowa, Missouri, Illinois, Indiana, and Ohio) are likely targets for development of the fuel-ethanol industry. Of the cornbelt states, only in Ohio should siting limitations for ethanol plants be severe.

Federal Water Pollution Control Act

It is the intent of the Clean Water Act to attain zero discharge of regulated pollutants such as BOD, solids, oil and grease by 1985. Technology-based effluent limitations on certain existing sources are predicated on the effectiveness of the best available control technology economically achievable by 1983.

Aqueous discharges from distilleries are relatively easy to monitor and regulate, as they generally involve point sources. There are no current technology-based effluent guidelines for fuel ethanol facilities or for beverage distilleries. Fuel-ethanol plants are under consideration by EPA, and a decision on the need for wastewater emission regulations for these facilities will be made in 1-2 years. The National Pollutant Discharge Elimination System (NPDES) distinguishes between major and minor dischargers, with a major source consisting of one discharging 100,000 gallons or more per day

Table 9-2. Nonattainment Areas of 48 Contiguous States*

STATE	NO. OF NONATTAINMENT COUNTIES (PARISHES)	TOTAL NO. OF COUNTIES (PARISHES)	% OF THE STATE'S COUNTIES IN NONATTAINMENT STATUS
Alabama	9	67	13
Arizona	13	14	93
Arkansas	1	75	1
California	43	58	74
Colorado	11	63	17
Connecticut	8	8	100
Delaware	1	3	33
Florida	7	67	10
Georgia	15	159	9
Idaho	5	44	11
Illinois**	34	102	33
Indiana**	14	92	15
Iowa**	13	99	13
Kansas	5	105	5
Kentucky	19	120	16
Louisiana	19	64	30
Maine	15	16	94
Maryland	10 + Baltimore	23	43
Massachusetts	14	14	100
Michigan	40	83	48
Minnesota	18	87	21
Mississippi	1	82	1
Missouri**	15 + St. Louis	114	13
Montana	8	57	14
Nebraska	4	93	4
Nevada	14 + Carson City	16	88
New Hampshire	10	10	100
New Jersey	21	21	100
New Mexico	7	32	22
New York	62	62	100
North Carolina	4	100	4
North Dakota	0	53	0
Ohio*	75	88	85
Oklahoma	4	77	5
Oregon	6	36	17
Pennsylvania	67	67	100
Rhode Island	5	5	100
South Carolina	6	46	13
South Dakota	1	67	1
Tennessee	17	95	18
Texas	18	254	7
Utah	6	29	21
Vermont	14	14	100
Virginia	12	96	13
Washington	9	39	23
West Virginia	9	55	16
Wisconsin	17	72	24
Wyoming	1	23	4

*EPA Data As of December, 1979.
**Included in the U.S. Department of Agriculture's "Corn Belt" Growing Region.

of wastewater. (Thus, commercial-scale ethanol plants would be classified as major sources.) Wastewater pretreatment regulations may have an impact on fuel-ethanol production. Final pretreatment regulations have been set governing water quality standards for nondomestic waste that is introduced to POTW with design flows of 5 million gallons per day or more. Between now and July 1, 1983, the provision for a pretreatment program will become a condition of all new or reissued NPDES permits, and industrial dischargers must be in compliance by that day.

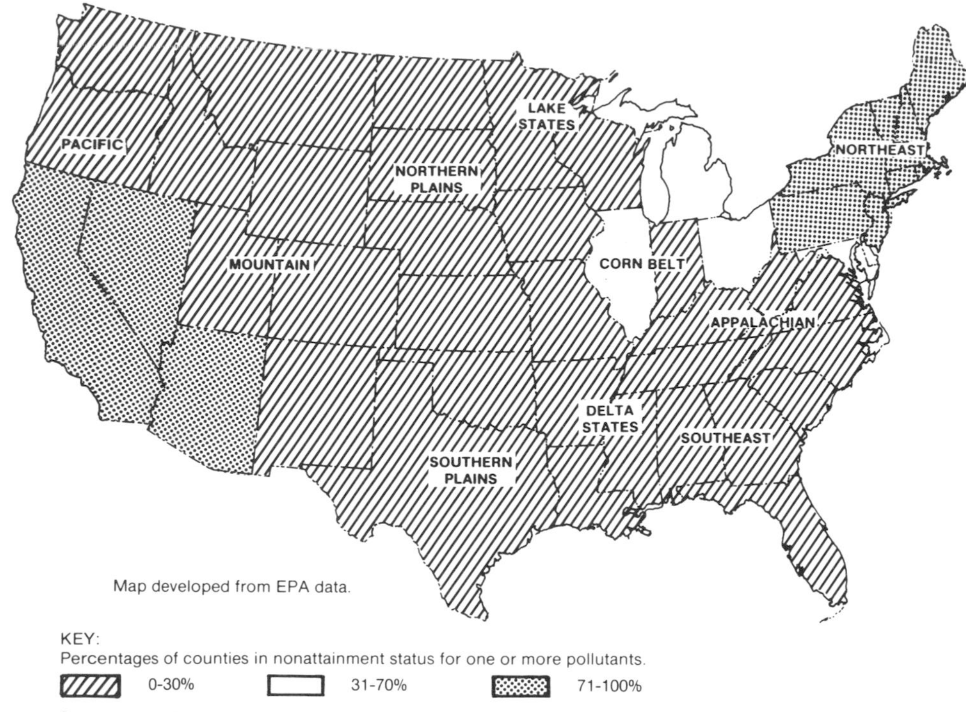

Source: *A Guide to Commercial Scale Ethanol Production and Financing*, SERI, Denver, 1980.

Figure 9-1. Extent of nonattainment of national ambient air quality standards in the 48 contiguous states.

Resource Conservation and Recovery Act (RCRA)

The primary areas of potential solid waste regulatory impacts applicable to ethanol-from-biomass activities are coal ash from boiler operation, grain elevator dust, particulates from the drying of distiller's grains (trapped during emission control efforts and disposed of as solid waste, unless recycled), and stillage, if it is applied to soil as fertilizer. If recycling of trapped grain and recovery of DDG from stillage are assumed, then the major federal legislation bearing on solid wastes from the biomass preparation and conversion steps is the RCRA. The extent of coverage and methods of implementation of this act are still being interpreted and developed by EPA and the courts. EPA has considered designating some energy-related hazardous wastes as "special waste" under RCRA. Interim standards would be established for special waste. Coal ash may be assigned to the special waste category until extensive analysis is completed.

ENVIRONMENTAL ASSESSMENT

Environmental assessments or impact statements may be required prior to construction of fuel-ethanol plants under certain conditions. The need for such assessment is triggered by federal participation in funding, whether by direct cost-sharing or by loan guarantee. For a large project, which may produce significant environmental effects, formal assessments will be required. The Office of Environmental Compliance and Overview within DOE rules on the need for such assessments.

Even in the absence of a federal requirement for an environmental assessment, it may be prudent for an investor to gather the pertinent environmental information, as such foresight may expedite the permitting process and facilitate public acceptance of a project. Although a federal agency must evaluate the environmental evidence, the fact-finding activity may be delegated to the applicant or its independent contractor.

Feedstock Production

In the production of grains as feedstocks for ethanol production, the single most critical environmental issue is erosion and its effects on water quality. Erosion serves as a transport mechanism for carrying nutrients, sediments, and pesticides to water resources, and it directly affects crop productivity and soil fertility.

Land and water use are also important issues. The need for irrigation in the western states puts increasing pressure on existing surface and groundwater resources, causing water tables to drop in some areas and the salinity content to increase in return flows to surface waters. The land use issue has two major components: (1) the loss of farmland to urban sprawl, and (2) the increasing pressure to put marginal farmland into use and convert fallow or natural areas to intensively cultivated farmlands.

EROSION

The rate of erosion is dependent on several factors, including type of soil, topography, amount and intensity of rainfall, ground cover, and control methods. It is, therefore, highly variable from site to site and over a period of time. The effects on soil are the direct loss of fertility and enhanced potential for water-quality degradation. The latter arises as the biological content of the soil is decreased, in turn reducing the rate at which toxic substances applied to the soil are broken down. Hazardous substances are thus more likely to reach waterways.

The loss of sediment to waterways results in silting. Runoff from agricultural lands can cause additional problems, since fertilizer and pesticide residues are carried to the waterways. The major soil nutrients—nitrogen and phosphates—promote algal growth and thereby accelerate eutrophication of surface waters. Corn requires relatively high fertilizer application rates; indeed, although corn comprises only 20 percent of the United States's cropland, it requires approximately 40 percent of the fertilizer used nationwide. Herbicide applications for corn are also higher than the national average.

WATER USAGE

The quantity of water used for irrigation can be substantial, and represents only one of several competing uses for a limited resource. Groundwater currently supplies approximately 40 percent of total U.S. irrigation needs, primarily in the western states. The importance of this issue for corn production is not as great as for other feedstocks, as minimal amounts of land are under irrigation in the major corn-producing states. Five of the six largest corn-producing states, for example, have less than 1 percent of total cropland under irrigation.

In addition to the use of water for agriculture, the quality of the returned water is of importance. Irrigation water picks up mineral salts, and the resulting increase in salinity of the water can have adverse effects on downstream users.

LAND USE

Increased demand for grain crops as ethanol feedstocks would result in the need to increase the amount of land in agricultural production. If marginal land is brought into production, more fertilizer and pesticides would be necessary to attain the average level of productivity for a given crop. These increases in chemical applications would intensify the water-quality impacts already associated with agricultural runoff; i.e., erosion from such marginal land would create more environmental impacts per acre, or per unit of output, than the average farmland in use today.

Another potential source of farmland is the conversion of natural areas such as forests to intensively managed crop production. Wildlife habitat would thus be destroyed, and erosion rates would be substantially increased.

The land use issue is further complicated by the loss of farmland to urban development. Annually, 3 million acres of farmland are lost to urban sprawl, and one-third of this total is considered prime farmland. As more of the existing farmland is lost to urban development, a greater proportion of marginal lands may be required for crop production.

A recent USDA cropland availability study reports that 78 million acres with high potential as cropland remain in this country, with much of it in the Cornbelt and Northern Plains regions. Of this high-potential land, 15 million acres have essentially no limitations on development. The latter acreage, if planted entirely in corn and used exclusively as ethanol feedstock, corresponds to approximately 4 billion gallons of fuel ethanol per year, or 80 of the reference-size, commercial-scale plants.

Use of Ethanol/Gasoline Blends in Highway Vehicles

Environmental impacts which could be associated with the use of ethanol/gasoline blends as fuel for highway vehicles are discussed briefly below. Not only users of the fuel but also the general population could be affected. Although potential impacts on the latter group are largely undefined, no barriers to the use of gasohol are anticipated.

IMPACTS ON USERS

Health

Ethanol is not highly toxic. Its major routes of entry into the human body are ingestion, inhalation, and absorption through the skin. Effects of exposure to the substance are primarily ones of discomfort, as described in an earlier section. Ingestion of ethanol/gasoline blends is unlikely, but the ethanol could be separated from the denaturing agents in some formulations. (For example, ethanol can be separated from gasoline denaturant.) A variety of physical symptoms can also arise from exposure to denaturants used with fuel ethanol.

Safety

Ethanol is a flammable liquid, with flammability limits of 3.3-19 percent by volume in air. Its flash point is above that of gasoline. Typical ethanol/gasoline blends pose explosion or fire hazards comparable to those of gasoline. In storage, however, ethanol presents one additional safety hazard—neat alcohols burn with invisible flames, thus making it possible for ethanol fires to remain undetected and/or to be approached too closely by workers.

IMPACTS ON GENERAL POPULATION

Air Quality

Both evaporative and exhaust emissions from ethanol/gasoline blends will differ slightly from those associated with unblended gasoline. Carbon monoxide and hydrocarbon emissions are not changed significantly. The exhaust is sensitive to the original air/fuel setting of the carburetor, whether or not the carburetor has been adjusted for the new fuel, the type of emission control system, and the engine configuration. A reduction of nitrogen oxide emissions is obtained with ethanol/gasoline blends, but aldehyde (currently unregulated) and evaporative emissions increase for the blend relative to gasoline. As ethanol contains no sulfur, sulfur dioxide is absent from the combustion products. Moreover, ethanol yields no particulates on combustion.

Ecosystems

Little information exists in the realm of potential effects of utilizing ethanol/gasoline blends in highway vehicles except for the above information on combustion. The major area of potential concern appears to be the possibility of introduction of ethanol to waterways through spills or other accidents. Ethanol dissolves readily in water and could, therefore, be transported easily in an aqueous medium.

Summary of Environmental Issues

Three major environmental problem areas associated with commercial-scale fuel-ethanol production are (1) safe disposal and treatment of distillery wastewaters; (2) exposure of workers to toxic and/or corrosive chemicals; and (3) atmospheric emissions and solid wastes resulting from combustion of coal to satisfy process steam requirements. Controls, in the form of hardware and procedures, are available commercially. The cost of direct controls amounts to approximately 13 percent of the capital investment for a 50-million-gallon-per-year fuel-ethanol plant.

The environmental regulatory climate can influence investment decisions. Current regulations may place limitations on siting of ethanol plants, particularly if they include coal-fired boilers. Besides the siting restrictions arising from air-quality considerations, the assimilative capacity of streams will play a role in siting decisions. Beyond the current generation of federal environmental regulations, investors need to consider the possibility of more stringent ones, as well as existing state and local regulations which may be more restrictive than their federal counterparts. Anticipated regulations could influence both the economics and the siting of future plants.

Availability of feedstocks for fuel-ethanol production is essentially unfettered by federal environmental laws. Environmental impacts associated with agricultural production tend to be non-point-source problems and are, therefore, difficult to monitor and regulate. The major concerns of increased agricultural production to meet fuel-ethanol feedstock requirements are erosion, which impacts water quality, and use of limited land and water resources.

When ethanol is blended with gasoline for use as a highway vehicle fuel, there are potential environmental implications involving both the user group and the general population. Ethanol is not highly toxic, but it does produce definite physical symptoms when inhaled or ingested. The same precautions need to be taken in handling ethanol as with any other explosive and flammable substance. The potential air-quality and ecosystem impacts of ethanol combustion in highway vehicles are not fully defined, but no barriers to deployment of ethanol for fuel use are currently foreseen.

Safety Considerations

Safety is of primary importance in a system producing a flammable product. In addition to the main product (ethanol), other hazardous compounds are consumed and generated in the process system. The chemicals considered hazardous are: sulfuric acid (H_2SO_4); calcium oxide (lime, CaO);

hexane (n-C_6H_{14}); calcium hydroxide [hydrated lime, $Ca(OH)_2$]; carbon dioxide (CO_2); and ethanol (grain alcohol, C_2H_5OH or EtOH). The hazards of these chemicals are briefly described in Table 9-3.

The codes and standards which are applicable to the production of fuel-grade ethanol are listed in Table 9-4.

Ethanol, 99.9 percent pure, is stored at ambient temperature and slight positive pressure. The flash point is 55°F (13°C) and flammability limits are 3.3-19.0 percent by volume. The saturated vapor/air mixture above the liquid ethanol is flammable between 50 and 110°F (10 and 43°C). Hence, a CO_2 blanket should be used in the ethanol storage tanks.

n-Hexane exists as a superheated vapor at 450°F (232°C) and 10 psig. The flash point is 0°F (-18°C). The flammability limits are 1.2-7.5 percent by volume. The saturated vapor/air mixture above the liquid hexane is flammable between 20 and 40°F (−7 and 4°C).

Benzene has flammability limits of 1.35-6.75 percent by volume in air, and its flash point is 12°F (−11°C). The saturated vapor/air mixture above the liquid is flammable between −10 and 64°F (−23 and 18°C).

Hazards

Fermentation ethanol plants are often located in rural areas in order to be close to their source of feedstock. Remoteness from city water supplies and fire departments places responsibility for fire protection almost entirely on the plant itself. Safety also depends on good construction and proper arrangement and safeguards for processes.

Because of the fire and explosion hazards inherent in handling large quantities of flammable liquids and also the potential for grain dust explosions in the grain storage areas, safety depends on supervision by well-trained operators, good maintenance, and process equipment safeguards.

Grain handling, milling, and feed preparation at distilleries present dust explosion hazards. Although grains and feeds are slow burning, fires in these materials may be deep-seated and difficult to extinguish. Wet grains will heat and sour if not dried promptly.

Process fire and explosion hazards are present during distilling, but are considered negligible during mashing and fermenting. Strict government regulations that require seals on every pipe joint, valve, and spigot reduce the probability of flammable liquid or vapor being released during distilling operations.

Flammable liquid hazards are also present in varying degrees in the various distilled-alcohol handling areas.

Because of alcohol's lower heat of combustion, radiant heat energy, and complete miscibility with water, lower sprinkler system demands are required than with other flammable liquids of equivalent flashpoint.

The quantity of water needed to extinguish fires in alcohol/water mixtures depends on the temperature of the liquid above its fire point and the

Table 9-3. Hazardous Liquids and Gases

HAZARDOUS MATERIAL	DESCRIPTION	FIRE AND EXPLOSION HAZARD	LIFE HAZARD	PERSONAL PROTECTION	EMERGENCY PROCEDURES	HANDLING TECHNIQUES
Sulfuric Acid (H_2SO_4)	Colorless (pure) to dark brown, oily dense liquid.	Not flammable but highly reactive. Reacts violently with water and organic materials with evolution of heat. Attacks many metals releasing hydrogen.	Causes severe, deep burns to tissue when contacting liquid. Vapors are extremely irritating to eyes and mucous membranes (nose and throat). AVOID CONTACT!	Wear rubber gloves, rubber apron, and indirectly ventilated, liquid-tight chemical goggles when transferring acid to storage tank or manually adding acid to any vessels.	In case of contact, Immediately flush skin or eyes (affected area) with large quantities of water for at least 15 minutes; FOR EYES, GET MEDICAL ATTENTION!!	Transfer acid by pump or gravity flow. Never use a compressed gas or air to pressurize an acid container. Always add acid to water, never water to acid, a violent reaction will occur causing acid to be ejected from container. Make addition slowly to minimize heating. Use proper carboy truck and tilter if acid is received in carboys. Small glass containers of acid shall be handled in impact-resistant chemical carriers.
Calcium Oxide (CaO)	Colorless crystal; a so known as unslaked, quick, or burnt lime.	Noncombustible but reactive. When wetted, it swells, gets hot, and becomes calcium hydroxide, $Ca(OH)_2$ (slaked lime, caustic lime).	Causes skin burns. Less corrosive than caustic soda (NaOH or sodium hydroxide) or caustic potash (KOH or potassium hydroxide). Dust is highly irritative to the eyes and mucous membranes and prolonged contact with skin can cause dermatitis. Avoid contact.	Wear rubber gloves, chemical goggles, and long-sleeved shirt or jacket.	In case of contact, immediately flush skin or eyes (affected area) with large quantities of water for at least 15 minutes. For eyes, get medical attention.	Transfer crystals from container to rubber pail. Pour required quantity into empty addition vessel. Close vessel cover and add required quantity of water to dissolve crystals.

Ethyl Alcohol (EtOH)	Clear, colorless, fragrant liquid, burning taste. Also, known as ethanol or grain alcohol.	Flammable liquid. Vapors form flammable mixtures with air. BURNS WITH INVISIBLE FLAME.*	Exposure to concentrations above 1000 ppm may cause headache and irritation of the eyes, nose, and throat. If continued for prolonged time, will cause drowsiness and stupor. Contact with liquid can cause defatting of skin. No known cumulative effect as is common with methyl alcohol.	Wear standard safety glasses. When breaking lines for maintenance, wear neoprene gloves and liquid-tight chemical goggles.	In case of body splash, flush with large quantities of water. Dilute liquid spills with large volumes of water. Attack small spill fires with ABC dry chemical extinquishers.	Product will normally be handled in a closed system. When transferring to transport vehicle, proper bonding and grounding procedures will be used.
Carbon Dioxide (CO_2)	A colorless, odorless, tasteless gas.	None. Is a fire extinguishing agent.	Is a simple asphyxiant. Symptoms include dizziness, headache, shortness of breath, muscular weakness, drowsiness, and ringing in ears. OSHA standards require oxygen concentration of 19.5% or greater before entry is made into tanks or vessels. All vessel entry is to be made under the Safe Work Permit system. Skin contact with CO_2 snow will cause frost burns.		Remove victim from oxygen-deficient atmosphere. Rescuers must wear self-contained breathing apparatus if vessel entry is necessary. If breathing has ceased, start mouth-to-mouth resuscitation. Call for medical assistance.	

*Note: Flammability Characteristics of Ethanol — Flash point, 55° F: Flammable limits, 3.3-19%; Ignition temperature, 685° F; Vapor density (air = 1), 0.8 (water soluble); Specific gravity (water = 1), 0.8 (water soluble); Boiling point, 173° F; Odor detectable at 5-10 ppm; Flash point of 5% alcohol/water solution, 144° F; Flash point of 10% alcohol/water solution, 120° F; Flash point of 15% alcohol/water solution, 110° F; Flash point of 20% alcohol/water solution, 97° F.

Source: *A Guide to Commercial Scale Ethanol Production and Financing,* SERI, Denver, 1980.

Table 9-4. Codes and Standards for the Production of Fuel-Grade Ethanol

TITLE	CODE*
Prevention of Dust Explosion in Industrial Plants	NFPA63
Basic Classification of Flammable and Combustible Liquids	NFPA321
Static Electricity	NFPA77
Flammable and Combustible Liquids Code	NFPA30
Occupational Noise Exposure	OSHA191094
Machinery and Machine Guarding	OSHA Subpart 0
Power Piping	ANSI B31.1
Standard for Steel Aboveground Tanks for Flammable and Combustible Liquids	UL142
Boiler and Pressure Vessel Code (B & PV)	ASME Code Section IV & VII Division I
All Electrical Instrumentation	NFPA70-1978
National Electric Code	Class II Division I

*Abbreviations:

NFPA	National Fire Protection Association
OSHA	Occupational Safety and Health Administration
UL	Underwriters Laboratory
ASME	American Society of Mechanical Engineers
ANSI	American National Standards Institute

Source: *A Guide to Commercial Scale Ethanol Production and Financing,* SERI, Denver, 1980.

effectiveness of mixing. The amount of water can be estimated from the following formula, assuming perfect mixing:

$$\frac{\text{\% alcohol in solution before fire}}{\text{\% alcohol at point of fire extinguishment}} - 1$$

Assume that a solution will be extinguished when the alcohol concentration is reduced to 20 percent. Applying the formula, a mixture containing 95 percent alcohol would require 3.75 volumes of water to extinguish each volume of burning liquid. A mixture containing 50 percent alcohol would require 1.5 volumes.

Insurance Loss Experience

A survey of industry losses for the years 1933-1972 indicated that approximately 75 percent of all property damage resulted from fires in 12 alcohol warehouses without sprinkler systems. However, several serious fire and explosion losses occurred in still-buildings. The most serious losses in still-buildings involved explosions with ensuing fires where sprinkler systems were damaged by the explosion. Several fires also occurred in driers which were processing dried grains from spent stillage or slops.

Plant Safety Considerations

GENERAL

Grain handling, milling and feed preparation facilities should be designed, arranged, and safeguarded in accordance with safety standards for grain storage and milling.

CONSTRUCTION AND LOCATION

Mashing and Fermenting

Mashing and fermenting areas should preferably be of fire-resistive or noncombustible construction.

Distilling

Distilling operations should be separated from other buildings by at least 100 feet (30 meters). Existing still-buildings that adjoin other buildings should be completely cut off by blank fire walls, parapeted above adjoining buildings. Basements, pipe trenches, and other spaces beneath still-buildings should be avoided.

Distilling equipment should be located with a minimum of enclosing structure. Structures should be of damage-limiting construction. Load-bearing steel members and exposed steel equipment supports should be fire-proofed with material having a minimum 2-hour fire-resistance rating. For existing buildings of substantial construction, explosion venting capacity should be provided through venting windows and roof panels in as high a ratio as practical.

Floor cutoffs are advisable at operating levels in high, enclosed buildings. If complete floor cutoffs are not practical, solid noncombustible mezzanines with curbs at levels supporting receivers or other equipment containing appreciable quantities of flammable liquids should be provided.

Unless the maximum possible spill can be extinguished by dilution while confined, emergency drainage facilities for the distilling area of buildings should be provided to prevent escaping liquids from exposing other areas of buildings.

Distilled Alcohol Handling

Alcohol-handling areas should preferably be of fire-resistive or non-combustible construction. Distilled-alcohol-handling areas should be cut off from surrounding occupancies. Vertical cutoffs should be provided in multistory buildings. Cutoffs should have at least a 1-hour fire-resistance rating.

Curbs, ramps, or trapped floor drains should be provided at doorways and other openings to prevent the spread of flammable liquids to other departments. Floor drains in each distilled-alcohol-handling area should be designed to handle expected sprinkler discharge unless the maximum possible spill can be extinguished by dilution while confined.

OCCUPANCY

Mashing and Fermenting

Grain meal should be discharged to precookers only through tight connections to prevent liberation of dust.

Distilling

Pressure vessels should be designed and constructed in accordance with applicable codes, standards, state and local laws, and regulations.

Stills should be equipped with vacuum and pressure relief devices piped to the outdoors. Any condenser vents also should be piped to the outdoors. Vents should be sized to discharge the maximum vapor generation possible at zero feed and maximum heating within the pressure limitations of the protected equipment. Vents should terminate at least 20 feet (6.1 meters) above the ground and preferably at least 6 feet (1.8 meters) above roof level and should be located so that vapor will not re-enter the building. Vent terminals should be equipped with flame arresters.

Equipment should be designed and maintained to eliminate or at least minimize any liquid and vapor leaks.

Where gauges are needed, Factory Mutual-approved gauging devices should be used. If ordinary gauge glasses are used, both connections normally should be kept closed and provided with weight-operated, quick-closing valves. The glass should be protected from mechanical injury. Where possible, tail boxes should be replaced with armored rotameters and specific gravity indicators, or with other instrumentation not subject to accidental breakage or leakage.

The steam supply for distillation should be thermostatically controlled and interlocked to shut down and sound an audible alarm on cooling-water failure. Alternately, powered standby pumps or gravity supplies of cooling water should be provided. Stills and other large equipment containing flammable liquids should be purged with steam or any inert gas (steam will be most generally available) before they are open for inspection or repair. Equipment should be washed with water following steaming.

Ventilation, designed and installed to ensure air movement throughout the entire structure, should be provided to prevent accumulation of explosive vapor-air concentrations within the building. The stack effect (i.e., natural ventilation) may suffice if the building is high; permanent openings are provided at grade and roof elevations; the equipment can be drained and cleared of vapors during shutdowns; and heat losses from the equipment maintain a temperature above that of the outdoors during all operating periods. If these operating conditions cannot be satisfied, or if blank walls or solid floors interfere with natural ventilation, mechanical exhaust ventilation should be designed to provide 1 cubic foot per minute per square foot (0.3 cubic meter per minute per square meter) of the floor area. Locate suction intakes near floor level to ensure a sweep of air across the area.

Electrical equipment, including wiring and lights, should be suitable for Class 1, Group D locations. Still-buildings should be considered Division 2 locations.

Distilled Alcohol Handling

Noncombustible, vapor-tight construction should be used for all tanks containing flammable concentrations of alcohol. Tanks should be kept tightly closed except when taking samples, and should be equipped with vents of adequate size terminating outdoors. Vents should be equipped with Factory Mutual-approved flame arresters if the flashpoint of the contents is less than 100°F (38°C).

Factory Mutual-approved liquid-level gauges should be installed on all tanks. If ordinary gauge glasses must be used, weight-operated, normally closed valves should be installed at both tank connections and the glass should be protected against physical damage. Wherever possible, top tank connections should be provided and liquids transferred by pumping through the top rather than by gravity flow. If draw-off stations are located in the same area as the supply tank, automatically operated emergency shutoff valves should be provided in gravity-feed lines. Flexible metallic hose should be used on all connections to scale tanks where fire exposure would release the tank contents or expose its vapor space.

Mechanical exhaust ventilation should be provided as needed, and arranged with suction near floor level to ensure air movement throughout the building. At dump troughs and similar installations, localized intakes are desirable. Careful attention should be given to below-grade installations, windowless buildings, sumps, pipe trenches, and similar installations. Usually, 0.25 cubic foot per minute of air per square foot (0.075 cubic meter per minute per square meter) of floor area will be adequate.

Electrical equipment, including wiring and lights, should be suitable for Class 1, Group D locations. Tank storage areas should be treated as Division 2 locations.

Fire Protection

Automatic sprinkler protection for distilleries, preferably of a type designed to flood the area, should be provided. Sprinkler control valves, dry-pipe valves, and riser drains should be readily accessible at all times to plant personnel. This is particularly important for areas under direct government supervision that may be locked during nonoperating periods.

Small hoses with combination shutoff nozzles should be provided throughout the distillery. Hose stream demand is a minimum of 500 gallons per minute (190 cubic meters per minute) for at least 60 minutes. Also, suitable portable fire extinguishers should be provided throughout the distillery.

The implementation of adequate safety procedures impacts the number and kinds of personnel needed to run an ethanol plant. It is estimated that a typical 50-million-gallon-per-year ethanol plant would require the services of a medical doctor half-time, assisted by four full-time nurses. Such a plant would also need at least one safety engineer to oversee safety procedures. Thus, ensuring adequate plant safety adds to the operating costs through the addition of 5.5 personnel for a 50-million-gallon-per-year plant.

Operating and Safety Training

The efficient operation of a still requires careful attention to many details with which most farmers are not familiar. Sanitation requirements are extremely high (similar to those of a Grade A dairy operation) to prevent the growth of undesirable organisms during fermentation that could, at best, reduce yields and, at worst, result in toxic, unusable coproducts.

Care must be taken to monitor pH and temperature levels and to control the conversion of starch to sugar, the conversion of sugar to alcohol, etc. In addition, safety requirements to prevent fire and explosion are greater than those for most farm operations. Alcohol vapors are as hazardous as those of gasoline.

To assure efficiency and safety, then, operator preparation should include one of the following:

Training by the equipment supplier at the factory or on the job

Training by the supplier of such materials as enzymes and yeast

Training by public agencies such as vocational schools and agricultural extension services.

The training of farmers or others probably will need to be done in each state where a significant number of plants is located. If public financing is used, then some of the first installations might be given additional assistance (grants) in return for their being available as training sites.

If training is to be done by the public sector, an intensive program should be inaugurated soon to train teachers. Training sites will be limited since only one or two relatively complete stills are known to be located in colleges or universities.

The training course should be held for at least 1 week. If a pot still is used, it will require most of this time to cook, ferment, and distill one batch.

In some areas and for installations which use larger boilers and/or steam pressures about 15 psi, it may be necessary to employ state licensed engineers.

The following specific training efforts must be stressed:

Prior to plant acceptance, detailed operating procedures (including safety procedures) should be documented in a technical operation manual. Detailed process diagrams, with color-coded piping illustrations, must be provided.

Operators must be trained to wear hardhats and be aware of plant safety features and location of fire equipment.

Small-Scale Plant Safety

As noted, alcohol is volatile, flammable, and potentially explosive in certain mixtures with air. Thus, great care must be exercised when distilling and storing ethanol. As any spark could ignite alcohol vapors, explosion-proof wiring, switches, and motors are mandatory in the area of the still. Open flames cannot be allowed in the area; and adequate ventilation, exit routes, and fire extinguisher locations must be planned. The potential for explosions and fire from alcohol vapor must be recognized and preventative measures incorporated in both plant design and operation procedures.

For example, the design of the still, especially the pot still, must be such that the likelihood of stoppage within the system is eliminated or, at least, greatly minimized. Plant design must eliminate obvious spark sources and other obvious hazards, and a well-trained knowledgeable operator is imperative to avert problems should any unusual situation arise.

Another potential source of an explosion or fire hazard is the boiler used for steam generation. Again, if the boilers are properly maintained according to the manufacturer's instructions, there should be little cause for concern; however, the use of biomass for boiler fuel may increase the likelihood of problems. Regulation of the biomass boilers may be somewhat more complicated than those using more easily metered liquid fuels. Proper maintenance and careful operating procedures can essentially eliminate the hazards associated with (commercially available) boilers.

There are a number of codes that should be adhered to in the building of the small distilling plants; most notably, the flammable liquids' code, the lighting protection code, the electrical code, and the life safety code should be considered. The codes are those of the National Fire Protection Association, the various states, or a combination of both. In many instances, the state adopts the NFPA codes. Specific area code information is available through a state's fire marshall's office or from reputable design engineers.

Chapter 10

Utilization of Gasohol

This chapter is composed of three sections. The first part describes the suitability of ethanol for fuel and the second part provides a profile of liquid fuel use. The final section describes in 10 steps the modification of a standard carburetor to utilize 100 percent ethanol.

Suitability of Ethanol for Fuel

Alcohols have been evaluated as fuels for internal combustion engines since the early 1900s and have proven to be satisfactory although subject to constraints.

BASIC FUEL PROPERTIES

The use of ethanol in agricultural engines is subject to a number of constraints that are a result of its basic properties. Thus, a comparison of ethanol to the more conventional sources of agricultural energy is in order. Table 10-1 shows a few of the properties of gasoline, octane (which is often used for research purposes), propane (representing agricultural LPG), ethanol, and diesel fuel (No. 1 and No. 2).

Both higher and lower heating values are shown in the table. Lower heating values assume that the water from combustion is exhausted in the gaseous state, a condition normally appropriate for engine usage. This is why the British thermal unit (Btu) value of anhydrous ethanol is established at 76,152 in this handbook. However, since higher heating values are often quoted in the popular press, these are also tabulated for comparative purposes.

Conventional hydrocarbon fuels contain 18,000-20,000 Btu per pound, with the differences in "per gallon" figures resulting primarily from specific gravity variations. Ethanol contains less energy per unit of weight than the conventional fuels. Compared to gasoline, ethanol contains about 61 percent as much energy per unit weight, or about 65 percent as much energy per unit volume. However, when compared to propane, ethanol contains 93 percent as much energy per unit volume.

Another key difference between ethanol and conventional fuels is the heat of vaporization. The conventional fuels require from 100 to 150 Btu in

Table 10-1. Fuel Properties

	Gasoline	Octane	Propane	Ethanol	No. 1 diesel	No. 2 diesel
Chemical formula	–	C_8H_{18}	C_3H_8	C_2H_5OH	–	–
Molecular weight	~126	114	44	46	~170	~184
Carbon percent by weight	–	84	82	52	–	–
Hydrogen percent by weight	–	16	18	13	–	–
Oxygen percent by weight	–	–	–	35	–	–
Heating value						
Higher, Btu/lb	20,260	20,590	21,646	12,800	19,240	19,110
Lower, Btu/lb	18,900	19,100	19,916	11,500	18,250	18,000
Btu/gal (lower)	116,485	111,824	81,855	76,152	133,332	138,110
Latent heat of vaporization, Btu/lb	142	141	147	361	115	105
Specific gravity	0.739	0.702	0.493	0.794	0.876	0.920
Research octane	85–94	100	112	106	10–30	–
Motor octane	77–86	100	97	89		
Cetane number	10–20	–	–	-20–8	~45	–
Stoichiometric mass air/fuel ratio	14.7	15.1	–	9.0	–	–
Distillation temperature (°F)	90–410	–	–	173	340–560	–
Flammability limits (volume percent)	1.4-7.6	–	–	4.3--19	–	–

Sources: J.L. Keller, *et al.*, *Use of Alcohol in Motor Gasoline—A Review,* American Petroleum Institute, 1971; L.C. Lichty, *Combustion Engines Processes,* McGraw-Hill, New York, 1967; Taylor and Taylor, *The Internal Combustion Engineer,* International Textbook Company, 2nd ed., 1966; J.L. Keller, *Methanol and Ethanol Fuels for Modern Cars,* Union Oil Company, presentation for World Federation of Engineering Organizations, November, 1979; J.H. Freeman, *et al.*, "Alcohol in Tractors and Farm Engines." *Agricultural Engineering* (February, 1941); W. Bandel, "Problems in the Application of Ethanol As a Fuel for Utility Vehicles," International Symposium on Alcohol Fuel Technology, Methanol and Ethanol, Novermber 21–23, 1977, Wolfsburg, Germany; *Ethanol Production and Utilization for Fuel,* Cooperative Extension Service, University of Nebraska, 1979.

order to vaporize each pound of fuel, with gasoline reported at 142 Btu per pound. By contrast, ethanol requires 361 Btu per pound, or about 2.5 times as much energy per unit weight. When combined with the heat of combustion, it is apparent that ethanol requires nearly 4.2 times as much vaporization energy per unit of heat input as does gasoline.

Ethanol is a single compound fuel; gasoline and diesel fuel contain a variety of compounds. As a result, ethanol boils at a specific temperature, rather than over a range of temperatures. The vapor pressure and flammability limits of ethanol, combined with its high heat of vaporization, can cause starting problems in engines operated on straight ethanol.

FARM FUEL USAGE AND ENGINE CHARACTERISTICS

The three basic fuels now in use in engines are gasoline, diesel fuel, and liquefied petroleum gas (LPG). The 1970 data (USDA-ERS, 1974) placed the average engine usage on U.S. farms (including transportation) of the three fuels at 18.6, 6.0, and 1.8 gallons per acre, respectively. However, when only farms over 500 acres were considered, the usage was estimated at 10.9, 7.9, and 2.3 gallons per acre, respectively.

The United States Department of Agriculture (USDA) conducted a detailed study of the energy used in farm production in 1974. Crop and livestock production operations used an average of 10.9 gallons of gasoline, 7.7 gallons of diesel fuel, and 4.3 gallons of LPG per acre of cropland harvested. When only crop activities were considered, the per acre inputs were 8.5 gallons of gasoline, 6.7 gallons of diesel fuel, and 3.4 gallons of LPG. That study did not include fuel for farm household use.

More recent information indicates that roughly 90 percent of the volume of fuel used for performing farm-field operations in Kansas consists of diesel fuel with gasoline and LPG filling the remainder.

Historically, there have been several shifts in tractor fuel types. Prior to World War II, about 35 percent of the models of tractors tested at the Nebraska Tractor Test Laboratory operated on gasoline and the remainder on kerosene and distillate (all were spark ignition). Immediately after the war, gasoline model tractors accounted for nearly 70 percent of the tests, but diesel and LPG models began to increase. By 1956 (the peak year for LPG), 50 percent of the units tested were gasoline, 34 percent diesel, and 26 percent LPG. In 1975, 8 percent of the models tested were gasoline, and 92 percent were diesel.

Retail sales document a rapid shift to diesel farm machinery (Table 10-2). In 1979, the USDA survey of farm production expenditures indicated that farmers purchased as much diesel fuel as gasoline for production purposes. The reasons for this shift included the efficiency and durability of the relatively low maintenance requirements of diesels and the comparatively low cost of diesel fuel over the period. In 1975, gasoline tractors tested at Nebraska averaged 23 percent efficiency (based on power take off horsepower), while

Table 10-2. Farm Machinery Stocks and Diesel As Percent of Sales of Tractors and Self-Propelled Combines

Farm machinery stocks[1]			Diesel as percent of sales[2]	
Year	Tractors (000)	Self-propelled combines (000)	Tractors	Self-propelled combines
1970	4,619	790	72	14
1973	4,518	701	86	50
1978	4,493	538	95	95

[1]U.S. Department of Agriculture, *Changes in Farm Production and Efficiency, 1978,* Economics, Statistics and Cooperative Service, Statistical Bulletin No. 628, January, 1980.

[2]U.S. Department of Agriculture, Economic Research Service, The U.S. Food Fiber Sector, *Energy Use and Outlook,* Senate Committee on Agriculture and Forestry, 1974, updated with sales data from the Farm and Industrial Equipment Institute.

diesels averaged 27.5 percent. Additionally, the total number of tractors on farms has declined gradually, since 1968, as agriculture uses fewer but larger machines. The number of tractors on farms as of January 1, 1979, was 4,350,000, the lowest number since 1955. However, total tractor power on farms increased to an all time high of 243 million horsepower.

In 1978, the most popular tractor size was 130-140 horsepower and no domestic major manufacturer has recently introduced diesel tractors in sizes less than 100 horsepower. Tractor engine rated speeds range from 1900 to 2800 rpm with 2000 to 2400 as the most common. Four hundred to 500 cubic inch displacements are common in the 100-175 horsepower range with up to 1500 cubic inch engines available in large articulated four-wheel drives. Engines above 110 horsepower are generally turbocharged, and aftercoolers are used on the more heavily boosted diesels in order to reduce intake temperatures.

ETHANOL IN SPARK IGNITION ENGINES

While spark ignition (SI) engines have largely been replaced by diesels for use in the field, the majority of highway vehicles are still gasoline powered. There are two fundamental ways in which ethanol can be used in SI engines: as a fuel mixture with gasoline, such as gasohol, and as ethanol alone. Since the engine performance, engine modifications, and usage problems are different for the two approaches, they will be discussed separately.

Gasoline-Ethanol Mixtures

The use of gasoline-ethanol mixtures is certainly not new. Investigations were reported in the 1910s and again in the 1930s and 1940s. Thus, there is a considerable history of research to be drawn upon in assessing the feasibility and properties of such mixtures.

Phase separation. One of the key, well-documented problems of gasoline-ethanol mixtures involves the phase separation of alcohol from gasoline in the presence of water. Such a phase separation results in the alcohol-water mix settling to the bottom of a vehicle's gasoline tank (or carburetor) and an extremely lean mixture (on a heat basis) being fed to the engine. This engine does not run on the gasoline-alcohol mixture but on alcohol alone. In all probability the engine will malfunction. Some experimenters have reported no separation problems encountered when mixing 160-190-proof ethanol with gasoline; but, in general, these trials involved little or no storage time between mixing and consumption and involved mixtures with relatively high levels of agitation and ambient air temperatures.

The water tolerance of an ethanol-gasoline blend depends on several factors including the ethanol's concentration and temperature, the presence of cosolvents, and the make up of the gasoline. In one test, water tolerance of a 10 percent ethanol blend was increased from 0.36 to 0.6 percent by heating it from -20 to 50°F (-29 to 10°C). Likewise increasing the aromatic content of the gasoline from 14 to 38 percent moved the tolerance from 0.52 to 0.65 percent at 50°F. The effects of cosolvents such as iso-butanol, *n*-butanol, amyl alcohol, and 2-ethyl hexyl alcohol have been investigated. The water tolerance of a 10 percent ethanol mix was improved from 0.6 to 0.98 percent at 50°F by the addition of 3.2 percent *n*-butanol.

The phase separation problem is important in considering on-farm options since most current on-farm still designs produce less than 200 proof ethanol. Even if the azeotrope of 95.6 percent alcohol is attained, a 10 percent mixture of ethanol in completely dry gasoline would still contain 0.44 percent water in the mixture. At this concentration phase, separation would be anticipated at lower temperatures.

Until satisfactory additives are developed to alleviate phase separation problems, mixing less than 200 proof ethanol with gasoline is impractical. Such additives, however, are under development and may be available on the market soon.

Heat content. Since ethanol contains only two-thirds the energy of gasoline, a mixture of ethanol and gasoline contains proportionately less energy than the original gasoline. In most cases, mixture precentages are given as a volume percent, so the calculated lower heating values are given in Table 10-3. The heat value of the mix is given in Btus per gallon and Btus per pound, as determined by brake specific fuel consumption (pounds per brake horsepower-hour). A small volume increase in mixing is neglected.

Octane. Owing to the relatively high octane number of ethanol, mixtures have a higher octane than straight gasoline. In fact, the improvement in octane number is generally greater than what would be calculated using the values in Table 10-1. The blending octane number of ethanol has been reported in the range of 128-136 for research octane and 95-112 for motor octane, when the ethanol is used in 10 percent blends. Older research,

Table 10-3. Calculated Heat of Combustion For Ethanol-Gasoline Mixtures

Volume percent ethanol	Btu/gal	Relative Btu/gal	Specific gravity	Btu/lb
0	116,485	100	0.739	18,900
10	112,452	96.5	0.745	18,099
20	108,418	93.1	0.750	17,333
30	104,385	89.6	0.756	16,556
100	76,152	65.4	0.794	11,500

Source: USDA, *Small-Scale Fuel Alcohol Production,* U.S. Government Printing Office, Washington, D.C., 1980.

dealing with low octane gasoline, placed the blending motor octane number of ethanol at 123-170. The end result of adding 10 percent ethanol to modern gasoline is typically a three-number increase in research octane and a two-number increase in motor octane.

Volatility. Ethanol is a single compound fuel which boils at 173°F (78°C); gasoline compounds boil at approximately 80-437°F (27-225°C). The addition of the ethanol to typical gasoline causes a marked dip in the distillation curve in the lower-temperature regions (Fig. 10-1). Similarly, Reid vapor pressure of gasoline is increased approximately 1 psi by the addition of 10 percent ethanol. Such changes can be expected to aggravate problems with carburetor evaporation losses, hot driveability, and vapor lock.

Driveability problems of alcohol blends have been assessed. As expected, the carburetor mixture of fuel and air had a major influence on total weighted driveability demerits and, since the alcohol mixtures had a leaning effect on the engine, they changed driveability accordingly. The conclusion of the tests was that driveability was a function of stoichiometry (the fuel-air ratio), and no additional problems associated with the volatility were reported. Volatility problems, like others associated with fuel alcohol, can be solved by appropriate fuel and ignition system modifications, and it is likely that future automotive engines will be modified.

Efficiency. Fuel economy is perhaps the single most controversial aspect of the combustion of ethanol-gasoline mixtures. Reports are varied, with many articles in the popular press citing improved gas mileage (on a volumetric basis) in vehicles. Although there is some basis for these claims in the results of one relatively uncontrolled study, reports of small mileage reductions have also been noted. In general, the mileage changes are small, either way.

The addition of ethanol affects fuel-air mixture and, therefore, motor efficiency. As mentioned previously, a 10 percent mix of ethanol will reduce the volumetric heat content by about 3.5 percent and result in leaning the mixture by a like amount. Since the main carburetor jets in nearly all vehicle carburetors are fixed and only the idle jets are adjustable, in normal vehicle

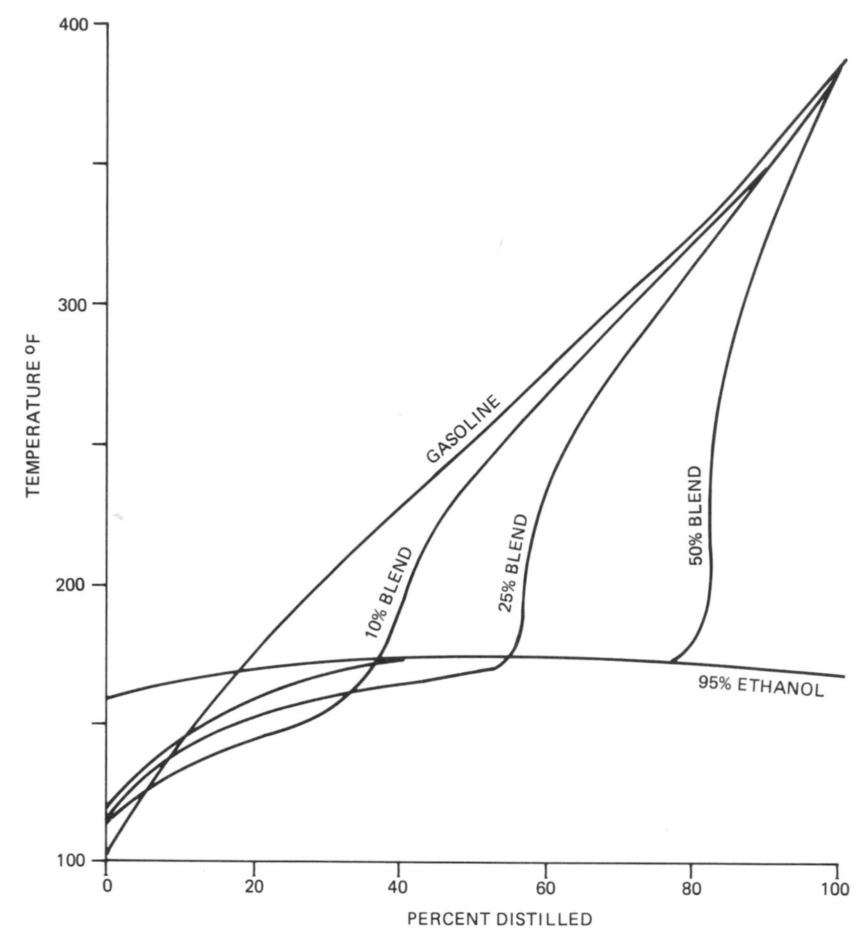

Source: A. R. Rogowski and C. F. Taylor, J. Aeron. Sci. 8:384 (1941) as reprinted in Lichty, L.C. *Combustion Engines Process*, McGraw-Hill, 1967.

Figure 10-1. ASTM distillation curves for gasoline, 95 percent ethanol, and various blends.

usage, the owner of the vehicle will not enrich the carburetor when using ethanol-gasoline blends. However, older SI tractors normally are equipped with adjustable main jets and owners could manually enrich the mixture to allow the use of higher proportions of alcohol. In addition, some 1980 cars use a closed-loop feedback carburetor system. This system uses an oxygen sensor in the exhaust manifold which provides a signal to a computer-controlled electromechanical carburetor to control the mixture. Within limits, this system automatically compensates for variations in fuel and reduces the driveability problems caused by the leanness of ethanol mixtures. Such carburetion systems are almost certain to become more popular in the future.

The effects of mixture on efficiency of an SI engine are well documented. Although the exact shapes and vertical position show some variation, the trends are usually similar to those shown in Fig. 10-2. That is, a minimum specific fuel consumption is usually evident at a fuel-air equivalence ratio of about 0.90-0.95 (or slightly lean). Enriching the mixture further leaves unburned fuel, while leaning below the minimum brake specific fuel consumption (BSFC) point reduces flame speed and ultimately leads to lean misfire.

Tests on a 1973 model, 7.5-liter V-8 vehicle produced the results shown in Fig. 10-3, expressed in terms of distance per unit energy, in volumetric fuel economy. As expected, the curves resemble the inverse of Fig. 10-2, with maximum mileage occurring at an equivalence ratio of about 0.95.

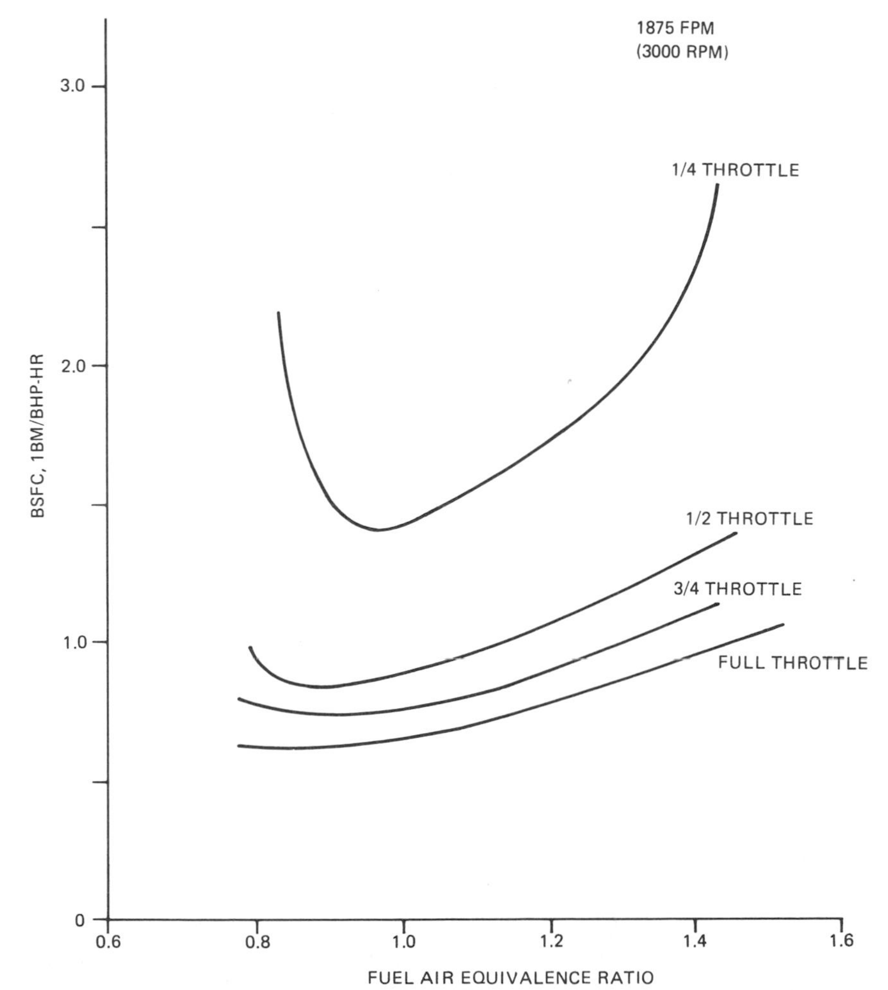

Source: Taylor and Taylor, *The Internal Combustion Engine*, International Textbook Company, 2nd Ed. 1966.

Figure 10-2. Effect of fuel-air equivalence ratio on brake specific fuel consumption at various throttle settings.

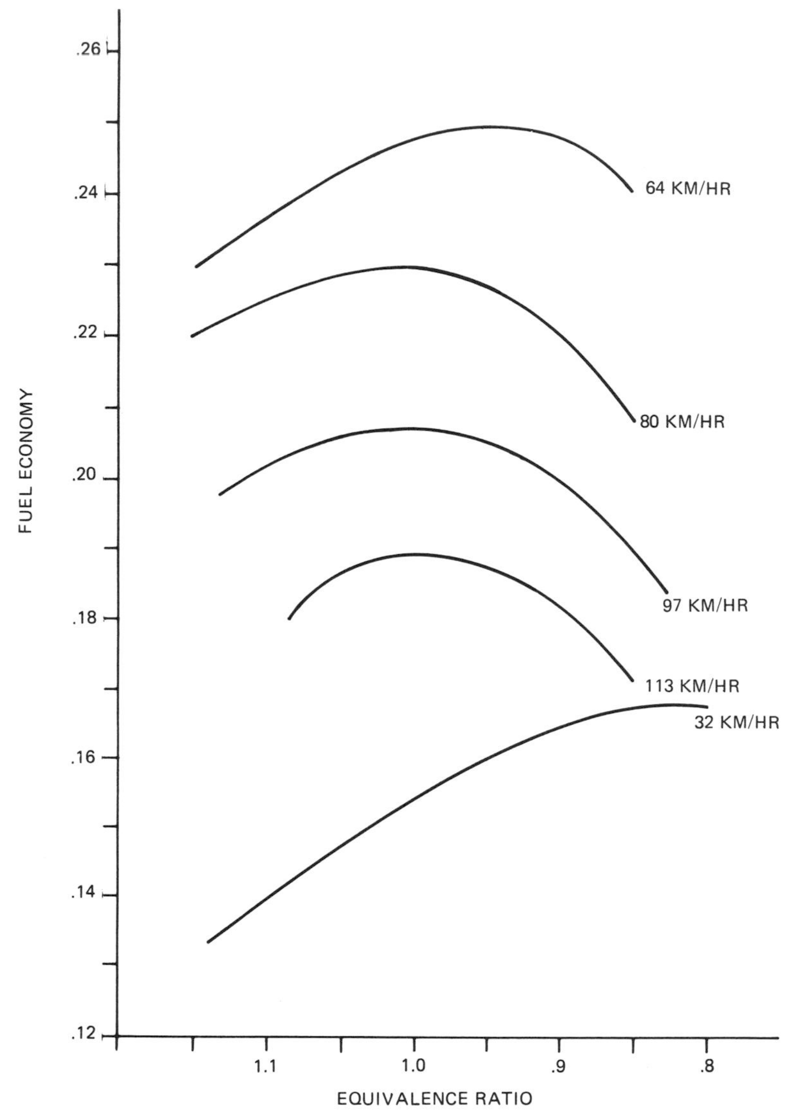

Source: Brinkman, N. D., N. E., Gallopoulos, and M. W. Jackson, Exhaust Emissions, Fuel Economy and Driveability of Volatiles Fueled with Alcohol-Gasoline Blends, SAE Paper 750120, February, 1975.

Figure 10-3. Effects of alcohol and carburetion on level road load fuel economy (energy basis).

The effect of leaning the mixture by 3.5 percent (corresponding to a 10 percent ethanol mixture) depends on the initial mixture supplied by the carburetor. If the carburetion is relatively rich (i.e., equivalence ratio greater than 0.95), then the leaning increases the thermal efficiency of the engine. If the mixture is initially lean (equivalence ratio less than 0.95), further leaning will reduce the thermal efficiency. This helps explain why some older cars that were carbureted rich may have shown an increase in fuel economy when operated on gasoline-ethanol mixtures.

In an effort to determine fuel economy on ethanol-gasoline blends, one study operated a vehicle at constant speeds as well as under highway, suburban, 1975 federal test procedure, and business district cycles. Both standard and modified carburetion were used. Standard carburetion used an average air-fuel ratio of 15.7 and gave an equivalence ratio of 0.91 on the test gasoline and 0.88 on the ethanol-gasoline mixture. The results of the standard carburetion tests showed a reduction in thermal efficiency for the ethanol blend at speeds above 36 miles per hour, with roughly equal efficiency below that speed.

The four test cycles were conducted at varying carburetor mixtures, and the results from ethanol blends generally fell on the same curves as the straight gasoline tests. Thus, the overall conclusion of the study was that ethanol blends are equal to gasoline when compared on an energy basis and represent roughly a 3 percent reduction in volumetric mileage.

Two Nebraska gasohol test vehicles were tested at the Energy Research and Development Administration (ERDA) Bartlesville Energy Research Center. The two-car average showed a reduction in volumetric fuel economy when operated on a 10 percent ethanol blend. The reductions were 1 percent for the urban test and 3.1 percent for the highway test, figures in general agreement with those performed by Brinkman *et al.*

A more recent study involved a paired test of gasohol- and gasoline-fueled cars at highway speeds. The cars were equipped for rapid fuel changes, and the remaining fuel in the car was weighed following each run. The results of the study were statistically significant at the 99 percent level, showing about 3 percent reduction in volumetric fuel mileage on gasohol.

Tests were made on a 1979 Toyota Supra equipped with a three-way catalyst and an oxygen-sensing, closed-loop feedback system. The mixture compensating characteristics of this vehicle allowed operation on up to 50 percent ethanol while still meeting federal exhaust and evaporative emission standards. Volumetric fuel economy declined on the blends, while energy-based fuel economy was unchanged.

These various fuel economy tests indicate that the mixture leaning caused by the addition of ethanol to gasoline will produce the greatest loss in volumetric gas mileage at highway speeds. From a judgmental standpoint, volumetric gasoline mileage of an average vehicle will probably be reduced about 2.5 percent by a 10 percent ethanol blend. The octane enhancement from the alcohol is an advantage, if the blend is used in engines that need the added antiknock properties.

Engine modifications. Although alcohol blends improve the octane number of gasolines, it is unlikely that farm vehicles will be modified (by raising compression ratio) to take advantage of the improved octane, since this would discourage the operation of the vehicle on the lower octane fuel available off the farm.

Presently, the operation of blends over approximately 25 percent ethanol may cause driveability and economy problems, particularly if the vehicle was originally carbureted lean. Carburetors can be re-jetted, but this (like modified compression ratio) would also compromise operation on straight gasoline, a condition solved only by the use of externally adjustable main jets or by the installation of a dual carburetor intake manifold.

The mixture-associated driveability problems of alcohol-gasoline blends will become less of a factor with the anticipated widespread introduction of feedback carburetion on 1981 automobiles. By the same token, fuel economy of blends will probably degenerate to coincide with their energy content even at lower highway speeds, since the feedback system would compensate for the slight leaning of the mixture.

Unblended Ethanol in Spark Ignition Engines

Considerable research and field experience information have been accumulated on the use of unblended ethanol in spark engines. The modifications associated with converting engines to ethanol are related to three general properties of ethanol.

Energy content. As discussed previously (Table 10-2), the volumetric heat content of ethanol is about two-thirds that of gasoline. Thus, the direct total substitution of ethanol for gasoline will produce such drastic mixture leaning that many engines will not run at all or will run very poorly. However, if exclusive alcohol operation is anticipated, the drilling of fixed jets can alleviate these difficulties.

Mingle suggests that the jet diameter should be increased by 27 percent when changing from gasoline to 190 proof ethanol; for lower proof, Brown suggests a 40 percent increase in jet size. However, another study involving automobiles suggests that more thorough modifications (including the enlargement of carburetor passageways) would be desirable in order to achieve proper mixtures over the required wide range of speed and load. If both gasoline and pure ethanol operations are desired, some form of adjustable jet or dual carburetion could be desired.

The amount of enrichment needed for ethanol may not be as much as would be expected, however, because of its ability to burn leaner (on a stoichiometric basis) than does gasoline. For example, one report placed the lean misfire limit of a test engine at 22 percent lean (from stoichiometric) for gasoline, but 36 percent lean for methanol. The report generalized a similar though somewhat leaner burn capability for ethanol.

Latent heat of vaporization. The high latent heat for ethanol is both an advantage and a disadvantage. On the positive side, the ethanol serves as a coolant and reduces intake temperature and improves volumetric efficiency. Furthermore, the increased likelihood of fuel being vaporized on the compression stroke tends to reduce compression work through lower temperatures and pressures. Both of these factors increase the performance of an engine, even when the compression ratio is fixed at gasoline levels. One study placed the power output for 95 percent ethanol (5 percent water) at 3 percent higher than for gasoline at the same compression ratio.

The high heat of vaporization can cause mixture distribution problems in multicylinder carbureted engines. Most engines show cylinder-to-cylinder variations in mixture even with gasoline, and a later vaporizing fuel like ethanol maintains more liquid in the intake manifold and worsens the situation. Potential cures for this problem are additional intake manifold heat from either water or exhaust, fuel injection to individual intake ports, or more careful manifold flow design to intensify higher mixture velocity and turbulence.

Perhaps the most noticeable effect of ethanol's low volatility is cold starting. Saturated vapors are too lean to ignite below about 50°F (10°C) and starting problems can become significant below about 40°F (4°C). Several cures are being developed. Conventional coolant heaters also have been used. Another approach uses the addition of 5-10 percent gasoline to the alcohol. Ether and acetone have also been tried, with 21 percent ether producing starts down to 0°F (-18°C). At any rate, some type of starting aid will be desirable for operation below a 40°F ambient temperature. Such aids are in common use with agricultural deisels, and the same general approach (commonly ether injection) may be a viable solution for alcohol-fueled SI engines.

Octane. The high octane rating of ethanol presents an opportunity to improve the efficiency of most SI engines. If an engine is intended solely for ethanol operation, it can be modified internally to raise the compression ratio well above that for gasoline. One review suggested that, for Brazilian gasohol, compression ratios of only about 7:1 are practical; while with ethanol, it is possible to raise the ratio to about 12:1. However, the relatively low speeds and high loads imposed by agricultural equipment use may dictate a practical limit somewhat below 12:1. For example, a common practice with engines using propane (which has an octane value above that of ethanol) was to use 8:1 to 10:1 compression ratios in farm tractor and irrigation engines.

Efficiency. Increasing an engine's compression ratio also increases its thermal efficiency. This is particularly true at lower compression ratios, such as those used in older farm tractors. Figure 10-4 shows the relative efficiency of ethanol, gasoline, and diesel fuel as a function of compression ratio. The graph arbitrarily defines an 18:1 compression ratio (CR) as 100 percent efficiency and compares the relative efficiencies of other ratios to it. For example, an 8:1 compression ratio engine would extract about 84 percent as much work from each unit of heat input as would the diesel. This is in close agreement with the results of the Nebraska tractor tests. By operating an alcohol SI engine at a 12:1 compression ratio, efficiency should be about 95 percent as high as diesel values.

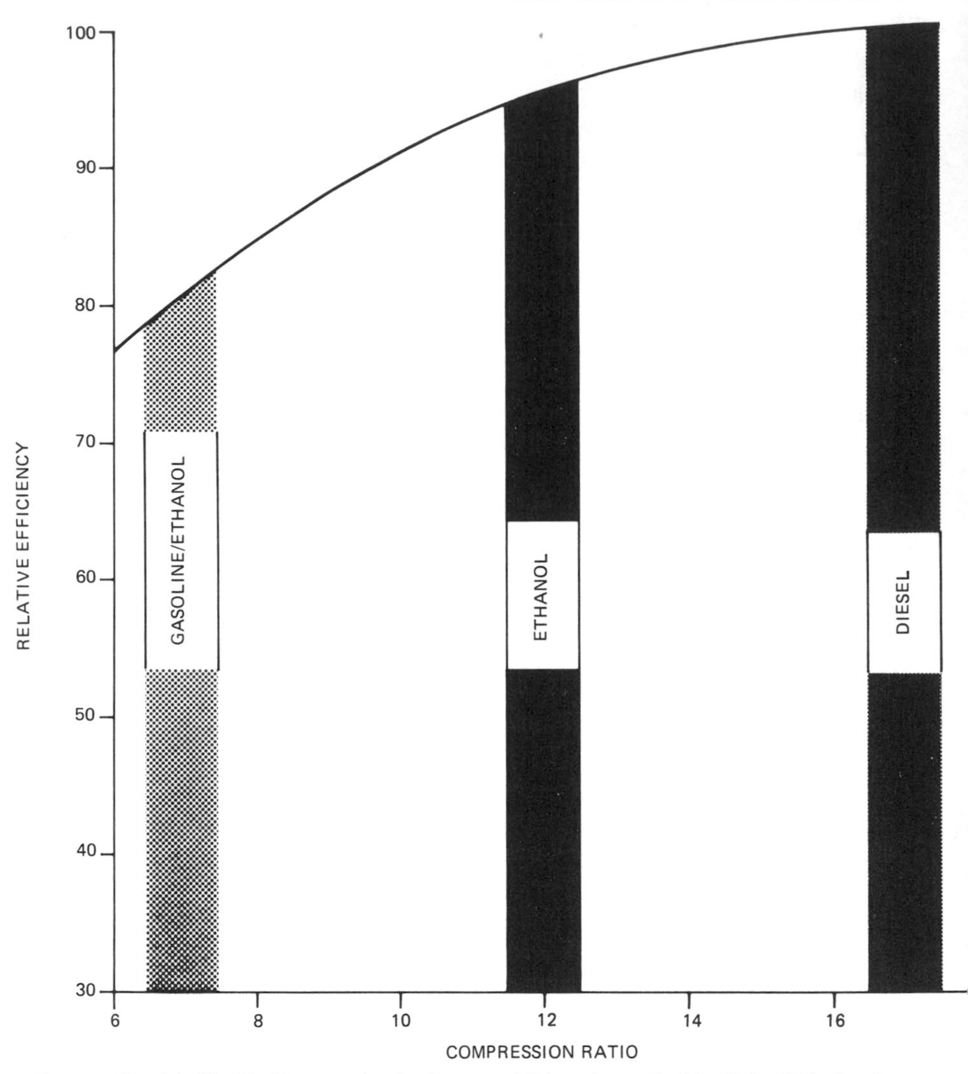

Source: Bandel, W., Problems in the Application of Ethanol as a Fuel for Utility Vehicles, International Symposium on Alcohol Fuel Technology: Methanol and Ethanol, November 21-23, 1977, Wolfsburg, Germany.

Figure 10-4. Relative degree of effective force in relation to compression ratio.

Other reports have suggested that additional factors improve the energy efficiency of ethanol combustion above what would be expected from compression ratio improvements alone. In one test, energy usage with ethanol was roughly 12 percent lower than with gasoline; power output was about 3 percent higher. Although the compression ratio was constant in these tests, the fuel-air ratio and ignition timing which were not reported may have been responsible for part of the variation.

There is not universal agreement in the test results comparing gasoline to ethanol at a constant compression ratio. The positive effects from ethanol can be summarized as:

Intake cooling leading to higher volumetric efficiency

Reduction in compression work due to internal cooling action

Slight improvement in mole ratio of combustion (1.047 for gasoline vs 1.065 for ethanol)

Ability to burn at leaner (relative to stoichiometric) mixtures than gasoline

The primary negative effect on engine efficiency of ethanol would be its poorer mixture distribution. In effect, then, comparative data imply that the positive factors outweigh the negative and that the *energy* (Btu) efficiency of ethanol operations at a constant compression ratio will average about 3 percent better than gasoline operations.

Proof. Most of the engine research that has been reported deals with pure or 95 percent ethanol operations. Little information on lower (i.e., 120-180) proof research results are available.

One study in 1945 operated a 1942 Plymouth engine on ethanol from 200 proof down to 70 proof. The tests were conducted at full throttle, and both ignition timing and mixture were adjusted to achieve best power. No changes were made to the intake manifold heat supply, but spark plug wetting required installation of hotter plugs. Smooth operation was obtained down to 80 proof, but 70 proof resulted in inconsistent operation. During the 70 proof tests, 1 hour of operation raised the crankcase oil level by 6-8 quarts from its dilution by water and alcohol. The results are summarized in Table 10-4.

Table 10-4. Engine Performance on Low-Proof Alcohol

Proof	Brake horsepower	Gal/hr blend	Ethanol (lb/bhp-hr)	Thermal efficiency[a]	Optimum spark advance
200	47.67	6.85	0.944	21.2	17.5
190	46.18	7.53	1.029	19.4	20.4
180	45.67	7.94	1.042	19.2	21.3
160	45.07	9.52	1.127	17.8	27.0
140	45.48	11.33	1.162	17.2	29.0
120	43.60	12.90	1.207	16.6	33.0
110	42.00	14.53	1.278	15.6	35.0
100	42.94	19.00	1.490	13.4	36.0
90	42.35	17.90	1.277	15.6	41.5
80	41.40	20.87	1.341	14.9	47.0
70	34.10	26.70	1.853	10.8	50.0

[a]Higher heating value used by this study (Duck, 1945) to calculate thermal efficiency.
Source: J.T. Duck, and C.S. Bruce, "Utilization of Non-Petroleum Fuels in Automotive Engines," *Natl Bur Stand J Res,* **35:** 439 (1945).

The steady reduction in thermal efficiency for lower proofs stems, in part, from the latent heat of the water and the poorer mixture distribution.

A more basic problem of low-proof ethanol is its low energy density as compared with that of conventional hydrocarbons. For example, a fuel tank that would carry a vehicle 200 miles on gasoline would last about 135 miles with 200 proof ethanol, 95 miles with 160 proof ethanol, and 60 miles on 120 proof. For most farm machines (i.e., tractors and combines), the added weight and space occupied by an auxiliary tank are not crucial; on conventional vehicles, it may be inconvenient and will reduce load capacity and lead to further reductions in fuel economy.

Materials Compatibility

Fuel systems in automobiles and farm machinery contain a number of different metal alloys and plastics, and several reports have mentioned difficulties from corrosion and chemical attack.

First, the solvent action of ethanol blends tends to loosen existing gum formations (and the rust they might hold) and to deposit them in the fuel filters. This is most likely to occur just after the vehicle or storage tank is converted to ethanol storage and would tend to be a relatively short-term problem. Cleaning or replacement of the fuel filter after a tank full of gasohol would be helpful in reducing this problem.

Ethanol blends have led to the cracking of polyamide filter housings and to the degradation of polyurethane and polyester bonded fiberglass. When enough water is present to cause phase separation of ethanol blends, severe corrosion of steel and terne alloys (fuel tank material) has resulted. At least one small engine manufacturer has warned that gasohol may cause corrosion in its engine. However, most major automobile manufacturers extend their new car warranty to include the use of gasohol.

Such incompatibility is apparently unpredictable, and some studies report no problems. Brazilian experience suggests that some new vehicle models (from countries not using ethanol mixtures) realize problems but that corrective action can be taken to "design in" ethanol compatibility. Thus, it is reasonable to expect a certain amount of corrosion or chemical attack to occur in current U.S. produced vehicles and farm machinery if they are converted to ethanol or ethanol blends.

Wear. As with several other conditions previously mentioned, no clear consensus concerning the effects of ethanol or ethanol blends on engine wear is apparent. Nebraska work done by Scheller reported no unusual wear from a 10 percent ethanol mixture in vehicle operation, and work with alcohol-water injection reported fewer deposits and less wear with the injection than without it.

In contrast, a more recent battery of tests using a Coordinating Lubricants Research engine in ASTM sequence II-C and V-C tests showed some negative results. The two test sequences approximated short-trip winter service and low-speed, low-temperature stop-and-go driving, respectively. Both denatured ethanol (BATF Formula 28A) and a 15 percent ethanol-gasoline mix were used. Ethanol showed a 180 percent increase in iron wear compared to that for gasoline use and increased oil viscosity and acid content. The 15 percent mixture tests were inconclusive, and no lubricant effects were seen at the 240 hour point of the test.

Thus, it appears that, at certain low-temperature conditions, excessive cylinder wear can occur when straight ethanol is used. Work is continuing to better define the causes and extent of the problem, but it does not appear serious at higher engine temperatures.

Valve recession. Following the switch to unleaded fuels in the late 1960s and early 1970s, many automobiles experienced a problem with rapid valve seat wear. The removal of tetraethyl lead reduced the level of lubrication between the valve face and seat and the resulting softer seat eroded and allowed the valve to work deep into the head.

The problem was solved in the early 1970s by most manufacters, mainly by induction hardening of the integral valve seats. However, vehicles produced before 1970 are still subject to the problem when operated for extended periods on unleaded fuels.

Since ethanol contains no tetraethyl lead, its use in older cars may result in valve recession problems. The problem is serious when it occurs, often requiring head replacement or the installation of seat inserts.

Since SI tractors were usually equipped with seat inserts, even the older models are unlikely to experience this problem.

ETHANOL IN DIESEL ENGINES

Since diesel engines are the prime movers of agricultural field operations, it is important that any prospective farm-produced fuel be adaptable to operation in diesels. Four basic approaches to using ethanol in a diesel have been mentioned:

Converting the diesel to a high compression SI engine

Modifying the diesel to tolerate straight ethanol injection

Mixing the ethanol with diesel fuel

Carbureting the ethanol separately

Diesel engines operate on an entirely different combustion system than SI engines. Diesel combustion is commonly divided into the following three phases.

Ignition delay is the period between the start of injection and actual combustion. During the first part of this phase (or physical delay), fuel is atomized, vaporized, and mixed with air. Next, a preflame oxidation reaction occurs (called chemical delay) and is followed by the localized ignition of the fuel. Since the total ignition delay time is less than the injection time, the first increments of fuel are made during ignition delay and additional fuel accumulates in the chamber.

The second phase is rapid pressure rise. Once ignition begins, the temperature and pressure in the chamber rises, a condition which shortens the ignition delay of the accumulated fuel. This causes a very rapid pressure rise and produces high stresses, the source of the well-known diesel "knock." Higher cetane fuels generally reduce the ignition delay and, thereby, control the rate of pressure rise within the combustion chamber. (Cetane rating measures the ease of self-ignition—a desirable property for diesel fuels.)

Controlled pressure rise, the third phase, occurs after the initial "pooled" fuel has burned, and combustion then continues to develop roughly at the rate of fuel injection.

For fuels in general, the autoignition properties that are desirable in a SI engine are highly undesirable in a diesel. The octane rating system measures a fuel's resistance to self-ignition, while cetane rating measures the ease of self-ignition. Fuels that have a high octane number (ON) have a low cetane number (CN). The approximate relationship is:

$$\text{Cetane Number (CN)} = \frac{104 - \text{Octane Number (ON)}}{2.75}$$

Injecting a high octane fuel into a diesel can, therefore, be expected to produce a long ignition delay, followed by a very rapid pressure rise. The accompanying knock and stresses may be objectionable from the standpoint of both noise and engine life.

Converting Methods

Most current production diesel tractors operate with compression ratios of between 14:1 and 19:1. As such, they usually feature robust construction in the block, crank, and rods, so a diesel might seem to be an ideal candidate for conversion to high compression ethanol operation. However, since only a few current agricultural engines are now built in both SI and diesel configurations, a ready source of SI heads, manifolds, pistons, etc., are not available for most models.

One engine for which such parts are available is the 855 Cummins diesel. This basic engine is currently being used in five models of four-wheel drive tractors, and it is also built in SI natural gas configuration. The cost of conversion from diesel to natural gas varies with the initial condition of the diesel, but a minimum net conversion cost is about $3000. Full conversion from diesel to ethanol would be very similar to that for natural gas conversion. It would require the following new or replacement parts: pistons, head assembly, intake and exhaust manifolds, ignition system, carburetor, governor.

Another approach to conversion would involve installing a spacer to raise the diesel head and reduce compression to about 12:1. The diesel head would be drilled and sleeved to allow the installation of a spark plug. Ethanol would be injected through the diesel injectors, but timing would be advanced so that vaporization would take place during the compression stroke. A few potential problems of this system are that:

the spacer would require careful machining to ensure that combustion seal as well as oil and water flow would be maintained;

the resulting combustion chamber would not be of ideal configuration;

longer push rods and head bolts would be needed;

drilling and sleeving a diesel head would probably compromise the spark plug location, structurally weaken the head, and present sealing problems;

a throttle plate (synchronized with the injector pump) would be needed to maintain fuel to air ratios at light loads;

cylinder wash down may result from the injection of the ethanol during the compression stroke;

injector pump lubrication may not be adequate without adding oils to the ethanol.

In essence, then, the problems suggest that this alternative will not achieve a satisfactory level of performance and durability.

Unblended Ethanol

Like gasoline and other low cetane fuels, ethanol is poorly adapted for direct use in diesels. Its use would require modifying the engine for enhanced multifuel capability and adding cetane improvers to the ethanol to improve its self-ignition properties.

Engine modification. A compression ratio of 25:1 in a direct injection diesel permitted full-load operation on ethanol and that part-load operation and starting on ethanol were not possible. For light-load operation, additional heat was applied to the intake by way of exhaust gas recirculation. This enabled light-load operation, but peak pressure and the rate of pressure rise were "far above" those of the original diesel design. Timing was retarded with the hope of reducing the rates of pressure rise, but it was thought that redesign with more rigid construction would be needed. Since exhaust gas recirculation would not ease starting difficulties, an additional modification would be necessary. The most practical solution would be to start the engine on diesel fuel and change to alcohol.

Some multifuel engines, developed for the military, use similar techniques to achieve satisfactory operation on fuels ranging from No. 2 Diesel (spec VVF-800) to combat gasoline (MIL-F-3056). However, ethanol is higher octane (and lower cetane) than gasoline, so it would lie considerably outside the "multifuel" design range. When GM diesels are provided in multifuel configuration, the compression ratio is raised to 23:1, and modified injectors are installed to compensate for the higher volatility of gasoline.

The MAN diesel (also known as the Meurer, M, or Whisper diesel) is a unique combustion chamber designed for smooth operation. As such, it can tolerate relatively low cetane fuels without excessive combustion noise. It accomplishes this by injecting a coarse fuel spray that impinges on the inside of a spherical piston cup. The high swirl then controls the rate at which the fuel is heated and leads to a lower rate of pressure rise. With a compression ratio of 19:1, MAN diesels have been operated on 80 octane gasoline with no apparent combustion noise. Additionally, the MAN diesel would seem to be the most alcohol tolerant of all the current diesel design. However, multifuel capability has not been a design goal in agricultural equipment, and apparently only two manufacturers (International Harvester and White) have offered MAN diesels in the agricultural market during the last 10 years, and then only for relatively short periods.

The multifuel concept has considerable appeal, since one engine fuel system could theoretically suffice for diesel, gasoline, and ethanol. However, rack adjustments would be necessary when changing fuels because of the lower volumetric heat content of gasoline and alcohol, the increased leakage past the pump, and the lower bulk modulus of gasoline and alcohol.

Ethanol modification. The second means of using ethanol directly in diesels is to improve the cetane number of the ethanol. Several compounds may be used for this purpose, but amyl nitrate seems most popular in the United States and cyclohexanol nitrate in Europe. One study of cyclohexanol nitrate found that 10 percent of the additive in ethanol improved the ethanol to a par with diesel fuel and required no major engine combustion modifications. Of course, the fuel setting was increased for the above-mentioned reasons, but the thermal efficiency was unchanged from that of diesel fuel operation. The cost of this specific additive is not readily available, but a related study reported that 20 percent of a similar additive was uneconomic. Amyl nitrate raises the cetane of hydrocarbon fuel about 15 points when used at a 1.5 percent additive level. However, no reports documenting its performance when added to ethanol have been found.

From the standpoint of ethanol usage in the short term, the direct injection of ethanol in diesels offers little potential for the following reasons:

Multifuel engines are not used to any great extent in agriculture.

Kits to convert agricultural engines to multifuel combustion systems are not currently available, and retrofit conversion would be expensive.

The probability of economical cetane improvement to allow ethanol use in conventional diesel combustion systems appears low.

In a longer term, the direct substitution of ethanol in diesels is perhaps the most practical for agricultural machinery. The following two elements appear to be necessary:

> The MAN diesel system could be developed for multifuel agricultural usage. The combustion system has already been produced for diesel-only use in agriculture, so it follows that production costs of the system itself must not be excessive. Additional changes to improve performance with ethanol would include installing a multifuel injector pump and pressurizing the fuel to avoid vapor blockage of the pump.
>
> If cetane improvers act on ethanol similarly to their action on hydrocarbons, the values in Table 10-5 can be calculated.

Since MAN diesel (or other multifuel) combustion with 80 octane fuel has been achieved, it appears that an economic compromise may be reached by using multifuel engine technology in addition to small amounts of cetane improvers in ethanol. This approach would have the added benefit of allowing a rapid conversion back to diesel fuel, using only revised rack settings and, perhaps, injection timing.

Ethanol and Diesel Fuel

The blending of ethanol, especially in low amounts (10-40 percent), with diesel fuel has been studied by several researchers.

The phase separation problem, discussed under SI engines, apparently becomes even more critical when diesel fuel is used as the base. For example, one study found that at 32°F (0°C) only 0.05 percent water could be tolerated by ethanol-diesel blends at the 10-30 percent ethanol level. At 70°F (21°C), the 10, 20, and 30 percent blends would tolerate 0.13, 0.20, and 0.27 percent water, respectively. Thus, it is apparent that azeotropic ethanol (95.6 percent alcohol-4.4 percent water) would separate from diesel fuel at a normal operating temperatures and that anhydrous ethanol would be desirable for diesel-ethanol mixtures.

Changes in distillation curves are far more marked for diesel-ethanol mixtures than for gasoline-ethanol mixtures. The 5 percent recovery point

Table 10-5. Theoretical Values of Octane and Cetane for Ethanol With Additives[a]

Fuel	Theoretical Octane	Calculated Cetane
Ethanol	97.5	2
Ethanol + 1 percent amyl nitrate	68	13
Ethanol + 2 percent amyl nitrate	46	21
Ethanol + 3 percent amyl nitrate	33	26

[a]Calculated from cetane enhancement values (Lichty, 1967) and cetane-octane conversion formula (Cummings, 1977).

Source: USDA, *Small-Scale Fuel Alcohol Production,* U.S. Government Printing Office, Washington, D.C., 1980.

falls at about 380°F (193°C) for No. 1 diesel fuel and at about 180°F (82°C) for a 10 percent ethanol blend. Essentially, the ethanol evaporates at 170-180°F (77-82°C), after which the remaining diesel fuel heats to 350-400°F (177-204°C), before continuing the curve. The drastic changes in distillation properties suggest that high evaporation losses in storage, as well as vapor lock and the cavitation of the injector pump, might be expected. Both of these problems were experienced in another study which found that it was necessary to store the ethanol-diesel blends in sealed containers to avoid ethanol evaporation. In addition, both chilling the blend and pressurizing it were investigated as ways to prevent stoppages caused by vaporization in the injector pump.

Tests of ethanol diesel mixtures in unmodified tractor engines in 1962, on a 173-cubic-inch Ford diesel, and in 1978 on a 219-cubic-inch John Deere, were used. Both were direct injection designs with 16.8:1 compression ratio and operated at 1800 and 2000 rpm, respectively. The performance for both engines was similar. For the Ford, brake specific fuel consumption on blends was increased, reflecting the lower heat content of the ethanol. The increase in fuel consumption was greatest at light (one-fourth) loads when the reduced cetane number of the blends markedly increased ignition delay. Brake thermal efficiency was essentially unchanged for ethanol concentrations up to 30 percent at full and three-fourths load, while at one-fourth load, efficiency decreased exponentially at over 15 percent ethanol. The John Deere engine showed a similar loss in thermal efficiency at both one-fourth and one-half load.

For both engines, increased noise was apparent even at the 10 percent ethanol levels, while at 30 percent mixtures, both engines were "extremely noisy." Additional combustion studies showed a definite delay in ignition with the blends, especially at light loads. The peak cylinder pressures were also higher with the blends.

Another study ran diesel-ethanol blends in both a GM-3-53 and a military LD-465-1 (multifuel) diesel. Blends containing from 20 to 45 percent ethanol were tested, and a commercial additive was used to form the 45 percent emulsion. With the 30 percent ethanol blend, power was reduced from 5 to 18 percent, depending on engine speed. Efficiency was measured in terms of relative vehicle range, a volumetric (not energy) basis. Again, with the 30 percent blend, range was reduced from 3 to 13 percent depending on speed.

The key problems with ethanol-diesel blends in unmodified diesels appear to be:

the water sensitivity of the blends, which required anhydrous ethanol;

the modified distillation curves, requiring fuel pressurization on some tractor models;

the reduced cetane number and energy content of the blends, leading to concerns about cylinder pressures and engine life.

Most current diesel injection pumps are lubricated by the diesel fuel that they supply to the engine. This and the lower viscosity of ethanol compared to diesel fuel cause concern for injector pump lubrication and life.

As part of a recent diesel fuel screening project, bench tests were conducted to determine the friction and wear characteristics of various substitute fuels. Conducted in accordance with ASTM method D2714-68, the tests measured the coefficient of friction as well as the width of the wear track on a standard test block. Table 10-6 shows the results for diesel fuel, unleaded gasoline, ethanol, and an 80 percent gasoline-20 percent diesel fuel mixture.

Although the lubrication performance of ethanol was somewhat poorer than that of diesel fuel, it is superior to gasoline's. Also of interest is the substantial improvement of gasoline's lubrication performance with the addition of 20 percent diesel fuel. No ethanol-diesel blends were tested, but if a similar improvement occurs, potential injector pump lubrication problems should be greatly reduced by the addition of even small amounts of diesel fuel to the ethanol.

Carbureted Ethanol

Carbureting (or "aspirating" or "fumigating") SI-type fuels into diesel engines has been practiced for many years using fuels such as natural gas, LPG, and gasoline. In some cases, the object was to reduce diesel smoke emissions; in others, more power and multifuel operation were sought.

Of the four options for using ethanol in diesels, this concept is currently receiving the most popular attention, due in large part to the availability of a conversion kit sold by the M&W Gear Company, Gibson City, Illinois. The device introduced 100 proof ethanol into the engine's air intake just upstream from the turbocharger. It uses the turbo boost pressure to pressurize the alcohol tank and, thereby, meter the alcohol as a function of engine load. Ethanol is not added at light loads. Advertising literature states that in company tests the device reduced the tractor diesel fuel consumption from 8½ gallons per hour to 6 gallons per hour while adding 2 gallons of 100 proof

Table 10-6. Friction and Wear Characteristics of Some Substitute Fuels

Fuel	Average wear track width (mm)	Maximum observed friction coefficient
Diesel fuel	1.17	0.15
80 percent gasoline – 20 percent diesel	1.32	0.18
Ethanol	1.58	0.27
Gasoline	1.99	0.45

Source: USDA, *Small-Scale Fuel Alcohol Production,* U.S. Government Printing Office, Washington, D.C., 1980.

ethanol. The tractor produced 125 horsepower in both cases, so in the pure diesel test, thermal efficiency was 27.1 percent. With the alcohol injection, the reported figures would correspond to a thermal efficiency of 35.2 percent.

Alcohol fumigation tests date from the early 1950s. In general, tests have shown that alcohol fumigation leads to poorer thermal efficiency at light loads and better thermal efficiency at overload. Fumigation is better adapted to direct injection, open-chamber engines.

The reduction in efficiency at part load was thought to be caused by the increased ignition delay brought on by the alcohol fumigation. The combustion was highly subject to the ratio of diesel flow to alcohol flow; at 30 percent of maximum diesel flow, only a small amount of alcohol could be burned before the engine began missing. Ignition delay with 40 percent of maximum diesel flow was 43.5; with 75 percent of maximum diesel flow, the delay was about 33.

The increase in thermal efficiency at high loads is apparently due to the better air utilization from the fumigation. The results of pure diesel and 75 percent of maximum diesel flow tests are summarized in Table 10-7.

It is apparent that, at overfueled conditions (when the thermal efficiency of a conventional diesel decreases), a comparable amount of power can be obtained by using fumigation while maintaining a high level of thermal efficiency. Under such conditions, peak cylinder pressures were about the same, but the rate of pressure rise was higher for the alcohol fumigation.

The tests also concluded that the advancement of injection timing would in part compensate for the increased ignition delay and allow higher levels of alcohol use. In one test, 80 percent of the total heat input was supplied by ethanol. Since the no-load diesel fuel flow was about 30 percent of the maximum flow, the proposed concept was that diesel flow be kept at no-load flow and that the amount of carbureted alcohol would control the engine's response to load. High rates of pressure rise were deemed unacceptable in some cases, and work on the concept is continuing.

A study using a special direct injection diesel with a single hole injector to study pilot injection concepts found that a minimum of 30 percent of the heat energy should be supplied by the pilot (diesel) fuel. The study reported a strong increase in thermal efficiency (from 14.7 to 19.9 percent at the 1.61 kilowatt reference condition), a condition again thought to be due to im-

Table 10-7. Thermal Efficiency of a Single Cylinder Test Engine at 1500 rpm

Horsepower	Thermal efficiency pure diesel (percent)	Thermal efficiency 75 percent diesel flow remainder ethanol (percent)
5	31	31
6	31	33
7	25	33

Source: USDA, *Small-Scale Fuel Alcohol Production,* U.S. Government Printing Office, Washington, D.C., 1980.

proved air utilization. It should be commented that the single hole injector used in this study is not representative of normal practice and may have been responsible for the rather low efficiencies of the straight diesel operation. It was also apparent from this study that the air-ethanol ratio must be held below certain levels to avoid the lean misfire of the bulk gases.

Barnes reported on a system remarkably similar to that using the M&W conversion kit. A pressurized alcohol tank and nozzle was added to the test engine—an Oliver F-310-DBLT—a turbocharged 6-cylinder unit used in the model 1855 tractor. Alcohol was introduced upstream of the turbocharger, and it served as an effective charge coolant to lower intake temperature by 70°F when the ratio A = mass alcohol/mass diesel fuel reached 1.0 during full load tests.

Changes in brake thermal efficiency from adding the ethanol at full load were negligible, with all points falling between 34 and 35 percent. A was varied between 0 and 1.2 in the tests. However, full load runs using isopropanol showed a gradual increase from 35 percent to about 37.5 percent thermal efficiency as A increased from 0 to 1.0.

In the final analysis, it appears that ethanol fumigation can improve thermal efficiency under certain conditions; specifically, if an overfueled condition exists while using straight diesel fuel, a comparable amount of power can be obtained at higher efficiency by reducing the amount of diesel fuel injected and replacing it with fumigated ethanol. The homogeneous nature of the air-ethanol mixture is believed to improve air utilization of the diesel, but one study suggests that the air-alcohol ratio must still be maintained below a certain level in order to avoid the incomplete burning of the alcohol. Ethanol fumigation seems relatively ineffective at engine loads below 50 percent but, at higher loads, a fairly large proportion of total energy can be supplied by the ethanol. As loads increase, some additional diesel fuel must be used. We estimate that, on the average, alcohol could replace 30 percent of the diesel fuel by volume.

Overfueling is generally not present in standard farm tractors, as evidenced by the fact that full load efficiency is almost always higher than 75 percent load fuel efficiency. Thus, as a general practice for farm tractors, ethanol fumigation will not be credited with an improvement in efficiency but should be valued on a heat content basis. In one review, it is mentioned that lower-proof ethanol also produced no improvement in thermal efficiency over that at 200 proof; therefore, low proofs should also be valued on a heat content basis.

APPLICATIONS

Engine Use

Gasoline-ethanol mixtures in SI engines. The acceptable use of ethanol in mixtures with gasoline such as gasohol is feasible and, since virtually no engine modifications would be necessary, easy conversion to and from gasoline would be possible. However, the phase separation problem must be dealt with either through the use of 200 proof ethanol or by the addition of cosolvents and other additives.

As discussed previously, volumetric fuel economy from a 10 percent ethanol mixture will probably be reduced by about 2.5 percent, a condition that indicates that ethanol replaces 75 percent of the gasoline heat value.

Ethanol in SI engines. Case 1 will consider an engine with carburetion and induction modifications designed to allow operation with either ethanol or gasoline. Changes would require either an adjustable carburetor or dual carburetors, additional intake manifold heat, a starting aid, and an easily adjusted ignition timing system. However, the compression ratio would remain standard (an assumed 8:1) to allow operation on gasoline.

The value of 200 proof ethanol used in this application would be its energy value plus a 3 percent efficiency improvement due to factors discussed previously. Combining this with the data on efficiency versus proof provides the results found in Table 10-8.

Notice that no credit is given for high octane, since the compression ratio was not raised to take advantage of it.

Case 2 considers an engine with induction and carburetion changes plus the raising of compression ratio from 8:1 to about 12:1. Starting aids are also needed. Such an engine could not be converted easily back to gasoline on a day-to-day basis.

The cost of such a conversion would vary greatly. For example, a high-compression, natural gas irrigation engine might require only minor heat planing, a replacement carburetor, a fuel pump, and a tank, at a total cost under $700. Other engines might not have adequate head material for planing and could require high compression pistons to increase the total overhaul and conversion cost to $2000 or more.

The value of ethanol in this application would be increased from case 1 in proportion to the engine efficiency due to its increased compression ratio (CR). From Fig. 10-3, the ratio "Efficiency at 12:1/Efficiency at 8:1" equals roughly 1.13, producing the data shown in Table 10-9.

Diesels converted to high compression SI. Since SI parts are not available for most currently used diesels, such components would have to be developed prior to the widespread retrofitting of existing engines. In addition, the costs of such a retrofit would be high and would discourage the conversion of older diesels.

According to Fig. 10-4, a high compression SI ethanol engine theoretically achieves 95 percent of the thermal efficiency of the diesel. Since this conversion adds a throttle plate and accompanying pumping losses at light loads, the previously proposed 3 percent efficiency advantage due to ignition advance is not applied in this case. The value of the ethanol is shown in Table 10-10.

Table 10-8. Results of Case 1 Study

Proof	Energy content (Btu)	Thermal efficiency relative to gasoline (percent)	Volumetric value relative to gasoline
200	76,152	103	0.67
190	72,344	94	0.58
160	60,921	86	0.45
120	45,691	81	0.32

Source: USDA, *Small-Scale Fuel Alcohol Production,* U.S. Government Printing Office, Washington, D.C., 1980.

Table 10-9. Results of Case 2 Study

Proof	Energy content (Btu)	Thermal efficiency relative to gasoline at 8:1 CR (percent)	Volumetric value relative to gasoline
200	76,152	116	0.76
190	72,344	106	0.66
160	60,921	97	0.51
120	45,691	91	0.36

Source: USDA, *Small-Scale Fuel Alcohol Production,* U.S. Government Printing Office, Washington, D.C., 1980.

Table 10-10. Thermal Efficiency Value

Proof	Energy content (Btu)	Thermal efficiency relative to diesel at 17:1 CR (percent)	Volumetric value relative to No. 2 diesel
200	76,152	95	0.52
190	72,344	87	0.46
160	60,921	79	0.35
120	45,691	75	0.25

Source: USDA, *Small-Scale Fuel Alcohol Production,* U.S. Government Printing Office, Washington, D.C., 1980.

Ethanol in diesels. Straight ethanol lies far outside most diesel engine manufacturers' fuel specifications, so the direct substitution of ethanol for diesel fuel cannot be seriously contemplated. Poor engine performance, knock, and severe engine damage are almost certain to occur as a result of such a substitution.

Additives to improve the cetane of ethanol are a distinct possibility, but the current costs and quantities required seem to discourage their use. A more probable long-term option would seem to be the combination of moderate amounts of additives and revised multifuel engine design. Although considerable development of this concept would be necessary and implementation would be slow, it merits consideration.

The value of ethanol in this application is not well documented, but multifuel research in general suggests an energy substitution. This leads to a value of 200 proof ethanol of 0.55 times that of diesel fuel.

Ethanol-diesel mixtures. Ethanol-diesel mixtures are generally subject to the same difficulties as ethanol-gasoline mixtures, plus some additional problems. Phase separation is apparently at least as critical as with gasoline, and the lower energy content of ethanol is reflected as well. The major additional problem stems from the low cetane rating of ethanol, a condition which tends to increase the ignition delay and reduce efficiency at light loads.

Because of these difficulties, ethanol will have a value of somewhat less than its heating value when used in this application. At three-fourths load (considered an average), thermal efficiency reductions varied from zero to 3 percent for a 30 percent ethanol blend. Thus, an average 1.5 percent thermal efficiency reduction for 30 percent blends is equivalent to an 8 percent reduction in the effective heating value of ethanol, giving it a value of 0.51 times that of diesel fuel in this application.

Carbureting ethanol into diesels. This approach appears to be the most feasible near-term technology for using ethanol in diesel engines. Retrofit hardware is currently available in limited quantities, efficiency is equal to pure diesel operation, and risk of engine damage appears low (provided the aspirated ethanol is used to replace diesel fuel rather than to boost power). The primary disadvantages of the approach are its moderate conversion costs and the inconvenience of its separate fuel tanks.

The value of ethanol in this application will be estimated at its heat value. (See Table 10-11.)

Other Uses

Potentially, ethanol could serve as a substitute for other farmstead energy requirements, including grain drying and livestock confinement heating. For these purposes, a lower-proof ethanol would be satisfactory.

In order to use ethanol in crop dryers, the major modification would be to the burners. The relative value of ethanol in grain drying compared to propane can be estimated based on the energy content of the two fuels. The energy that can be obtained from burning a gallon of propane is 81,855 Btu compared to 76,152 Btu for ethanol. If substitution were strictly on a Btu basis, then 1.07 gallon of ethanol would be required to replace 1 gallon of propane.

If ethanol is to be substituted for fuel oil in the heating of buildings, then, based on energy content (76,152 Btu for ethanol; 138,690 Btu for fuel oil),

Table 10-11. Value of Ethanol in Diesels

Proof	Energy content (Btu)	Volumetric value relative to No. 2 diesel
200	76,152	0.55
190	72,344	0.52
160	60,921	0.44
120	45,691	0.33

Source: USDA, *Small-Scale Fuel Alcohol Production,* U.S. Government Printing Office, Washington, D.C., 1980.

1.8 gallons of ethanol would be required to replace 1 gallon of fuel oil. The flash point of ethanol [55°F (13°C)] is considerably lower than that for fuel oil, making it less attractive than fuel oil for heating homes and commercial buildings.

The feasibility of the seven engine applications discussed in this chapter are summarized in Table 10-12. The "percent fuel replaced" column refers to the amount of petroleum fuel (either gasoline or diesel fuel) that ethanol might replace for those engines that are actually converted to ethanol.

Near- and long-term potentials are highly subjective, and the introduction of new additives or engine technologies could alter this assessment. Only carbureted diesel operation is given a high near-term potential, while the other three diesel applications are judged to have low near-term potential. All SI applications are given moderate to high ratings.

The applications are given high long-term (5 years hence) ratings. If properly implemented, both the standard CR spark engine and the multifuel diesel could be switched between ethanol and petroleum fuel very quickly, giving them added flexibility. Ethanol-diesel mixtures would probably require both cetane improvers and cosolvents, so this application was given a low long-term rating.

Liquid Fuel Use Profile

Ethanol can be used as a liquid fuel for mobile and stationary engines, grain driers, and boilers. In the near term, its primary use will be as a fuel for vehicle engines. Ethanol may be used either as a blend of 200 proof with gasoline, and perhaps diesel fuel, straight at 200 or lower proofs, or aspirated with diesel fuel. The quantities and specific applications of each type of ethanol fuel depend to some extent on the rate at which necessary modifications or adaptations of present equipment and the rate of replacements of equipment currently in use occur. The present liquid fuel consumption on farms and in the transportation sector for which ethanol fuels may be a complete or partial substitution or an extender are discussed below.

ON-FARM UTILIZATION

On-farm fuel usage by type of operation has been compiled by USDA for 1978. These data are subdivided to show energy usage associated with crop-production operations from preplant through harvest to grain drying and those associated with livestock operations. For each of the separate operations, the quantity of fuel used is compiled. These data, shown in Table 10-13, indicate that for the United States, 83 percent of the gasoline used on farm (exclusive of household use) was used for crop operations and 17 percent for livestock operations. Crop operations accounted for 85 percent of diesel fuel use and livestock operations for 13 percent. Crop operations

Table 10-12. Summary of Ethanol Application in Engines

Application	1) Ethanol-gasoline mixtures	2) Ethanol in Spark Ignition, Standard Compression Ratio	3) Ethanol in Spark Ignition High Compression Ratio	4) Ethanol in Spark Ignition converted diesels high Compression Ratio	5) Ethanol in diesels	6) Ethanol-diesel mixtures	7) Carbureted ethanol C.I.
Engine population	Vehicles and older Spark Ignition tractors	Vehicles, natural gas tractors, engines, old Spark Ignition tractors	Natural gas Irr engines old Spark Ignition tractors	C.I. tractors, combines	C.I. tractors, combines	C.I. tractors, combines	C.I. tractors, combines
Approx % fuel replaced[a]	10	100	100	100	100	10	30
Utilize low proof?	No	Yes	Yes	Yes	Yes[b]	No	Yes
Near term potential	Moderate[c]	Moderate	Moderate	Low	Low[c]	Low[c]	High
Retrofit costs	Very Low[c]	Low	Moderate	High	High[c]	Low	Moderate
Long term potential	Moderate	High	Moderate	--	High	Low[c]	Moderate
200 proof ethanol value	.75 x gasoline	.67 x gasoline	.76 x gasoline	.52 x diesel	.55 x diesel	.51 x diesel	.55 x diesel
Phase separation	X					X	
Driveability	X						
Valve recession		X	X				

Application	1) Ethanol-gasoline mixtures	2) Ethanol in Spark Ignition, Standard Compression Ratio	3) Ethanol in Spark Ignition High Compression Ratio	4) Ethanol in Spark Ignition converted diesels high Compression Ratio	5) Ethanol in diesels	6) Ethanol-diesel mixtures	7) Carbureted ethanol C.I.
Starting (Below 40° F)		X	X	X	X	X	
Vapor Lock	X				X	X	
Unavailability of retrofit hardware				X	X		
Materials compatability	X	X	X	X	X	X	
Injector pump Lubrication					X		
Oil dilution @ light loads		X	X	X			
Combustion knock					X	X	
Inconvenience							X

a Assuming 100% adoption

b Speculative

c Would become more favorable with the development of low cost additives

Source: U.S. Department of Agriculture. *Small-Scale Fuel Alcohol Production,* 1980. Washington, D.C. GPO No. 001-000-04124-0.

Table 10-13. United States Energy and Agriculture, 1974 Data Base, Summary by Operation

	Gallons of gasoline (000)	Gallons of diesel (000)	Gallons of fuel oil (000)	Gallons of LPG (000)	Cubic feet of natural gas (mil.)	Tons of coal (mil.)	Kilowatt hours of electricity (mil.)
Operations—crops							
Preplant	61,424	923,331	—	22,719	—	—	—
Plant	25,581	242,480	—	9,461	—	—	—
Cultivate	24,476	280,338	—	9,053	—	—	—
Fertilizer application	27,023	64,044	—	9,995	—	—	—
Pesticide application	32,711	74,754	—	12,099	—	—	—
Irrigation	70,553	177,144	—	236,331	132,318	—	19,264
Frost protection	46,443	39,326	218,549	1,458	—	—	200
Harvest	495,052	443,729	—	183,102	—	—	—
Farm truck	538,405	5,605	—	—	—	—	—
Grain handling	16,314	—	—	—	—	—	33
Crop drying	—	—	76,564	664,440	27,182	—	858
Farm pickup	1,036,304	1,058	—	—	—	—	—
Farm automobile	442,722	—	—	—	—	—	—
Electrical overhead	—	—	—	—	—	—	1,705
Miscellaneous	64,283	35,354	—	—	—	—	—
Total—crops	2,881,291	2,287,163	295,113	1,148,658	159,500	—	22,060
Operations—livestock							
Lighting	—	—	—	—	—	—	1,653
Feed handling	149,633	195,283	—	—	—	—	1,073
Waste disposal	98,036	44,645	688	6,754	435	—	123
Water supply	—	25,511	—	—	—	—	1,605
Livestock handling	19,391	1,033	—	—	—	—	—
Space heating	—	1	—	55,957	10	—	174
Ventilation	—	—	—	—	—	—	2,014
Water heating	—	—	—	70,737	—	—	975
Milking	—	—	—	—	—	—	820
Milk cooling	—	—	—	—	—	—	1,344
Egg handling	—	—	—	—	—	—	31
Brooding	—	—	8,129	188,120	4,181	32,725	—
Farm vehicles	233,310	69,330	—	469	—	—	—
Farm automobile	198,417	—	—	—	—	—	—
Other	118,578	16,613	—	10,848	—	—	215
Total—livestock	817,365	352,416	8,817	332,885	4,626	32,725	10,027
Total—agriculture	3,698,656	2,639,579	303,930	1,481,543	164,126	32,725	32,087

Source: U.S. Department of Agriculture/Federal Energy Administration, *Energy and Agriculture, 1974 Data Base,* 1977.

accounted for 78 percent of LPG use and livestock operations for 22 percent. Usage in individual states varies from these national averages and are shown in Table 10-14 for Minnesota and Illinois.

Fuel usage per farm varies widely by geographic area. Preliminary 1978 data compiled by USDA show variations by region in the fuels and lubrication cost per acre for grains. The values for corn given in Table 10-15 exemplify these variations. Corn production in the southwest is predominantly irrigated and therefore has high energy cost per acre.

Variations also occur by type of farm. Table 10-16 shows the fuel, oil, and grease expenditures per tillable acre for several types of Illinois farms—both grain and livestock. The livestock farms use about 50 percent more petroleum products than do the grain farms. Fuel usage for farms in two areas of Minnesota are about the same as for grain farms in Illinois. Fuel usage on Kansas farms is less than on either Illinois or Minnesota farms, a rate which probably reflects the lower tillage energy requirements for wheat and milo compared to those for corn and soybeans.

Fuel usage per acre is relatively independent of farm size as shown in Fig. 10-5, where fuel expenditures are shown for grain and hog farms in Illinois. There seems to be little economy of scale as far as fuel usage is concerned.

Table 10-14. On-Farm Use of Gasoline, Diesel Fuel, and LPG

	Gasoline[a]	Diesel[a]	LPG[a]
Minnesota			
Crop	87	86	60
Livestock	13	14	40
Illinois			
Crop	93	83	92
Livestock	7	17	8

[a]Percent of total.
Source: USDA, *Small-Scale Fuel Alcohol Production,* U.S. Government Printing Office, Washington, D.C., 1980.

Table 10-15. Value of Corn in the United States

Region	Corn-production fuels and lubrication cost per acre (dollars)
Northeast	5.92
Lake states/cornbelt	6.11
Northern plains	16.06
Southeast	6.95
Southwest	28.30
United States (average)	8.41

Source: USDA, *Small-Scale Fuel Alcohol Production,* U.S. Government Printing Office, Washington, D.C., 1980.

Table 10-16. Fuel, Oil, and Grease Expenditures Per Tillable Acre

Type of operation	Fuel, oil, and grease ($/tillable acre)
N. Illinois, high-production soil, grain[a]	8.02
N. Illinois, low-production soil, grain	8.11
S. Illinois, grain	8.19
N. Illinois, high-production soil, hog	13.52
N. Illinois, low-production soil, hog	13.58
S. Illinois, hog	11.90
N. Illinois, dairy	13.73
S. Illinois, dairy	14.17
N. Illinois, beef	11.70
S.E. Minnesota[b]	9.80
S.W. Minnesota[c]	8.78
Kansas[d]	6.78

[a]Summary of Illinois Farm Business Records, 1978, Univ. of Illinois.

[b]Southeastern Minnesota Farm Management Association, 1978 Report, Univ. of Minnesota.

[c]Southwestern Minnesota Farm Management Association, 1978 Report, Univ. of Minnesota.

[d]Farm Management Summary and Analysis, 1978, Kansas State University.

Source: USDA, *Small-Scale Fuel Alcohol Production,* U.S. Government Printing Office, Washington, D.C., 1980.

Table 10-17 shows the consumption of gasoline, diesel, and LPG fuels per cropped acre in the United States. The large amount of LPG in the cornbelt area probably reflects that area's large energy requirements for drying of grain. Relatively large amounts of LPG in drier regions probably reflect the use of LPG for irrigation pump engines. Alcohol could be used for crop dryers with little modification of burners. LPG engines used for irrigation would also require minimum modification and would probably have higher compression ratios and use alcohol more efficiently.

More than 3 billion gallons of gasoline are used annually on farms for business use in trucks and automobiles as well as combines (which use a higher percentage of gasoline engines than do the larger tractors).

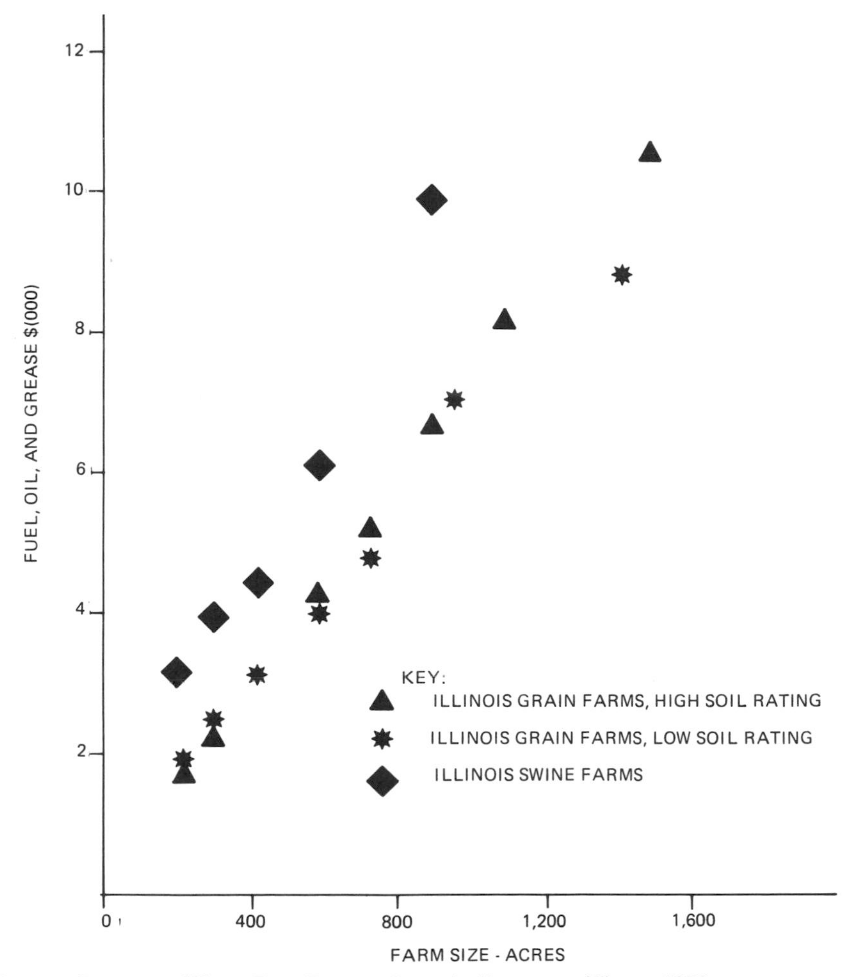

Source: Summary of Illinois Farm Business Records. University of Illinois, 1978.

Figure 10-5. Fuel, oil, and grease expenditures for selected Illinois farms.

More than 3.3 billion gallons of diesel fuel are used annually on farms. Conversion of diesel tractors to alcohol fuels can occur through retrofitting or replacement. This conversion will be more diffucult than that for gasoline or LPG, but the existing use of gasoline and LPG indicate that much alcohol of either 190 or 200 proof can be absorbed by the agricultural sector without waiting for the conversion of diesel tractors.

States with intensive agricultural production, including livestock and truck crops, use significantly larger amounts of fuel per crop acre. These data should be useful in estimating fuel usage in small areas of a particular state.

In addition, these data may be used to calculate the quantities of 200 proof ethanol required to replace the fuels used on an average farm. Using data for fuel usage per acre (Table 10-17) and an average-size farm in the

Table 10-17. Agricultural Fuel Use by State[a]

State	Crop area[b]	Fuel use[c]			Fuel use[d]		
		Gasoline	Diesel	LP	Gasoline	Diesel	LP
Alabama	3.66	38.6	37.6	24.0	10.5	10.3	6.56
Alaska	0.023	0.282	0.048	0.064	12.3	2.1	2.78
Arizona	1.52	24.1	16.5	1.30	15.9	10.9	0.86
Arkansas	7.82	71.3	94.6	57.4	9.12	12.1	7.34
California	9.74	157.4	166.2	24.9	16.16	17.06	2.56
Colorado	9.67	50.1	45.8	8.53	5.18	4.74	0.88
Connecticut	0.178	4.05	1.65	2.00	22.75	9.27	11.24
Delaware	0.515	10.3	5.70	7.08	20.0	11.1	13.8
Florida	3.11	78.8	88.5	22.0	25.3	28.5	7.0
Georgia	5.32	68.7	62.6	54.9	12.9	11.8	10.3
Hawaii	0.320	9.00	6.49	0.173	28.1	20.3	0.5
Idaho	5.52	42.1	41.3	6.77	7.6	7.5	1.2
Illinois	23.1	268.0	130.0	139.0	11.6	5.6	6.0
Indiana	12.7	144.0	67.0	73.9	11.3	5.3	5.8
Iowa	24.5	328.0	165.0	143.0	13.4	6.7	5.8
Kansas	27.7	149.0	136.0	45.0	5.4	4.9	1.6
Kentucky	4.91	59.4	34.4	15.9	12.1	7.0	3.2
Louisiana	4.64	48.0	68.6	10.5	10.3	14.8	2.3
Maine	0.555	7.98	3.06	0.844	14.4	5.5	1.5
Maryland	1.63	25.4	11.5	13.5	15.6	7.1	8.3
Massachusetts	0.209	4.22	1.89	0.985	20.2	9.0	4.7
Michigan	7.43	85.4	48.6	21.8	11.5	6.5	2.9
Minnesota	21.9	244.0	109.0	89.9	11.1	5.0	4.1
Mississippi	6.23	56.7	70.9	19.8	9.1	11.4	3.2
Missouri	13.8	140.0	75.8	46.1	10.1	5.5	3.3
Montana	14.8	58.5	29.4	5.86	4.0	2.0	0.4
Nebraska	20.1	141.0	241.0	149.0	7.0	12.0	7.4
Nevada	0.600	8.30	8.53	1.99	13.8	14.2	3.3
New Hampshire	0.131	1.85	0.591	0.436	14.1	4.3	3.3
New Jersey	0.631	15.1	6.89	1.76	23.9	10.9	2.8
New Mexico	1.76	26.8	19.3	18.1	15.2	11.0	10.3
New York	4.86	74.9	29.0	11.6	15.4	6.0	2.4
North Carolina	5.48	66.1	56.7	143.0	12.1	10.3	26.0
North Dakota	28.0	124.0	98.4	5.79	4.4	3.5	0.2
Ohio	11.2	119.0	51.6	38.7	10.6	4.6	3.5
Oklahoma	11.2	103.9	62.0	21.5	9.3	5.5	1.9
Oregon	4.46	30.7	23.7	6.50	6.9	5.3	1.5
Pennsylvania	5.00	74.0	32.9	18.7	14.8	6.6	3.7
Rhode Island	0.024	0.507	0.169	0.097	21.1	7.0	4.0
South Carolina	2.86	29.8	30.2	26.2	10.4	10.6	9.2
South Dakota	16.9	89.0	104.4	31.3	5.3	6.2	1.8
Tennessee	4.87	50.4	37.8	10.7	10.3	7.8	2.2
Texas	24.6	264.0	182.0	67.3	10.7	7.4	2.7
Utah	1.54	14.8	11.8	3.01	9.6	7.7	1.9
Vermont	0.570	8.55	1.54	2.32	15.0	2.7	4.1
Virginia	3.00	33.3	27.4	29.6	11.1	9.1	9.9
Washington	7.68	43.0	31.7	7.30	5.6	4.1	0.9
West Virginia	0.878	9.13	4.28	2.82	10.4	4.9	3.2
Wisconsin	10.1	178.0	45.1	46.6	17.6	4.5	4.6
Wyoming	2.79	17.6	14.2	1.76	6.3	5.1	0.6
Total		3,697.1	2,639.0	1,481.3			

[a]Excludes fuel oil, which is mainly used for frost protection.

[b]U.S. Department of Agriculture, *Agricultural Statistics,* 1978. Figures in millions of acres.

[c]U.S. Department of Agriculture/Federal Energy Administration, *Energy and Agriculture, 1974 Data Base,* 1977. Figures in millions of gallons.

[d]Figures in gallons per acre.

Table 10-18. Calculations for Replacement of Gasoline, Diesel Fuel, and LPG by 200 Proof Alcohol

Fuel	Gallons per acre	Total fuel (gal)	Volumetric value relative to 200 proof ethanol	Ethanol for total replacement (gal)
Gasoline	11.6	3,160	1.5	4,740
Diesel	5.6	1,520	1.8	2,740
LPG	6.0	1,630	1.8	2,930

Source: USDA, *Small-Scale Fuel Alcohol Production,* U.S. Government Printing Office, Washington, D.C., 1980.

state, the fuel requirements for the farm are estimated. By combining this information with the relative volumetric values for specific engines or other uses, the quantity of alcohol required for replacement may be obtained. The calculations for an average Illinois farm of 272 acres requiring a total of 10,410 gallons of 200 proof ethanol (to replace the gasoline, diesel fuel, and LPG) are shown in Table 10-18.

Similarly, individuals could, by using data for their own farming operations, determine the ethanol required for liquid fuel replacements for their own farming operations and thus sizing requirements for an ethanol-production facility.

COMMERCIAL MARKET FOR 200 PROOF ETHANOL

The current and near-term potential market for 200 proof ethanol is probably greatest as a blending agent in the production of gasohol—10 percent alcohol and 90 percent unleaded gasoline.

Ten Steps to Carburetor Modification

Modification of the carburetor (Fig. 10-6) or fuel injection fuel-air metering system will be required to convert a spark-ignited engine to run on straight ethanol. In some engines, this might simply involve enlarging the carburetor fuel-air metering main jet to compensate for differences in the energy content and stoichiometry between ethanol and gasoline. However, the carburetor should be properly calibrated for the fuel being used in order to obtain completely satisfactory engine performance at all operating ranges.

The consequences of adding alcohol can be better comprehended by referring to Fig. 10-7. This figure shows the general variations in engine efficiency and emissions with air-fuel ratio, which is represented by the equivalence ratio. Equivalence ratios less than 1 are lean mixtures; those greater than 1 are rich mixtures. At the top of the figure, the approximate operating range for the different catalytic systems is indicated. On adding 5 percent alcohol, the shift in equivalence ratio to the left will not exceed 0.04 units. Depending on the starting point, the effects of adding small percentages of alcohol will vary in terms of fuel economy and emissions, i.e., there may be a small increase in hydrocarbons and a reduction in NO_x—or the converse.

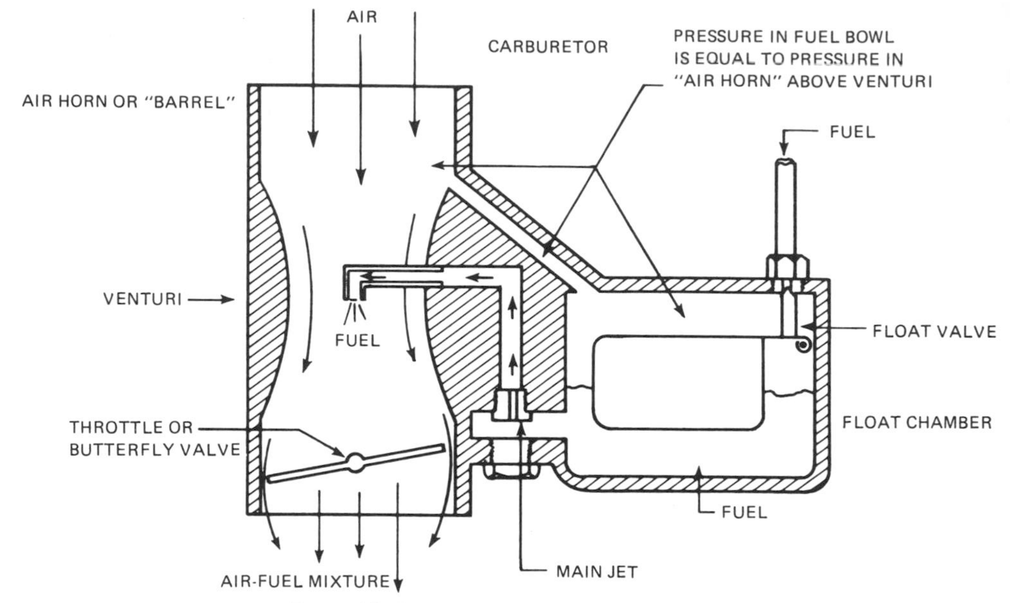

Figure 10-6. Diagram of a typical carburetor.

Ethanol vaporizes less readily than gasoline. This often results in difficult cold engine start-up at temperatures below 40°F (4°C) and poor distribution of the air-fuel mixture to the engine cylinders. A more volatile fuel may be needed to start the engine, and a system for heating the intake manifold may be needed to improve air-fuel distribution.

Because ethanol has a higher octane value than gasoline, raising the compression ratio can enhance vehicle performance. This major alteration should not be performed if the conversion is not intended for the lifetime of the engine. Engine conversions should be done by, or with the assistance of, qualified mechanics.

This section provides information on the carburetor modification required to permit the use of 100 percent ethanol. This information is provided through the courtesy of Desert Publications and their book entitled *Brown's Alcohol Motor Fuel Cookbook* by Michael H. Brown, and an article in the September/October 1979 issue of the *Mother Earth News* entitled "Mother's Experimental Alcohol-Powered Truck." This material is provided for the reader's information; liability is not assumed by the author or publisher for such engine modification.

The basic modification required to run your automobile on 100 percent alcohol involves changing a small piece of brass called the main jet. All carburetors designed to use liquid fuel have a main jet as shown in Fig. 10-6. In the care of carburetors designed to work on V-8 engines, there will normally be jets in multiples of even numbers; for example, a "two-barrel" carburetor will have two main jets and two venture tubes. There are several

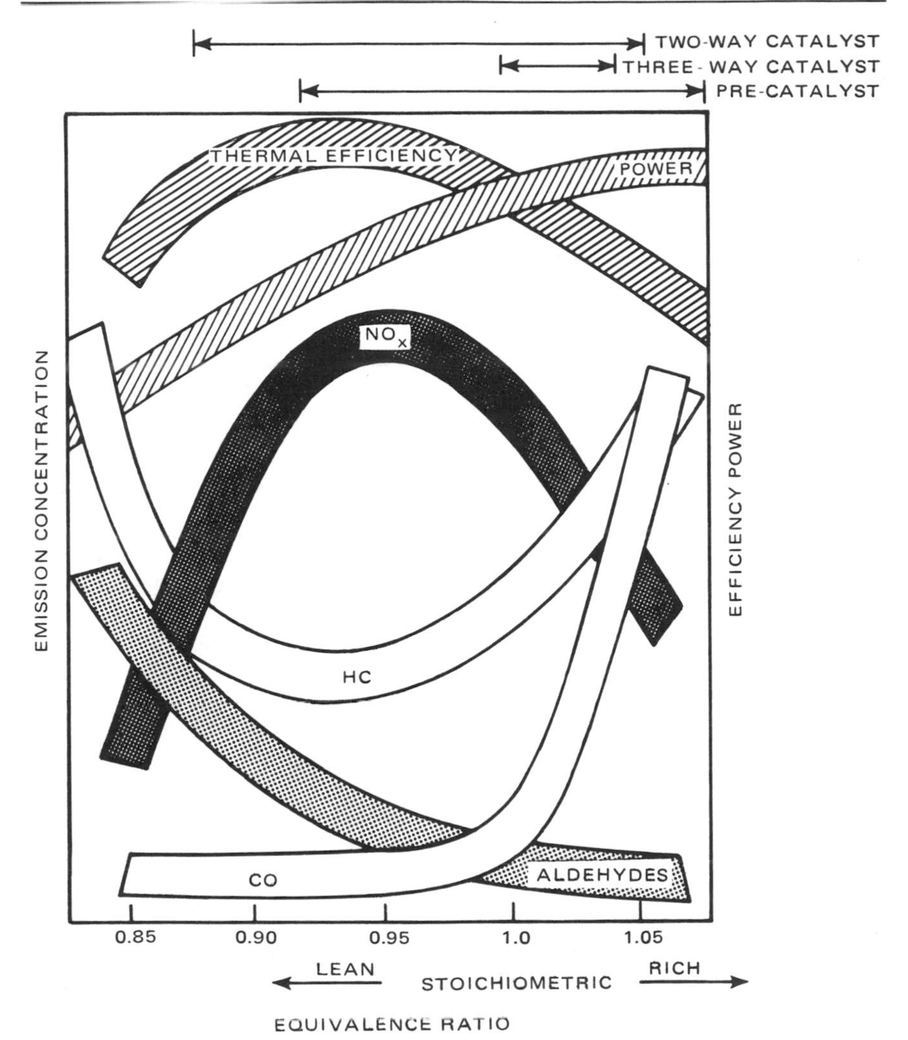

Source: *Mechanical Engineering*/November 1979

Figure 10-7. Efficiency, power, and emissions are functions of the equivalence ratio.

reasons for enlarging the main jet(s) orifice to accommodate an alcohol-powered engine. The first is that alcohol requires a fuel-air or air-fuel ratio of 9:1 to ignite in the engine (i.e., 9 pounds of air for every 1 pound of alcohol). Gasoline requires a 15:1 ratio of air to fuel. The difference between alcohol and gasoline is about 40 percent. The alcohol carries oxygen in the fluid, where gasoline does not, which may explain why alcohol requires less air to detonate or ignite.

Another reason is the difference in weight of gasoline and alcohol. Alcohol weighs 6.6 pounds per gallon; gasoline weighs 6.1 pounds per gallon. Even at 6.6 pounds (anhydrous or pure 100 percent alcohol), alcohol is a thicker fluid than gasoline and will not flow readily through a gasoline-sized jet.

Although the difference in the fuels by weight is only 10 percent, alcohol has only two-thirds the energy (Btu's) that gasoline has. To compensate for this, one must add one-third for energy loss and 10 percent for difference in fuel weight; this gives you the 40 percent (approximate) enlargement necessary.

The venturi tube is a restricted passage in the body of the carburetor which causes the air to speed up, lose some of its pressure, and create a partial vacuum. Since gasoline must be vaporized before it can be ignited and liquid cannot exist in a vacuum, the partial vacuum created in the venturi tube helps vaporize the gasoline.

The first step (Fig. 10-8) in carburetor conversion is getting ready. Gather all the tools and hardware you need to complete the job. In most cases a screwdriver, a pair of needlenose pliers, assorted end wrenches, vise-grip pliers, and a power drill—with bits ranging in size from 0.050 inches (No. 55) to 0.089 inches (No. 43)—are the basic tool list. To make the modification easier to follow you could take Polaroid photographs of the carburetor disassembly, purchase a carburetor rebuilding kit, or use standard auto manuals like Motor, Glenn, or Chilton to help visualize the modification.

Courtesy Brown's Alcohol Motor Fuel Cookbook, Desert Publications, Cornville, AZ.

Figure 10-8. Step 1—Gather tools and prepare working diagram of carburetor assembly. The carburetor shown in this series of illustrations is a two-barrel Rochester from a 307 Chevrolet V8 Engine.

These auto repair manuals provide excellent exploded illustrations to guide you through the necessary carburetor disassembly. The purchase of a carburetor rebuilding kit for your particular make and model automobile would provide working diagrams and a supply of new gaskets and other parts that may get damaged during the stripdown process or may be in poor condition. You probably will want to purchase several of the removable main metering jets for your engine (at a cost of approximately $1.00 apiece) so you can easily convert your car back to gasoline fuel, if desired.

The second step is to remove the carburetor air filter housing—and all its hoses, tubes, and paraphernalia—from the engine. Next, disconnect the throttle linkage (Fig. 10-9) from the carburetor, and if your automobile is so-equipped, any choke-linkage rods that are not self-contained on the carburetor body.

Now unscrew the fuel line from the carburetor inlet fitting and remove any other hoses that fasten to the unit, including vacuum and other air control

Courtesy Brown's Alcohol Motor Fuel Cookbook, Desert Publications, Cornville, AZ.

Figure 10-9. Step 2—disconnect the throttle linkage.

lines. You should label all of these cables and fittings before removal to assure their correct reconnection.

The third step is to remove the carburetor from the manifold after it is completely free from all external attachments. Single-barrel units will have only two fastening nuts or bolts, while two- or four-barrel units use four-point mounts. Once the carburetor is off the engine, drain the gasoline from it by turning it upside down (Fig. 10-10). If covered with grime, the carburetor should be cleaned with an automotive degreasing solvent (do not use a carburetor cleaner).

The fourth step is to enlarge the opening in the carburetor's main jet (or jets, if your carburetor is a multithrust model). Start by removing the air horn

Courtesy Brown's Alcohol Motor Fuel Cookbook, Desert Publications, Cornville, AZ.

Figure 10-10. Step 3—remove the carburetor from the manifold, drain gasoline, and clean unit.

from the float bowl (Fig. 10-11). In most cases there will be a choke step-down linkage rod—and possibly some other mechanical connection—between these components. Disconnect these, if you can, before unthreading the air horn's fastening screws.

The fifth step is to locate the main jet (Fig. 10-12). Some carburetors have the main jet installed in a main well support (a towerlike mount fastened to the air horn), while others mount the metering device directly in the float bowl body. In any case you should not have any trouble identifying the removable main jet: It is a round brass fitting—with a hole in its center and a slot in its top—that threads into place.

Now remove the float assembly as the sixth step. Unscrew the main jet (Fig. 10-13), and measure the diameter of its central orifice (Fig. 10-14). The simplest way to do this is to find a drill bit that fits snugly into the hole, then determine the size of the bit by matching its drill number to its diameter—in thousandths of an inch—using a machinist's conversion chart.

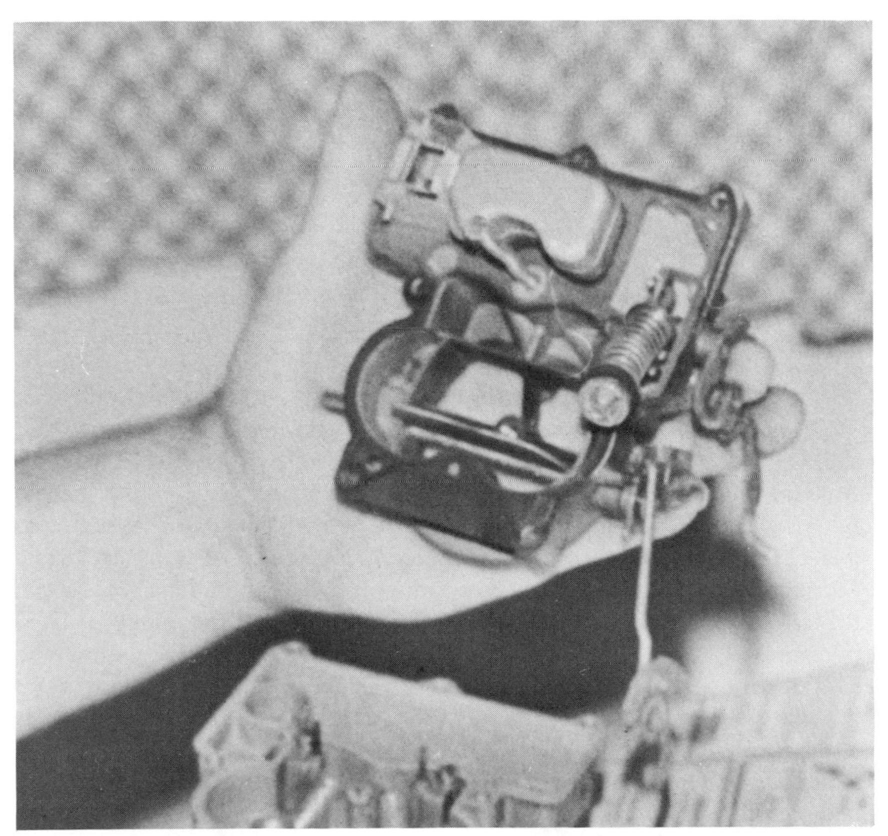

Courtesy Brown's Alcohol Motor Fuel Cookbook, Desert Publications, Cornville, AZ.

Figure 10-11. Step 4—set the top half of the carburetor aside.

Courtesy Brown's Alcohol Motor Fuel Cookbook, Desert Publications, Cornville, AZ.

Figure 10-12. Step 5—locate the main jets (shown at the end of the screwdriver).

Step 7 is to increase the size of the main jet orifice (Fig. 10-15). Once you have determined in step 6 the "normal" size of the main jet orifice, prepare to increase that dimension by about 40 percent. Remember that this isn't a fixed percentage for every engine; you may have to experiment with several different size main jets. Multiply the "normal" size of the main jet by 40 percent and add the product of the multiplication to the size of the main jet. This determines the approximate new drill bit size. As an example, suppose that the main jet "normal" size is 0.0595 according to your measurement. Multiplying 0.059 by 0.40 gives 0.0236. Next, add 0.0236 and 0.059 which sum is 0.83. This is the approximate new drill bit size required to modify the carburetor.

Step 8 (Fig. 10-16) is to enlarge the main jet orifice. The jets can be drilled out while still in the carburetor or the jet could be held with vise-grip

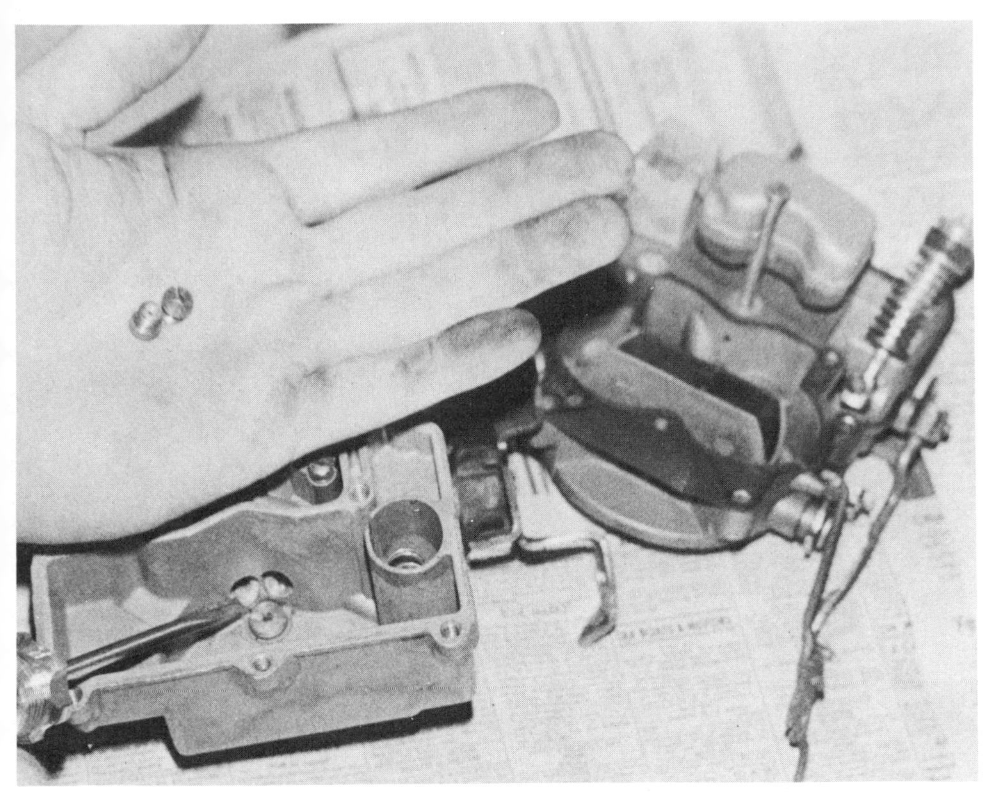

Courtesy Brown's Alcohol Motor Fuel Cookbook, Desert Publications, Cornville, AZ.

Figure 10-13. Step 6—remove the main jets (screwdriver points to where they came from).

pliers while you carefully bore out its central orifice. Be sure to clean any brass residue out of the carburetor and its components with compressed air.

With the main jet modification completed, the ninth step is to replace the main jet in the carburetor, install the float, and reassemble the carburetor. Position a fresh gasket on the manifold (if you bought a refurbishment kit) making sure that both metal surfaces are clean, and bolt the carburetor assembly in place. Reconnect all fittings.

The last step is to road test the vehicle. Three problems might occur after it is reassembled and the car is road tested. First, the car may not idle if the screws on the bottom of the carburetor are not set for alcohol; they have to be opened up (see Fig. 10-17 for location of idle screws). Next, because alcohol burns slower than gasoline, the car may not run smoothly at low speeds. The slower burning alcohol, however, lessens engine vibration considerably.

Third, slightly more ethanol than gasoline is required to run the car. If the hole in the jet is too large, there will be additional fuel consumption. Because of the large amount of oxygen in alcohol, it will keep burning in an engine

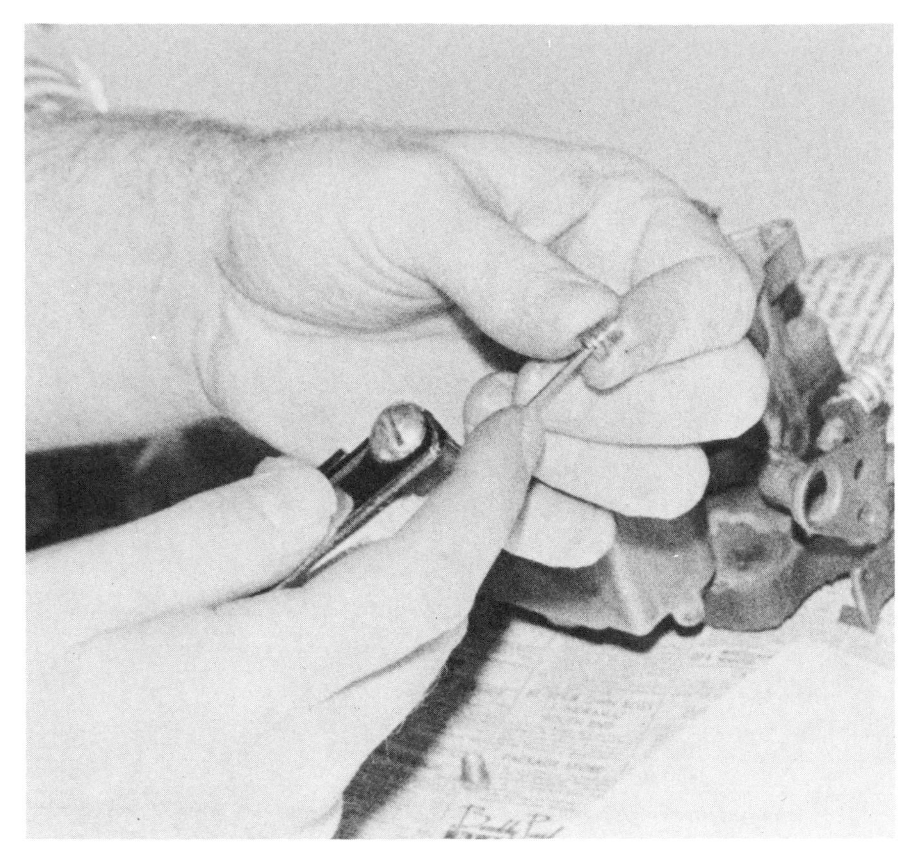

Courtesy Brown's Alcohol Motor Fuel Cookbook, Desert Publications, Cornville, AZ.

Figure 10-14. Step 7—measure the size of the main jet.

long after the same percentage of gasoline would have flooded and stalled the engine.

If the jet size is too small, the valves may burn. A gasoline-fueled engine will sputter and misfire when the jets are too small; an alcohol-fueled engine with undersized jets will burn hotter which, in turn, burns the valves. Most American cars are designed to use unburned fuel and tetraethyl lead to lubricate the exhaust valves. Engines designed to use unleaded gas usually have hardened valve sets, so this problem may not occur.

Valve burning can be prevented in several ways. One is to put one-half cup of vegetable oil into the fuel tank with the alcohol. (One-half cup of diesel fuel could be used, but it is not recommended since alcohol and petroleum products will not mix if there is any water in the alcohol.) If the engine runs cool enough, the alcohol will sometimes act as a valve lubricant.

Two other alternatives for preventing burning valves are to install stellite or stainless-steel racing valves with hardened seats, or to buy a spare car-

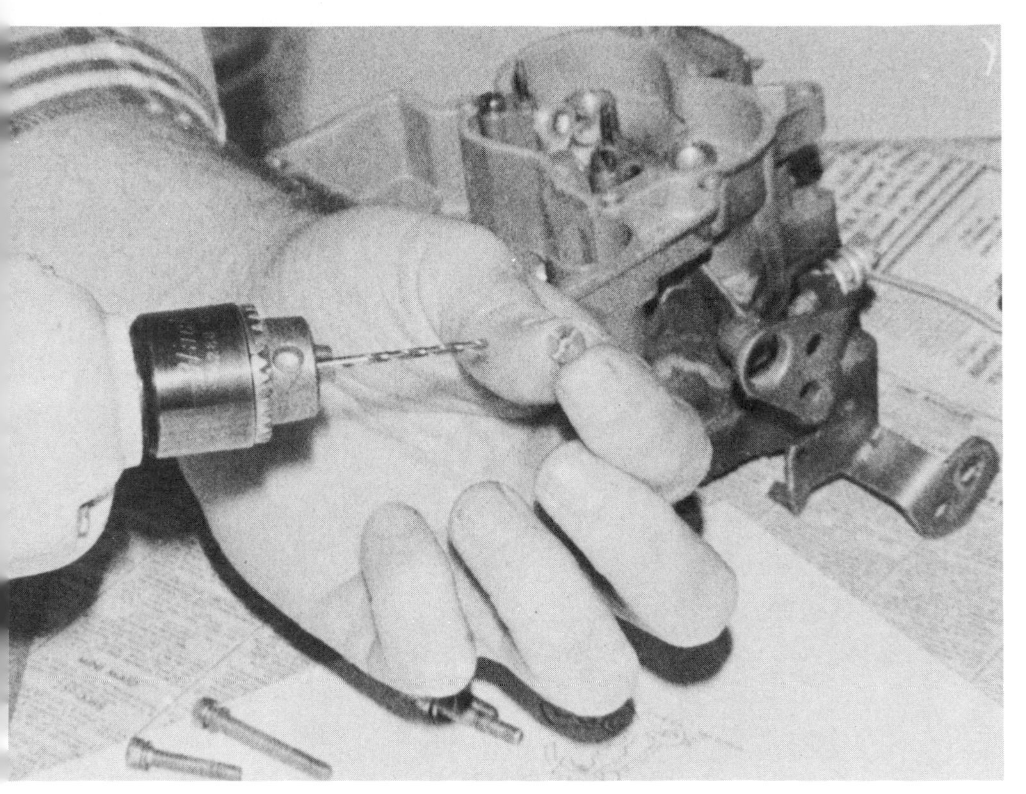

Courtesy Brown's Alcohol Motor Fuel Cookbook, Publications, Cornville, AZ.

Figure 10-15. Step 8—drill out the main jets for alcohol using the appropriate size drill.

buretor. In some instances, with the aid of adaptor plates, a Ford carburetor may replace a Chevrolet carburetor, and vice versa. This is frequently an advantage because some carburetors are much easier to work on than others.

Gasoline and Alcohol Modification

If you want to alternate between gasoline and alcohol, only a needle valve is required. All carburetors used to be equipped with them. A tapered shaft was inserted into the main jet by means of a screw thread adjustment. The screw thread adjustment allowed the tapered shaft to be moved in and out of the hole in the jet. The further the shaft was inserted into the hole, the smaller the hole became. As it was unscrewed or withdrawn, the opening became larger.

There are several means for setting the opening: using an rpm gauge, a vacuum gauge, or emission control gauge. The emission control gauge measures unburned hydrocarbons which, in turn, result in better mileage because more fuel is being used by the engine. Also, you can listen to the

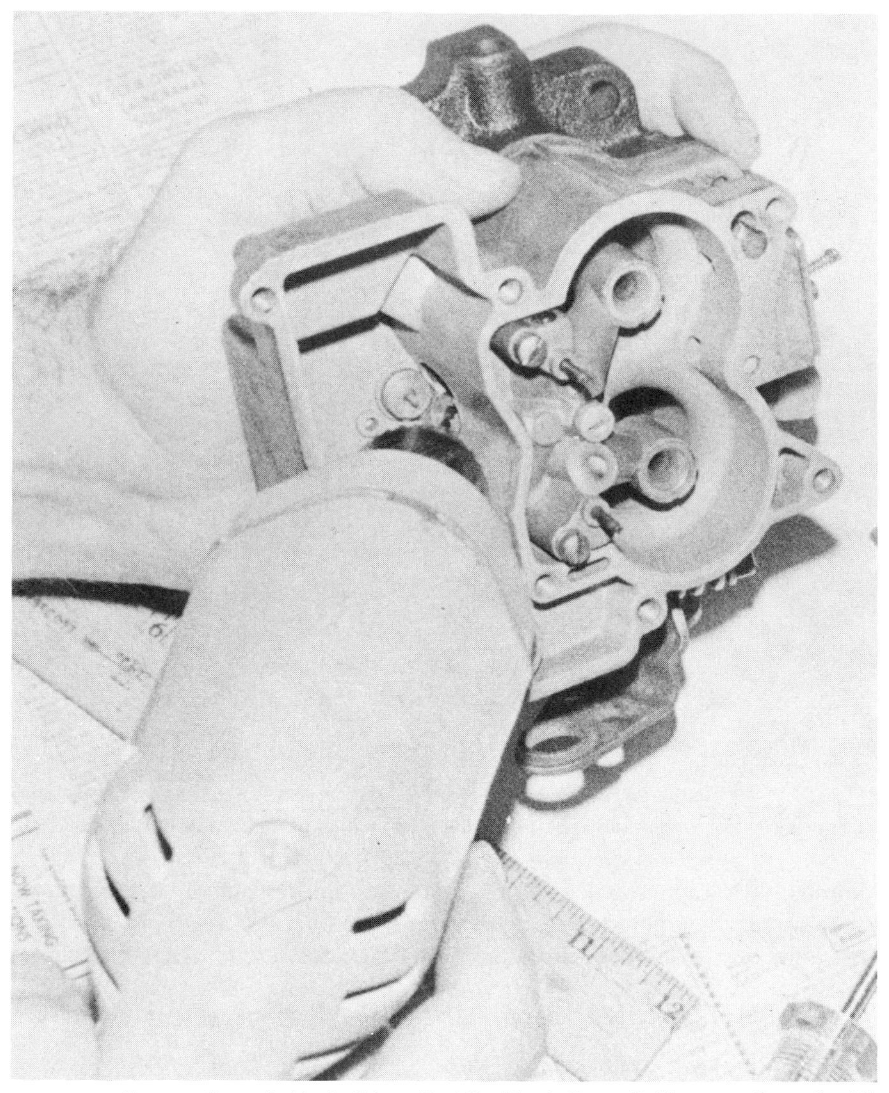

Courtesy Brown's Alcohol Motor Fuel Cookbook, Desert Publications, Cornville, AZ

Figure 10-16. Step 9—enlarge main jet orifice. Jet could be drilled while in carburetor.

engine as you unscrew the needle valve when you switch from gasoline to alcohol.

With the advent of the downdraft carburetor, most American car manufacturers used the fixed-jet principle. Most of the old updrafts had adjustable needle valves (Fig. 10-18).

A few engines and carburetors are factory equipped with a main jet adjustment (Figs. 10-19 and 10-20). Facet Carburetor Corporation markets a number of carburetors with main jet adjustments for stationary power

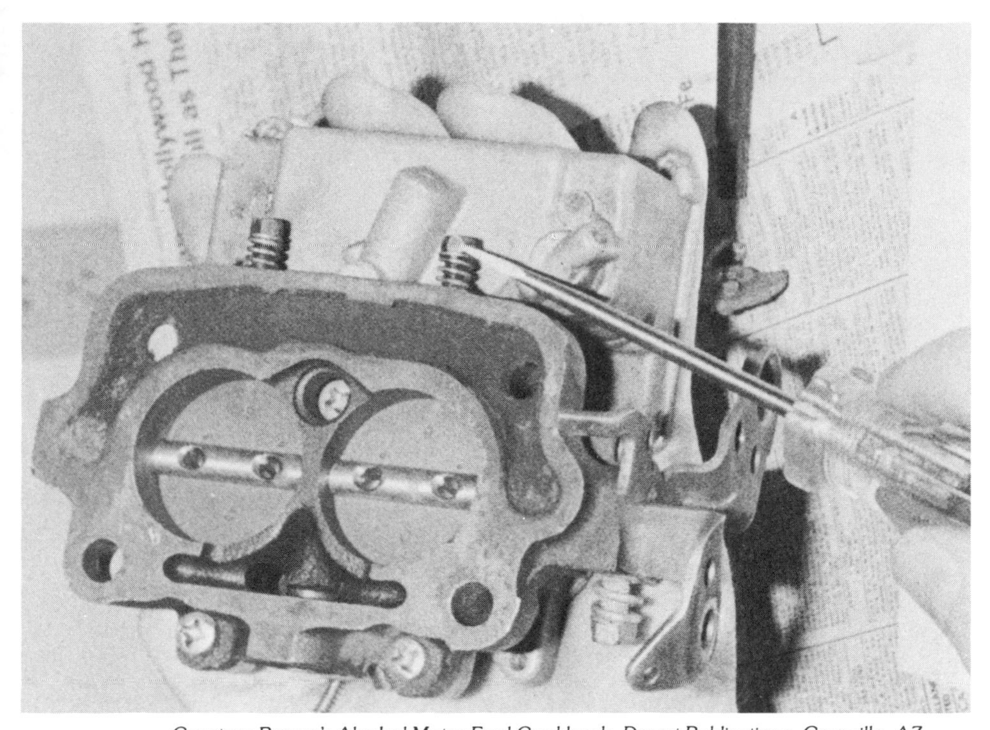

Courtesy Brown's Alcohol Motor Fuel Cookbook, Desert Publications, Cornville, AZ.

Figure 10-17. The idle adjusting screw is under the carburetor. The carburetor in this view is upside down. The screwdriver points to the idle adjusting screw.

plants. Many small air-cooled engines also have main jet adjustments. Main jet adjustments allow altitude adjustments, but they have no accelerator pump.

Most passenger car carburetors have a screw at their base which is the needle valve. This valve controls the idle adjustment (Fig. 10-17). At idle, an engine is normally running on a 6:1 air to gasoline mixture instead of the normal 15:1.

It is not necessary to disassemble the carburetor each time that fuels are changed. On most four-barrel carburetors, the back two barrels are usually used only for passing or accelerating. Normally no gasoline will flow through the back two jets. For example, on a Rochester, the back two jets are opened and closed by two metering rods that function almost the same way a needle valve does. By pressing on the accelerator pedal, the metering rods should rise up out of the jets. This means that, if the main jets are blocked off in the front two barrels, the metering rods are remachined to a taper, the jets are bored out, and the linkage is reset, the car should run on either gasoline or alcohol.

Another method is to buy two Rochesters, cut off the back two barrels with the metering rods in them. Then, turn one barrel backwards and weld

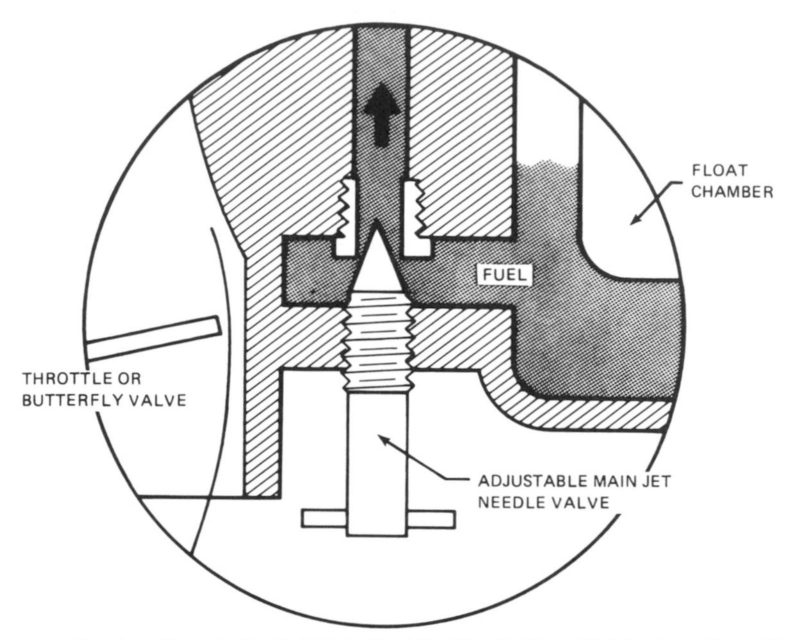

Courtesy Brown's Alcohol Motor Fuel Cookbook, Desert Publications, Cornville, AZ.

Figure 10-18. Enlarged section of carburetor drawing showing an adjustable main jet in place of the fixed jet.

them back together. Install a gasoline fuel line to either the front or back half of the carburetor. Enlarge the jets and install an alcohol fuel line to the other half of the carburetor. Run the alcohol and gasoline fuel lines to their respective fuel tanks with shutoff valves. An adaptor plate will have to be fabricated for the intake manifold.

Until 1959, John Robert Fish manufactured a carburetor that operated on a self-adjusting principle; it would run on alcohol, gasoline, or kerosene. Some manufacturers are trying to get it back into production since it usually produced 40 percent better gas mileage.

This carburetor had six jets in the butterfly (or throttle) valve of the one barrel. The fuel passage from the jets made a right-angle turn down a radical arm that swung at the bottom of the fuel bowl and picked up fuel from a groove cut in the side of the float chamber. The groove was deeper and wider at one end than it was at the other. The further the throttle was opened, the deeper and wider the groove became under the hole in the radical arm, which increased the size of the main jet as the gas pedal was pushed further down. This meant that the accelerator pedal had to be pushed down more when the fuel was alcohol than it did for gasoline.

The float adjustment can also alter fuel economy. For maximum economy, weigh the float on a gram scale, then braze 10 percent of the total weight to the top middle of the float. (Alcohol as a fluid is 10 percent heavier than gasoline.)

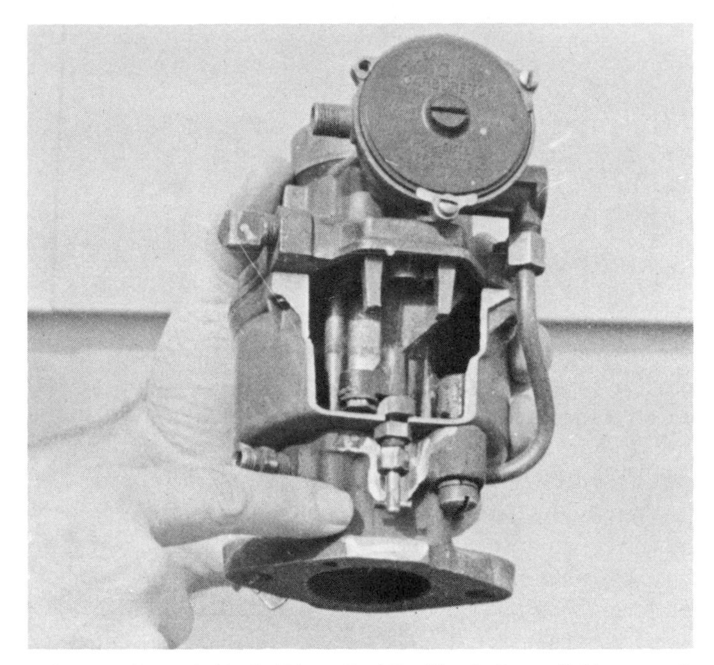

Courtesy Brown's Alcohol Motor Fuel Cookbook, Desert Publications, Cornville, AZ.

Figure 10-19. Carburetor with cut-a-way section to show the addition of an adjustable main jet.

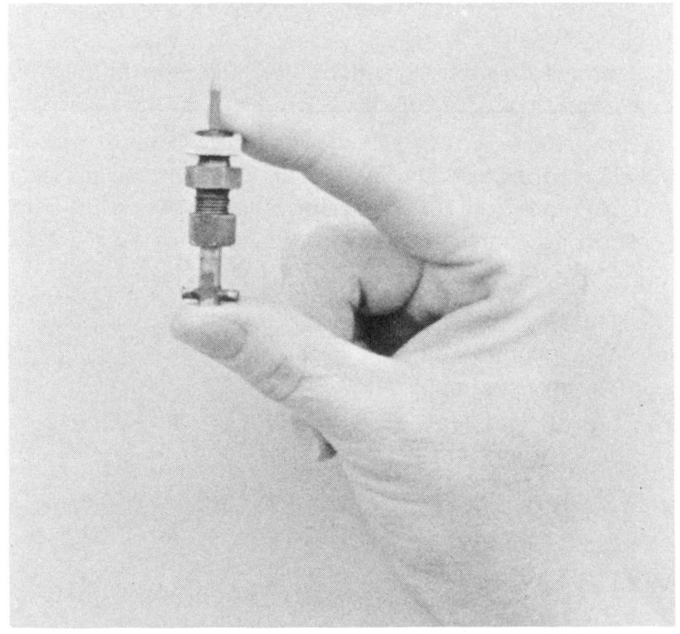

Courtesy Brown's Alcohol Motor Fuel Cookbook, Desert Publications, Cornville, AZ.

Figure 10-20. Adjustable main jet assembly removed from carburetor.

Chapter 11

Business Planning

Entry into the alcohol industry requires equal planning from both business and technological perspectives. Many technologically sound projects have never gotten off the ground for lack of business acumen, lack of adequate financing, and lack of a cohesive team capable of transforming ideas and strategies into operational results. These concerns are of particular significance in the alternative energy field where there is insufficient financial and management history to ensure an adequate return on investment. The purpose of a business plan is to blend innovative ideas into a detailed plan of action. The principal objectives of a business plan include:

- The establishment of a project's goals and objectives, such as monetary return, investment criteria; protection of supply, image, diversification, and public or national spirit. As a general rule, the more owners or investors in a project, the more diverse a project's objectives become.
- The development of a decision matrix for "Go Ahead," "No Go," and modification of plan decisions.
- The assembly of a task force team including both internal and external members capable of accomplishing the project's objectives, and likewise having the ability to advise the investors when to abort an undesirable project. This team should include:

  —investors, whose function is to establish the objectives of the project;

  —internal management and staff personnel, whose function is to carry out the investors' objectives;

  —external consultants and vendors, who will usually include:

    attorneys

    certified public accountants

    engineering consultants

    investment bankers

    commercial bankers

contractors

insurance agents and underwriters.

Once the project team is assembled, the goals are defined, and a decision matrix is established, a detailed schedule or work program should be developed for project control. This detailed schedule or work program should include, as a minimum, the following:

- Designation of a project manager or director, and description of all project members' functions.
- Specific assignment of tasks to project members with definite dates for completion.
- Definition of the project's objectives and goals.
- A method for interim reporting on the status of each assignment.
- A calendar of expected completion dates for key areas in the project which will require decisions.

The following discussion addresses the key requirements of studying, organizing, financing, constructing, and operating an ethanol plant.

Seed Money Stage

"Seed money" is usually the most difficult to raise for any investment. By definition, seed money pays for expenses incurred during the planning stage of a project, before decisions are made on how best to proceed; consequently, seed money is considered high risk by investors. Because of their greater risk, investors who enter the venture at this time will receive an ownership position in the project greater than their proportional dollar share of investment. Seed money is particularly important to alcohol-fuel projects to finance the prefeasibility study. The financing and economic feasibility study requires the accumulation of marketing, legal, economical, technical, and financial information by engineers, attorneys, and CPA's, to evaluate the viability of the project.

In planning an ethanol plant, one must obtain funds to cover the various project costs. These costs and possible funding sources are listed in Table 11-1.

Once a project team is assembled, seed money is acquired, and a financing plan is developed, the organizational form of the entity must be considered.

Organizing the Venture

Alcohol plants may be established in various organizational structures depending on the objectives of investors and applicable regulations. The

Table 11-1. Project Costs and Possible Funding for an Ethanol Plant

COST ITEM	FUNDING SOURCE		
	SEED MONEY OR EQUITY	MAY BE DEBT FINANCED	POSSIBLE GRANT
Prefeasibility Cost	X		
Underwriting the Cost of Raising Equity	X		
Feasibility Study	X		X
Exploring Sources of Financing	X		
Preliminary Engineering	X		X
Final Engineering		X	
Site Options	X		
Acquiring a Site		X	
Plant and Equipment Cost (Construction Period)		X	X
Plant and Equipment Cost (Permanent)		X	X
Working Capital Requirement for the Plant	X	X	

Source: *A Guide to Commercial Scale Ethanol Production and Financing,* SERI, Denver, 1980.

organizational basis is the legal and business framework for the ethanol-production facility. Broadly speaking, there are three principal kinds of business structures: proprietorship, partnership, and corporation.

A proprietor is an individual who operates without partners or other associates and, consequently, has total control of the business. A proprietorship is the easiest type of organization to begin and end, and has the most flexibility in allocating funds. Business profits are taxed as personal income and the owner/proprietor is personally liable for all debts and taxes. The cost of formation is low, especially in this case, since licensing involves only the BATF permit to produce ethanol and local building permits.

A partnership is two or more persons contractually associated as joint principals in a business venture. This is the simplest type of business arrangement for two or more persons to begin and end and has good budgetary flexibility (although not as good as a proprietorship). The partners are taxed separately, with profits as personal income, and all partners are personally liable for debts and taxes. A partnership can be established by means of a contract between two or more individuals. Written contractual agreements are not legally necessary, and therefore oral agreements will suffice. In a general partnership, each partner is personally liable for all debts of the partnership, regardless of the amount of equity which each partner has contributed.

A corporation is the most formalized business structure. It operates under the laws of incorporation of a state; it has a legal life all its own; it has its scope, activity, and name restricted by a charter; it has its profits taxed separately from the earnings of the executives or managers, and makes only the company (not the owners and managers) liable for debts and taxes. A Board of Directors must be formed and the purposes of the organization must be laid out in a document called "The Articles of Incorporation." Initial taxes and certain filing fees must also be paid. Finally, in order to carry out the business for which the corporation was formed, various official meetings must be held. Since a corporation is far more complex in nature than either a proprietorship or a partnership, it is wise to have the benefit of legal counsel. A corporation has significant advantages as far as debts and taxes are concerned. Creditors can only claim payments to the extent of a corporation's assets; no shareholder can be forced to pay off creditors out of his or her pocket, even if the company's assets are unequal to the amount of the debt.

There are often differences in the ease with which a business may obtain start-up or operating capital. Sole proprietors stand or fall on their own merits and worth. When a large amount of funding is needed, it may be difficult for one person to have the collateral necessary to secure a loan or to attract investors. Partnerships have an advantage in that the pooled resources of all the partners are used to back up the request for a loan and, consequently, it is often easier to obtain a loan because each and all of the partners are liable for all debts. Corporations are usually in the best position to obtain both initial funding and operating capital as the business expands. New shares may be issued; the company's assets may be pledged to secure additional funding; and bonds may be issued, backed up by the assets of the corporation.

A nonprofit cooperative is a special form of corporation. Such a cooperative can serve as a type of tax shelter. While the cooperative benefits each of its members, they are not held liable, either individually or collectively, for taxes on the proceeds from the sale of their products. The different entity forms which may be considered include:

Proprietorship

General partnership

Limited partnership

Joint venture

New subsidiary of an existing corporation

New division of an existing corporation

New corporation

Acquisition or merger of an existing corporation

Because of the complicated legal, accounting, financial, and tax considerations involved in choosing an organization structure, a decision should not be made until detailed discussions are held with the project's attorney and CPA to determine which entity format is best suited to a given situation.

The key factors and pros and cons of the alternative entity forms are shown in Table 11-2.

Table 11-2. Alternative Entity Options

Factor of comparison	Individual/partnership or consolidated corporate subsidiary	New nonconsolidated corporate ownership
Start-up cost, depreciation or losses may be deducted when incurred by investor	Usually passed out to the investor for a current deduction	Carried forward to offset against future income
Tax credit such as investment tax credit and energy tax credit	Usually passed out to the investor for a current deduction	Carried forward to offset against future income
Rates of taxation	The personal tax rate of the individual or partners; the corporate rate of corporate parent	From 17 percent of the first $25,000 to 46 percent on income in excess of $100,000
Liability of investor	Unlimited to proprietor or general partner, others usually limited	Usually limited
Taxation	To the individual or corporate owner when incurred; no taxation of dividends to individuals and an 85 percent exemption for corporate parents	Taxed at corporate level when earned; dividends also taxed to the individual when distributed

Source: *A Guide to Commercial Scale Ethanol Production and Financing,* SERI, Denver, 1980.

Corporate Options

Establishing a project office within an existing corporation has the following advantages:

Ability to select from a large, tried, and proven management staff

Lower administrative burden

Established sources of internal funds

Existing credit rating

Existing credit sources

Reliance on internal development has been a rule for many firms, yet the costs and risks involved in such action should be weighed against the opportunities created by acquisition which can often provide more rapid and predictable results. The addition of new corporate capabilities by this method should be investigated as early as possible for any conceivable antitrust implications. The public interest, as reflected in the antitrust laws, prohibits any acquisition or merger that substantially lessens competition or creates a monopoly. In favor of this alternative is the immediate acquisition of trained management, a new source of working capital, and an expanded credit base.

A new corporation is an obvious organizational alternative. A charter defines its purpose and scope, and formation of a new corporation has significant debt and tax advantages. Financing may be obtained by issuing new shares; and company assets may be pledged as security for loans and bonds.

A joint venture is a special form of a partnership or corporation created by two or more existing legal entities to accomplish a specific, limited objective. The necessary seed money, management, labor, and board of directors are provided by the sponsoring organizations, but their operational expenses thereafter are borne by the new group.

Once the legal and management structures have been determined, the organization should set regular business sessions and stockholders' meetings. It is not important at this point to fine-tune the organization; the most pressing detail is building the project team which is capable of structuring the financial arrangements for the project.

Criteria in Selection of the Project

Unless the organization is a joint venture among highly sophisticated firms with sufficient resources available, or an organization with an internal source of expertise in all fields necessary for operation of an alcohol project, it will be necessary to turn to external sources for expertise. Outside consultants will probably be needed.

It is important to choose the right professionals to handle the organization's needs. Arrangements or contracts should be negotiated and signed with a recognized engineering design firm, a legal firm, a CPA firm, a construction contractor, and, most likely, an investment banker. Because of the rapid development of the alcohol industry, there are a number of young, highly competitive, and aggressive companies in the marketplace offering services in process design, engineering, construction, and financing. Unfortunately, the rapid growth has also enticed a number of marginally qualified firms into this marketplace. Promoters and investors should take great care in selecting qualified firms in order to assure quality work and to protect their investment. It is advisable to obtain personal and business references, and to utilize firms with experience not only in the field of alcohol distilling, but also to choose those who have structured projects of a similar dollar magnitude compared to the project involved.

In evaluating the engineering firm, it is important to recognize that there are no monopolies on alcohol-production processes; many choices are available. As each process undoubtedly has strengths and weaknesses, both price and quality should be considered in the selection process. Several specific questions are important in the determination of an engineering firm. The organization should request:

A detailed economic analysis and description of the process which the firm recommends and an estimated manufacturing cost per gallon.

Specific projects which the firm has engineered in the alcohol industry.

A copy of the process and performance guarantees used by the firm.

A copy of the legal contracts which the firm uses in contracting for services.

The current financial statements and bank references of the engineering firm.

The above information should be reviewed by the project's attorneys and CPA's, and the attorney should finalize all contracts with the engineering firm.

In selecting attorneys, CPA's, and investment bankers, it is essential to consider the following:

Experience in the industry

Location of the professionals, due to the almost daily contact required in the development stage

The attorney's specific knowledge in areas of taxation, contracts (possibly securities law and underwriting), federal regulations, patents, etc.

The CPA firm's special knowledge in project financing, forecasting, taxation, feasibility studies, management analysis, and, possibly, in SEC regulations and underwriting

The investment banker's experience in successfully placing project financing and, if necessary, the equity portion of a project

The criteria in selecting the contractor are identical to those of the engineering firm, with the addition that the contractor be bondable.

Project Financing

Once the team is assembled and the feasibility study is underway, the detailed financing package must be assembled by the CPA's and investment bankers. This package, in addition to the feasibility study, must include:

Personal résumés of applicants

Personal résumés of selected plant managers

Contracts for supply of raw materials

Contracts for energy supply

Contracts for sale of production

Contracts for sale of coproducts

Corporate/partnership history and historical financial statements

Summary of collateral offered

Summary of projected use of funds

Summary of projected source of funds

It is important to estimate accurately the time required to accumulate all of the legal and financial documentation required. This is especially true where government forms and applications are necessary. It is also important, when requesting federal financial assistance, to consider aspects which address the government's objectives. For example, some government agencies are presently encouraged to give priority assistance to projects which use a farm cooperative to supply raw materials and thus contribute to the enrichment of the local economy, or to projects which use municipal wastes or coal to fuel the plant. If such options are reasonable, then the loan package should emphasize their expected effects on the local economy.

Identification of federal involvement has another dimension; i.e., a request for financial assistance should give reference to such federal incentives as the Crude Oil Windfall Profits Tax Act of 1980. The alcohol-fuel sections of the Act may provide long-term market support for gasohol, cut some of the red tape involved in licensing and regulatory procedures, and offer tax incentives for investors in alcohol-fuel plants.

Finally, a banking agent for the project should be identified and the bank itself should make some commitment to the project to act as a supplier of funds. This commitment indicates a solid relationship with the bank and gives further creditability to the project. In virtually all instances, with the exception of the costs of documenting the loan package, financing services should be based on results. Investment banking commitments will normally cost 1-2 percent of the total loan requested. CPA firms, which professionally handle government applications for financing, will normally charge a negotiable fee for assisting in the feasibility study and preparation of the financing package. This fee, based on the time and billing rates of the individuals involved in the preparation of the financing package, will generally amount to about 1-1.5 percent of the loan amount requested.

Plant Construction and Operation

In the final phase of the business plan, management must become actively involved in the project. The owners will need to hire management and operations personnel to work with the engineering and construction companies to gain in-depth knowledge of the plant operations. It is extremely important that cost controls be effected early in the project development. Thus, the company must implement its internal accounting procedures at the earliest possible date.

Regular meetings among management, engineers, and construction personnel should be maintained throughout the construction phase and into start-up operations. Management should locate necessary skilled and semi-skilled operations personnel, and plan the initial logistics of start-up operations during the project construction period. A time-phased approach will prove most effective in overall business planning.

Small-Scale Plant Business Plan

This business plan is for the 25-gallon-per-hour small-scale ethanol plant described in SERI's *Fuel from Farms—A Guide to Small-Scale Ethanol Production*. The case study was prepared by Jim Smrcka of Galusha, Higgins, and Galusha of Glascow, Montana.

The case study of the Johnson family demonstrates the process for determining the feasibility of a small farm-sized fermentation ethanol plant by developing a business plan. It is a realistic example, but the specific factors are, of course, different for every situation. This process may be used by anyone considering small-scale ethanol-plant development, but the numbers must be taken from one's own situation. Table 11-3 delineates the assumptions used in the case study.

BACKGROUND INFORMATION

The Johnson family operates a 1280-acre corn farm, which they have owned for 15 years. They feed 200 calves in their feedlot each year. The family consists of Dave, Sue, and three children: Ted, 24 years old and married, has been living and working at the farm for 2 years, and he and his wife have a strong commitment to farming; Sara is 22 years old, married, and teaches in a town about 25 miles away; Laura is 15 years old, goes to high school in town 25 miles away, and also works on the farm.

Table 11-3. Case Study Assumptions

Corn is the basic feedstock.
25-gal-per hour ethanol production rate.
Operate 24 hours/day; 5 days/week; 50 week/year.
Feed whole stillage to own and neighbors' animals.
Sell ethanol to jobber for $1.74/gal.
Sell stillage for 3.9¢/gal.
Corn price is $2.30/bu (on-farm, no delivery charge, no storage fees).
Operating labor is 4 hours/day at $10/hour.
Corn stover cost is $20/ton.
Equity is $69,000.
Debt is $163,040; at 15 percent per annum; paid semiannually.
Loan period is 15 years for plant; 8 years for operating capital and tank truck.
Miscellaneous expense estimated at 12¢/gal ethanol produced.
Electricity costs estimated at 2¢/gal ethanol produced.
Enzymes estimated at 4¢/gal ethanol produced.

Source: *Fuel From Farms—A Guide to Small-Scale Ethanol Production,* SERI, Denver, 1980.

The Johnsons are concerned about the future cost and availability of fuel for their farm equipment. The Johnsons have known about using crops to produce ethanol for fuel for a long time, and recent publicity about it has rekindled their interest.

They have researched the issues and believe there are five good reasons for developing a plan to build a fermentation ethanol plant as an integral part of their farm operation:

To create another market for their farm products.

To produce a liquid fuel from a renewable resource.

To gain some independence from traditional fuel sources and have an alternate fuel available.

To gain cost and fuel savings by using the farm product on the farm rather than shipping it, and by obtaining feed supplements as a coproduct of the ethanol-production process.

To increase profit potential by producing a finished product instead of a raw material.

They first analyzed the financial requirements in relation to their location, farm operations, and personal financial situation.

ANALYSIS OF FINANCIAL REQUIREMENTS

The local trade center is a town of 5000 people, 35 miles away. The county population is estimated at 20,000. Last year 7 million gallons of gasoline were consumed in the country according to the state gasoline tax department. A survey of the Johnsons' energy consumption on the farm for the last year shows:

Gasoline = 13,457 gallons
Diesel = 9,241 gallons
LPG = 11,487 gallons

They decided to locate the plant close to their feedlot operation for ease in using the stillage for their cattle. They expect the plant to operate 5 days a week, 24 hours a day, 250 days a year. It is designed to produce 25 gallons of anhydrous ethanol per hour or 150,000 gallons in 1 year, using 60,000 bushels of corn per year. In addition to ethanol, the plant will produce stillage and carbon dioxide as coproducts at the rate of 230 gallons per hour.

After researching the question of fuel for the plant, this family has decided to use agricultural residue as the fuel source. This residue will be purchased from the family farm. The cost of this fuel is figured at $20 per ton.

They will have to purchase 1 year's supply of fuel since it is produced seasonally in the area. The tonnage of residue per acre available is 6 tons per acre based on measurements from the past growing season.

The water source is vitally important. They will need about 400 gallons of water each hour. To meet this demand, a well was drilled and an adequate supply of water was found. The water was tested for its suitability for use in the boiler, and the test results were favorable.

The family has determined that they can operate both the plant and the farm without additional outside labor. The ethanol they produce will be picked up at the farm as a return load by the jobbers making deliveries in the rural area. The Johnsons will deliver stillage in a tank truck to neighbors within 5 miles.

At the present there are no plans to capture the carbon dioxide, since the capital cost of the equipment is too high to give a good return on their investment. There is no good local market for the carbon dioxide; but there are many uses for carbon dioxide, and selling it as a coproduct may prove to be profitable in other situations.

The family's plans are to market their products locally. They have contacted local jobbers who have given them letters of intent to purchase the annual production of ethanol. They plan to use the distiller's grains in their own feedlot and to sell the rest to their neighbors. The neighbors have given them letters stating they would purchase the remainder of the distiller's grains produced. These letters are important in order to accurately assess the market.

ORGANIZATIONAL FORM

The Johnsons chose to establish a closely held corporation for this business. Other possibilities they considered included partnerships, sole proprietorships, and profit and nonprofit corporations. If additional equity had been needed, a broader corporation or a partnership would have been selected. However, their financial status was sufficient to allow them to handle the investment themselves, as shown by their balance sheet shown in Fig. 11-1.

The Johnsons formed a corporation because it afforded the ability to protect themselves from product liability and gave them the option to give stock to all family members as an incentive compensation package. Also, the use of a corporation avoids an additional burden on their credit line at the bank for their farming operation since they were able to negotiate a loan with no personal guarantee of the corporate debt. In a partnership they would have had personal liability for the product, the debt, and the actions of the partners in the business. The recordkeeping requirements of the corporation and the limited partnership were equal, and the former afforded greater security. Co-ops and nonprofit corporations were considered also, but these two options were discarded because of operating restrictions.

After formation of the corporation, they transferred (tax-free) one-half of a year's supply of corn (30,000 bushels) in exchange for stock in the cor-

GLASGOW, MONTANA

November 6, 1979

National Bank of Golden Rise
Golden, Colorado

We have assisted in the preparation of the accompanying projected balance sheet of Johnson Processors, Inc. (a sample company), as of December 31, 1980 and 1981, and the related projected statements of income and changes in financial position for the years then ended. The projected statements are based solely on management's assumptions and estimates as described in the footnotes.

Our assistance did not include procedures that would allow us to develop a conclusion concerning the reasonableness of the assumptions used as a basis for the projected financial statements. Accordingly, we make no representation as to the reasonableness of the assumptions.

Since the projected statements are based on assumptions about circumstances and events that have not yet taken place, they are subject to the variations that will arise as future operations actually occur. Accordingly, we make no representation as to the achievability of the projected statements referred to above.

The terms of our engagement are such that we have no obligation or intention to revise this report or the projected statements because of events and transactions occurring after the date of the report unless we are subsequently engaged to do so.

James Sourcka

G, H & G
Certified Public Accountants

Source: *Fuel From Farms—A Guide to Small-Scale Ethanol Production,* SERI, Denver, 1980.

Figure 11-1. Projected financial statements for Johnson Processors Inc.

JOHNSON PROCESSORS, INC.
PROJECTED BALANCE SHEET
UNAUDITED

(NOTES 1 THROUGH 4 ARE AN INTEGRAL PART OF THESE PROJECTED FINANCIAL STATEMENTS AND PROVIDE AN EXPLANATION OF ASSUMPTIONS USED IN THIS REPORT)

	FOR YEAR ENDED			
		December 31, 1980		December 31, 1981
Assets				
Current assets				
Cash		$24,617		$18,452
Accounts receivable		55,350		55,350
Raw materials and supplies		22,214		26,785
Work in process inventory		1,362		1,601
Finished goods inventory		18,765		22,041
Marketable securities		30,000		30,000
Total current assets		$152,308		$154,229
Plant, equipment, and structures				
Plant and equipment	$107,000		$107,000	
Building	17,280		17,280	
Total plant and equipment	$124,280		$124,280	
Less accumulated depreciation	8,605		17,210	
Net plant and equipment		115,675		107,070
Total assets		$267,983		$261,299
Liabilities and Capital				
Current liabilities				
Accounts payable		$3,800		$4,130
Current portion of loans		2,997		3,500
Total current liabilities		$6,797		$7,630
Long-term liabilities				
Bank loan	$156,043		$113,821	
Less current portion	2,997		3,500	
Total long-term liabilities		153,046		110,321
Total liabilities		$159,843		$117,951
Capital		108,140		143,348
Total liabilities and capital		$267,983		$261,299

Figure 11-1. ***(continued)***

JOHNSON PROCESSORS, INC.
PROJECTED INCOME STATEMENT
UNAUDITED

(NOTES 1 THROUGH 4 ARE AN INTEGRAL PART OF THESE PROJECTED FINANCIAL STATEMENTS AND PROVIDE AN EXPLANATION OF ASSUMPTIONS USED IN THIS REPORT)

		FOR YEAR ENDED		
		December 31, 1980		**December 31, 1981**
Revenue				
Alcohol		$229,680		283,500
Stillage		55;525		55,973
Total sales		$285,205		$339,473
Cost of goods sold				
Beginning finished goods inventory		0		$18,765
Cost of goods manufactured		$225,634		266,634
Cost of goods available for sale		$225,634		$285,399
Ending finished goods inventory		18,765		22,041
Cost of goods sold		$206,869		$263,358
Gross profit		$78,336		$76,115
Selling expenses				
Marketing and delivery expenses (scheduled)	$25,545		$29,074	
Total selling expenses		$25,545		$29,074
Net operating profit		$52,791		$47,041
Income taxes		13,651		11,833
Net income		$39,140		$35,208

Figure 11-1. *(continued)*

JOHNSON PROCESSORS, INC.
PROJECTED STATEMENT OF CHANGES IN FINANCIAL POSITION (Cash Basis)
UNAUDITED

(NOTES 1 THROUGH 4 ARE AN INTEGRAL PART OF THESE PROJECTED FINANCIAL STATEMENTS AND PROVIDE AN EXPLANATION OF ASSUMPTIONS USED IN THIS REPORT)

	FOR YEAR ENDED	
	December 31, 1980	December 31, 1981
CASH GENERATED		
Net income	$39,140	$35,208
Add (deduct) items not requiring or generating cash during the period		
Trade receivable increase	(55,350)	(0)
Trade payable increase	3,800	330
Inventory increase	(42,341)	(8,086)
Depreciation	8,605	8,605
Subtotal	$(46,146)	$36,057
Other sources		
Contributed by shareholders	69,000	
Bank loan	163,040	
Total cash generated	$185,894	$36,057
CASH APPLIED		
Additional loan repayment	$ 4,000	$38,722
Purchase of plant and equipment	107,000	
Purchase of building	17,280	
Reduction of bank loan	2,997	$3,500
Total cash applied	$131,277	$42,222
Increase in cash	$ 54,617	$(6,165)*

*Net decrease in cash is caused by an accelerated pay-off of the operating capital rate in the amount of $38,722.

Figure 11-1. ***(continued)***

JOHNSON PROCESSORS, INC.
EXHIBIT I
UNAUDITED

(NOTES 1 THROUGH 4 ARE AN INTEGRAL PART OF THESE PROJECTED FINANCIAL STATEMENTS AND PROVIDE AN EXPLANATION OF ASSUMPTIONS USED IN THIS REPORT)

	FOR YEAR ENDED	
	December 31, 1980	**December 31, 1981**
ETHANOL		
Projected Production Schedule		
Projected gallons sold	132,000	150,000
Projected inventory requirements	18,000	18,000
Total gallons needed	150,000	168,000
Less inventory on hand	0	18,000
Projected production	150,000	150,000
Sales price per gallon	$1.74	$1.89
Projected Cost of Goods Manufactured		
Projected production costs:		
Labor	$20,805	$20,328
Corn	138,000	180,000
Electricity	3,000	3,300
Straw	10,714	11,785
Miscellaneous (scheduled)	18,000	19,800
Depreciation	6,730	6,730
Interest	23,747	18,330
Enzymes	6,000	6,600
Total costs of production	$226,996	$266,873
Add beginning work-in-process inventory		1,362
Subtotal	$226,996	$268,235
Less ending work-in-process inventory	1,362	1,601
Projected cost of goods manufactured	$225,634	$266,634

Figure 11-1. *(continued)*

JOHNSON PROCESSORS, INC.
NOTES TO THE PROJECTED FINANCIAL STATEMENTS
UNAUDITED

1. SIGNIFICANT ACCOUNTING POLICIES

Following is a summary of the significant accounting policies used by Johnson Processors, Inc. in the projected financial statements.

- Assets and liabilities, and revenues and expenses are recognized on the accrual basis of accounting.
- Inventory is recorded at the lower value (cost or market) on the first-in, first-out (FIFO) basis.
- Accounts receivable are recorded net of bad debts.
- Depreciation is calculated on the straight line basis.

2. ASSETS

Current Assets

- Accounts receivable are projected at each balance sheet date using 30 days of sales for ethanol and 90 days of sales for stillage.
- Inventory—raw materials—is made up of corn and corn stover. Thirty days in inventory is used for corn and one year's supply is used for stover.
- Inventory—work in process—consists of 1½ days' production.
- Inventory of finished goods consists of raw materials and cost of production. Thirty days in inventory is used for ethanol and 2 days is used for stillage.

The estimates of number of days in accounts receivable and finished goods inventory are higher than those quoted in Robert Morris Associates averages for feed manufacturers and wholesale petroleum distributors.

Fixed Assets

Management anticipates purchasing the equipment for production of ethanol. Consulting engineers contacted verified that the equipment and plant costs listed in Table VI-2 were reasonable.

REPRESENTATIVE PLANT COSTS

Equipment and Materials	$ 71,730
Piping	4,000
Electrical	1,500
Excavation and Concrete	2,000
Total Equipment and Materials	79,230
10% Contingency	7,923
Total	87,153
Tank Truck	14,847
Erection Costs	5,000
Grand Total	$107,000

Figure 11-1. *(continued)*

Investments

Investments consist of the amount of excess cash accumulated from operation during the first and second year of operation.

3. LIABILITIES AND CAPITAL

Management estimated accounts payable using 30 days in payables for conversion costs. It is anticipated that corn will be paid for a month in advance.

Management estimates that a bank loan in the amount of $163,040 will be required, payable semiannually at 15% interest. Anticipated payback period for the portion of the loan covering plant and equipment is 15 years. The payback period for the portion covering working capital is 8 years. The payback period for the truck is 8 years. An additional $4,000 the first year is projected to be paid on the equipment loans and to repay the working capital loan in the second year. The loan will be used to finance plant and equipment and working capital. The anticipated plant and equipment and working capital for the first year is estimated as follows.

Plant and equipment	$124,280
Working capital	$107,760
Total	$232,040

Cash could be very lean during the first year that the plant operates at capacity because of dramatic increases in working capital resulting from accounts receivable and inventory requirements. Inadequate financing would make maximum production impossible because of inability to fund working capital demands.

4. INCOME STATEMENT

Sales

Sales volume was estimated at maximum production (150,000 gallons of ethanol and 1,380,000 gallons stillage) for the first year. Ethanol price was taken to be $1.74 per gallon (the actual delivered price at Council Bluffs, Iowa on November 6, 1979). The price of ethanol is projected to increase by 9.1% for the entire period covered by the projections. The increase of 9.1% is the projected price increase by a marketing firm from Louisiana.

It is conceivable that as the price of gasoline increases to a point greater than the price of ethanol, producers could raise the price of ethanol to equalize the prices of the two liquid fuels. In order to be conservative, management did not consider this effect .

The stillage sales price was taken to be 3.9 cents per gallon for the 2 years. This sales price was based on the sales price charged by Dan Weber in Weyburn, Saskatchewan, Canada. The local stillage market is, however, worthy of a thorough study before a decision is made to enter the fermentation ethanol business. If a large brewery or distillery is located in the area, the price of stillage can be severely depressed. For example, Jack Daniels and Coors sell their stillage for 0.4 cents per gallon and 0.8 cents per gallon, respectively.

Cost of Sales

Management has projected cost of sales to include raw materials and production costs. The cost of corn is projected at $2.30 per bushel during the first year of operation (the price received by farmers in Iowa on November 6, 1979). Management has the total amount of corn available from the corporate shareholder. To demonstrate the effect of substantial increases in corn prices on profitability and cash flow, management projected that the cost of corn would rise to $3.00 per bushel for year two.

Management anticipates that depreciation will remain constant using the straight line method. A 15-year life for the plant was used with a salvage value of $4,000, while an 8-year life and 20-year life were used for the truck and building respectively. No other salvage values were taken into consideration.

Labor cost was computed allowing 4 hours per day for work necessary in the processing of the ethanol, based on the engineer's time requirement estimates. The labor was valued at $10 per hour, including a labor overhead factor. Bookkeeping labor was computed at $6,000 per year, assuming this plant would only require part-time services. It is anticipated that some additional time may be required the first year. For this, $2,325 has been added to the labor cost as a contingency.

Figure 11-1. ***(continued)***

Enzyme cost was estimated at $6,000 per year by the engineers working on the project. Electricity was estimated at 2 cents per gallon of ethanol produced.

Stover cost was computed based on a cost of $20 per ton. A Btu value of 7,000 Btu per pound (as estimated by the engineers) was used. An 80% efficiency for the boiler was assumed, so anticipated Btu values were 5,600 Btu per pound of straw, and a total Btu requirement of 40,000 per gallon of ethanol produced.

Miscellaneous Expenses

Miscellaneous expenses were estimated at 12 cents per gallon. These expenses are shown in the detail schedule at the end of this report. To be conservative, figures are included in the miscellaneous expenses for shrinkage due to the grain handling and a contingency for any minor items that may have been overlooked.

Interest expense is for the bank loan. Interest expense is calculated at 15%. In year two of the operation, it is projected that the working capital portion of the notes payable will be paid off.

Management projects that other projected costs will increase 10% per year because of inflation. To be conservative, management did not estimate the cost savings potential of improved technology. Research is currently being performed in crops that have the potential of producing several times the amount of ethanol as does corn. Use of such crops could produce substantial cost savings in ethanol production. The process is very new in design, so improvements in the production process are also probable. Such improvements could further reduce the cost of producing ethanol.

Selling and Administrative Expenses

To be conservative, management estimated marketing expenses at 5% of sales. It is anticipated that this expense may not actually be necessary. Delivery expenses take into consideration the following items:

- interest was computed at 15% on the bank loan for the truck based on semi-annual payments;
- the time necessary to deliver the stillage was estimated based on 10 hours per week at $10 per hour including labor overhead;
- maintenance for the truck was estimated at $1,000; and
- fuel for the truck was computed based on 75 miles per week and a fuel consumption of 4 miles per gallon and a fuel cost of $1.025 per gallon. The cost is estimated to increase 36.5% for the second year of operation.

Income Taxes

The shareholders of Johnson Processors, Inc. plan to elect to have income taxed to the shareholder rather than to the corporation, under Internal Revenue Code Section 1372(a). The shareholders anticipate changing the election after the first year of operation. Taxes have been estimated based on a 6% state tax rate and a 6.75% federal tax rate in effect during 1979. For purposes of these projections, the projected financial statements (assuming a conventional corporation and a full 12 months of operation in each period) have been shown to demonstrate the projected results of operation that could be anticipated.

The shareholders of Johnson Processors, Inc. anticipate contributing $69,000 to the corporation. This amount is 30% of the total project. This will be contributed by transferring corn inventory equal to the ½-year supply necessary for processing. For purposes of this illustration, the contribution of corn is treated as cash to demonstrate to the bank the payback potential of the plant.

Figure 11-1. *(continued)*

JOHNSON PROCESSORS, INC.
DETAIL SCHEDULES
UNAUDITED

(NOTES 1 THROUGH 4 ARE AN INTEGRAL PART OF THESE PROJECTED FINANCIAL STATEMENTS AND PROVIDE AN EXPLANATION OF ASSUMPTIONS USED IN THIS REPORT)

	FOR YEAR ENDED	
	December 31, 1980	**December 31, 1981**
Schedule of Marketing and Delivery Expenses		
Marketing 5% of sales	$14,260	$16,974
Interest on truck	2,211	2,045
Depreciation	1,875	1,875
Labor	5,200	5,720
Maintenance	1,000	1,100
Fuel	1,000	1,360
Total	$25,546	$29,074
Schedule of Miscellaneous Expenses		
Property taxes	$2,250	$2,475
Insurance	2,100	2,310
Chemicals and supplies	600	660
Yeast	450	495
Shrinkage	4,200	4,620
Other	2,550	2,805
Contingencies	5,850	6,435
Total	$18,000	$19,800

Figure 11-1. ***(continued)***

poration. They elected Subchapter S treatment upon incorporation and had the first year of operation reported on a short-period return. Generally, Subchapter S has many of the advantages of a partnership but not the liabilities. (Consult an accountant or lawyer for a detailed description.) They could pass through the investment credit which is proposed to be 20 percent of the capital cost, assuming that the Internal Revenue Service would authorize a fuel-grade ethanol plant to qualify for the additional investment credit for being a producer of renewable energy. After the first short-period return is filed, the stockholders can then elect not to be a Subchapter S corporation. This plan helps the cash flow as they would personally recover some tax dollars through the investment credit.

The corporation will lease from the family, on a long-term basis, 2 acres of land on which to locate the plant. They considered transferring this land to the corporation, but the land is pledged as security for the Federal Land Bank so it would be cumbersome to get the land cleared of debt. Also, the 2 acres would require a survey and legal description, thereby adding additional cost, and there are no local surveyors who could do this work.

The corporation will purchase corn and agricultural residue from the family farm and damaged corn from neighbors when there is a price advantage to do so. The family will also purchase the distiller's grains and ethanol used on the farm from the corporation as would any other customer. All transactions between the family farm and the corporation will use current prices that would be paid or received by third parties.

STATUS

The initial visit with the bank was encouraging. The local banker was acquainted with ethanol production through the publicity it had been receiving. The bank was receptive to the financing, saying they would consider it an equipment loan. The bank required a schedule of production, funds-flow projections, projected income statement, and projected balance sheets for the next 2 years. The bank was primarily concerned that these statements demonstrate how the plant could be paid for.

Before meeting with their accountant, the Johnsons prepared decision and planning worksheets as described in Chapter 2. This work on their part saved them some accountant's fees and gave them an idea as to the feasibility of such a plant. The projected financial statements were then prepared with the assistance of their accountant for the bank's use.

The projected financial statement shown in Fig. 11-1 is based on decisions made about the operations and management of the plant. It served the Johnsons as a tool in deciding whether or not the plant would be a good investment for them and also as a final presentation to the bank for loan approval. The assumptions used in preparing these financial statements are included with the financial statements and represent an integral part of the business plan.

After the financial projections were completed and the bank had reviewed them, there was one more area of concern. The bank wanted to know whether the system as designed was workable and could produce what it was projected to do from a technical feasibility standpoint. The family furnished the bank with the engineers' report which documented systems that were in operation and that were successfully using their proposed technology. The bank contacted some of the people operating these plants to verify their production. The bank then completed their paperwork and made a loan to the family's corporation secured only by the equipment. They also approved the line of credit for the working capital required based on the projected financial statements.

Chapter 12

Preparation of Feasibility Analysis

The purpose of this chapter is to identify the significant aspects in performing an effective feasibility study and suggest how best to address them. The form and format proposed should not be taken literally, rather as guidelines for consideration as needed. Knowledge of the audience is mandatory. If the study is to be prepared for a group of investors with varied backgrounds, then more-detailed technical explanations of the basic manufacturing process and a historical overview of the industry are probably required. An enunciation of the potential and inherent risks of the proposed venture must be clearly stated, regardless of the audience addressed. If the study is for use by more-informed and experienced lending institutions, then too much preamble may dilute the primary points of the proposal.

The basic points to be addressed in the study are how successful the venture will be and how much risk is involved to all parties. The length of the analysis and the amount of detail required are governed by the size of the project and the amount of control the involved parties (the proposed lender as well as the proposed borrower) have over the decision variables. For example, less information need be presented if the process raw materials are already owned or contracted for, or if the sale of all production of the product and coproducts is guaranteed. Discretion should be used as to the volume of information to be provided.

First, the potential investor must be convinced that the project is worth his time and effort, and a few major issues must be addressed for this prefeasibility phase: namely, the availability of markets, feedstock cost and availability, water, power, and status of legislation.

Prefeasibility Phase

The initiator of the project must be satisfied that (1) there is a market and it can be reached (conduct market and marketing analysis); (2) resources and means of acquiring them have been identified, e.g., feedstock, water, sewage access, power, and labor (evaluate factors of production); and (3) all applicable federal, state, and local requirements can be met (determine environmental, health, and safety factors). With respect to the existence of a market for ethanol and coproducts, the entrepreneur must determine how firm the need, at what price, and at what risk to the project. With respect to the necessary feedstocks and chemicals, he must determine their availability, price, and assurance for each. With respect to the availability of needed water, the same considerations need to be faced. For fuel, the potential investor should check the availability and cost of alternative resources, particularly cogeneration, and coal, and natural gas. With respect to laws, regulations, and rules, he should determine which are prohibitive and which are supportive of the project.

Feasibility Phase

Once the potential entrepreneur is convinced that the project deserves an in-depth analysis, a feasibility study should be performed in sufficient detail to define the means by which financing, construction, and operation will be accomplished. The analysis should address:

Technical and economic feasibility

Resource assessment and availability

Financing alternatives

Ability to construct and operate an ethanol plant in an environmentally acceptable manner on the selected site

The marketing, materials, and legal sections must be reviewed to provide more specificity, and sections on technical and engineering design, site selection and suitability, economic and financial analysis, and management planning should be added. Sufficient detail should be provided to facilitate a lender's decision on whether or not to provide the support required.

The following discussion, and outline information at the end of this chapter, only provide a guide to preparing a feasibility analysis; the entrepreneur must do the convincing. The basic goal of the analysis should be considered at all times: an answer to the question, "Can this specific project make a profit over the anticipated life of the investment?" The feasibility analysis should critically examine the proposed project to uncover problem areas and consider all possible risks. Solutions should be offered to reduce risks and unknown economic factors that may cause the venture to fail.

Key Elements of Feasibility Analysis

INTRODUCTION

____ ✓ Purpose
____ ✓ Scope
____ ✓ Limitations
____ ✓ Summary and opinion

This section sets the foundation for the reader, focusing his or her attention on the most important conclusions and business factors and their effects on the proposed project. It emphasizes the positive side of the findings, but also indicates where help is needed. The summary and opinion are very important because many investor/banks want to know the bottom line before reviewing the project in greater detail.

PART 1. MARKET AND MARKETING ANALYSIS

A. Ethanol

- ____ ✓ Market description
- ____ ✓ Gasoline demand
- ____ ✓ Alcohol demand
- ____ ✓ Product analysis
- ____ ✓ Competitive factors
- ____ ✓ Presecured markets—contracts for sale of product
- ____ ✓ Distribution analysis

The objective of this analysis section is to determine if the total proposed output of the plant can be sold at a profit over the life of the venture. The nature and detail of the analysis will identify and quantify specifics of the market composition, size, and description (e.g., refiners, marketers, consumers, geographic area); competitive factors, both current and future; gasoline and alcohol, current and future pricing and anticipated demand. The intent is to develop a realistic evaluation of immediate market conditions and to prepare realistic market projections for the future.

B. Coproducts

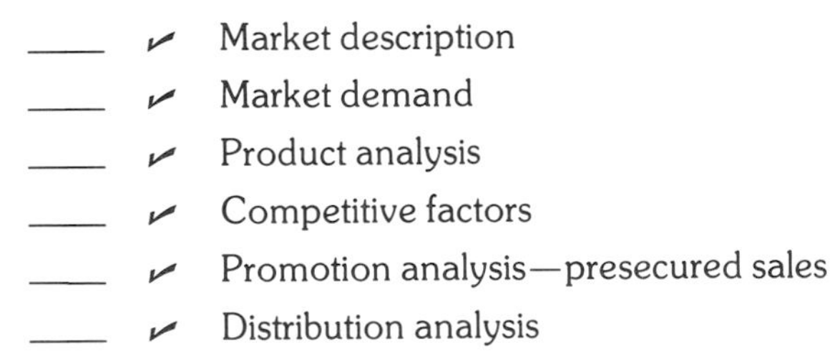

- ____ ✓ Market description
- ____ ✓ Market demand
- ____ ✓ Product analysis
- ____ ✓ Competitive factors
- ____ ✓ Promotion analysis—presecured sales
- ____ ✓ Distribution analysis

The sale of coproducts such as CO_2, yeast, stillage (both wet and as distiller's dried grains), and gluten can be a large determinant of the profitability of plant operations. It is likely that, for such products as DDGS, the future market will be totally new because of the unprecedented volumes of supply and the untested demand. Price fluctuations of the DDGS have been rapid and irregular with the supply drying up during the summer months. Any long-term contracts for selling the coproducts that can be established in advance are invaluable in establishing the viability of the proposal.

PART 2. TECHNICAL AND ENGINEERING DESIGN

____ ✓ Technological state of development
____ ✓ Process description
____ ✓ Production schedule
____ ✓ Resource requirements
____ ✓ Equipment procurement plan
____ ✓ Construction management plan
____ ✓ Process guarantee/service warranty
____ ✓ Assessment of uncertainties/contingency planning

The objective of this analysis is to determine the technological feasibility of the proposed project. The potential investor must provide evidence that the proposed venture has available a viable process ready for construction and installation according to well-considered timetables. This section provides the rationale for process selection and should analyze the uncertainties surrounding commercial application of the process. It should clearly identify all major areas of project design and operation that represent significant new technological innovation and should delineate what steps (if any) remain to be taken before commercial readiness is achieved.

The process description should include process flow diagrams, energy and mass balances, mechanical specifications, process specifications, major equipment requirements, plant layout sketches, plot plans and off-site requirements, waste-handling procedures, and other general design requirements. Resource requirements include all necessary raw materials such as water and power. All plans should be time-phased and indicate manpower and subcontractor requirements, where appropriate. Some engineering companies will provide a process guarantee and subcontractors and suppliers will provide service warranties. An assessment of the uncertainties surrounding the commercial application of the proposed process must be presented, including presentation of your strategy for dealing with unanticipated events.

PART 3. FACTORS OF PRODUCTION

A. Feedstock

____ ✓ Alternative feedstocks
____ ✓ Selection criteria
____ ✓ Cost and availability of selected feedstock
____ ✓ Contingency plan

The choice of feedstock for ethanol production varies from region to region, and even from site to site. The financial success of the venture is dependent on the steady procurement of low-cost feedstock. Present availability as well as future agricultural projections should be evaluated for each proposal. The results of current crop development work will influence future feedstock choices and the investor should have full knowledge of the available alternatives.

All available feedstocks need to be catalogued and compared to provide as many options as reasonable. Purchase prices at which the use of each alternative feedstock becomes less economical than another feedstock should be constantly updated to provide a basis for purchasing decisions. Factors affecting feedstock supply and availability should be reviewed in this section of the study, including natural factors as well as likely political or competitive events. During your feasibility study, you should identify sources of supply that can be contracted at favorable prices, particularly those products of substandard grade or classified as waste from some other process.

B. Power

____ ✓ Alternative sources
____ ✓ Selection criteria
____ ✓ Cost and availability of selected source
____ ✓ Contingency plan

Potential energy sources for process fuel should be analyzed and the best source selected. The criteria usually are availability, cost, and the resulting net energy balance. The better sources of energy, from a national viewpoint, are waste and other renewable resources; and, depending on the site, probably coal and possibly natural gas. Some financing sources, particularly the government, may have explicit guidelines on the use of different fuels. Any likely potential for shortages due to labor shutdowns in other fields should be addressed.

C. Water

____ ✓ Availability of sources
____ ✓ Requirement
____ ✓ Cost
____ ✓ Sewage disposal
____ ✓ Contingency plan

Significant amounts of water are used in the ethanol-production process (about 16 gallons of water per gallon of ethanol produced). This demand includes requirements for generating steam, cooling, and preparing mashes.

The actual amount of water needed will be significantly less due to the recapture and recycling of water used, but in some regions this may be a critical consideration. The source, process requirements, and cost should be analyzed, and a plan developed for acquisition, showing a complete understanding of potential problems. Correction of inadequate sewer facilities can rapidly increase the overall cost for a small-scale plant.

D. Labor

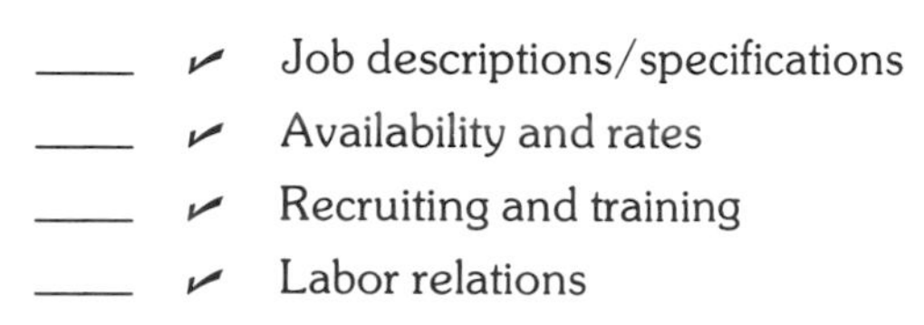

In this new industry, production manpower requirements are not well known. Different processes should require specific numbers and types of workers, yet, as experience is acquired, the work force required will probably change. The first estimate of the number of workers necessary can be obtained from the engineering design contractor. The feasibility analysis should deal with the quality and characteristics of the available labor pool; how much they will be paid; how they will be recruited, selected, and trained; and an outline of promotion policies and retention means. Any potential for labor disputes or shutdowns should be addressed for each site considered.

PART 4. SITE SELECTION AND SUITABILITY

The primary site requirement for location of an ethanol plant is the availability of an abundant water supply. Other site requirements are similar to those for any chemical manufacturing facility, except for the particular material handled and the storage of agricultural and liquid products. In comparing alternative sites, project or anticipate future problems such as the availability/cost of railroad service, regional and local political forces, and distance from the raw materials and sewage disposal facilities.

A potential site should be near several sources of different raw materials and several different markets. Multiple modes of transportation are likewise desirable. The site should be located where there will be no unusual circumstances preventing licensing and permits. The costs of these factors should be used in evaluating the best site. The recommended site should be readily available for acquisition.

Any firm agreement to purchase or lease the site should be subject to engineering approval, receipt of required permits, and completion of other legal and environmental requirements.

PART 5. ECONOMIC ANALYSIS

A. Cost of Production

- ____ ✓ Plant size
- ____ ✓ Fixed costs
- ____ ✓ Variable costs
- ____ ✓ Economies of scale
- ____ ✓ Sensitivity analysis

A detailed economic analysis should be presented for the proposed plant. As described in Chapter 6 of this handbook, the size chosen for the plant can mean the difference between purchasing raw materials and developing, storing, and treating them, thus incurring a different cost structure for the plant. Detailed assumptions concerning the cost of operation, i.e., fixed and variable costs, need to be identified for the near and long terms as a function of output. An accounting model of the plant should be developed, and the effects of changing the cost of the key variables should be evaluated. This sensitivity analysis will illustrate the range of possible consequences and help the project developer organize a plan to reduce dependence of the project on uncontrollable variables.

B. Investment Analysis

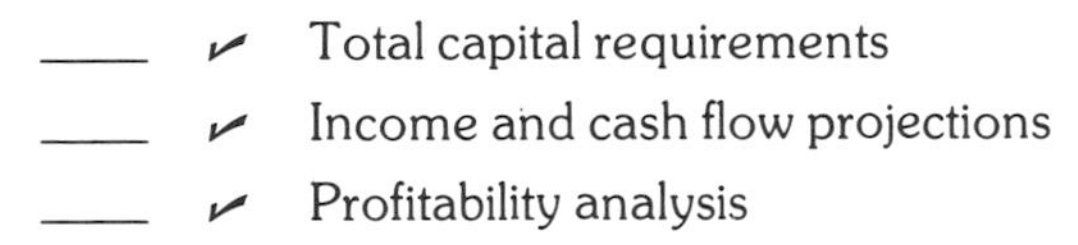

- ____ ✓ Total capital requirements
- ____ ✓ Income and cash flow projections
- ____ ✓ Profitability analysis

An analysis of the total capital requirements must be prepared, including appropriate breakouts of the capital components such as the elements of fixed capital requirements, and others, such as working capital. Of course, the final projections must wait until many of the project characteristics are finalized, such as site-specific construction requirements and the size, terms, and interest rates of the necessary loans. However, assumptions should be made based on findings from other sections of this feasibility study in order to determine bottom-line estimates as early as possible.

Working capital requirements should be estimated in conjunction with operating characteristics of the engineering design. Actual amounts would differ, depending on the inventory requirements, the lines of credit opened with a commercial lender, and the amount of product produced. Less than full-capacity operation should be planned, especially during the start-up period, and most probably thereafter. Allow for sufficient funds to cover contingencies, as well.

As the income and cash flow projections are made, the least favorable conditions also should be considered. Profitability analysis should include projections of the classical ratios, including return on investment, payback period, discounted cash flows, net present value, and so on.

The funding options available to the entrepreneur should be a fundamental part of these analyses. The investor should include the likelihood of federal, state, and local government support as well as taxes, and discuss their effects on the project's financial picture.

C. Risk Analysis

____ ✓ Government intervention
____ ✓ Competitive reaction
____ ✓ Product/process obsolescence

Examination of the potential risks involved with alcohol-fuel production is of primary importance to both the entrepreneur and the investor. Not only will it help to identify the possibilities for failure but, if the project is undertaken, it will identify potential problem areas for management. This early identification of economic risk also will aid in determining the optimum project configuration to allow for the most flexibility in dealing with changing economic and technological climates. In conjunction with the economic analysis, an examination of the available margins at various market conditions should be made to determine the effects of changing markets. The purpose of this section is to highlight potential pitfalls to prevent producing a product which costs too much, or having to sell a product at too low a price.

PART 6. MANAGEMENT PLAN

____ ✓ Background and experience
____ ✓ Responsibilities
____ ✓ Management systems

As in any new venture, and especially a pioneering effort such as alcohol-fuel production, management capabilities are a most critical component. The management plan should indicate the involvement and responsibilities to be assumed for the entire life-span of the project. Responsibilities and authorities must be clearly identified and each position filled with an experienced individual. The technical and business backgrounds are critical, with agribusiness, as well as refinery and plant management experience, preferred. The lending institution must be assured that top-rated managers are being hired.

PART 7. ENVIRONMENTAL HEALTH AND SAFETY AND OTHER REGULATORY REQUIREMENTS

____ ✓ Statutory and regulatory requirements

____ ✓ Public concerns

____ ✓ Management responsibilities

The purpose of this assessment is to examine the issues likely to have the most impact on the permitting and scheduling of the proposed facility. The potential investor should research all applicable local, state, and federal statutory and regulatory requirements in order to determine whether the environmental, health, and safety requirements can be met, and if so, on what schedule.

The baseline environmental quality of the proposed site should be defined; a determination of applicable environmental standards and constraints should be presented; a technical description and cost evaluation of mandated environmental control measures should be prepared; the quantities of air, water, solid, and radiation residuals of the proposed process assuming applicable control technology should be assessed; and an analysis of the impact of residuals on the baseline environmental quality of the site should be presented, using appropriate predictive models.

Assessments of particular concern to the community or state should be included, quantitative analyses should be presented of worker and community exposures to, and disposal of, known or suspected hazardous materials.

Finally, the entrepreneur should describe control options, costs and risks, and control measures that will be instituted to meet requirements. The process of making alcohol fuels should be environmentally safe, and the investor should be certain that no moral, ethical, or legal issue will tie up the project for long periods, or even kill it after it is started.

Key Elements of Feasibility Study

This outline contains elements upon which the entrepreneur can expand his feasibility analysis.

1. Title Page
   - Name of firm submitting the proposal
   - Mailing addresses
   - Telephone numbers
   - Person to contact if further information is needed
2. Table of Contents
   - All chapters
   - All appendixes
   - All tables, charts, and maps

3. The Objective
   Who/what/when/where/why/cost/annual capacity
4. Background
   History of this particular venture and alcohol fuels
5. Market Description of the Feedstock(s) and Products
   Summary of the specific feedstocks for the production of alcohol and the market for the products of the particular plant
   Who are the sellers and buyers?
6. Organization and Founders
   Briefly describe all the owners
   Outline the organization that will direct this project
   Indicate the distribution of ownership
7. Project Components
   Outline the major tasks to accomplish the objective
   Delineate the management responsibility for each task
   Indicate the integration of appropriate tasks
8. Time Schedule
   List all tasks in summary form in time sequence
   Depict the tasks on a chart (see example in Fig. 12-1)
9. Project Cost and Sources of Funding
   Summarize the costs
   List all sources of funding

The Appendixes to Feasibility Study

A. MARKET AND MARKETING ANALYSIS

The procurement and retailing of products with all internal and external impacts are estimated. The historical and projected value and cost are determined along with the phases of development and operation. The relationship of the project to regional, state, and local energy policy should be noted. Suggested considerations include:

Feedstock(s)
- Sources and locations (gathering radius)
- Cost of collection and transportation
- Quantity and quality for alcohol production
- Historical, current, and projected cost
- Availability over the life of the plant

Alcohol
- Market price for expected proof
- Transportation cost
- Current and projected cost

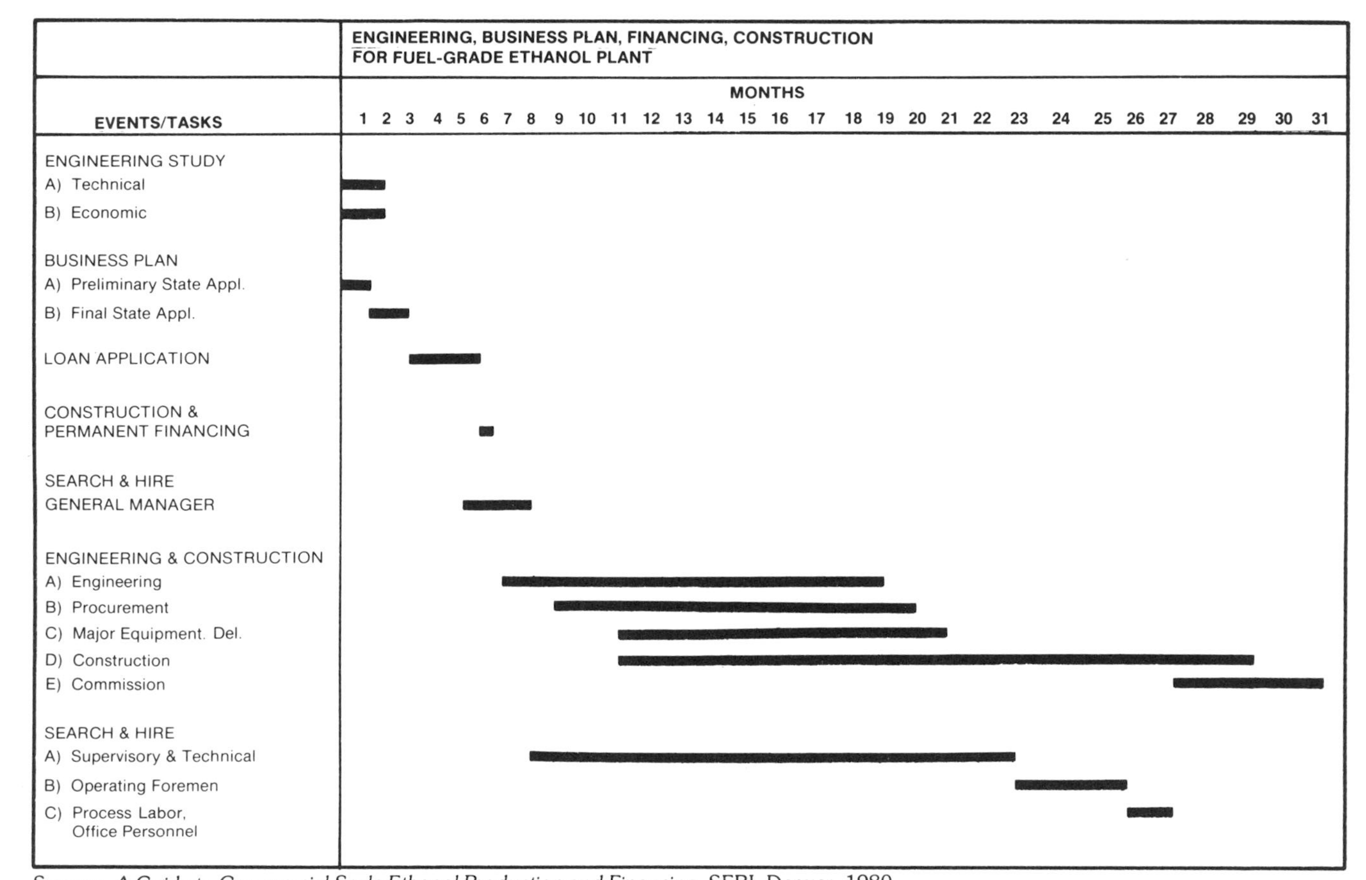

Source: *A Guide to Commercial Scale Ethanol Production and Financing*, SERI, Denver, 1980.

Figure 12-1. Likely schedule of events/tasks.

Coproducts
    Value on-site
    Market price

Commitments from sellers, customers, and brokers

Location of the plant
    Include map(s)
    Description of adjacent industries, utilities, general population, and the potential complementary or supplementary relationship with the project

Commitment from utilities, energy suppliers, fire, and police

All written support which has been elicited from the community should be provided. This may take the form of: letters; endorsements; resolutions and ordinances from businesses and nonprofit organizations; town, city, county, and state governments; Jaycees and other community-based organizations; planners; mayors; councils; governors; legislators; or congressmen. If you want federal financing, you need all the political clout possible.

B. TECHNICAL AND ENGINEERING DESIGN AND FACTORS OF PRODUCTION

The production cycle design will come from the selection of one particular process. The following items need to be taken into account to calculate economic and energy costs of production. Each should be described in the process with particular attention to the production capacity through quantitative descriptions in each step.

Feedstock(s)
    Primary/alternate

Collection (harvest and transportation)
    Includes methods and distances

    Storage of feedstock
    Preparation (cleaning and grinding)
    Preliminary conversion (cooking)
    Saccharification
    Fermentation
    Distillation
    Dehydration
    Separation of solids
    Denaturing
    Drying of solids
    Storage of stillage
    Use of stillage
    Water recycling
    Heat source—primary/secondary

Integrated energy generation on-site

Heat recovery

Use of other coproducts (CO_2, etc.) and wastewater

Detailed energy use calculation on a Btu-per-gallon basis in the plant

Inputs are all nonrenewable sources for cooking, distillation, grinding, augering, site transportation

Outputs include alcohol plus the coproducts

Equipment (instruments/controls/boiler/engines/pumps), lead time, availability, and costs

Laboratory facility and equipment on-site for testing and analyzing parts of the process

Preliminary process flow diagrams

C. ECONOMIC ANALYSIS AND FINANCIAL PLAN

A pro forma, or projected, sources and uses of funds statement can be constructed to show how the project plans to acquire and employ funds. This is the basic summary document to show the flow of resources. A projected balance sheet and income statement is required for the financial plan.

An itemized estimate of all project costs including the basis for the estimates is needed for the development of the project.

Cost estimates for the project
- Engineering design
- Site preparation, construction, and installation
- Labor
- Production
  - Feedstocks
  - Utilities
  - Chemicals
  - Bonding, insurance, and taxes
- Maintenance
- Depreciation
- Other overhead

All private and public financing options that will be explored should be listed in as much detail as possible.

The complete financial plan includes the following time periods:

Development and construction of plant (with a subtotal for each task in the work plan)

First year of operation, month by month

Five-year quarterly projection

Debt service on a quarterly basis

Previous 3 years' profit and loss statement and financial statement of existing businesses

Documentation should be complete for deeds, leases, contracts, agreements, options, appraisals, and insurance.

D. MANAGEMENT AND STAFFING PLAN

The organization of the entire project should be detailed with job titles and descriptions. An organization chart provides clarity and brevity for project management structure. List everyone by name, if possible, from the Board of Directors through the supervisory and technical personnel.

The education, technical training, employment, and related business experience of each participant in the project should be detailed on one page in a resume format for each person. Salaries, fees, and overhead cost for labor are included here.

Engineering and construction companies and subcontractors are included with scheduling and reporting requirements. Cost and project management responsibilities need to be clearly indicated.

Provisions for a permanent, trained labor force must be quite specific. This training could be accomplished before the beginning of the operation of the particular plant.

E. SAMPLE INFORMATION ELEMENTS

Sample information elements are shown in Figs. 12-2, 12-3, and 12-4. The following information should be provided.

1. With respect to each partner or shareholder (e.g., over 5 percent of total shares outstanding):
   - Name
   - Address
   - Citizenship
   - Principal occupation
   - Percentage of ownership
   - Personal financial statement
   - Credit reliability/analysis
   - Personal interest in any concern which shall have some relationship to project
2. With respect to any corporate organization:
   - Parent companies, subsidiaries, affiliates
   - Charter of organization
   - Financial statement of organization
   - SEC 10K report for companies listed on stock exchange
   - Tax status

SAMPLE INFORMATION ELEMENTS

	SALES FORECASTS — GOALS				
	81	82	83	84	85
A. ETHANOL					
JOBBER: ______					
JOBBER: ______					
JOBBER: ______					
OTHER: ______					
TOTAL ______					

	SALES FORECAST — TONS				
	81	82	83	84	85
B. COPRODUCTS					
BUYER: ______					
BUYER: ______					
BUYER: ______					
TOTAL ______					

COPRODUCT CONTENT: PROTEIN ______ %

FAT ______ %

FIBER ______ %

ASH ______ %

MOISTURE ______ %

Source: *A Guide to Commercial Scale Ethanol Production and Financing*, SERI, Denver, 1980.

Figure 12-2. Sales forecasts.

F. REGULATORY COMPLIANCE REQUIREMENTS

Obtaining building permits and other necessary certifications, and meeting compliance requirements of the appropriate regulatory agencies having jurisdiction over the project, should indicate the following:

Environmental (federal and state)

Fire and safety

Building codes

Zoning

Utilities

BATF (federal and state)

National Historic Preservation Act

A.
- PLANT CAPACITY: ____________ GAL/HOUR
- PRODUCTION ETHANOL: ____________ GAL/YR
- PRODUCTION COPRODUCT: ____________ TON/YR
- ETHANOL PRICE, F.O.B. PLANT: ____________ $/GAL
- COPRODUCT PRICE, F.O.B. PLANT: ____________ $/TON
- PLOT SIZE: ____________ FT. X ____________ FT.
- PLANT SIZE: ____________ FT. X ____________ FT.
- SOIL STRENGTH: ____________ PSF @ ____________ FT.
- PLASTICITY INDEX: ____________
- ____________ pH RATING ____________ RESISTIVITY LEVEL
- RAILROAD SIDING: ____________ FEET
- ACCESS ROAD QUALITY: ____________

B.
- FEEDSTOCK TYPE: ____________
  FEEDSTOCK CONTENT: STARCH ____________ % WEIGHT
  PROTEIN ____________ % WEIGHT
  OTHER SOLIDS ____________ % WEIGHT
  MOISTURE ____________ % WEIGHT
  AMOUNT REQUIRED: ____________ BUSHELS/YR
  JOBBER: ____________
  NUMBER DAYS INVENTORY: ____________
- FEEDSTOCK TYPE: ____________
  FEEDSTOCK CONTENT: STARCH ____________ % WEIGHT
  PROTEIN ____________ % WEIGHT
  OTHER SOLIDS ____________ % WEIGHT
  MOISTURE ____________ % WEIGHT
  AMOUNT REQUIRED: ____________ BUSHELS/YR
  JOBBER: ____________
  NUMBER DAYS INVENTORY: ____________
- ENZYME TYPE: ____________
  AMOUNT REQUIRED: ____________ GAL/DAY
  JOBBER: ____________
  NUMBER DAYS INVENTORY: ____________
- ENZYME TYPE: ____________
  AMOUNT REQUIRED: ____________ GAL/DAY
  JOBBER: ____________
  NUMBER DAYS INVENTORY: ____________

C.
- POWER SOURCE: COAL
  AMOUNT REQUIRED: ____________ TONS/HR
  JOBBER: ____________
  NUMBER DAYS INVENTORY: ____________
- POWER SOURCE: ELECTRIC
  AMOUNT REQUIRED ____________ kW/HR
  JOBBER: ____________

D.
- WATER SOURCE: ____________
  AMOUNT REQUIRED: ____________ GALS/HP
  JOBBER: ____________
  NUMBER DAYS INVENTORY: ____________

E.
- LABOR ____________

  ____________ NUMBER PLANT OPERATION AND MAINTENANCE
  ____________ NUMBER ADMINISTRATION AND SUPERVISION
  ____________ NUMBER GENERAL ADMINISTRATION AND SALES

Source: *A Guide to Commercial Scale Ethanol Production and Financing*, SERI, Denver, 1980.

Figure 12-3. Plant characteristics.

A. COST OF PRODUCTION

	$/YR	$/GAL
a. CAPITAL COSTS		
• EQUIPMENT	______	______
• LAND	______	______
• INVENTORY	______	______
• TAXES	______	______
• INSURANCE	______	______
• DEPRECIATION	______	______
• INTEREST	______	______
b. OPERATING COSTS		
• FEED MATERIAL	______	______
• SUPPLIES	______	______
• FUEL	______	______
• WASTE DISPOSAL	______	______
• OPERATING LABOR	______	______
c. MAINTENANCE COSTS		
• LABOR	______	______
• SUPPLIES	______	______
• EQUIPMENT	______	______
UNSCHEDULED MAINTENANCE		
• LABOR	______	______
• SUPPLIES	______	______
• EQUIPMENT	______	______

B. INVESTMENT ANALYSIS

• REVENUES		$______
ETHANOL	______	
STILLAGE	______	
• LESS COST OF SALES		
FEEDSTOCK	______	
DEPRECIATION	______	
• GROSS PROFIT	$______	
• LESS G&A AND O.H.	$______	
• NET PROFIT BEFORE TAXES	$______	
• BREAK EVEN QUANTITY	______	GALS.
• PAYBACK PERIOD	______	YEARS
• RETURN ON INVESTMENT	______	%

C. RISK ANALYSIS

PROCESS GUARANTEE:	______	PROOF
	______	FUSEL OIL CONTENT
PERFORMANCE GUARANTEE:	______	PRODUCTION RATE
	______	QUANTITY ENERGY
	______	QUANTITY PER INPUT

Source: *A Guide to Commercial Scale Ethanol Production and Financing*, SERI, Denver, 1980.

Figure 12-4. Economic analysis.

Occupational Safety and Health Administration

Other permits, ordinances, and regulations

Assuming productive use of the alcohol, stillage, and CO_2 produced, there still may be a number of environmental factors to be considered. These might include:

Odors released during the process.

Disposal of wet thin stillage and wastewater. How much backset (recycling of spent thin stillage into fresh fermentation) is built into the plant design? Is there any concentration of thin stillage into syrup? Is there any digestion of thin stillage? What is the projected discharge of wastewater in volume or quality? Does the plant site encompass additional buffer area that could be used for waste disposal lagooning or irrigation? Will a discharge permit be required for the plant?

Air pollution in boiler operations. Compliance with local and federal standards. Are fuel alternatives available? Cogeneration prospects? What is the quality of the stack gases from the boiler? Can they be used for stillage drying by direct contact or indirectly?

Removal of crop residuals causing erosion. Does crop residual removal for fuel feedstock or alcohol feedstock exceed 50 percent of total crop biomass? Will the production of alcohol result in farm and soil management shifts which are very different from current practices?

Ethanol use of fuel for vehicles and resulting emissions. Seasonal air flow patterns and local air quality history.

In conjunction with the considerations developed under the sections which contain marketing and economic analyses, and the analysis of the production process, alternatives to the final decision and their consequences should be considered. In addition, the environment to be affected should be described (e.g., quality of soil, air, water) prior to undertaking this project.

Completing forms similar to those shown in Figs. 12-2 through 12-4 will provide the basic information elements necessary for the preparation of the feasibility analysis.

Chapter 13

International Development

Introduction

The search for biomass-derived fuels is not limited to the United States. Many countries of the world, both developed and less developed, suffer from an energy deficit which forces them to import expensive petroleum, thereby jeopardizing their potential for industrial development. A number of foreign countries have initiated ethanol-fuel programs to reduce their dependence on imported oil. Most notable of these is the alcohol-fuel program presently underway in Brazil. Some other foreign countries are engaged in research and development activities to analyze the potential of alcohol-fuel production and use within their own context of biomass resources and projected demand for liquid fuels. Some of these countries which have the biomass resources to support an ethanol-fuel program may offer opportunities for investors interested in foreign activities.

The objective of this section is to give an overview of ethanol-fuel activities in foreign countries.

Background

Ethanol-fuel production in foreign countries follows the same principles as those described earlier in this handbook. However, some differences in process and project implementation may occur because of national goals, circumstances, and environment.

FEEDSTOCKS

The preferred feedstocks may be different from those mentioned in Chapter 4 of this handbook because of climate, soil, and nutritional habits of the country.

Sugarcane is an important potential feedstock in Brazil, and other South and Central American countries, and the Caribbean Islands. Cane molasses, also known as blackstrap molasses, is a coproduct of sugar production from cane; it has been and still is a common feedstock for ethanol production. The availability and cost of these sugar feedstocks, however, is closely related to world demand and prices for sugar. With the exception of some countries such as Brazil, where enough land is available for sugarcane feedstock production, a sugarcane-based ethanol-fuel industry could be deprived of its feedstock when higher economic return can be obtained by exporting sugar on world markets than by producing ethanol fuel internally. As is the case in the United States, sweet sorghum is considered an attractive biomass feedstock for ethanol production in the future in many countries. Sugar beets, a relatively little used feedstock in the United States, have been an important feedstock for ethanol production in some countries, for example, France.

Potential starchy feedstocks for ethanol-fuel production include the grains and tubers mentioned earlier, as well as regional crops such as cassava, also called maniot. Cassava is a root crop grown as a subsistence crop in many developing countries. Cassava is an attractive feedstock for ethanol-fuel production because (1) it is an efficient converter of solar energy and therefore can yield high quantities of ethanol fuel per unit land area; (2) it can be grown on marginal lands; (3) it withstands adverse weather conditions; (4) it requires little commercial energy inputs; and (5) it can be harvested throughout most of the year. One disadvantage of cassava is that it produces no residual energy source for the generation of process energy. (The residue of sugarcane processing to sugar and alcohol, bagasse, is a lignocellulosic material which can be burned to produce steam and electricity.) One ton of cassava produces about 43.1 gallons of ethanol as compared to 15.2 gallons per ton of sugarcane and 89.4 gallons per ton of corn.

In Brazil, the average productivity of cassava is about 5.5 tons per acre per year, which translates in a potential annual yield of about 237 gallons of ethanol per acre. For comparison, the average annual yield of ethanol from sugarcane is about 338 gallons per acre in Brazil, and the average annual yield of ethanol from corn is about 228 gallons per acre in the United States.

Other regional feedstocks, such as the babassu palm in Brazil, are also being investigated. Complete processing of this coconut can result in a variety of coproducts including charcoal for ore refining, oil for food use and soap production, alcohol by fermentation, and process heat from the residues of the process.

PROCESSES

Some variations in the production process may occur because of local conditions. Process heat requirements may be lowered in tropical areas, while cooling requirements may be increased in these areas. Also, in tropical areas, solar-produced process energy generated by concentrating collectors for example could be an economically viable alternative to other sources of process energy.

END USES, MARKETS, AND POLICIES

Several factors may impact the alcohol-fuel policy of developing countries. The distribution network for the alcohol fuels produced in agricultural and sometimes remote areas is less developed than it is in the

United States. As a result, the market for the alcohol fuels may be a local/regional market. This circumstance offers the advantage of a captive market but has the disadvantage of a lack of flexibility for the marketing of the output of a large-scale production plant. The same problems must be faced when marketing the coproducts (if any) of ethanol production.

In contrast to the situation in the United States, where emphasis is placed on the production of anhydrous alcohol for mixing with unleaded gasoline, the option of using hydrated ethanol directly in automotive engines is explored in some countries. The selection of this option may result from the existence of numerous small local production units (on-farm units, village-scale units) for which the production of anhydrous alcohol may not be justified technically or economically. The adoption of hydrated alcohol rather than anhydrous alcohol as fuel will impact the marketing of the fuel as well as the end-user sector. In Brazil, for example, about 13 percent of the 1978-1979 production of ethanol was in the form of hydrated fuel.

Considerations other than the desire of producing a substitute fuel for oil-derived fuels may influence the decision to produce or not to produce ethanol fuel in some countries. One reason for encouraging ethanol production is to create an alternative, steady internal market for commodities with prices that are extremely volatile on world markets. Sugar is one of these commodities. Another reason for encouraging alcohol production is to promote employment in rural areas and to reduce the migration to cities. The latter is one of the reasons for Brazil's alcohol-fuel policy. The production of ethanol may be less attractive than other uses of the biomass feedstock. The case in Jamaica is illustrative. Molasses converted to ethanol for fuel use has a value of $70 per ton. Molasses shipped to New Orleans has a value of about $86 per ton (average 1979 price), while the same molasses converted to rum has an equivalent value of $200 per ton. For a country short on foreign exchange, producing beverage alcohol appears economically more attractive in the near term than producing ethanol fuel.

The remarks made above, therefore, indicated that significant differences in policies may exist in various countries which should be carefully reviewed before considering to produce ethanol fuel in areas offering apparently attractive resource and market conditions.

Prospects for Ethanol-Fuel Production in Foreign Countries

The implementation of a biomass-derived ethanol-fuel industry in a country depends heavily on the present status and prospects of the agricultural, industrial, and energy sectors of the countries concerned. The options are complex and will have to be addressed on a country-by-country basis. The overall potential for alcohol production from biomass, however, can tentatively be assessed by first identifying the countries which offer an agricultural/energy balance that would favor a biomass energy program and then determining which of these countries offer the economic background required to launch an economically successful alcohol production program.

In a recent analysis, Norman Rask has addressed this problem of identification of countries offering good prospects for ethanol production. The approach used by Rask is summarized below.

AGRICULTURAL AND ENERGY SELF-SUFFICIENCY

The agricultural and energy self-sufficiency of a sample number of countries was analyzed and is shown in Fig. 13-1. (Agricultural self-sufficiency is defined as the total value of agricultural production divided by the value of agricultural production consumed in the country; it is a measure of trade balance for agricultural products. A similar definition is used for energy self-sufficiency.) The data in the figure show that four major country situations can be considered.

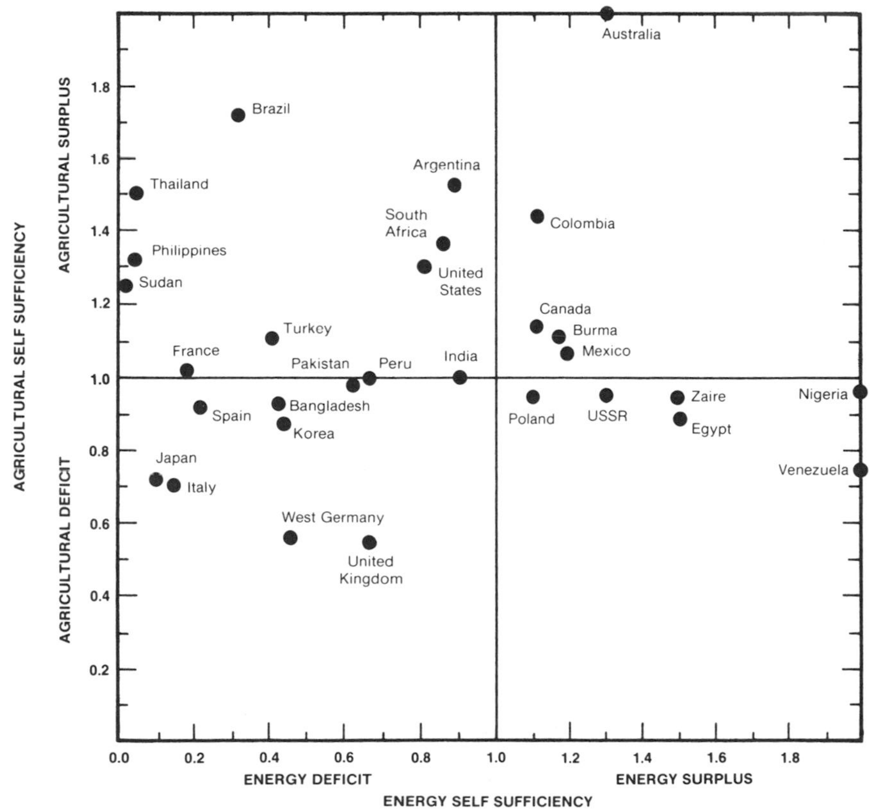

*The results indicate country situations where biomass energy programs may be feasible. The underlying analysis is being refined by taking into account averages for several years and focusing directly on **food** as compared to **agricultural** self-sufficiency measure used in this frame.

Source: Developed by Dr. N. Rask from FAO and World Bank Data.

Figure 13-1. Alcohol production from biomass energy and agricultural self-sufficiency ratios for selected countries — 1976.

The first category includes countries having surplus agricultural production and being net importers of energy (upper-left-hand segment of Fig. 13-1). In these countries, government policies will probably promote domestic energy production and conservation.

The second category includes countries that are surplus producers of both agricultural products and energy (upper-right-hand segment of Fig. 13-1). Biomass-to-energy programs are not expected to be encouraged in these countries and strong economic viability will have to be demonstrated prior to developing a fuels-from-biomass industry.

The third category includes countries with deficits both in agricultural products and energy production (lower-left-hand segment of Fig. 13-1). These countries will generally pursue policies to support agriculture but may support biomass energy programs if these tend to use feedstocks with low economic/food value. Surplus biomass material (such as molasses or sugarcane during periods of world sugar surpluses) or processing wastes may be used as feedstocks for biomass-fuel production in these countries. The key question in countries such as these is how to best make economic and social use of their agricultural resources for food use (domestic and export) and biomass-fuel production.

The fourth category includes countries with agricultural deficits and energy surpluses (lower-right-hand segment of Fig. 13-1). Emphasis will be placed on development of agriculture, while biomass energy programs will only be encouraged if the biomass feedstock is a low (food) value product.

The analysis of Fig. 13-1 focuses mostly on food rather than total agricultural production. As an example, Egypt is reported in the figure as a country showing a food deficit. Egypt, however, has significant surpluses of molasses which can be used as feedstock for ethanol production.

ECONOMIC FACTORS

A number of economic and related factors will impact the attractiveness of a biomass-fuel industry. Countries with surpluses of easily converted feedstocks (molasses or sugarcane as opposed to coffee, tea, or even grains) will be prime candidates for an alcohol-fuel industry. Similarly, alcohol-fuel production will be more attractive in countries where biomass feedstocks are available at low cost and where the industrial infrastructure needed to develop an alcohol industry is available. The high cost of delivery of petroleum products in remote and inaccessible regions/countries may make alcohol-fuel production economically attractive even under high raw materials or high plant costs.

CANDIDATE COUNTRIES FOR ETHANOL FROM BIOMASS INDUSTRIES

On the basis of the preceding discussions, two categories of countries may be considered as candidates for viable ethanol-fuel programs: those countries which have surplus biomass such as molasses that can be converted to fuels with little impact on the country's food balance, and those countries which have large biomass-production potential where some agricultural resources could be devoted to fuel production. In the former case, the decision to produce ethanol fuel will be based mostly on economic reasons; in the latter case, economic and social factors will have a bearing on the decision to produce ethanol fuels. Table 13-1 shows a partial list of candidate countries for ethanol-fuel production. Further studies may, of course, result in the addition or deletion of some countries listed in the table.

International Alcohol-Fuels Production

Although Brazil is the only country that has made a full commitment to developing an ethanol-fuel industry, other countries have shown interest in ethanol fuel or are engaged in research and development (R&D) or pilot ethanol projects. A brief description of some of these projects is given below. The countries are listed in alphabetical order.

AUSTRALIA

R&D programs dealing with feedstock production and process improvement are underway in Australia. The production of sweet sorghum and cassava, their storage and handling requirements, and the bioconversion of cellulose are being investigated. The biology of fermentation and continuous fermentation processes are also the object of research.

A pilot ethanol-fuel plant using wheat feedstock and having a capacity of about 670,000 gallons/year (2000 metric tons/year) is under construction in

Table 13-1. Sample List of Candidate Countries for Ethanol-Fuel Production

COUNTRIES WITH SURPLUS BIOMASS[1]	COUNTRIES WITH LARGE BIOMASS POTENTIAL[2]
Colombia	Argentina
Dominican Republic	Brazil
Ecuador	Papua New Guinea
Egypt	Philippines
Guatemala	Sudan
India	Thailand
Ivory Coast	
Jamaica	
Kenya	
Mali	
Peru	
Sri Lanka	
Swaziland	
Other Central American and Caribbean Countries	

[1]i.e., Molasses

[2]i.e., Sugarcane, Cassava, Wood

Source: N. Rask, "Using Agricultural Resources to Produce Food or Fuel — Policy Intervention or Market Choice", Paper presented at the first Inter-American Conference on Renewable Source of Energy, New Orleans, LA, November 1979.

Sydney and is scheduled for operation by 1981. The plant will use a continuous fermentation process developed by Biotechnology Australia Property, Ltd.

Larger plants with outputs of 80 to 150 million gallons/year (250,000 to 500,000 metric tons/year) are planned for the future. Australia hopes to produce 15 percent of its liquid fuel needs from alcohol fuels by 1985.

BRAZIL

Because of its alcohol program, Brazil has been singled out as an example of the successful use of energy derived from biomass. The rapid development of the Brazilian program is due mainly to two facts. First, liquid fuels are the main energy problem facing Brazil in the foreseeable future; and second, Brazil is a major sugar and alcohol producer. Brazil is able to rely on an existing agricultural and industrial base which has helped to speed up the implementation of the program. There is no major concern about a global energy shortage in Brazil. Brazil's hydroelectric potential is about 208,000 megawatts, its coal reserves are estimated at 22 billion tons, and its uranium reserves have been evaluated at 215,000 tons of yellowcake. The rate of energy consumption has risen by 12 percent each year in the last 10 years. This rapid increase in energy consumption is related to the gross national product growth, which has reached 6 percent per year in the last 20 years. During this period, however, oil's share of the energy budget has grown from 9 percent in 1940 to 42 percent in recent years. This has been due to the fact that, besides heavy industrialization, the transport system, which increased 10 times in 30 years in terms of roads built, has relied almost exclusively on oil for its operation. As far as liquid fuels are concerned, gasoline and diesel now account for 60 percent of overall petroleum products consumption.

This expansion in petroleum demand can also be judged by Brazil's automobile fleet which increased from 235,000 vehicles in 1950 to more than 8 million vehicles in 1979. Brazil's automobile industry is delivering more than 1 million cars a year, which places the country as the ninth major car market in the world. Unfortunately, this expansion in petroleum consumption has not been followed by a parallel growth in domestic oil production. Oil output, which represented 80 percent of the demand in the early 1950s, is now reduced to only 17 percent. The dependence of imported petroleum and the rapid increase in oil price puts strong pressure on the Brazilian balance of payments.

Within this broad framework, and taking into account the profile of its energy balance, the Brazilian Government has decided to tackle these questions within the context of the national capability to produce alternative sources of energy. The Government has set an oil import ceiling of about 1 million barrels per day, which will mean hard currency expenditures of almost 10 to 12 billion U.S. dollars in 1980, depending on the behavior of petroleum prices. In parallel with the pressures resulting from the increased cost of oil, equipment, and petrochemical products, the high rate of gross national product growth implies a continuous increase in oil consumption.

In view of the oil price outlook and the limitation of world reserves, the Brazilian government has decided to replace petroleum derivatives by renewable sources of energy. In this context, the conversion of biomass into energy was given the highest priority.

The choice of the renewable source of energy had to be made taking into account the country's potentialities that would minimize the risk of such a program.

The main energy problem in Brazil being that of liquid fuel availability, interest has focused on the photosynthetic process and the rich organic material, biomass, which may be converted into alcohol.

The principal biomass feedstock options in Brazil are discussed below.

Sugarcane

This grass species can be grown almost all over the country. Because of its sensitivity to freezing, it is cultivated from a latitude 25° south of the Equator. With exception of Paraiba State, sugarcane is grown as a rainfall crop. São Paulo, Pernambuco, and Paraiba are the main producing areas.

Yields vary from about 20 to about 45 tons per acre-year (45 to 100 tons per hectare) because of the different ecological and social conditions in the sugarcane regions. This range is based on a first harvest followed by three or four extra crops. The sugar-producing stalks constitute about 70 percent of the total biomass; water content represents about 75 percent of the stalks' weight. The fiber content ranges from 11 to 14 percent according to the region and the varieties, and the soluble solids amount to 10 to 16 percent.

This biomass demands the selection of different genotypes to cover the range of conditions in the country. A great number of highly productive varieties with resistance to viruses, bacteria, smut, and other diseases are available. The mechanization of such a crop varies from region to region. In some areas, like São Paulo, nearly all phases of sugarcane farming, including harvest, can be mechanized. On the other hand, some areas in the northeast are based on intensive labor use. For this crop, there is sufficient technological information and an excellent research structure to improve the production of alcohol.

Cassava

The cassava, native to Brazil, has been used for centuries as staple food and for animal feeding. Being a tropical crop it can be grown from the south to the north of Brazil with few limitations. Despite its broad adaptation, cassava yields range from about 4 to about 11 tons of roots per acre-year (8 to 25 tons of roots per hectare). It should be pointed out that the potential yields are high. Varietal tests have shown productivity of 20-31 tons per acre-year (45-69 tons per hectare). Under favorable soil and climatic conditions, yields of 22-31 tons per acre-year (50-70 tons per hectare) have been reached. The starch content ranges from 20 to 40 percent, depending on the variety.

Despite its being an important staple crop covering more than 2 million hectares, there is not sufficient experience on plantation-type fields. Cassava is usually produced on small farms. Consequently, it has been recommended as an energy crop for allowing better participation of small- and medium-size farms in the production of alcohol. Although it can be grown in low fertility soil, experimental areas have indicated that cassava responds economically to mineral fertilization. Bacterial blight is the most critical disease affecting the plant. Cultivation technology is well established, but there is less experience on mechanization of cultivation and harvest. Several prototypes have recently been developed for partial mechanical harvest.

Sorghum

Sweet sorghum is being tested in many areas of Brazil. It can be cultivated all over the country; however, its biological and management limitations must be studied. Sorghum can be planted twice during the same agricultural season, and it has high tolerance to drought. For this reason, it is being evaluated in semiarid areas in the northeast or in regions with high probability of drought during the growing season.

The yield obtained in several experiments or pilot plantings ranges from 18 to 31 tons per acre-year (40 to 70 tons per hectare). The reducing sugar total varies from 11 to 15 percent, and the fiber content is around 11 percent. The alcohol conversion is about 17 gallons per ton of stalks (70 liters per ton of stalks). Besides the sugar produced for alcohol conversion, about 0.9 tons of sorghum kernel can be obtained per acre. This additional output of starch material makes sorghum an interesting energy crop to be evaluated. Because of its climatic adaptability, sorghum has been suggested as a secondary crop to expand the milling period in sugarcane distilleries from 150 to 300 days.

Elements of Brazil's PROALCOOL program were conceived as early as 1931: a government decree in Brazil required that 5 percent ethanol be added to gasoline. (In Brazil, a 10 percent mixture of ethanol in gasoline refers to a mixture of 10 gallons of ethanol and 100 gallons of gasoline, i.e., about a 9.1 percent mixture under U.S. definitions. Gasohol, a 10 percent mixture of ethanol and gasoline, refers to a mixture of 1 gallon of ethanol in 9 gallons of gasoline.) It was hoped that this decree would help stabilize the country's sugar industry. However, as recently as 1977, this goal of a 5 percent ethanol/gasoline mixture had not been reached. In 1975, the government launched the National Alcohol Program (PROALCOOL) which consists of two phases.

In the first phase (1975-1979), the objective is the partial replacement of gasoline consumption through the blending of anhydrous ethanol in gasoline up to a limit of 20 percent (16.67 percent gasohol) which requires no technical change in existing vehicle engines or fuel distribution facilities. A target production capacity of about 800 million gallons per year was set for 1980. In this phase, the program took advantage of existing idle capacity in the sugar industry and of the depressed world sugar prices. The increase in ethanol-fuel output capacity resulted mostly from the implementation or expansion of distilleries associated with the existing sugar mills. The production target of the program was achieved by the end of 1979. By that year, about 17 percent of the country's consumption of gasoline had been replaced by alcohol.

It is important to note that all of the ethanol produced is not used for fuel. In 1976-1977, about 33 percent of the production was used for fuel, about 57 percent was used by industry, and about 10 percent was exported. Export of ethanol is important to Brazil because, for each barrel of ethanol sold, the country can buy about two barrels of oil. As reported by the *Wall Street Journal* (June 18, 1980), Brazilian exports of ethanol to the United States were 29 million gallons in the first quarter of 1980 and 76 million gallons in April, 1980. The selling price was $1.40 per gallon.

Various Brazilian government incentives and subsidies tend to encourage the development of an ethanol industry. These include a government guarantee to purchase all anhydrous ethanol at an equity price with crystal sugar (equity is based on a ratio of 42 liters of alcohol to 60 kilograms of crystal sugar or 1 gallon per 12 pounds of crystal sugar), no taxation on anhydrous fuel used as fuel until 1979, and various low interest loan programs for agricultural and industrial development.

During the first phase of the program, the main feedstock for ethanol production is sugarcane. The pricing of ethanol at equity with world sugar prices does seriously impact the economic competitiveness of ethanol versus gasoline. When sugar prices were about $0.07/per pound ($0.15 per kilogram), the equity price of ethanol was about $0.81 per gallon as compared to world prices for gasoline of about $0.91. In Brazil, because of taxes, gasoline was retailing for about $2.65 per gallon. Recently, sugar prices have risen to about $0.30 per pound ($0.66 per kilogram) and the equity price of ethanol is about $3.75 per gallon, which is above the price of gasoline. As is discussed below, economics is not the only justification for PROALCOOL.

The second phase of PROALCOOL has a production capacity target of about 2.8 billion gallons of ethanol per year. This target includes about 800 million gallons of anhydrous ethanol for blending with gasoline, about 1.6 billion gallons of hydrated gasoline for direct use as fuel, and about 400 million gallons to be used as raw material for the chemical industry. Most of the additional production capacity will come from independent distilleries not associated with sugar mills. At present, sugarcane is the primary raw material, and the most frequent capacity of ethanol plants is about 3200 gallons per day (about 1 million gallons per year for a 300-day production year). These plants require an average investment of about $6 million (about $6 per gallon per year capacity) and supporting investments of about $3 million in the agricultural sector.

Currently, 259 projects have already been approved by PROALCOOL. This represents an additional capacity of 4.8 billion liters above the capacity before 1985. This capacity is distributed between north/northeast (34 percent) and center-southern (66 percent) regions.

A significant R&D program also supports PROALCOOL particularly in the areas of new raw materials (cassava and sweet sorghum in the near term and wood in the long term), improvement of sugarcane yields, reduction of production costs, and use of the fuel.

This technological program has already resulted in the development of proven technology for the use of hydrated ethanol as a direct fuel for the replacement of gasoline.

Accordingly, the Brazilian government signed agreements with the automotive industry for the production of full-alcohol powered cars and the conversion of existing ones for the use of alcohol as single fuel. The schedule for the next three years is:

1980:	New cars 250,000; converted cars 80,000; total 330,000
1981:	New cars 300,000; converted cars 90,000; total 390,000
1982:	New cars 350,000; converted cars 100,000; total 450,000
1980-1982 Total:	New cars 900,000; converted cars 170,000; total 1,070,000

The implementation of PROALCOOL is fully carried out by private enterprise. The government provides the overall program coordination, grants financing at special interest rates, and guarantees the purchase of the production of alcohol from PROALCOOL-approved projects.

The level of subsidy is basically designed to establish the following types of projects:

- Projects located in the north/northeastern regions
- Independent distilleries, not associated with sugar mills
- Using other raw materials instead of sugarcane (for example, cassava, babussu nuts, or wood)

The funding of PROALCOOL is obtained from the following sources:

- Commercialization of alcohol
- Taxes on oil imports
- Tax on the annual renewing of car licenses
- Local and foreign loans

All program phases, from the production of raw materials, alcohol, production and equipment manufacturing to the use of alcohol as a fuel in vehicles designed specifically for this purpose, have been fully developed with Brazilian technology. Additional efforts are being pursued to substitute other fractions of oil derivatives, such as diesel oil and fuel oil.

Concerning diesel oil, several options have been technologically proven in experiments conducted by the Brazilian government. Their economics are being evaluated for use on a commercial scale:

- Use of hydrated alcohol with additives as replacement for diesel oil
- Blending of 4 percent of anhydrous alcohol into diesel oil
- Double injection system using 40 percent alcohol and 60 percent diesel oil
- Use of vegetable oils (peanut, palm, soybean, etc.) blended with diesel oil
- Replacement of diesel engines by explosion engines fed with hydrated alcohol

In the area of fuel oil, a major effort is being concentrated in its replacement by mineral coal and charcoal.

These programs, together with PROALCOOL, represent Brazil's major contribution to the development of alternative energy sources.

As was pointed out earlier, the decision to launch the PROALCOOL program is not solely based on economic considerations. Other factors supporting the decision to proceed include: reduction or minimization of the country's foreign debt, support to the sugar industry, availability of large land areas for feedstock production, availability of domestic suppliers of ethanol-production equipment, lack of significant petroleum reserves, and desire to subsidize agriculture to provide employment in rural areas and reduce migration to the cities. To illustrate the potential impact that PROALCOOL could have, it has been estimated that the implementation of the second phase of the program would replace all growth of gasoline consumption as compared to that of 1973 by alcohol, result in industrial/agricultural investments of the order of $5 billion, result in the development of about 4.2 million acres of new agricultural areas, create about 350,000 new direct jobs, and make available a quantity of energy equivalent to the total present Brazilian production of petroleum (180,000 barrels per day).

Attention is being focused on the pollution effect of the distilleries' effluents (stillage) to the hydrological basins in Brazil. The stillage has been shown to have a high BOD and high amount of minerals, particularly potassium, and can therefore be highly polluting. The stillage makes an excellent fertilizer or feedstock and can be converted into methane or be used as substrate for single-cell-protein production. The application of stillage on the soil as fertilizer has been shown to be one of the most economical methods of simple disposal. Application of 0.2-0.5 acre-inch per year (50-100 cubic meters) on crops led to increase in the tonnage of more than 50 percent without any additional fertilizer requirement. Other possible uses cannot be ignored as alternative methods of disposal because of the tremendous volume of stillage produced yearly (about 5.3 billion cubic feet or more than 150 million cubic meters). In order to guarantee that the alcohol program will not affect the environment because of the high volume of stillage, strict measures must be taken by the government. No existing or new project will be allowed to operate without a plan for using the stillage.

The replacement of gasoline by ethanol has created the problem of disposing of a surplus of low-octane gasoline. This situation results from the fact that, while ethanol has expanded the supply of gasoline, the demand for diesel and fuel oil has increased, which has led to further imports of crude oil. Because the refineries were designed to process crude to methane for fuel oil, diesel oil, and gasoline, a surplus of the latter has resulted. Redesigning of the refineries is underway.

CAMEROON

The World Bank is considering loans to Cameroon to finance ethanol plants. Sugar mills are found in this country and, because of difficulty and cost of transportation, molasses are wasted. The basis for the loan is that it would be easier to transport and find a market for ethanol fuel than for molasses. (Conversion of molasses to ethanol reduces the volume of products by 2.5.)

CANADA

Although not directly involved in biomass-derived ethanol-fuel production, Canada maintains an R&D effort focusing on the conversion of lignocellulosic material (wood is a major biomass resource) to ethanol.

COSTA RICA

Two distilleries of about 32,000-gallons-per-day capacity are scheduled to start operations by 1981. The feedstock is sugarcane. The objective is to develop an industry capable of substituting about 20 percent of the imported oil or about 11 million gallons per year. Larger quantities of sugarcane than are available at present will, however, be required to reach this goal.

EGYPT

Blackstrap molasses from sugarcane processing is the feedstock used to produce ethanol. The total annual production is on the order of 16 gallons per year of 90-95 proof alcohol. About 9 million gallons are used by the food, beverage, perfume, chemical, and pharmaceutical industries; about 3.7 million gallons are used for heating and lighting; and about 3.3 million gallons are available for export.

To date, ethanol, alone or blended, has not been used for fuel in Egypt, but this option is being explored.

GUATEMALA

Research is being conducted on various aspects of ethanol production such as packed bed fermenters and new strains of yeast. Small pilot units (10-15 gallons) are being operated.

INDIA

In 1978, India produced about 132 million gallons of ethanol from molasses. About 32 million gallons were used as chemical feedstocks (production of acetaldehyde, acetic acid, DDT, acetates, PVC, acetone, styrene, and others); about 45 million gallons were used for beverage; and the remainder was exported. The high price of molasses in relation to the government-controlled price for ethanol does not make ethanol production an attractive commercial proposition. Since the mid-1950s when the use of some ethanol in automobile engines was discontinued, there has been no systematic attempt toward the use of ethanol fuel in India.

IRELAND

After World War II, legislation requiring that alcohol derived from foodstuffs be blended with gasoline was passed. The legislation was designed to help support the price of farm products. A government facility was built to produce ethanol from agricultural surpluses, originally potatoes. Other feedstocks were used when potatoes became unavailable at competitive prices. Agricultural surpluses became scarce after the entry of Ireland in the Common Market, so the plant was converted to use molasses as feedstock. Because of the high price of molasses, Ireland is currently faced with the problem of disposing of very expensive ethanol for blending with gasoline.

ITALY

Research and development work to test the performance of internal combustion engines running on various gases and alcohols is being pursued in Italy. For example, a modified Fiat engine coupled with an asynchronous electric motor has been reported to deliver 15 kilowatts of electricity and about 120,000 Btu of heat per hour, resulting in 90 percent utilization of the biogas fed to the system. Emphasis appears to be placed on developing low cost systems presumably for use in rural areas and in developing countries.

JAMAICA

Jamaica has a plant for producing anhydrous ethanol from molasses. This plant has been closed down because it was not competitive. As was pointed out in an earlier section, molasses converted to rum has over two times the value of molasses sold to the international market and over three times the value of molasses converted to ethanol fuel.

JAPAN

About 60 percent of Japan's ethanol production (total production was about 63 million gallons in 1977) is obtained through fermentation of biomass feedstocks such as molasses, potatoes, sweet potatoes, and sugar beets. This ethanol is used almost exclusively for industrial applications such as food preserves, cosmetics, and solvents. A special council has been established in the Ministry of International Trade and Industry to develop an ethanol-fuel industry in Japan. This program is developed in cooperation with Brazil, Indonesia, the Philippines, and Thailand. Feedstocks are expected to be cassava, seaweed, and garbage. Plans call for using gasohol as an automotive fuel by 1985. Research on methanol and ethanol use in automotive engines and on emissions from such engines has been reported to be underway in Japan.

NEW ZEALAND

The government has established a target of 50 percent self-sufficiency in transport fuels by 1987. Research is being pursued in the areas of compressed natural gas and liquid petroleum gas use, methanol and ethanol use, energy farming of sugar and fodder beets for ethanol production, and pine trees for methanol production. Vehicle utilization tests include fleet tests of 15 percent methanol and 85 percent gasoline blends, methanol use in diesel engines, and an assessment of modifications to existing gasoline engines for conversion to alcohol-fuel use.

PAPUA, NEW GUINEA

Construction of a 530,000 gallons ethanol per year facility using cassava as feedstock was begun in 1980. The fuel produced will be used locally in the remote area where it is produced.

PARAGUAY

A large-scale ethanol plant using sugarcane as feedstock is under construction and is expected to be operational by 1981. The ethanol will be blended with gasoline.

PHILIPPINES

The government has made a commitment to develop an ethanol-fuel production capability. The ethanol will be blended (15 percent) with gasoline (alcogas). A 16,000 gallons per day plant (about 5 million gallons per year at 300 days of production per year) that uses molasses as feedstock has been producing hydrous ethanol fuel since 1974. A 26,000 gallons per day plant (about 8 million gallons per year) is in the planning stage.

One-half of this new plant output will be shipped to Japan (hydrous form), and one-half will be blended with gasoline into alcogas for consumption in the Philippines. Current exports of ethanol to Japan are about 5 million gallons per year. Some of the ethanol plants currently in operation are integrated with sugar mills and use bagasse as fuel for process heat generation. Animal feed coproducts are often recovered.

Plans call for the implementation of eight distilleries of about 48,000 gallons per day capacity (about 14 million gallons per year capacity) in the next 10 years. Blended at 15 percent with gasoline, this alcogas would displace about 5.8 million barrels of oil per year.

SENEGAL

A plan similar to that described for Cameroon is under consideration.

SOUTH AFRICA

South Africa is in a situation similar to that of Brazil. Large areas of underutilized land capable of growing ethanol feedstock crops (corn, sugarcane, or sweet sorghum) are available. The development of these areas as feedstock-producing regions would provide employment in rural areas and would reduce migration to the cities. Ethanol from maize has been used for several years as an industrial chemical. A limited quantity of ethanol derived from sugarcane is currently used as a fuel extender. Expanding the production of ethanol for blending with gasoline is under consideration.

SWEDEN

A program to test pure methanol as automotive fuel was initiated by Volvo and the Swedish Government in 1978. Methanol would be derived from wood. A program is also underway to test ethanol fuel. Feedstocks in this case would be grain initially and cellulosic materials in the long run. A decision concerning a commitment to ethanol fuel is expected to be reached by 1981.

WEST GERMANY

A test program of a 15 percent methanol and 85 percent unleaded gasoline blend is underway. Simultaneously, Hoechst A.G. and Uhde,

GmbH., have been operating a 264 gallons per day (about 80,000 gallons per year) pilot ethanol unit since 1978. This unit uses molasses as feedstock. A starch-based ethanol plant using continuous fermentation is reportedly scheduled for operation by 1981. Research is also pursued to improve the ethanol distillation process.

Resource People and Organizations

INTRODUCTION

The list of resource people and organizations is provided for your information. In addition, a brief survey of the federal, state, and local governments' responsibilities that are concerned with any aspect of alcohol-fuel production is given. The reader is referred to the particular governmental entity for specific information and assistance.

ASSOCIATIONS AND ORGANIZATIONS

American Agriculture Foundation
P.O. Box 57
Springfield, CO 81073
(303) 523-6223

The Bio-Energy Council
1625 Eye Street, NW
Suite 825A
Washington, DC 20006
Contact: Carol Canelio
(202) 833-5656

Brewers Grain Institute
1750 K Street, NW
Washington, DC 20006

Distillers Feed Research Council
1535 Enquirer Building
Cincinnati, OH 45202
Contact: Dr. William Ingrigg
(513) 621-5985

International Biomass Institute
1522 K Street, NW
Suite 600
Washington, DC 20005
Contact: Dr. Darold Albright
(202) 783-1133

National Alcohol Fuel Institute
5716 Jonathan Mitchell Rd.
Fairfax Station, VA 22039
(703) 250-5136

National Alcohol Fuels Commission
492 1st St., S.E.
Washington, DC 20003
Contact: Jim Childress
(202) 426-6490

National Alcohol Fuels Information Center
Solar Energy Research Institute
1617 Cole Boulevard
Golden, CO 80401
Contact: Cecil Jones
(800) 525-5555

National Gasohol Commission, Inc.
521 South 14th St., Suite 5
Lincoln, NE 68508
Contact: Myron Reamon
(402) 475-8044 or 8055

National Oil Jobbers Council
11th Floor
1707 H Street, NW
Washington, DC 20006
(202) 331-1078

Nebraska Agricultural Products Industrial Utilization Committee
P.O. Box 94831
Lincoln, NE 68509
(402) 471-2941

Solar Energy Research Institute
1617 Cole Boulevard
Golden, CO 80401
Contact: Cecil Jones
(303) 231-1205

BIOLOGISTS

Paul Middaugh
South Dakota State University
Brookings, SD 57007
(605) 688-4116

Robert Middaugh
1704 Third Street
Brookings, SD 57006
(605) 692-5760

Micro-TEC Lab, Inc.
Route 2, Box 19L
Logan, IA 51546
Contact: John W. Rago
(712) 644-2193

Leo Spano
The Army/Navy Lab
Natick, MA 01760
(617) 653-1000, Ext. 2914

BLENDING TERMINALS

GATX Terminals Corporation
P.O. Box 409
Argo, IL 60501
(312) 458-1330

CHEMISTS

Lance Crombie
Route 1
Webster, MN 55088
(507) 652-2804

Dr. Harry P. Gregor
Columbia University
353 Seeley West Mudd Building
New York, NY 10027
(212) 280-4716

Dr. Richard Spencer
Southwest State University
Marshall, MN 56258
(507) 537-7217

COMPONENT MANUFACTURERS

Abbeon Industrial Lab
and Plant Buying
123-52A Gray Avenue
Santa Barbara, CA 94119
(805) 966-0810

Agri-Tech
P.O. Box 26581
Denver, CO 80226
Contact: Tony Ball
(303) 989-3343

Alco-Fuels, Inc.
5118 Valley Road
Fairfield, AL 35064
Contact: Celia Gasque
(205) 787-3835

ALCOGAS
220 Equitable Building
730 17th Street
Denver, CO 80203
Contact: Evan L. Goulding
(303) 572-8300

Alternative Energy Sources, Inc.
752 Duvall
Salina, KS 67401
(913) 825-8218

Alternative Energy Systems
305 Basseron
Vincennes, IN 47524
Contact: Dave Snider
(812) 882-7320

Arlon Industries, Inc.
P.O. Box 347
Sheldon, IA 51201
Contact: David Vander Griend
(712) 324-3305

W. A. Bell
P.O. Box 105
Florence, SC 29503

Better Way, Inc.
2697 Casco Point Road
Wayzata, MN 55391
Contact: Dick Large
(612) 471-7088

Bienergy Organizers, Inc.
P.O. Box 5715
Baltimore, MD 21208
Contact: Paul Gibson
(301) 484-8913

Cole-Parmer Instruments Co.
7425 N Oak Park Ave.
Chicago, IL 60648
(312) 647-7600

Conrad Industries
Box 130
Bonapart, IA 52620
(319) 592-3131

Dale Devermon
3550 Great Northern Ave.
Route 4
Springfield, IL 62707
(217) 787-9870

Energy Products of Idaho
P.O. Box 153
Coeur d'Alene, ID 83814
Contact: Mike Oswald
(208) 667-2418

Energy Restoration, Inc.
1201 J Street
Century House Suite 403
Lincoln, NE 68508
(402) 475-9237

Ernst Gage Co.
250 South Livingston Ave.
Livingston, NJ 07039
(201) 992-1400

Ethanol International, Inc.
1372 South Fillmore
Denver, CO 80210
Contact: Stephen M. Munson

Ferguson-Poer, Inc.
Box 217
Dunreith, IN 47337
(317) 465-5108

Flexi-Liners
5940 Reeds Rd.
Mission, KS 66202
(913) 265-1234

Fluid Metering, Inc.
29 Orchard Street
Oyster Bay, NY 11771
(516) 922-6050

Fort Wayne Dairy Equipment Co.
Box 269
Fort Wayne, IN 46801
(219) 424-5517

Grau Enterprises, Inc.
Sioux Rapids, IA 50585
Contact: Ted Grau
(712) 283-2316

D. N. Gray
Biotechnology and Toxicology
Toledo, OH 43666
(419) 247-9206

Hot Energy Shoppe
490 W 300 South
Provo, UT 84601
Contact: Hal Foutz
(801) 377-8130

Industrial Innovators, Inc.
P.O. Box 387
Ashford Plaza Shopping Center
Ashford, AL 36312
Contact: Vernon Heines
(205) 899-3314

Jacobson Machine Works, Inc.
2445 Nevada Ave. N
Minneapolis, MN 55427
(612) 544-8781

Kargard Industries
Marinette, WI 54143
(715) 735-9311

KBK Industries
East Hwy 96
Rush Center, KS 67575
(913) 372-4331

Rochelle Development Inc.
Box 356
Rochelle, IL 61068
Contact: John Askvig
(815) 562-7372

SEI Corp
54-45 44th Street
Maspeth, NY 11378
Contact: Dr. Tica

Seven Energy Corporation
3760 Vance
Wheat Ridge, CO 80033
(303) 425-4239

Sioux Equipment
1310 East 39th St. N
Sioux Falls, SD 57104
Contact: Elmer Biteler
(605) 334-1653

Slarton, Inc.
2617 Sibley
St. Charles, MO 63301
Contact: Ron Williams
(314) 926-1816

S & S Galvanizing Co.
P.O. Box 37
Clay Center, NE 68933
Contact: Eldon L. Shetler
(402) 762-3884

Still Company
115 East 9th Street
P.O. Box 9570
Panama City Beach, FL 32407
(907) 769-4269

Stone & Webster
One Penn Plaza
250 West 34th St.
New York, NY 10001

3T Engineering Inc.
Box 80
Arenzville, IL 62611
Contact: Wm. C. Talkemeyer
(217) 997-5921

Tomco Equipment
1862 South Duth
Hidden Hills Parkway
Stone Mountain, GA 30088
Contact: Morton Bradlyn
(404) 466-2235

United Industries
P.O. Box 11
Buena Vista, GA 31803
Contact: John Daniel
(912) 649-7444

Vendome Copper and Brass
153 North Shelby Street
Box 1118
Louisville, KY 40202
(502) 587-1930

E. Dale Waters
Double "L" Mfg. Inc.
P.O. Box 533
American Falls, ID 83211
(208) 226-5592

Wenger Alko-Vap System
1220 Rochester Boulevard
Rochester, IN 46975
Contact: Oscar Zehier
(219) 223-3335

Wenger Manufacturing
714 Main Street
Sabetha, KS 66534
Contact: Kenneth Sanderude
(913) 284-2133

Weslipp
Franklin, NE
Contact: Brian Hayers
(308) 425-3101

WESTCO
P.O. Box 1236
Evergreen, CO 80439
Contact: Wayne Bethrum
(303) 674-6099

Archy Zeithamer
Route 2, Box 63
Alexandria, MN 56308
(612) 763-7392

CONSULTANTS FOR REPORTS[1]

Alltech, Inc.
271 Goldrush Rd.
Lexington, KY 40503
Contact: Dr. Pearse Lyons
(606) 276-3414

Battelle Memorial Institute
505 King Ave.
Columbus, OH 43201
Contact: Dr. Billy Allen
Mr. David Jenkins
(614) 424-6424

Bartlesville Energy Technology Center
Bartlesville, OK
Contact: Jerry Allsup
(918) 336-4268

Center for the Biology of Natural Systems
Washington University
St. Louis, MO 63130
Contact: Mr. David Freedman
(314) 889-5317

Department of Energy
Office of Alcohol Fuels
Forrestal Building
Room 6A-211
1000 Independence Ave.
Washington, DC 20585
(202) 252-9487

Development Planning and Research Associates
200 Research Drive
P.O. Box 727
Manhattan, KS 66502
Contact: Mr. Milton David
(913) 539-3565

The Eakin Corporation
401 Delphine St.
Baton Rouge, LA 70806
Contact: Mr. Sam Eakin
(504) 346-0453

EG&G Idaho, Inc.
P.O. Box 1625
Idaho Falls, ID 83415
Contact: Don LaRue
(208) 526-0509

Energy Inc.
P.O. Box 736
Idaho Falls, ID 83401
Contact: Steve Winston
(208) 524-1000

Galusha, Higgins and Galusha
P.O. Box 751
Glascow, MT 59230
Contact: Jim Smrcka
(406) 228-9391

William S. Hedrick
844 Clarkson
Denver, CO 80218
(303) 832-1407

E. F. Hutton
3340 Peachtree St. Rd., N.E.
Atlanta, GA 30026
Contact: Mr. Strud Nash
(404) 262-2110

Pincas Jawetz
Independent Consultant on Energy Policy
425 East 72nd Street
New York, NY 10021
(212) 535-2734

[1]Consultants for *Fuel From Farms* and *A Guide to Commercial Scale Ethanol Production and Financing*

Ralphael Katzen Associates
1050 Delta Ave.
Cincinnati, OH 45208
Contact: Dr. Raphael Katzen
(513) 871-7500

A. T. Kearney, Inc.
699 Prince St.
Alexandria, VA 22314
Contact: Dr. Cathryn Goddard
(703) 836-6210

PEDCo International
1149 Chester Rd.
Cincinnati, OH 45246
Contact: Dr. William Stark
(513) 782-4717

Quaintance and Swanson
Suite 120, Van Brunt Building
226 N. Phillips Ave.
Sioux Falls, SD 57101
Contact: Mr. Robert Mabee, Attorney
(605) 335-1777

Solar Energy Research Institute
1617 Cole Blvd.
Golden, CO 80401
Contact: Cecil Jones
(303) 231-1205

Arthur Young and Company
235 Peachtree St., N.E.
Atlanta, GA 30303
Contact: Mr. Michael Thomas
(404) 577-8773

CONSULTANTS/DESIGN ENGINEERS

ACR Process Corp.
808 South Lincoln, #14
Urbana, IL 61801
Contact: Robert Chambers
(217) 384-8003

Alltech, Inc.
271 Goldrush Rd.
Lexington, KY 40503
Contact: Dr. Pearse Lyons
(606) 276-3414

Bartlesville Energy Technology Center
Bartlesville, OK 74003
Contact: Jerry Allsup
(918) 336-4268

Battelle Columbus Laboratories
505 King Avenue
Columbus, OH 43201
Contact: Dr. Billy Allen (Small Scale Systems)
(614) 424-6424

Biomass Suchem
Clewiston, FL 33440
Contact: Dr. Ron DeStephano
(813) 983-8121

Center for Biology of Natural Systems
Washington University
St. Louis, MO 63130
Contact: David Freedman
(314) 889-5317

Chemapec, Inc.
230 Crossways Park Drive
Woodbury, NY 11797
Contact: Dr. Ing. Hans Mueller
(516) 364-2100

Development Planning and Research Associates
200 Research Drive
P.O. Box 727
Manhattan, KS 66502
Contact: Milton David
(913) 539-3565

Raphael Katzen Associates
1050 Delta Avenue
Cincinnati, OH 45208
Contacts: Dr. Raphael Katzen
George Moon
(513) 871-7500

Vulcan Cincinnati, Inc.
2900 Vernon Place
Cincinnati, OH 45219
Contacts: Donald Miller
Ted D. Tarr
(513) 281-2800

Lew Weiner Associates
10134 N. Port Washington Rd.
Mequon, WI 53092

CO-OPS IN OPERATION

Ted Landers
New Life Farm
Drury, MO 65638
(417) 261-2553

DDG PURCHASERS

The Pillsbury Company
3333 South Broadway
St. Louis, MO 63118
(314) 772-5150

ECONOMISTS

Center for the Biology of Natural Systems
Washington University
Box 1126

St. Louis, MO 63130
Contact: David Freedman
(314) 889-5217

ENGINE MODIFICATION ENGINEERS

Michael H. Brown
Desert Publications
Cornville, AZ 86325
(602) 634-4650

M & W Gear Company
Gibson City, IL 60936
(217) 784-4261

Dick Pefley
University of Santa Clara
Santa Clara, CA 95053
(408) 984-4325

Southwest Research Institute
San Antonio, TX

ENGINEERING FIRMS

Bechtel Corporation
50 Beale Street
15th Floor, Room B15
San Francisco, CA 94119
Contact: Jackson Yu
(415) 768-2971

Bohler/Vogelbusch
1625 West Belt North
Houston, TX 77043
Contact: Jerry Korff
(713) 465-3373

C & I Girdler Co.
1721 South 7th Street
P.O. Box 32940
Louisville, KY 40234
Contacts: Steve Knapp
William McGammon
(502) 637-8701

Chemapec Inc.
230 Crossways Park Drive
Woodbury, NY 11797
Contact: Rene Loser
(516) 364-2100

Day & Zimmerman, Inc.
1818 Market Street
Philadelphia, PA 19103
Contact: William J. Jones
(215) 299-8193

Honeywell, Inc.
Corp. Tech. Center
10701 Lyndale Avenue South
Bloomington, MN 55420
Contact: Ulrich Bonne
(612) 887-4477

Hydrocarbon Research Inc.
134 Franklin Corner Rd.
P.O. Box 6047
Lawrenceville, NJ 08648
Contact: Maurice Jones
(609) 896-1300

I. E. Associates
3704 11th Avenue South
Minneapolis, MN 55407
Contact: Tom Abeles
(612) 825-9451

Raphael Katzen Associates
1050 Delta Ave.
Cincinnati, OH 45208
Contacts: Dr. Raphael Katzen
George Moon
(513) 871-7500

Keep Chemical Company
Box 441
Cornwall, NY 12518
Contact: Peter J. Ferrara
(914) 534-4755

A. G. McKee Corporation
10 South Riverside Plaza
Chicago, IL 60606
Contact: Edward A. Kirchner
(312) 454-3685

PEDCo International, Inc.
11499 Chester Road
Cincinnati, OH 45246
Contact: Timothy Devitt
(513) 782-4717

Power Engineering Co.
1313 S.W. 27th Ave.
Miami, FL 33145
Contact: John Scopetta

Seven Energy Corporation
3760 Vance
Wheat Ridge, CO 80030
Contact: Ernie Barcell
(303) 989-7777

Stone & Weber Eng. Corp.
One Penn Plaza
New York, NY 10001
Contact: Miles Connors
(212) 760-2000

3-T Engineers
Arenzville, IL 62611
Contact: R. E. Talkemeyer
(217) 997-2188

Vulcan Cincinnati, Inc.
2900 Vernon Plaza
Cincinnati, OH 45219

Contacts: Donald Miller
Ted D. Tarr
(513) 281-2800

ENZYME AND YEAST PRODUCERS

Alltech, Inc.
271 Goldrush Road
Lexington, KY 40503
(606) 276-3414

Anheuser-Busch, Inc.
721 Pestalozzi Street
St. Louis, MO 63118
(314) 577-2000

Biocon, Inc.
2348 Palumbo Drive
Lexington, KY 40509
(606) 269-6351

Chemapec, Inc.
230 Crossways Park Drive
Woodbury, NY 11797
Contact: Rene Loser
(516) 364-2100

Enzyme Development Corporation
210 Plaza
New York, NY 10001
Contact: Randy Ross
(212) 736-1580

Fermco Biochemicals, Inc.
2638 Delton Lane
Elk Grove Village, IL 60007
(312) 595-3131

G. B. Fermentation Industry, Inc.
One North Broadway
Des Plaines, IL 60061
(312) 827-9700

Grove Engineering
1714 Gervais Avenue
North St. Paul, MN 55109
(612) 777-8545

Norbert Haverkamp
Compost Making Enzymes
Rural Route 1, Box 114
Horton, KY 66439
(913) 486-3302

Miles Laboratories, Inc.
Enzyme Products Division
P.O. Box 932
Elkhart, IN 56515
(219) 564-8111

Nova Laboratory, Inc.
59 Danbury Road
Wilton, CT 06897
(203) 762-2401

Rohm & Haas
Independence Mall West
Philadelphia, PA 19105
Contact: Bob Broadbent
(215) 592-2517

Schwartz Service International Ltd.
230 Washington Street
Mt. Vernon, NY 10551
Contact: Mr. Aberdein
(914) 664-1100

Scientific Products Co.
North Kansas City, MO 64116
(816) 221-2533

ETHANOL PRODUCERS/ DISTRIBUTORS

Amoco Production Company
P.O. Box 5340a
Chicago, IL 60680
(312) 856-2222

ARCO Oil Corporation
Houston Natural Gas Building
Houston, TX 77002
(713) 658-0610

Archer-Daniels-Midland Company
P.O. Box 1470
Decatur, IL 62525
(217) 424-5200

Big D & W
Refining and Solvent Co., Inc.
705 South 23rd Street
Vanburen, AR 72956
(501) 474-5258

Encore Energy Resources, Incorporated
11951 Mitchell Road, Mitchell Island
Richmond, British Columbia
(604) 327-8394

Exxon Corporation
Exxon Research and Engineering
Public Relations
P.O. Box 639
Linden, NJ 07036

Georgia-Pacific
900 Southwest 5th Avenue
Portland, OR 97204
(503) 222-5561

Georgia Pacific Company
Bellingham Division
P.O. Box 1236
Bellingham, WA 98225
(206) 733-4410

Grain Processing Corporation
Muscatine, IA 52761
(918) 264-4211

Marcam Industries
527 North Easton Road
Glenside, PA 19038
(215) 885-5400

Midwest Solvents Company
1300 Main Street
Atchison, KS 66002
(913) 367-1480

Milbrew, Incorporated
330 South Mill Street
Juneau, WI 53039
(414) 462-3700

Ronald C. Mode
Box 682
Glen Alpine, NC 28628
(704) 584-1432

Publicker Industries
777 W. Putnam Avenue
Greenwich, CT 06830
(203) 531-4500

Quaternoin Chemical Industries
72026 Livingston Street
Oakland, CA 94604
Contact: Louis Nagel
(415) 535-2311

Sigmor Corporation
P.O. Box 20267
San Antonio, TX 78220
(512) 223-2631

Texaco, Incorporated
135 East 42nd Street
New York, NY 10017
(212) 953-6000

TIPCO
9000 N. Pioneer Rd.
Peoria, IL 61614
(309) 692-6543

Hiram Walker
31275 Northwestern Highway
Farmington Hills
Detroit, MI 48018

Worum Chemical Company
2130 Kasoto Avenue
St. Paul, MN 55108
Contact: Mr. Ritt
(612) 645-9224

FEEDSTOCK CONTACTS

General

Distillers Feed Research Council
1435 Enquirer Bldg.
Cincinnati, OH 45202
(513) 621-5985

Sorghum

Grain Sorghum Producers Association
1708 - A 15th Street
Lubbock, TX 79401
Contact: Elbert Harp
Executive Director
(806) 763-4425

Producers Grain Corp.
P.O. 111
Amarillo, TX 79105
(806) 374-0331

Potatoes

National Potato Council
45th & Peoria
Denver, CO 80239
(303) 373-5639

Grains

National Council of Farmer Co-ops
1800 Massachusetts Ave. NW
Washington, DC 20036
(202) 659-1525

MOLECULAR SIEVE MATERIAL

W. R. Grace
P.O. Box 2117
Davison Chemical Division
Baltimore, MD 21203
Contact: Paul E. Cevis
(301) 659-9000

ON-FARM DEMONSTRATION PLANTS

Phil Apple
Apple Agri-Sales
Rural Route #6
Crawfordsville, Indiana 47933

Beckman Construction Company
Alcohol Engineering Division
7201 West Vickery Street
Fort Worth, TX 76116

Randy Butters
4257 Two-and-one-half-mile Road
Homer, MI 48245

Alan and Archie Zeithamer
Route 2, Box 63
Alexandria, MN 56308
(612) 762-1798 (Alan)
(612) 763-7392 (Archie)

PLANT OPERATION CONSULTANTS

ACR Process Corporation
808 South Lincoln, #14
Urbana, IL 61801
Contact: Robert Chambers
(217) 384-8003

A. G. McKee Associates
10 South Riverside Plaza
Chicago, IL 60606
Contact: E. Kirchner
(312) 454-3685

Agri-Stills of America
3550 Great Northern Avenue
Route 4
Springfield, IL 62707
Contact: Al Mavis
(217) 787-4233

Alltech, Inc.
271 Goldrush Road
Lexington, KY 40503
Contact: Dr. Pearse Lyons
(606) 276-3414

Biomass Suchem
Clewiston, FL 33440
Contact: Dr. Ron DeStephano
(813) 983-8121

Chemapec, Inc.
230 Crossways Park Drive
Woodbury, NY 11797
Contact: Dr. Ing. Hans Mueller
(516) 364-2100

Development Planning and Research Associates
2000 Research Drive
P.O. Box 727
Manhattan, KS 66502
Contact: Milton David
(913) 539-3565

Enerco, Inc.
139A Old Oxford Valley Road
Langhorne, PA 19047
Contact: Miles J. Thomson
(215) 493-6565

GB Fermentation Industries, Inc.
One North Broadway
Des Plaines, IL 60016
(312) 827-9700

REGIONAL OFFICES OF BUREAU OF ALCOHOL, TOBACCO, AND FIREARMS

Central Region: Indiana, Kentucky, Michigan, Ohio, West Virginia
Regional Regulatory Administrator
Bureau of Alcohol, Tobacco, and Firearms
550 Main Street
Cincinnati, OH 45202
(513) 684-3334

Mid-Atlantic Region: Delaware, District of Columbia, Maryland, New Jersey, Pennsylvania, Virginia
Regional Regulatory Administrator
Bureau of Alcohol, Tobacco, and Firearms
2 Penn Center Plaza, Room 360
Philadelphia, PA 19102
(215) 597-2248

Midwest Region: Illinois, Iowa, Kansas, Minnesota, Missouri, Nebraska, North Dakota, South Dakota, Wisconsin
Regional Regulatory Administrator
Bureau of Alcohol, Tobacco, and Firearms
230 S. Dearborn Street, 15th Floor
Chicago, IL 60604
(312) 353-3883

North-Atlantic Region: Connecticut, Maine, Massachusetts, New Hampshire, New York, Rhode Island, Vermont,
Puerto Rico, Virgin Islands
Regional Regulatory Administrator
Bureau of Alcohol, Tobacco, and Firearms
6 World Trade Center, 6th Floor
For letter mail
P.O. Box 15, Church Street Station
New York, NY 10008
(212) 264-1095

Southeast Region: Alabama, Florida, Georgia, Mississippi, North Carolina, South Carolina, Tennessee
Regional Regulatory Administrator
Bureau of Alcohol, Tobacco, and Firearms
3835 Northeast Expressway
For letter mail
P.O. Box 2994
Atlanta, GA 30301
(404) 455-2670

Southwest Region: Arkansas, Colorado, Louisiana, New Mexico, Oklahoma, Texas, Wyoming
Regional Regulatory Administrator
Bureau of Alcohol, Tobacco, and Firearms
Main Tower, Room 345
1200 Main Street
Dallas, TX 75202
(214) 767-2285

Western Region: Alaska, Arizona, California, Hawaii, Idaho, Montana, Nevada, Oregon, Utah, Washington
Regional Regulatory Administrator

Bureau of Alcohol, Tobacco, and Firearms
525 Market Street, 34th Floor
San Francisco, CA 94105
(415) 556-0226

SMALL-SCALE PLANT SERVICES

Agri Products
Hillsboro, OH
Contact: Jerry Garrison

American Standard
Cincinnati, OH
Contact: Jack Connor

Aqua Chem
Milwaukee, WI
Contact: Anna Zoltak

Beard Industries, Inc.
Frankfort, IN
Contact: Ronald Noyes

Bird Machine Company
South Walpole, MA
Contact: Glen Reirstad

Burnham Boiler
Gahanna, OH
Contact: Len Myers

Carrier Cooling Towers
Cincinnati, OH
Contact: Michael Shurr

Chem-Pro Equipment Corp.
Fairfield, NJ
Contact: Bill Schweitzer

Circle Steel Corporation
Taylorville, IL
Contact: Dennis Neal

Cleaner Brooks
Cincinnati, OH
Contact: Joe Bollinger

C. E. Raymond Impact Mills & Dryers
Chicago, IL
Contact: Ralph Rogers

Corporate Equipment Co.
Cincinnati, OH
Contact: Ron Kastner

Davenport Machine and Foundry Co.
Davenport, IO
Contact: David J. Aldayse

Easy Engineering
3351 Larimer Street
Denver, CO 80205
Contact: Richard Stewart
(303) 893-8936

Energy Independence Corporation
Montrose, MN 56308
Contact: Jack Harmon
(612) 625-3131

Ethanol International, Inc.
1372 South Fillmore
Denver, CO 80210
Contact: Stephen M. Munson
(303) 744-8355

Expert Industries, Inc.
80 19th Street
Brooklyn, NY 11232
(212) 788-2896

Farm Fan, Inc.
Indianapolis, IN
Contact: Joe Zimmer

William S. Hedrick
844 Clarkson
Denver, CO 80218
(303) 832-1407

S. Howes Co.
Buffalo, NY
Contact: W. E. Bantle

Hunter Mfg. Co.
Winchester, OH
Contact: Bill Pfeiffer

Hydro-Thermal Heaters
Milwaukee, WI
Contact: Paul Gum

Lectrodryer
Richmond, KY
Contact: Hassan Mahdara

Major Refrigerating Engineers
Cincinnati, OH
Contact: Steve Stein

McEver Engineering, Inc.
6363 Richmond Avenue
Houston, TX 77057
(713) 780-3465

Middle State Manufacturing Co.
Box 788
Columbus, NE 68601
Contact: Bill Fritz
(402) 564-1411

Moier Construction
511 North Cedar
P.O. Box 381
Owatonna, MN 55060
(507) 451-5254

Philip-Rahn Inc.
Houston, TX
Contact: Jim Bond

Process Equipment Corp.
Cincinnati, OH
Contact: James Hockett

Proctor & Schwartz Inc.
Philadelphia, PA
Contact: Larry Harris

A. K. Robins Washers
Baltimore, MD
Contact: A. S. Brodsky

Rotex Screens
Cincinnati, OH
Contact: Nash McCulley

Schmitt Energy Systems
Route 2
Hawkeye, IA 52147
(319) 427-3497

Sharples Centrifugals
Cleveland, OH
Contact: Bob Cotab

Silver Engineering Works, Inc.
3309 Blake Street
Denver, CO
Contact: Richard D. Smith
(303) 623-0211

Solar Fuel Company
Box 423
Dubuque, IA 52001

United Evaporator
Union, NJ
Contact: Anthony Diflippo

Urschel Laboratories Inc.
Columbus, OH
Contact: B. E. Pritchard

United International, Inc.
United International of Buena Vista
P.O. Box 271
Buena Vista, CA

Weil McLain Boilers
Cincinnati, OH
Contact: Bob Finnell

Wenger Manufacturing
Sabetha, KA
Contact: Gary Edelman

Werner and Pfleiderer Corp.
Ramsey, NJ
Contact: Richard Ziminski

Wylain Filters Inc.
Cincinnati, OH
Contact: Tom Fritz

York Cooling Towers
Cincinnati, OH
Contact: Richard T. Yerian

FINANCIAL ASSISTANCE PROGRAMS

The Federal Government affects the distribution of national resources by direct loan and loan guarantee programs aimed, directly or indirectly, at private industry. Tax credits, tax exemptions, and price support programs are other forms this distribution can take. The purpose of loan programs is to channel funds to those sectors of the economy which could not otherwise obtain credit or could obtain it only at very high costs. Sometimes the channel is indirect, going through state or local governments or other nonprofit organizations. The specific goal, in this case, is to enable private ventures to make more alcohol-production capability available than if left to normal business dealings.

In addition to the Department of Energy (DOE), the federal agencies with such authority are the Small Business Administration (SBA); the Farmers Home Administration (FmHA) of USDA; Urban Development Action Grants, the Department of Housing and Urban Development (HUD); and Economic Development Administration (EDA) of the Department of Commerce. The existing legislation in all cases has already been interpreted to apply to alcohol-fuel production. Several legislative efforts are underway that will specifically address alcohol fuels.

REGULATIONS AND PERMITS

Regulations concerning the use of distilled spirits for fuel, as covered by Public Law 96-223, can be obtained from the Bureau of Alcohol, Tobacco, and Firearms of the U.S. Department of Treasury.

Most plants designed for the production of fuel alcohol from grain will be designed to produce coproducts in the form of grain residues suitable for use as animal feeds. Most states have adopted commercial feed laws governing the production and marketing of animal feed; therefore, the regulations in each state should be checked for particular requirements. In addition, the sale of feeds on an interstate basis is regulated at the federal level by various agencies, primarily the Food and Drug Administration, and the surface transportation of raw materials and finished product may be regulated by state and federal agencies, depending on the location of the plant and its proximity to the source of materials and its markets. State and local building codes should also be examined. Finally, the following laws also have some alcohol-fuel aspects:

The Energy Security Act: P.L. 96-294.

Department of Interior Appropriations for FY 1980: P.L. 96-126.

Department of Energy, Alternative Fuels Production: P.L. 96-304.

Energy Tax Act of 1978: P.L. 95-618.

Crude Oil Windfall Profit Tax Act of 1980: P.L. 96-223.

Crude Oil Entitlements Program.

Consolidated Farm and Rural Development Act of 1972: P.L. 92-419, amended. Farm Loans.

Consolidated Farm and Rural Development Act of 1972: P.L. 92-419, amended. Business and Industrial Loans.

Food and Agricultural Act of 1977: P.L. 95-113 Section 1420.

Public Works and Economic Development Act of 1965: P.L. 89-136.

Small Business Act, amended: P.L. 95-315.

Housing and Community Development Act of 1974: P.L. 93-383; Urban Development Action Grants: P.L. 93-128.

Appendix B

Reference Sources

CHAPTER 1. INTRODUCTION TO GASOHOL

DOE, *Report of the Alcohol Fuels Policy Review,* Department of Energy Report No. DOE/PE-0012, Superintendent of Documents, Stock Number 061-000-00313-4, June, 1979.

DOE, *Preliminary Design Report Small-Scale Fuel Alcohol Plant,* DOE Office of Alcohol Fuels, October, 1980.

Katzen, Raphael Associates, *Grain Motor Fuel Alcohol Technical and Economic Assessment Study,* DOE Report Number HCP/J6639-01, dated June, 1979.

SERI, *Fuel From Farms—A Guide to Small-Scale Ethanol Production,* SERI Report No. SP-451-519, dated May, 1980.

SERI, *A Guide to Commercial Scale Ethanol Production and Financing,* dated November, 1980.

USDA, *Small-Scale Fuel Alcohol Production,* U.S. Department of Agriculture, dated March, 1980.

CHAPTER 2. DECISION TO PRODUCE GASOHOL

"Ethanol." *Chemical and Engineering News,* p. 12, October 29, 1979.

Jawetz, P., "Improving Octane Values of Unleaded Gasoline Via Gasohol." Proceedings of the 14th Intersociety Energy Conversion Engineering Conference. Volume I, pp. 301-302; abstract Volume II, p. 102; Boston, MA, August 5-10, 1979. Available from the American Chemical Society, 1155 Sixteenth Street NW, Washington, DC 20036.

Panchapakesan, M.R. et al. "Factors That Improve the Performance of an Ethanol Diesel Oil Dual-Fuel Engine." International Symposium on Alcohol Fuel Technology—Methanol and Ethanol. Wolfsburg, Germany; November 21-23, 1977. CONF-771175.

Reed, T. "Alcohol Fuels." Special Hearing of the U.S. Senate Committee on Appropriations; Washington, D.C.: January 31, 1978; pp. 194-205. U.S. Government Printing Office. Stock Number 052-070-04679-1.

Ribeiro, F. F. A., "The Ethanol-Based Chemical Industry in Brazil." Workshop on Fermentation Alcohol for Use as Fuel and Chemical Feedstock in Developing Countries. United Nations Industrial Development Organization, Vienna, Austria, March 1979. Paper No. ID/WG.293/14 UNIDO.

SERI, *Fuel From Farms—A Guide to Small-Scale Ethanol Production,* SERI Report No. SP-451-519, dated May, 1980.

SERI, *A Guide to Commercial Scale Ethanol Production and Financing,* dated November, 1980.

Sharma, K. D., "Present Status of Alcohol and Alcohol Based Chemical Industry in India." Workshop on Fermentation Alcohol for Use as Fuel and Chemical Feedstock in Developing Countries. United Nations International Development Organization, Vienna, Austria, March, 1979. Paper No. ID/WG.293/14 UNIDO.

USDA, *Small-Scale Fuel Alcohol Production,* U.S. Department of Agriculture, dated March, 1980.

CHAPTER 3. BASIC PRODUCTION PROCESS

Alfa-Laval, *Fermenter Cooling System for Alcohol Fermentations,* NOTB40536E 3 undated.

Department of the Treasury, Bureau of Alcohol, Tobacco and Firearms, *Ethyl Alcohol for Fuel Use*, ATFP 5000, 1 (9-78), available from U.S. Government Printing Office, Stock Number 048-012-00045-1.

DOE, *Report of the Alcohol Fuels Policy Review*, Department of Energy Report No. DOE/PE-0012, Superintendent of Documents, Stock Number 061-000-00313-4, June, 1979.

DOE, *Preliminary Design Report Small-Scale Fuel Alcohol Plant*, DOE Office of Alcohol Fuels, October, 1980.

Heinz, D. J., *Technology of Ethanol Production*, Hawaiian Sugar Planters' Association, Media Briefing on Energy, September 11, 1979.

Katzen, Raphael Associates, *Grain Motor Fuel Alcohol Technical and Economic Assessment Study*, DOE Report Number HCP/J6639-01, dated June, 1979.

SERI, *Fuel From Farms—A Guide to Small-Scale Ethanol Production*, SERI Report No. SP-451-519, dated May, 1980.

SERI, *A Guide to Commercial Scale Ethanol Production and Financing*, dated November, 1980.

Shelton, D., and A. R. Rider, "Ethanol Production Equipment and Processes," *Ethanol Production and Utilization for Fuel*, Cooperative Extension Service, Institute of Agriculture and Natural Resources, University of Nebraska, Lincoln, October, 1979.

Shreve, R. Norris (ed.), *Chemical Process Industries*, McGraw-Hill Book Company, N.Y., 1967.

Underkoffer, L. A. and R. J. Hickey (eds.), *Industrial Fermentation*, Chemical Publishing Co. Inc., N.Y., 1954, Vol. I.

Universal Foods Corporation, *Fermentation Alcohol For Use In "Gasohol,"* Milwaukee, Wisconsin, 1979.

USDA, *Small-Scale Fuel Alcohol Production*, U.S. Department of Agriculture, dated March, 1980.

CHAPTER 4. FUNDAMENTALS OF FEEDSTOCKS

DOE, *Report of the Alcohol Fuels Policy Review*, Department of Energy Report No. DOE/PE-0012, Superintendent of Documents, Stock Number 061-000-00313-4, June, 1979.

DOE, *Preliminary Design Report Small-Scale Fuel Alcohol Plant*, DOE Office of Alcohol Fuels, October, 1980.

Federal Energy Administration and U.S. Department of Agriculture, *Energy and U.S. Agriculture: 1974 Data Base, Vol. 2: Commody Series of Energy Tables*, 1977.

Katzen, Raphael Associates, *Grain Motor Fuel Alcohol Technical and Economic Assessment Study*, DOE Report Number HCP/J6639-01, dated June, 1979.

Lipinski, E. S., D. R. Jackson, S. Kresovich, M. F. Arthur, and W. T. Lawkon, "Carbohydrate Crops as a Renewable Resource for Fuel Production," Vol. I, *Agricultural Research*, Battelle Columbus Division, May 15, 1979.

Morrison, F. B., *Feeds and Feeding*, Abridged, 9th ed., The Morrison Publishing Co., 1961.

Nathan, R. A., *Fuels from Sugar Cane: Systems Study for Sugarcane, Sweet Sorghum and Sugar Beets*, Technical Information Center, U.S. Department of Energy, 1978.

SERI, *Fuel From Farms—A Guide to Small-Scale Ethanol Production*, SERI Report No. SP-451-519, dated May, 1980.

SERI, *A Guide to Commercial Scale Ethanol Production and Financing*, dated November, 1980.

USDA, *Motor Fuels from Farm Products*, Miscellaneous Publication No. 327, Washington, DC, December, 1938.

USDA, *Agricultural Statistics*, U.S. Government Printing Office, Washington, DC, 1977 and 1978.

USDA, *Energy in Agriculture*, 1978.

USDA, *Small-Scale Fuel Alcohol Production*, U.S. Department of Agriculture, dated March, 1980.

CHAPTER 5. MARKETS FOR ETHANOL AND COPRODUCTS

Barber, R. S., R. Braude, and K. G. Mitchell. "Further Studies on the Water Requirements of the Growing Pig." *Anim. Prod.* **5,** 277-282 (1963).

Brady, S., A. C. Ragsdale, H. J. Thompson, and D. M. Worstell. "The Effect of Wind on Milk Production, Feed and Water Consumption and Body Weight in Dairy Cattle." *Res. Bul. 45,* Univ. of Mo. Ag. Expt. Sta., 1954.

Braude, R. and J. G. Rowell. "Comparison of Dry and Wet Feeding of Growing Pigs," *J. Agric. Sci. Camb.,* **68,** 325-330 (1967).

Gardner, R. H. and H. G. Sanders. "The Water Consumption of Suckling Sows." *J. Agric. Sci.* **27,** 638-643 (1937).

Garrigus, W. P., "Wet Stillage for Beef Cattle," *Distillers Feed Res. Corn. Proc.* 3 (1948).

Holme, D. W. and K. L. Robinson. "A Study of Water Allowances for the Bacon Pig." *Anim. Prod.* **7,** 377-384 (1965).

Ittner, N. R., C. F. Kelley, and N. R. Guilbert, "Water Consumption of Hereford and Braham Cattle and the Effect of Cooled Drinking Water in a Hot Climate." *J. Anim. Sci.* **10,** 742-751 (1962).

Keinholz, E. W. and D. L. Rossiter. *Grain Alcohol Fermentation By-Products for Feeding in Colorado.* Colo. State Univ., Ft. Collins, CO, 1979.

Leitch, I. and J. S. Thompson, "The Water Economy of Farm Animals." *Nutrition Abstracts and Reviews.* **16**(2), 197-223 (1944).

Lipinsky, E. S., D. A. Scantland, and T. A. McClure. *Systems Study of the Potential Integration of U.S. Corn Production and Cattle Feeding with Manufacture of Fuels via Fermentation.* Battelle, Columbus, Ohio, 1979.

Miller, T. B., "Evaluation of Whiskey Distillery By-products, I. Chemical Composition and Losses during Transport and Storage of Malt Distillers Grains." *J. Sci. FD. Agric.* **20,** 477-481 (1969).

Mount, L. E., C. W. Holmes, W. H. Close, S. R. Morrison, and I. B. Start, "A Note on the Consumption of Water by the Growing Pig at Several Environmental Temperatures and Levels of Feeding." *Anim. Prod.* **13,** 561-563 (1971).

Owen, J. B., E. L. Miller, and P. S. Bridge. "A Study of the Voluntary Intake of Food and Water and the Lactation Performance of Cows Given Diets of Varying Roughage Content *ad libitum,*" *J. Agr. Sci.* **70,** 223-235 (1968).

SERI, *Fuel From Farms—A Guide to Small-Scale Ethanol Production,* SERI Report No. SP-451-519, dated May, 1980.

SERI, *A Guide to Commercial Scale Ethanol Production and Financing,* dated November, 1980.

Steckley, J. D., D. G. Grieve, G. K. MacLeod, and E. T. Moran, "Brewers Yeast Slurry, I. Composition as Affected by Length of Storage, Temperature and Chemical Treatment." *J. Dairy Sci.* **62,** 941-946 (1979).

USDA, *Small-Scale Fuel Alcohol Production,* U.S. Department of Agriculture, dated March, 1980.

Wilford, E. J., "Distillery Slop for Hogs," *Ky. Ag. Res. Bul.* 408 (1944).

Winchester, C. F. and M. J. Morris, "Water Intake Rates of Cattle." *J. Anim. Sci.* **15,** 722-740 (1956).

CHAPTER 6. ECONOMICS OF ETHANOL PRODUCTION

David, M. L., *et al., Gasohol: Economic Feasibility Study,* U.S. Department of Energy, SAN 1681-T1, July, 1978.

Katzen, Raphael, *Grain Motor Fuel Alcohol Technical and Economic Assessment Study,* DOE Report No. HCP/J6639-01, dated July, 1979.

USDA, *Small-Scale Fuel Alcohol Production,* U.S. Department of Agriculture, dated March, 1980.

U.S. National Alcohol Fuels Commission, *Farm and Cooperative Alcohol Plant Study—Technical and Economic Assessment as a Commercial Venture,* By Raphael Katzen and Associates, dated October, 1980.

CHAPTER 7. OVERVIEW OF PROCESS TECHNOLOGY

Katzen, Raphael, *Grain Motor Fuel Alcohol Technical and Economic Assessment Study,* DOE Report No. HCP/J6639-01, dated June, 1979.

U.S. National Alcohol Fuels Commission, *Alcohol Fuels From Biomass: Production Technology Overview,* prepared by The Aerospace Corporation, July, 1980.

CHAPTER 8. ETHANOL PLANT DESIGNS

DOE, *Report of the Alcohol Fuels Policy Review,* Department of Energy Report No. DOE/PE-0012, Superintendent of Documents, Stock Number 061-000-00313-4, June, 1979.

DOE, *Preliminary Design Report Small-Scale Fuel Alcohol Plant,* DOE Office of Alcohol Fuels, October, 1980.

Katzen, Raphael Associates, *Grain Motor Fuel Alcohol Technical and Economic Assessment Study,* DOE Report Number HCP/J6639-01, dated June, 1979.

SERI, *Fuel From Farms—A Guide to Small-Scale Ethanol Production,* SERI Report No. SP-451-519, dated May, 1980.

SERI, *A Guide to Commercial Scale Ethanol Production and Financing,* dated November, 1980.

USDA, *Small-Scale Fuel Alcohol Production,* U.S. Department of Agriculture, dated March, 1980.

CHAPTER 9. ENVIRONMENTAL AND SAFETY IMPACTS

Aerospace Corporation, "Environmental Control Perspective for Ethanol Production from Biomass." Draft. Germantown. MD, June 1980. ATR-80 (7848-01)-1.

Brown, D., R. McKay, and W. Weir, "Some Problems Associated with the Treatment of Effluents from Malt Whiskey Distilleries." *Progress in Water Technology* **8** (2/3), 291-300 (1976).

Council on Environmental Quality, "Environmental Quality: The Tenth Annual Report of the Council of Environmental Quality." Washington, DC, December, 1979.

DOE, *Preliminary Design Report Small-Scale Fuel Alcohol Plant,* DOE Office of Alcohol Fuels, October, 1980.

Hagey, *et al.*, "Methanol and Ethanol Fuels—Environmental, Health and Safety Issues." U.S. Department of Energy and Mueller Associates, Inc., Presented at the International Symposium on Alcohol Fuel Technology, Wolfsburg, FRG. November 21-23, 1977.

Jackson, E. A. "Distillery Effluent Treatment in the Brazilian National Alcohol Programme." *The Chemical Engineer* 239-242 (April, 1977).

Kant, F. H., *et al. U.S. Environmental Protection Agency, Feasibility Studies of Alternative Fuels for Automotive Transportation.* Report No. EPA-460/374-009, 1974. Available from NTIS, PB235 581/GGI, $4.50 paper copy, $3.00 microfiche.

Raphael Katzen Associates, *Grain Motor Fuel Alcohol Technical and Economic Assessment Study*. U.S. Department of Energy, 1979.

Lee, L. K., *A Perspective on Cropland Availability,* Economics, Statistics, and Cooperatives Services, U.S. Department of Agriculture, Report No. 406.

Lowry, S. P. and R. S. Devoto, "Exhaust Emissions from a Single-Cylinder Engine Fueled with Gasoline, Methanol, and Ethanol." *Combustion Science and Technology* **12**(4, 5, and 6), 177-182 (1976).

Moriarity, A. J. "Toxicological Aspects of Alcohol Fuel Utilization." *Proceedings of the International Symposium on Alcohol Fuel Technology: Methanol and Ethanol.* Wolfsburg, FRG, November, 1977.

Radian Corporation, *Source Test and Evaluation Report: Alcohol Synthesis Facility for Gasohol Production.* McLean, VA, January, 1980.

Scarberry, R. M. and M. P. Papai, Radian Corp., *Source Test and Evaluation Report Alcohol Synthesis Facility for Gasohol Production, ibid.*

SERI, *Fuel From Farms—A Guide to Small-Scale Ethanol Production,* SERI Report No. SP-451-519, dated May. 1980.

SERI, *A Guide to Commercial Scale Ethanol Production and Financing,* dated November, 1980.

Sittig, M. *Hazardous and Toxic Effects of Industrial Chemicals.* Noyes Data Corp., pp. 196-197, 1979.

S. G. Unger, *Environmental Implications of Trends in Agriculture and Silviculture, Vol. 1: Trend Identification and Evaluation.* EPA-600/3-77-121, Development Planning and Research Associates, Manhattan, KS, October, 1977.

U.S. Congress, Office of Technology Assessment, *Gasohol—A Technical Memorandum.* GPO #052-003-00706-1, September 1979.

USDA, *Small-Scale Fuel Alcohol Production,* U.S. Department of Agriculture, dated March, 1980.

U.S. Environmental Protection Agency, *Report to Congress, Industrial Cost Recovery.* December, 1978.

CHAPTER 10. UTILIZATION OF GASOHOL

Allsup, J. R. and D. B. Eccleston, *Ethanol-Gasoline Blends as Automotive Fuel,* Bartlesville Energy Technology Center, 1979.

Bailey, B. K. and J. A. Russel, "Emergency Fuels Composition and Impact, Phase II Formulation and Screening Diesel Emergency Fuels," MED Report 101, Southwest Research Institute, June, 1979.

Bandel, W., "Problems in the Application of Ethanol As a Fuel for Utility Vehicles," International Symposium on Alcohol Fuel Technology—Methanol and Ethanol, November 21-23, 1977, Wolfsburg, Germany.

Barger, E. L., "Power Alcohol in Tractors and Farm Engines." *Agricultural Engineering,* February, 1941.

Barnes, K. D., D. B. Kittelson, and T. E. Murphy, "Effect of Alcohol as Supplemental Fuel for Turbocharged Diesel Engines," SAE paper 750469, 1975.

Bernhardt, W., "Possibilities for Cost Effective Use of Alcohol Fuels in Otto Engine-Powered Vehicles," International Symposium on Alcohol Fuel Technology—Methanol and Ethanol, November 21-23, 1977, Wolfsburg, Germany.

Bro, K. and P. S. Pedersen, "Alternative Diesel Engine Fuels: An Experimental Investigation of Methanol, Ethanol, Methane, and Ammonia in a D.I. Diesel Engine with Pilot Injection," SAE paper 770794, 1977.

Brinkman, N. D., N. E. Gallopoulos, and M. W. Jackson, "Exhaust Emissions, Fuel Economy, and Driveability of Vehicles Fueled with Alcohol-Gasoline Blends," SAE paper 750120, February, 1975.

Courtney, R. L., and H. K. Newhall, "A Primer on Current Automotive Fuels," *Automotive Engineering,* December, 1979.

Cummings, D. R. and W. M. Scott, "Dual Fueling the Truck Diesel with Methanol," International Symposium on Alcohol Fuel Technology—Methanol and Ethanol, Nov. 21-23, 1977, Wolfsburg, Germany.

Duck, J. T., and C. S. Bruce, "Utilization of Non-Petroleum Fuels In Automotive Engines," *Natl Bur. Std. J. Res.* **35**, 439 (1945).

"Engineering Highlights of the 1980 Automobiles," *Automotive Engineering,* November, 1979.

Ethanol Production and Utilization for Fuel, Cooperative Extension Service University of Nebraska, Lincoln, 1979.

Farm Management Summary and Analysis, 1978, Kansas State University.

Freeman, J. H., *et al., Alcohols—A Technical Assessment of Their Application as Fuels,* American Petroleum Institute, July, 1976.

"Gasohol and Diesohol Tested in Farm Tractors," *Doanes Agricultural Report* **42**(42-5) (October, 1979).

Kaufman, K. R. and H. J. Klosterman, "A Highway Test of Gasohol." *Farm Research* **37**(1) (July-August, 1979).

Keller, J. L., *Methanol and Ethanol Fuels for Modern Cars,* Union Oil Company, presentation for World Federation of Engineering Organizations, November, 1979.

Keller, J. L., *et al., Use of Alcohol in Motor Gasoline—A Review,* American Petroleum Institute, August, 1971.

Kirik, M., *Alcohol as Motor Fuel,* Ontario Ministry of Agriculture and Food, 1979.

Leviticus, L. E., *Nebraska Tractor Test Data, 1979,* University of Nebraska, Tractor Testing Lab.

Lichty, L. C., *Combustion Engines Processes,* McGraw-Hill, New York, 1967.

Lichty, L. C., and C. W. Phelps, "Gasoline-Alcohol Blends in Internal Combustion Engines." *Industrial and Engineering Chemistry,* **30**(1) (1938).

Lucke, C.E., "THe Use of Alcohol and Gasoline in Farm Engines," Farmers Bulletin No. 277, USDA, February 17, 1907.

Meyer, A. J. and R. E. Davis, "Burning Alcohol in a Gasoline Tractor and Thermodynamic Properties of an Ethyl Alcohol-Air Mixture and Its Products," *Univ. Kentucky Engg. Exp. Sta. Bull.* **2,** 4 (1948).

Meiners, E., *M&W Water/Alcohol Injection in Diesel Engines,* M&W Gear, August, 1979.

Mingle, J. G., "Converting Your Car to Run on Alcohol Fuels," Bulletin 56, Oregon State University Experiment Station, October, 1979.

Mueller Assoc., *Status of Alcohol Fuels Utilization Technology for Highway Transportation,* U.S. Dept. of Energy, June, 1978.

Obert, E. F., *Internal Combustion Engines,* International Textbook Company, 1968.

Owens, E. C., "Effects of Alcohol-Containing Fuels on Spark Ignition Engine Wear," U.S. Army Fuels and Lubricants Research Laboratory, Southwest Research Institute, April, 1979.

Panchapakesan, N. R., K. V. Gopalakrishnan, and B. S. Murthy, "Factors that Improve the Performance of an Ethanol-Diesel Oil Dual-Fuel Engine," International Symposium on Alcohol Fuel Technology—Methanol and Ethanol, November 21-23, 1977, Wolfsburg, Germany.

Pefley, R. K., *Alcohol Fueled Automobiles Performance—Emissions—Environment,* University of Santa Clara, October, 1979.

Private Communication, Cummins Rio Grande, January, 1980.

Scheller, W. A., *The Use of Ethanol-Gasoline Mixtures for Automotive Fuel,* Univ. of Nebraska, Lincoln, January, 1977.

Scheller, W. A., "Tests on Unleaded Gasoline Containing 10% Ethanol—Nebraska Gasohol," International Symposium on Alcohol Fuel Technology—Methanol and Ethanol, November 21-23, 1977, Wolfsburg, Germany.

Scheller, W. A., and B. J. Mohr, *Gasohol Road Test Program Second Progress Report,* Univ. of Nebraska, Lincoln, July, 1975.

Schrock, M. D., S. J. Clark, and J. A. Kramer, "Fuel Requirements for Field Operations in Kansas," Extension Bulletin (to be published, 1980).

Southeastern Minnesota Farm Management Association, 1978 Report, University of Minnesota.

"Spectrum," *Grounds Maintenance,* July 1979.

Strait, J., J. J. Boedicker, and K. C. Johansen, *Diesel Oil and Ethanol Mixtures for Diesel Powered Farm Tractors,* Univ. of Minn., 1978.

"Summary of Illinois Farm Business Records," 1978, University of Illinois.

Taylor and Taylor, *The Internal Combustion Engine,* 2nd ed., International Textbook Company, 1966.

USDA/FEA, *Energy and Agriculture, 1974 Data Base.*

USDA, *Agricultural Statistics,* 1978.

U.S. Department of Agriculture, *Energy in Agriculture,* 1978.

U.S. Department of Agriculture, Economics, Statistics, and Cooperative Service, "Costs of Producing Selected Crops in the United States—1977, 1978 and Projection for 1979 for the Committee on Agriculture, Nutrition and Forestry," United States Senate, June 15, 1979.

USDA, *Changes in Farm Production and Efficiency, 1978, Economics, Statistics and Cooperative Service,* Statistical Bulletin No. 628, January, 1980.

USDOE, Energy Information Administration, *Monthly Energy Review,* September, 1979.

The U.S. Food and Fiber Sector: Energy Use and Outlook, U.S. Department of Agriculture, Economic Research Service, Senate Committee on Agriculture and Forestry, 1974.

Wiebe, R. and J. D. Hummell, "Practical Experiences with Alcohol-Water Injection in Trucks and Farm Tractors," *Agricultural Engineering,* May, 1954.

Wrage, K. E. and C. E. Goering, "Technical Feasibility of Diesohol," *Agricultural Engineering,* October, 1979.

Yahya and Goering, "Some Trends in Fifty-five Years of Nebraska Tractor Tests," University of Missouri, ASAE paper no. MC-77-503.

CHAPTER 11. BUSINESS PLANNING

SERI, *Fuel From Farms—A Guide to Small-Scale Ethanol Production,* SERI Report No. SP-451-519, dated May, 1980.

SERI, *A Guide to Commercial Scale Ethanol Production and Financing,* dated November, 1980.

USDA, *Small-Scale Fuel Alcohol Production,* U.S. Department of Agriculture, dated March, 1980.

CHAPTER 12. PREPARATION OF FEASIBILITY ANALYSIS

SERI, *Fuel From Farms—A Guide to Small-Scale Ethanol Production,* SERI Report No. SP-451-519, dated May. 1980.

SERI, *A Guide to Commercial Scale Ethanol Production and Financing,* dated November, 1980.

USDA, *Small-Scale Fuel Alcohol Production,* U.S. Department of Agriculture, dated March, 1980.

CHAPTER 13. INTERNATIONAL DEVELOPMENT

Lampreia, Minister Luis Felipe. *The Brazilian Experience in Alcohol Fuels.* Chicago: May 1-2, 1980.

Proceedings of the Bio-Energy Council, 1980.

Rask, Norman. *Using Agricultural Resources to Produce Food or Fuel—Policy Intervention or Market Choice.* Paper presented at the First Inter-American Conference on Renewable Sources of Energy, New Orleans, LA, November, 1979.

SERI, *Gasohol Reference Guide,* October, 1980.

Ward, Roscoe F. *Power Alcohol,* United Nations, 1980.

World Bank Report No. 3021, International Biomass Fuel Report, June, 1980.

Appendix C

Bibliography

DATA BASE SOURCES

Alcohol Fuels (Citations from the NTIS Data Base), Volume 1, 1964-1977. Washington, DC: NTIS, July 1979. Report No. NTIS/PS-79/0712. Available from the National Technical Information Service, 5285 Port Royal Road, Springfield, VA 22161 (hereinafter referred to as NTIS).

Alcohol Fuels (Citations from the NTIS Data Base), Volume 2, 1978-June 1979. Washington, DC: NTIS, July 1979. Report No. NTIS/PS-79/0713. Available from NTIS.

Alcohol Fuels (Citations from the NTIS Data Base), Report for 1979-June 1979. Washington, DC: NTIS, July 1979. Report No. NTIS/PS-79/0714. Available from NTIS.

MAGAZINES/NEWSLETTERS

Alcohol Update. August 1980. Available from Alcohol Update, P. O. Box 35211, Minneapolis, MN 55435.

Alcohol Week. December 1980. Available from Alcohol Weekly, P. O. Box 7167, Benjamin Franklin Station, Washington, DC 20042.

Biomass Digest. Available from Technical Insights, P. O. Box 1304, Fort Lee, NJ, 07024.

Biotimes. January 1979. Available from International Biomass Institute, 1522 K St., N. W., Suite 600, Washington DC 20005.

Gasohol USA. June 1979. Available from Box 9547, Kansas City, MO 64133.

BACKGROUND MATERIAL

Baratz B., R. Ouellette, W. Park, and B. Stokes, *Survey of Alcohol Fuel Technology, Volume I.* Mitre Corporation, McLean, VA. November 1975. Report No. PB-256007. Available from NTIS.

Cheremisinoff, N. P., *Gasohol for Energy Production.* Ann Arbor Science, Ann Arbor, MI, 1979.

Freeman, J. H. *et al.*, *Alcohols—A Technical Assessment of Their Application as Fuels.* American Petroleum Institute, Washington, DC, July, 1976. Publication No. 4261. Available from American Petroleum Institute, 2101 L Street, N. W., Washington DC 20037.

Office of Technology Assessment, *Gasohol—A Technical Memorandum.* OTA, Washington, DC, U.S. Congress, 1979. Stock No. 052-003-00706-1. Available from Superintendent of Documents, U.S. Government Printing Office, Washington, DC 20402.

Park, W., E. Price, and D. Salo, *Bimass-Based Alcohol Fuels: The Near Term Potential for Use with Gasoline.* Mitre Corporation, McLean, VA, August, 1978. Report No. HCP/T4101-3.

Paul, J. K., *Ethyl Alcohol Production and Use as a Motor Fuel.* Chemical Technology Review. No. 144. Noyes Data Corporation, Park Ridge, NJ, 1979.

Pimental, D., *et al.*, "Energy and Land Constraint in Food Production," *Science* **190** (4126), 754-761 (November 21, 1975).

Proceedings of the Third International Symposium on Alcohol Fuels Technology, Asilomar, California, May 29-31, 1979. U.S. Department of Energy, Washington, DC, 1980. Report No. CONF-79052. Available from NTIS.

Solar Energy Research Institute, *A Guide to Commercial Scale Ethanol Production and Financing*. SERI, Golden, CO: 1980.

Solar Energy Research Institute, *Fuel from Farms—A Guide to Small-Scale Ethanol Production*. Golden, CO: SERI, 1980. Stock No. 061-000-00372-0. Available from Superintendent of Documents, U.S. Government Printing Office, Washington, DC 20402.

U.S. Department of Agriculture, *Small-Scale Fuel Alcohol Production*. USDA, Washington, DC, 1980. Stock No. 001-000-04124-0. Available from Superintendent of Documents, U.S. Government Printing Office, Washington DC 20402.

U.S. Department of Energy, *The Report of the Alcohol Fuels Policy Review*. U.S. Department of Energy, Washington, DC, June, 1979. Report No. DOE/PE-0012. Available from NTIS.

U.S. National Alcohol Fuels Commission, *Farm and Cooperative Alcohol Plant Study: Technical and Economic Assessment as a Commercial Venture*. Prepared by Raphael Katzen Associates International, Inc., 1980. Available from NTIS.

U.S. National Alcohol Fuels Commission, *Fuel Alcohol: Report and Analysis of Plant Conversion Potential to Fuel Alcohol Production*. 1980. Prepared by Davy McKee Corporation. Available from NTIS.

U.S. National Alcohol Fuels Commission, *Ethanol: Farm and Fuel Issues*. Prepared by Schnittker Associates, 1980. Report No. PB 80-215692. Available from NTIS.

U.S. National Alcohol Fuels Commission, *Energy Balances in the Production and End-Use of Alcohols Derived from Biomass*. Prepared by TRW, 1980.

CONGRESSIONAL HEARINGS

U.S. House of Representatives, Ninety-fifth Congress, Second Session, *Alcohol Fuels: Hearings Before the Subcommittee on Advanced Energy Technologies and Energy Conservation Research, Development and Demonstration of the Committee on Science and Technology*. U.S. House of Representatives, Washington, DC, 11-13 July 1978. Stock No. 35-520. Available from House Committee on Science and Technology, Room 3154, House Annex # 2, Washington DC 20515.

U.S. House of Representatives, Ninety-sixth Congress, First Session, *National Fuel Alcohol and Farm Commodity Production Act of 1979: Hearings Before the Subcommittees on Conservation and Credit Livestock and Grains, of the Committee on Agriculture*. U.S. House of Representatives, Washington, DC, 15-16 May 1979. Stock No. 052-070-05071-3. Available from Superintendent of Documents, U.S. Government Printing Office, Washington, DC 20402.

U.S. House of Representatives, Ninety-sixth Congress, First Session, *Oversight-Alcohol Fuel Options and Federal Policies. Hearings Before the Subcommittee on Energy Development and Applications of the Committee on Science and Technology*. U.S. House of Representatives, Washington, DC, 4 May, 12 June 1979. Stock No. 49-650. Available from Superintendent of Documents, U.S. Government Printing Office, Washington, DC 20402.

U.S. Senate, Ninety-fifth Congress, Second Session, *Alcohol Fuels: Hearing Before the Committee on Appropriations*. U.S. Senate, Washington, DC, 13 January 1978. Stock No. 052-070-04679-1. Available from Superintendent of Documents, U.S. Government Printing Office, Washington DC 20402.

U.S. Senate, Ninety-fifth Congress, Second Session, *The Gasohol Motor Fuel Act of 1978: Hearings Before the Subcommittee on Energy Research and Development of the Committee on Energy and Natural Resources*. U.S. Senate, Washington, DC, 7-8 August 1978. Publication No. 95-165. Available from Superintendent of Documents, U.S. Government Printing Office, Washington, DC 20402.

CONVERSION

Brushke, H., "Direct Processing of Sugarcane into Ethanol." *Proceedings of the International Symposium on Alcohol Fuel Technology: Methanol and Ethanol*. Wolfsburg, Federal Republic of Germany, November 21-23, 1977, Report No. CONF-771175, paper 5-5. Available from NTIS.

Nathan, R. A., *Fuels from Sugar Crops: Systems Study for Sugar Cane, Sweet Sorghum, and Sugar Beets.* Technical Information Center, Oak Ridge, TN, U. S. Department of Energy, 1978. Report No. TID-22781. Available from NTIS.

COPRODUCTS

Colorado State University, *Analysis of Alcohol Fermentation By-Products for Livestock and Poultry Feeding in Colorado.* Colorado State University, Fort Collins, CO, 1979. Available from Department of Animal Sciences, Colorado State University, Fort Collins, CO 80523.

Paturau, J. M., *By-products of the Cane Sugar Industry.* Elsevier Publishing Co., Amsterdam, 1969.

Reilly, P. S., *Conversion of Agricultural By-Products to Sugars, Progress Report.* Iowa State University, Ames, IA, 1978. Available from Department of Chemical Engineering, Engineering Research Institute, Iowa State University, Ames, Iowa 50010.

Wisner, R. N. and J. O. Gidel, *Economics Aspects of Using Grain Alcohol as a Motor Fuel with Emphasis on By-product Feed Markets.* Iowa Agriculture Experiment Station, June 1979. Report No. 9. Available from Agriculture Engineering Extension, Davidson Hall, Ames, IA 50010.

DESIGN

Brackett, A. T., *et al., Indiana Grain Fermentation Alcohol Plant.* Department of Commerce, Indianapolis, IN, 1976. Available from Indiana Department of Commerce, State House, Room 336, Indianapolis, IN 46204.

Chambers, R. S., *The Small Fuel-Alcohol Distillery: General Description and Economic Feasibility Workbook.* ACR Process Corporation, Urbana, IL, 1979. Available from ACR Process Corporation, 808 S. Lincoln Ave., Urbana, IL 61801.

Grain Motor Fuel Alcohol Technical and Economic Assessment Study. Raphael Katzen Associates, Cincinnati, OH, December, 1978. Report No. HCP/J6639-01. Available from NTIS.

DISTILLATION

King, C. J., *Separation Processes.* McGraw-Hill Book Co., New York, 1971.

Tassios, D. P., "Rapid Screening of Extractive Distillation Solvents." *Extractive and Azeotropic Distillation.* Advances in Chemistry Series, No. 115. American Chemical Society, Washington, DC, 1972.

McCabe, W. and J. C. Smith, *Unit Operations in Chemical Engineering,* 3rd ed., McGraw-Hill Book Co., New York, 1976.

ECONOMICS

David, M. L., *et al., Gasohol: Economic Feasibility Study—Final Report.* Development Planning and Research Associates, Inc., Manhattan, KS, July, 1978. Report No. SAN-1681-T1. Available from NTIS.

Gasohol from Grain—The Economic Issues. Economics, Statistics and Cooperative Service, Washington, DC, January 19, 1978. Report No. PB-280120/75T. Available from NTIS.

ENERGY BALANCE

Alich, J. A., *et al., An Evaluation of the Use of Agricultural Residues as an Energy Feedstock; A Ten Site Survey.* Stanford Research Institute-International, Palo Alto, CA, January, 1978. Report No. TID-27904/2. Available from NTIS.

Commoner, B., *Testimony before United States Senate Committee on Agriculture, Nutrition and Forestry, Subcommittee on Agricultural Research and General Legislation on "The Potential for Energy Production by U. S. Agriculture.*" Center for the Biology of Natural Systems, Washington University, St Louis, MO, July 23, 1979.

Ladisch, M. R. and K. Dyck, "Dehydration of Ethanol: New Approach Gives Positive Energy Balance." *Science* **205** (31), 898-900 (August 3, 1979).

Lewis, C. W., "Fuels from Biomass—Energy Outlays vs. Energy Returns: A Critical Appraisal." *Energy* **2** (3), 241-248 (September, 1977).

ENVIRONMENTAL CONSIDERATIONS

Aerospace Corporation, *Environmental Control Perspective for Ethanol Production from Biomass* (Draft). Germantown, MD, June 1980. Report No. ATR-80(7848-01)-1.

Brown, D., R. McKay, and W. Weir, "Some Problems Associated with the Treatment of Effluents from Malt Whiskey Distillers." *Prog. Water Tech.* **8** (2/3), 291-300 (1976).

Council on Environmental Quality, *Environmental Quality: The Tenth Annual Report of the Council of Environmental Quality.* Washington DC, December, 1979.

Grain Motor Fuel Alcohol Technical and Economic Assessment Study. Raphael Katzen Associates, Cincinnati, OH, December 1978. Report No. HCP/J6639-01. Available from NTIS.

Hagey, G., *et al.*, "Methanol and Ethanol Fuels: Environmental, Health and Safety Issues." *Proceedings of the International Symposium on Alcohol Fuel Technology: Methanol and Ethanol.* Wolfsburg, Federal Republic of Germany, November 21-23, 1977. Report No. CONF-771175, paper 8-2. Entire report available from NTIS.

Jackson, E. A., "Distillery Effluent Treatment in the Brazilian National Programme." *Chem. Eng.* **319,** 239-242 (April, 1977).

Kant, F. H., *et al., Feasibility Studies of Alternative Fuels for Automotive Transportation.* Environmental Protection Agency, Washington, DC, 1974. Report No. EPA-460/374-009. Available from NTIS.

Lee, Linda K., *A Perspective on Cropland Availability.* Economics, Statistics, and Cooperatives Services, U. S. Department of Agriculture, Washington, DC, 1978, Report No. 406. Available from Superintendent of Documents, U. S. Government Printing Office, Washington, DC 20402.

Lowrey, S. P. and R. S. Devoto, "Exhaust Emissions from a Single Cylinder Engine Fueled with Gasoline, Methanol, and Ethanol." *Combustion Science and Technology* **12** (4, 5, 6), 177-182 (1976).

Moriarty, A. J., "Toxicological Aspects of Alcohol Fuel Utilization." *Proceedings of the International Symposium on Alcohol Fuel Technology: Methanol and Ethanol.* Wolfsburg, Federal Republic of Germany, November 21-23, 1977. Report No. CONF-771175, paper 8-1. Entire report available from NTIS.

Scarberry, R. M. and M. P. Papai, *Source Test and Evaluation Report: Alcohol Synthesis Facility for Gasohol Production.* Radian Corporation, McLean, VA, 1980. Available from Radian Corporation.

Sittig, M., *Hazardous and Toxic Effects of Industrial Chemicals,* Noyes Data Corporation, Park Ridge, NJ, 1979.

Office of Technology Assessment, *Gasohol—A Technical Memorandum.* OTA, U. S. Congress, Washington, DC, 1979. Stock No. 052-003-00706-1. Available from Superintendent of Documents, U. S. Government Printing Office, Washington, DC 20402.

Unger, S. G., *Environmental Implications of Trends in Agriculture and Silviculture, Volume 1: Trend Identification and Evaluation.* Development Planning and Research Associates, Manhattan, KS, October, 1977. Report No. PB.-274-233. Available from NTIS.

U. S. Environmental Protection Agency, *Report to Congress, Industrial Cost Recovery.* Office of Water Program Operations, Washington, DC, December 1978. Report No. PB80-204746. Available from NTIS.

FEEDSTOCKS

Atchison, J. E., *Preliminary Investigation of New Process for Separation of Components of Sugar Cane, Sweet Sorghum, and Other Plant Stalks.* Battelle, Columbus, OH, 1977. Report

No. TID-28734. Available from NTIS.

Atlas of Nutritional Data on United States and Canadian Feeds. National Academy of Science, Washington, DC, 1971.

Chubey, B. B. and D. G. Dorrell, "Jerusalem Artichoke, A Potential Fructose Crop for the Prairies." *J. Canad. Institute Food Science Tech.* **7** (2), 98-100 (1974).

Jones, J. L., *Mission Analysis for the Federal Fuels from Biomass Program. Volume 1: Summary and Conclusions.* Stanford Research Institute International, Menlo Park, CA, December, 1978. Report No. SAN-0115-T2. Available from NTIS.

Hertzmark, D., *Agricultural Sector Impacts of Making Ethanol from Grain.* Solar Energy Research Institute, Golden, CO, March, 1980. Report No. SERI/TR-352-554. Available from NTIS.

Lee, L. K., *A Perspective on Cropland Availability.* U. S. Department of Agriculture, Washington, DC, 1978. Report No. 406. Available from Superintendent of Documents, U. S. Government Printing Office, Washington, DC 20402.

Lipinsky, E. S., *et al.*, *Systems Study of Fuels from Sugarcane, Sweet Sorghum, and Sugar Beets, Volume 1: Comprehensive Evaluation.* Battelle, Columbus, OH, March 1976. Report No. BMI-1957. Available from NTIS.

Lipinsky E. S., *et al.*, *Systems Study of Fuels from Sugarcane, Sweet Sorghum, and Sugar Beets. Volume 2: Agricultural Considerations.* Battelle, Columbus, OH, December 31, 1976. Report No. BMI-1957. Available from NTIS.

Lipinsky, E. S., *et al.*, *Systems Study of Fuels from Sugarcane, Sweet Sorghum, and Sugar Beets. Volume 3: Conversion to Fuels and Chemical Feedstocks.* Battelle, Columbus, OH, December 31, 1976. Report No. BMI-1957. Available from NTIS.

Lipinsky, E. S., *et al.*, *Systems Study of Fuels from Sugarcane, Sweet Sorghum, and Sugar Beets. Volume 4: Corn Agriculture.* Battelle, Columbus, OH, March 31, 1977. Report No. BMI-1957A. Available from NTIS.

Lipinsky, E. S., *et al.*, *Systems Study of Fuels from Sugarcane, Sweet Sorghum, Sugar Beets, and Corn. Volume 5: Comprehensive Evaluation of Corn.* Battelle, Columbus, OH, March 31, 1977. Report No. BMI-1957A. Available from NTIS.

Lipinsky, E. S., *et al.*, *Fuels from Sugar Crops: First Quarterly Report.* Battelle, Columbus, OH, July 19, 1977. Report No. TID-28414. Available from NTIS.

Lipinsky, E. S., *et al.*, *Fuels from Sugar Crops—Second Quarterly Report.* Battelle, Columbus, OH, October 31, 1977. Report No. TID-27834. Available from NTIS.

Lipinsky, E. S., *et al.*, *Fuels from Sugar Crops—Third Quarterly Report.* Battelle, Columbus, OH, 1978. Report No. TID-28191. Available from NTIS.

Nathan, R. A., *Fuels from Sugar Crops. Systems Study for Sugar Cane, Sweet Sorghum, and Sugar Beets.* Technical Information Center, U. S. Department of Energy, Oak Ridge, TN, 1978. Report No. TID-22781. Available from NTIS.

U. S. Department of Agriculture, *Agricultural Statistics, 1979.* USDA, Washington, DC, 1979. Stock No. 001-000-03775-7. Available from Superintendent of Documents, U. S. Government Printing Office, Washington, DC 20402.

FERMENTATION

Engelbart, W., "Basic Data on Continuous Alcoholic Fermentation of Sugar Solutions and of Mashes from Starch Containing Raw Materials." *Proceedings of the International Symposium on Alcohol Fuel Technology: Methanol and Ethanol.* Wolfsburg, Federal Republic of Germany, November 21-23, 1977. Report No. CONF-771175, paper 5-3. Entire report available from NTIS.

Lipinsky, E. S., *et al.*, *Systems Study of the Potential Integration of U. S. Corn Production and Cattle Feeding with Manufacture of Fuels via Fermentation.* Battelle, Columbus, OH, June 1979. Report No. BMI-2033. Available from NTIS.

Miller, D. L., "Ethanol Fermentation and Potential." C. R. Wilke (ed.) *Biotechnology and Bioengineering Symposium No. 5: Cellulose as a Chemical and Energy Resource Conference, Berkeley, CA.* John Wiley, New York, 1974.

INTERNATIONAL

Ribeiro, F. F. A., "The Ethanol-Based Chemical Industry in Brazil." *Workshop in Fermentation Alcohol for Use as Fuel and Chemical Feedstock in Developing Countries.* Vienna, Austria, 26-30 March 1979. Paper No. ID/WG. 29 3/4UNIDO. Available from UN Publications, Room A 3315, New York, NY 10017.

Sharma, K. D., "Present Status of Alcohol and Alcohol Based Chemicals Industry in India." *Workshop on Fermentation Alcohol for Use as a Fuel and Chemical Feedstock in Developing Countries.* Vienna, Austria, 26-30 March 1979. Paper No. IDWG. 293/14 UNIDO. Available from UN Publications, Room A 3315, New York, NY 10017.

REGULATORY

Abeles, T. P. and J. R. King, *Parameters for Legislature Consideration of Bioconversion Technologies.* Minnesota Legislature Science and Technology Project, St. Paul, MN, February 1978. Report No. PB 284742/45T. Available from NTIS.

Bureau of Alcohol, Tobacco, and Firearms. *Ethyl Alcohol for Fuel Use.* BATF, Washington, DC, July, 1978. Brochure. Available from BATF Distribution Center, 3800 S. Four Mile Run Drive, Arlington, VA 22206.

Bureau of Alcohol, Tobacco, and Firearms. *Alcohol Fuel and ATF.* BATF, Washington, DC, August, 1979. Brochure No. ATF 5000.2. Available from ATF Distribution Center, 3800 S. Four Mile Run Drive, Arlington, VA 22206.

Denaturants for Ethanol/Gasoline Blends. Mueller Associates, Baltimore, MD, April, 1978. Report No. HCP/M2098-01. Available from NTIS.

"Fuel Use of Distilled Spirits—Implementing a Portion of the Crude Oil Windfall Profit Tax Act of 1980 (P.L.96-223); Temporary and Proposed Rule." *Federal Register* **45** (No. 121), 41837-41858 (20 June 1980).

TRANSPORTATION USE

Adt, R. R., Jr., *et al.*, "Effects of Blending Ethanol with Gasoline on Automotive Engines and Steady State Performance and Regulated Emissions Characteristics." Troy, MI: 1978. Report No. CONF-7805102. Entire report available from NTIS.

Allsup, J. R. and D. B. Eccleston, *Ethanol/Gasoline Blends as Automotive Fuel.* International Alcohol Fuels Technology, Asilomar, CA, May 1979. Report No. BETC/R1-79/2. Available from NTIS.

Bernhardt, W. "Future Fuels and Mixture Preparation Methods for Spark Ignition Automobile Engines." *Prog. Energy Combustion Sci.* **3** (3), 139-150 (1977).

Bushnell, D. J. and J. M. Simonsen, "Alcohol Assisted Hydrocarbon Fuels: A Comparison of Exhaust Emissions and Fuel Consumption Using Steady State and Dynamic Engine Test Facilities." *Energy Commun.* **2** (2), 107-132 (1976).

Ecklund, E. E., *Comparative Automotive Engine Operation when Fueled with Ethanol and Methanol.* U. S. Department of Energy, Washington, DC, May, 1978. Report No. HCP/W1737-0. Available from NTIS.

Panchapakesan, N. R. and K. V. Gopalakrishnan, "Factors that Improve the Performance of an Ethanol-Diesel Oil Dual Fuel Engine." *Proceedings of the International Symposium on Alcohol Fuel Technology: Methanol and Ethanol.* Wolfsburg. Federal Republic of Germany, November 21-23, 1977. Report No. CONF-771175, paper 2-2. Entire report available from NTIS.

Rutan, Al. *Alcohol Car Conversion.* Rutan Publishing Co., Minneapolis, MN, 1980.

Scott, W. M. "Alternative Fuels for Automotive Diesel Engines." J. M. Colucci and N. E. Gallopoulos (eds.), *Future Automotive Fuels: Prospects, Performance, Perspective.* Plenum Press, New York, 1977.

Appendix D

Glossary and Acronyms

GLOSSARY

absolute alcohol. Completely dehydrated ethyl alcohol of the highest proof obtainable (200 proof); also "neat" alcohol. (See **anhydrous.**)

acid hydrolysis. Decomposition or alteration of a chemical substance by water in the presence of acid.

acidity. The measure of how many hydrogen ions a solution contains per unit volume; may be expressed in terms of pH.

aflatoxin. The substance produced by some strains of the fungus *Aspergillus flavus;* the most potent carcinogen yet discovered; a persistent contaminant of corn that renders crops unsalable.

agitator. Device such as a stirrer that provides complete mixing and uniform dispersion of all components in a mixture. Agitators are generally used continuously during the cooking process and intermittently during fermentation.

alcohol. The family name of a group of organic chemical compounds composed of carbon, hydrogen, and oxygen; a series of molecules that vary in chain length and are composed of a hydrocarbon plus a hydroxyl group, $CH_3\text{-}(CH_2)_n\text{-}OH$; includes methanol, ethanol, isopropyl alcohol, and others. (See **ethanol.**)

Defined in the Crude Oil Windfall Profit Tax Act of 1980 (26 USC 44E P.L. 96-223) to include "methanol and ethanol but does not include alcohol produced from petroleum, natural gas, or coal or alcohol with a proof less than 150."

Defined in the Energy Security Act (42 USC 8802, P. L. 96-294) as "alcohol (including methanol and ethanol) which is produced from biomass and which is suitable for use by itself or in combination with other substances as a fuel or as a substitute for petroleum or petrochemical feedstocks."

alcohol fuel plant. Under BATF regulations, a distilled spirits plant established solely for producing, processing, and using or distributing distilled spirits to be used extensively for fuel use.

alcohol fuel producer's permits. The document issued by BATF pursuant to the Crude Oil Windfall Profit Tax Act (26 USC 5181, P. L. 96-223) authorizing the person named to engage in business as an alcohol-fuel production facility.

aldehydes. Any of a class of highly reactive organic chemical compounds obtained by controlled oxidation of primary alcohols, characterized by the common group CHO, and used in the manufacture of resins, dyes, and organic acids.

alkali. Soluble mineral salt of a low-density, low-melting point, highly reactive metal; characteristically "basic" in nature.

alpha-amylase. An enzyme that acts specifically to accelerate the hydrolysis of starch by conversion to dextrin.

ambient. The prevalent surrounding conditions usually expressed as functions of temperature, pressure, and humidity.

amino acids. The naturally occurring, nitrogen-containing building blocks of protein.

amylaceous feedstocks. Materials, such as cereal grains and potatoes, that are composed of saccharides in the form of starches.

amylase. Any of the enzymes that accelerate the hydrolysis of starch and glycogen.

amylodextrins. See **dextrins.**

anaerobic digestion. A type of bacterial degradation of organic matter that occurs in the absence of air (oxygen) and produces primarily carbon dioxide and methane.

anhydrous. Devoid of water; refers to a compound that does not contain water either absorbed on its surface or as water of crystallization.

anhydrous ethanol. 100 percent alcohol, neat alcohol, 200-proof alcohol.

apparent proof. The proof indicated by a hydrometer after correction for temperature but without correction of the obscuration caused by the presence of solids.

atmospheric pressure. Pressure of the air (and atmosphere surrounding us) which changes from day to day; it is equal to 14.7 psia.

auger. Rotating, screw-type device that moves material through a cylinder. In alcohol production, it is used to transfer grains from storage to the grinding site and from the grinding site to the cooker.

azeotrope. The chemical term for two or more liquids that, at a certain concentration, boil as though they are a single substance; alcohol and water cannot be separated further than 194.4 proof because at this concentration, alcohol and water form an azeotrope and vaporize together. For ethanol-water the azeotrope of 95.6 percent ethanol and 4.4 percent water boils at 172.2°F at 1 atmosphere.

azeotropic distillation. Distillation in which a substance is added to the mixture to be separated in order to form an azeotropic mixture with one or more of the components of the original mixture; the azeotrope formed will have a boiling point different from the boiling point of the original mixture and will allow separation to occur.

backset (*Also called* **set back**). The liquid portion of the stillage recycled as part of the process liquid in mash preparation.

bacterial spoilage. Occurs when bacterial contaminants take over the fermentation process in competition with the yeast.

bagasse. The cellulosic residue left after sugar is extracted from sugarcane.

balling hydrometer or Brix hydrometer. A triple-scale wine hydrometer designed to record the specific gravity of a solution containing sugar.

bankable debt. Debt which is sufficiently collateralized to allow financing through normal commercial bank channels.

barrel. A liquid measure equal to 42 American gallons or about 306 pounds of crude oil; one barrel equals 5.6 cubic feet or 0.159 cubic meters.

basic hydrolysis. Decomposition or alternation of a chemical substance by water in the presence of alkali.

batch distillation. Process in which the liquid feed is placed in a single container and the entire volume is heated, in contrast to continuous distillation in which the liquid is fed continuously to the still.

batch fermentation. Fermentation of a specific quantity of material conducted from start to finish in a single vessel.

BATF. Bureau of Alcohol, Tobacco, and Firearms, under the U. S. Department of Treasury; responsible for the issuance of permits, for both experimental and commercial facilities, for the production of alcohol.

beer. The product of fermentation of mash by microorganisms; the raw fermented mash, which contains about 6-12 percent alcohol; usually refers to the alcohol solution remaining after yeast fermentation of sugars.

beer still. The stripping section of a distillation column for concentrating ethanol, or the first column of a two- (or more) column system, in which the first separation from the mash takes place.

beer well. The surge tank used for storing beer prior to distillation.

beta-amylase. Enzyme which converts dextrins into glucose.

biomass. Organic matter, such as trees, crops, manure, and aquatic plants, that is available on a renewable basis.

Defined in the Energy Security Act (42 USC 8802, P. L. 96-294) as "any organic matter which is available on a renewable basis, including agricultural crops and agricultural wastes and residues, wood and wood wastes and residues, animal wastes, municipal wastes, and aquatic plants."

biomass energy project. Defined in the Energy Security Act (42 USC 8802, P. L. 96-294) as "any facility (or portion of a facility) located in the United States which is primarily for (a) the production of biomass fuel (and byproducts); or (b) the combustion of biomass for the purpose of generating industrial process heat, mechanical power, or electricity (including cogeneration).

biomass fuel. Defined in the Energy Security Act (42 USC 8802, P. L. 96-294) as "any gaseous, liquid, or solid fuel produced by conversion of biomass."

BOD (biochemical oxygen demand). A measure of organic water pollution potential.

boiler. Unit used to heat water to produce steam for cooking and distillation processes.

boiling point. The temperature at which the transition from the liquid to the gaseous phase occurs in a pure substance at fixed pressure.

bond. A type of insurance which gives the government security against possible loss of distilled spirits tax revenue; not required for alcohol-fuel plants producing less than 10,000 proof gallons per year.

British thermal unit (Btu). The amount of heat required to raise the temperature of one pound of water one degree fahrenheit under stated conditions of pressure and temperature (equal to 252 calories, 778 foot-pounds, 1055 joules, and 0.293 watt-hours); it is a standard unit for measuring quantity of heat energy.

bulk density. The mass (weight) of a material divided by the actual volume it displaces as a whole substance, expressed in pounds per cubic feet, kilograms per cubic meter, etc.

calorie. The amount of heat required to raise the temperature of one gram of water one degree centigrade.

carbohydrate. A chemical term describing certain neutral compounds made up of carbon, hydrogen, and oxygen; includes all starches and sugars; a general formula is $C_x(H_2O)_y$.

carbon dioxide. A gas produced as a coproduct of fermentation; chemical formula is CO_2.

cassava. A starchy root crop used for tapioca; can be grown on marginal croplands along the southern coast of the United States.

catalysis. The effect produced by a small quantity of a substance (catalyst) on a chemical reaction, after which the substance (catalyst) appears unchanged.

cell recycle. The process of separating yeast from fully fermented beer and returning it to ferment to a new mash; can be done with clear worts in either batch or continuous operations.

cellulase. An enzyme capable of decomposing cellulose into simpler carbohydrates.

cellulose. The main polysaccharide in living plants, forms the skeletal structure of the plant cell wall; can be hydrolyzed to glucose.

cellulosic feedstocks. Materials, such as wood, crop stalks, and newsprint, containing sugar units linked by bonds that are not easily ruptured.

celsius (centigrade). A temperature scale commonly used in the sciences; at sea level, water freezes at 0°C and boils at 100°C. [$°C = \frac{5}{9}(°F - 32)$.]

centrifuge. A rotating device for separating liquids of different specific gravities or for separating suspended colloidal particles according to particle-size fractions by centrifugal force.

cetane number (cetane rating). Measure of a fuel's ease of self-ignition; the higher the number, the better the fuel for a diesel engine.

COD (chemical oxygen demand). A measure of water pollution.

collateral value. The resale value of alcohol equipment and/or plants; specifically, the ability of the equipment to derive from sale the amount necessary to pay off the debt borrowed against the unit.

column. Vertical, cylindrical vessel containing a series of perforated plates or packed with materials through which vapors may pass, used to increase the degree of separation of liquid mixtures by distillation or extraction.

completely denatured alcohol (CDA). Ethyl alcohol which is at least 160 proof blended, pursuant to formulas prescribed by BATF, with sufficient quantities of various denaturants to make it unfit for and not readily recoverable for beverage use; this may then be distributed through retail outlets without permits. (Compare to **specially denatured alcohol.**)

compound. A chemical term denoting a specific combination of two or more distinct elements.

concentration. The quantity of ethyl alcohol (or sugar) present in a known quantity of water. Weight percent is the weight of alcohol (or sugar) per weight of water. Volume percent is the volume (or sugar) per volume of water.

condenser. A heat-transfer device that reduces a fluid substance from its vapor phase to its liquid phase by reducing its temperature as it contacts cooling surfaces in its path.

continuous fermentation. A steady-state fermentation system that operates without interruption; each stage of fermentation occurs in a separate section of the fermenter, and flow rates are set to correspond with required residence times.

cooker A tank or vessel designed to cook a liquid or extract or digest solids in suspension; the cooker usually contains a source of heat, and is fitted with an agitator; its purpose is to aid in breaking down starches into fermentable sugars.

cooking. The process that breaks down the starch granules in the grain making the starch available for the liquefaction and saccharification steps.

coproducts. The resulting substances and materials that accompany the production of ethanol by fermentation processes.

corporate bonds. Bonds issued and sold to the public which are backed by the corporation which issues them; instruments which provide debt financing to the private sector from institutions such as insurance companies, mutual funds, pension funds, etc.

DCF-IROR. Discounted cash flow—interest rate of return.

DDG. See **distiller's dried grains.**

DDGS. See **distiller's dried grains with solubles.**

DDS. See **dried grains with solubles.**

dehydration. The process of removing water from any substance by exposure to high temperature or by chemical means.

denaturant. A substance added to ethanol to make it unfit for human consumption so that it is not subject to alcohol beverage taxes.

desiccant. A substance having an affinity for water; used for drying purposes.

dewatering. Removal of the free water from a solid substance.

dextrins. A polymer of D-glucose which is intermediate in complexity between starch and maltose and formed by partial hydrolysis of starches.

dextrose. The same as glucose.

disaccharides. The class of compound sugars which yield two monosaccharide units upon hydrolysis; examples are sucrose, maltose, and lactose.

dispersion. The distribution of finely divided particles in a medium.

distillate. That portion of a liquid which is removed as a vapor and condensed during a distillation process.

distillation. The process of separating the components of a mixture by differences in boiling point; a vapor is formed from the liquid by heating the liquid in a vessel, and the vapor is successively collected and condensed into liquids.

distiller's dark grains. See **distiller's dried grains with solubles (DDGS).**

distiller's dried grains (DDG). The water-insoluble, dried distiller's grains' coproduct of the grain fermentation process which may be used as a high-protein (28 percent) animal feed. (See **distiller's grain.**)

distiller's dried grains with solubles (DDGS). A grain mixture obtained by mixing distiller's dried grains and distiller's dried solubles.

distiller' dried solubles (DDS). A mixture of water-soluble oils and hydrocarbons obtained by condensing the thin stillage fraction of the solids obtained from fermentation and distillation processes.

distiller's feeds. Coproducts resulting from the fermentation of cereal grains by the yeast *Saccharomyces cerevisiae;* the nonfermentable portion of grain mash.

distiller's grain. The nonfermentable portion of a grain mash comprised of protein, unconverted carbohydrates and sugars, and mineral material.

DOE. The U. S. Department of Energy.

drawback. A refund of part of the tax given when tax-paid alcohol is used to produce approved products unfit for beverage purposes.

dry milling. A process of separating various components of grains, such as germ, bran, and starch without using water.

DSP (distilled spirits plant). A plant, including fuel-alcohol plants, authorized by the Bureau of Alcohol, Tobacco, and Firearms to produce, store, or process ethyl alcohol in any of its forms.

Economic Regulatory Administration (ERA). A regulatory agency within the U. S. Department of Energy administering petroleum pricing and allocation programs, oil and gas fuel conversion programs, and other programs as assigned by the Secretary of Energy.

EDA. Economic Development Administration.

energy crops. Includes such agricultural crops as corn and sugarcane; also nonfood crops such as poplar trees. (See **biomass.**)

Energy Security Act (42 USC 8701, et seq., P. L. 96-294). June 30, 1980 legislation authorizing, inter alia, a U. S. biomass and alcohol-fuel program; established independent Office of Alcohol Fuels within DOE; and authorized program including loan guarantees, price guarantees, and purchase agreements with producers of fuel alcohol.

enrichment. The increase of the more volatile component in the condensate of each successive stage above the feed plate.

ensilage. Immature green forage crops and grains which are preserved by alcohol formed by an anaerobic fermentation process.

Entitlement Program. A DOE program administered by the Economic Regulatory Administration which pays producers of ethanol for gasohol an entitlement per gallon through September, 1981.

enzymes. The group of catalytic proteins that are produced by living microorganisms; enzymes mediate and promote the chemical processes of life without themselves being altered or destroyed.

EPA. Environmental Protection Agency.

equity capital. That portion of the total debt of a corporation in which stock is given in return for invested capital.

ethanol. Chemical formula C_2H_5OH; the alcohol product of fermentation that is used in alcoholic beverages and for industrial purposes; blended with gasoline to make gasohol; also known as ethyl alcohol or grain alcohol.

ethyl alcohol. Flammable organic compound (CH_3CH_2OH) formed during sugar fermentation. It is called ethanol, grain alcohol, or simply alcohol. (See **ethanol.**)

evaporation. Conversion of a liquid to the vapor state by the addition of latent heat of vaporization; usually refers to vaporization into the atmosphere.

excise tax, gasoline. A tax collected at the pump to support the construction and maintenance of highways. Gasohol is exempt through 1992 from the $0.04 federal excise tax and in some states from the state excise tax.

facultative (anaerobe). A microorganism that grows equally well under aerobic and anaerobic conditions.

fahrenheit scale. A temperature scale in which the boiling point of water is 212° and its freezing point 32°; to convert °F to °C, subtract 32, multiply by 5, and divide the product by 9 (at sea level). $°C = (°F - 32) \times \frac{5}{9}$.

FDA. The U. S. Food and Drug Administration.

feed plate. The theoretical position in a distillation column above which enrichment occurs and below which stripping occurs.

feedstock. The raw material, such as grain, fruit, or other agricultural products, used as the sugar source in the fermentation process.

fermentable sugar. Sugar (usually glucose) derived from starch or cellulose that can be converted to ethanol (also known as reducing sugar or monosaccharide).

fermentation. A biological sequence of enzymatic reactions that convert sugars to CO_2 and alcohol in the absence of oxygen. Generally refers to metabolism in the absence of oxygen.

fermentation ethanol. Ethyl alcohol produced from the enzymatic transformation of organic substances.

flash heating. Very rapid heating of material by exposure of small fractions to high temperature and using high flow rates.

flash point. The temperature at which a combustible liquid will ignite when a flame is introduced; anhydrous ethanol will flash at 51°F, 90-proof ethanol will flash at 78°F.

flocculation. The aggregation of fine suspended particles to form floating clusters or clumps.

FmHA. Farmers Home Administration.

fossil fuel. Any naturally occurring fuel of an organic nature which originated in a past geologic age (such as coal, crude oil, or natural gas).

fractional distillation. A process of separating alcohol and water (or other mixtures) by boiling and drawing off vapors from different levels of the distilling column.

fructose. A fermentable monosaccharide (simple) sugar of chemical formula $C_6H_{12}O_6$; fructose is a ketohexose.

fuel-grade alcohol. Usually refers to ethanol of 160-200 proof.

fusel oil A clear, colorless, poisonous liquid mixture of alcohols obtained as a coproduct of grain fermentation; major constituents are amyl, isoamyl, propyl, isopropyl, butyl, and isobutyl alcohols.

gasohol (gasahol). A registered trademark held by the State of Nebraska for a fuel mixture of agriculturally derived 10 percent anhydrous fermentation ethanol and 90 percent unleaded gasoline; it is often incorrectly used to mean any mixture of alcohol and gasoline to be used for motor fuel.

gasoline. A volatile, flammable liquid obtained from petroleum that has a boiling range of approximately 200-216°C and is used as fuel for spark-ignited internal combustion engines.

gelatinization. The rupture of starch granules by heat to form a gel of soluble starch and dextrins.

glucose. Simple sugar containing six carbon atoms ($C_6H_{12}O_6$). A sweet colorless sugar that is the most common sugar in nature and the primary component of starch and cellulose. The sugar most commonly fermented by yeast to produce ethyl alcohol.

glucoamylase. An enzyme that acts specifically to convert, by hydrolyses, dextrins to glucose.

glucosidase. An enzyme that hydrolyzes polymers of glucose monomers (glucoside); specific glucosidases must be used to hydrolyze specific glucosides; e.g., β-glucosidases are used to hydrolyze cellulose; α-glucosidases are used to hydrolyze starch.

grain alcohol. See **ethanol.**

guaranteed debt. 100 percent versus 90 percent. A distinction made to the structure of government-guaranteed debt; i.e., under the FmHA program the government guarantees 90 percent of the debt which the lender loans to the corporation; contrast to a 100 percent guarantee which eliminates risk from the bank altogether.

heat exchanger. Unit that transfers heat from one liquid (or vapor) to another without mixing the fluids. A condenser is one type of heat exchanger.

heat of condensation. The same as the heat of vaporization, except that the heat is given up as the vapor condenses to a liquid at its boiling point.

heat of vaporization. The heat input required to change a liquid at its boiling point to a vapor at the same temperature (e.g., water at 212°F to steam at 212°F).

heating value The amount of heat obtainable from a fuel and expressed, for example, in Btu's per pound.

hexose. Any of various simple sugars that have six carbon atoms per molecule

HHV (higher heating value). The heat released during combustion of fuel if all products are cooled to room temperature and water is condensed to liquid.

HUD. Housing and Urban Development.

hydrated. Chemically combined with water.

hydrocarbon. A chemical compound containing hydrogen and carbon.

hydrolysis. The decomposition or alteration of a substance by chemically adding a water molecule to the unit at the point of bonding.

hydrometer. A long-stemmed glass tube with a weighted bottom; it floats at different levels depending on the relative weight (specific gravity) of the liquid; the specific gravity or other information is read where the calibrated stem emerges from the liquid.

indolene. A standard mixture of chemicals used in comparative tests of automotive fuels.

industrial alcohol. Ethyl alcohol produced and sold for other than beverage purposes; depending on the use, may or may not be denatured.

industrial revenue bonds (IDR Bonds). Debt incurred through industrial revenue authorities in numerous states; since such authorities are permitted to issue bonds for sale to the public which are payable in tax-exempt interest rates, IDR or tax-exempt bonds normally reflect an interest rate substantially below the prime rate.

INEL. Idaho National Engineering Laboratory.

inoculum. A small amount of bacteria produced from a pure culture which is used to start a new culture.

inulin. A polymeric carbohydrate comprised of fructose monomers found in the roots of many plants, particularly Jerusalem artichokes.

inventory. Refers to the supplies of all factors of production which must be maintained in storage, i.e., grain, wood, enzymes, etc.

lactic acid. $C_3H_6O_3$, the acid formed from milk sugar (lactose) and produced as a result of fermentation of carbohydrates by bacteria called *Lactobacillus*.

lactose. A crystalline disaccharide made from whey and used in pharmaceuticals, infant foods, bakery products, and confections; also called "milk sugar," $C_{12}H_{22}O_{11}$.

leaded gasoline. Gasoline containing tetraethyl lead, added to raise the octane value.

lean fuel mixture. An excess of air in the air/fuel ratio; gasohol has a leaning effect over gasoline because the alcohol adds oxygen to the system.

leasing. A form of financing whereby debt and ownership are retained by a third party who is normally not involved in the management of the operation; allows for the loan of both debt and equity capital to the market.

LHV (lower heating value). The heat released during combustion of fuel if all products are cooled to room temperature while water remains as steam; the lower heating value is more representative of typical fuel burning than HHV.

lignified cellulose. Cellulose polymer wrapped in a polymeric sheath and extremely resistant to hydrolysis because of the strength of its linkages.

lignin. A polymeric, noncarbohydrate constituent of wood that functions as a binder and support for celluose fibers.

lime. White powder composed of calcium oxide that forms a highly alkaline solution when mixed with water. Lime is used to increase the pH of mash.

limited partnership. Legal mechanism in which investors can limit their liability by investing in a "managing partner" who operates the company; the managing partner is usually a corporation and limited partners are allocated tax credits and depreciation in a ratio which may exceed their pro rata share; basically used for tax-shelter-type investments, since the limited partners can be allocated depreciation, tax credits/losses which exceed their original investments.

linkage. The bond or chemical connection between constituents of a molecule.

liquefaction. The change in the phase of a substance to the liquid stage; in the case of fermentation, the conversion of water-insoluble carbohydrate to water-soluble carbohydrate.

LPG. Liquid petroleum gas.

malt. Barley softened by steeping in water, allowed to germinate, and used especially in brewing and distilling as a source of amylase.

maltose. A disaccharide of glucose.

mash. A mixture of grain and other ingredients with water to prepare wort that can be fermented to produce ethanol.

meal. A granular substance produced by grinding.

membrane. A sheet polymer which separates components of solutions by permitting a passage of certain substances but preventing passage of others.

methanol. A light, volatile, flammable, poisonous, liquid alcohol, CH_3OH, formed in the destructive distillation of wood or made synthetically and used especially as a fuel, a solvent, an antifreeze, or a denaturant for ethyl alcohol, and in the synthesis of other chemicals; methanol can be used as fuel for motor vehicles; also known as methyl alcohol or wood alcohol.

methyl alcohol. Also known as methanol or wood alcohol. (See **methanol.**)

molecular sieve. A compound which separates molecules by selective penetration into the sieve space on the basis of size, charge, or both.

molecule. The chemical term for the smallest particle of matter that is the same chemically as the whole mass.

monomer. A simple molecule which is capable of combining with a number of like or unlike molecules to form a polymer.

monosaccharides See **fermentable sugar.**

multiple-effect evaporator. A series of evaporators in which the vapors removed from each unit are used to supply heat to the next unit in the series.

municipal solid waste (MSW). Combined residential and commercial wastes generated within a municipal area and consisting of any materials, including food wastes, that are discarded or rejected as spent, worthless, or in excess and that are not wet enough to be free-flowing.

municipal waste. Defined in the Energy Security Act (42 USC 8802, P. L. 96-294) as "any organic matter, including sewage, sewage sludge, and industrial or commercial waste, and mixtures of such matter and inorganic refuse (i) from any publicly or privately operated municipal waste collection or similar disposal system, or (ii) from similar wastes flows (other than such flows which constitute agricultural wastes or residues, or wood wastes or residues from wood harvesting activities or production of forest products)."

municipal waste energy project. Defined in the Energy Security Act (42 USC 8802, P. L. 96-294) as "any facility (or portion of a facility) located in the United States primarily for (i) the production of biomass fuels (and byproducts) from municipal waste; or (ii) the combustion of municipal waste for the purpose of generating steam or forms of useful energy, including industrial process heat, mechanical power, or electricity (including cogeneration)."

NAFI. National Alcohol Fuels Institute.

net energy balance. The amount of energy available from fuel when it is burned, less the amount of energy it takes to produce the fuel.

octane number. A rating which indicates the tendency to knock when a fuel is used in a standard internal combustion engine under standard conditions.

offering statement A prospectus of an investment in the form of an offering of stock or partnership in the investment in return for cash invested into the company.

Office of Alcohol Fuels. An independent office within the Department of Energy responsible for administration of all alcohol-fuel programs in the department and coordination of related programs in other federal agencies.

OSHA. Occupational Safety and Health Administration.

osmotic pressure. Applied pressure required to prevent passage of a solvent across a membrane which separates solutions of different concentrations.

overhead. The relatively low-boiling-point liquids removed from the top of a distillation unit.

Over-the-Counter (OTC) Stock Market. Market for new or speculative stocks which are traded in the public sector but not to the extent which New York Stock Exchange or American Stock Exchange require for active trading.

packed column. Type of column or pipe, used in the distillation process, that is filled with large-surface-area materials, such as metal filings, ceramic saddles, or plastic or glass beads.

personal versus corporate debt. Under any loan agreement which involves high-risk capital speculation or unknown collateral values, personal endorsements are normally required by the lender, which allows the lender to have lien against other assests of the owners; corporate debt restricts the liability of the debt to the corporation and the assets in the corporation.

pH. A term used to describe the free hydrogen ion concentration of a system; a solution of pH less than 7 is acid; pH of 7 is neutral; pH over 7 is alkaline.

plate distillation column (seive tray column). A distillation column constructed with perforated plates or screens.

polymer. A substance made of molecules comprised of long chains or cross-linked simple molecules.

pounds per square inch absolute (psia). The measurement of pressure referred to a complete vacuum or zero pressure.

pounds per square inch guage (psig). The measurement of pressure expressed as a quantity measured from above atmospheric pressure.

pound of steam. One pound (mass) of water in the vapor phase, not to be confused with the steam pressure which is expressed in pounds per square inch.

practical yield. The amount of product that can actually be derived under normal operating conditions; i.e., the amount of sugar that normally can be obtained from a given amount of starch or the amount of alcohol that normally can be obtained is usually less than theoretical yield.

preculture. Process in which the alcohol-producing yeasts are propagated prior to introduction into the fermentation tank. Preculturing ensures a high concentration of active yeast, thus reducing the time required for fermentation and controlling the production of unwanted bacteria.

private placements. A security offering which is limited to a small group of investors and hence is not in the public markets or under the full security of the SEC; such investments may be in Subchapter S corporations, standard corporations, limited partnerships, etc.

process guarantee. Refers to the financial ability of an engineering company to successfully make its process perform, within given tolerance levels, in the event that the process does not meet production levels originally agreed to in the contract.

proof. A measurement of the alcohol concentration in an alcohol-water mixture, equal to twice the percentage by volume of the alcohol; e.g., 80 percent alcohol equals 160 proof, 100 percent alcohol equals 200 proof.

proof gallon. A U. S. gallon of liquid which is 50 percent ethyl alcohol by volume or the alcohol equivalent thereof; also one tax gallon.

protein. High-molecular-weight compounds composed of amino acids. Proteins are an essential ingredient in the diet of animals and humans.

pure ethyl alcohol. Ethyl alcohol that has not been denatured and is usually sold as 190 proof and 200 proof (absolute).

quad. One quadrillion (10^{15} or 1,000,000,000,000,000) Btu's (British thermal units).

rectification. With regard to distillation, the selective increase of the concentration of a component in a mixture by successive evaporation and condensation.

rectifying column. Section of the distillation column in which the alcohol concentration is increased by repeated interaction of the rising vapor with the liquid distillate. This section is above the beer injection point.

reflux. Liquid alcohol that is condensed in the distillation column or reintroduced into the column from the alcohol-product stream. Reflux results in more efficient separation and is mandatory, or only one stage of separation occurs.

relative density. See **specific gravity.**

renewable resources. Renewable energy; resources that can be replaced after use through natural means; examples: solar energy, wind energy, energy from growing plants.

road octane. A numerical value for automotive antiknock properties of a gasoline; determined by operating a car over a stretch of level road.

S-18. A relatively new program of the Securities and Exchange Commission which is designed to allow small stock offerings to go to the public sector through private placement, or through brokers up to $5 million on interstate placements; formerly, this was limited to $750,000 within a single state.

saccharide. A simple sugar or a compound that can be hydrolyzed to simple sugar units.

saccharification. Conversion process using acids, bases, or enzymes in which carbohydrates are broken down into fermentable sugars. Results in the conversion of short chains of glucose formed in the liquefaction step into single molecules of glucose.

saccharify. To hydrolyze a complex carbohydrate into simpler soluble fermentable sugars, such as glucose.

Saccharomyces. A class of single-cell yeasts which selectively consume simple sugars.

SBIC. Small Business Investment Company. Venture capital companies which are licensed by the Small Business Administration (SBA) for the purpose of investing in small businesses; requires a minimum of $500,000 in capital for which the SBA will lend an additional $1.5 million in investment capital to help small businesses; SBIC's can either lend to a small business or buy up to 49 percent of the stock in a small business, or a combination thereof.

scrubbing equipment. Equipment for countercurrent liquid-vapor contact of flue gases to remove chemical contaminants and particulates.

secondary market. Market of institutional investors who buy governm guaranteed debt for their portfolio.

SEIDB. Solar Energy Information Data Bank.

set back. The liquid portion of the stillage that is recycled as a portion of the process liquid in the mash preparation.

settling time. In a controlled system, the time required for entrained or colloidal material to separate from the liquid.

seive plate column. Type of distillation column that uses a series of perforated plates to promote the contact of liquid and vapor in the column.

sight gauge. A clear calibrated cylinder through which liquid level can be observed and measured.

simple sugars. See **fermentable sugars.**

small-scale biomass energy project. Defined in the Energy Security Act (42 USC 8802, P. L. 96-294) as a "biomass energy project with an anticipated annual production capacity of not more than 1,000,000 gallons of ethanol per year, or its energy equivalent of other forms of biomass energy."

special fuel. Defined in the Crude Oil Windfall Profit Tax Act of 1980 (26 USC 44E, P. L. 96-223) as "any liquid fuel (other than gasoline) which is suitable for use in an internal combustion engine."

specially denatured alcohol (SDA). Ethyl alcohol to which sufficient quantities of various denaturants have been added, pursuant to formulas prescribed by federal regulations, to render it unfit for beverage purposes without impairing its usefulness for other purposes; specially denatured alcohol may be distributed only to persons holding BATF permits.

specific gravity. The ratio of the mass of a solid or liquid to the mass of an equal volume of distilled water at 4°C.

specific performance bonds. Refers to the ability of a contractor, through a third-party insurer, to eliminate the risk of nonperformance on a construction job.

spent grains. The nonfermentable solids, remaining after fermentation of a grain mash.

standard corporation. A corporation in which the investment tax credits, depreciation, or corporate losses stay within the corporate shell and cannot be deducted from personal income taxes.

starch. A carbohydrate polymer comprised of glucose monomers linked together by a glycosidic bond and organized in repeating units; starch is found in most plants and is a principal energy storage product of photosynthesis; starch hydrolyzes to several forms of dextrin and glucose.

still. An appartus for distilling liquids, particularly alcohols; it consists of a vessel in which the liquid is vaporized by heat, and a cooling device in which the vapor is condensed.

tillage. The nonfermentable residue from the fermentation of a mash to produce ohol.

chiometric ratio. The ratio of chemical substances necessary for a reaction to completely.

stover. The dried stalks and leaves of a crop remaining after the grain has been harvested.

stripping column. The section of the distillation column in which the alcohol concentration in the starting beer solution is decreased. This section is below the beer injection point.

Subchapter S corporation. A corporation which is limited to 15 or fewer investors who can use the tax credits, depreciation, or losses which the corporation incurs in such a manner as to decrease personal income tax liability; the Sub-S election can be rescinded to convert the corporation at a later date to a standard 1244 corporation in which any loss investors incur can be deducted from personal income taxes

sucrose. A crystalline disaccharide carbohydrate found in many plants, mainly sugarcane, sugar beets, and maple trees; $C_{12}H_{22}O_{11}$.

sulfuric acid. A strong acid (H_2SO_4) used to lower the pH (increase the acidity) of a solution. Also known as battery acid, it can be obtained from most automotive stores.

surety bond. A type of insurance which satisfies the government's bonding requirements on distilled spirits production (see **bond**); obtainable from U. S. Treasury-authorized insurance companies, surety bonds usually carry an annual premium of 1 to 2 percent of face value.

surfactant. Surface-active agent, a substance that alters the properties, especially the surface tension, at the point of contact between phases; e.g., detergents and wetting agents are typical surfactants.

tapered auger. A rotating, screw-type device used to dewater the stillage or mash in preparing this material for animal feed.

tax-free alcohol. Pure ethyl alcohol withdrawn free of tax for government, for hospital use, for science, or for humanitarian reasons; it cannot be used in foods or beverages; all purchasers must obtain BATF permits, post bonds, and exert controls upon storage and use of tax-free alcohol.

tax-paid alcohol. Pure ethyl alcohol which has been released from federal bond by payment of the federal tax of $21.00 per gallon at 200 proof or $19.95 per gallon at 190 proof.

tetraethyl lead (TEL). An octane enhancer for gasoline now under environmental restriction.

thermal efficiency. Energy heating value; the ratio of energy output to energy input.

thermophilic. Capable of growing and surviving at high temperatures.

thin stillage. The water-soluble fraction of a fermented mash plus the mashing water.

tray. One of several types of horizontal pieces in a distillation column.

UDAG. Urban Development Action Grant.

USDA. U. S. Department of Agriculture.

vacuum distillation. The separation of two or more liquids under reduced vapor pressure; reduces the boiling points of the liquids being separated.

vaporization. The process of converting a compound from a liquid or solid state to the gaseous state. Alcohol is vaporized during the distillation.

vaporize. To change from a liquid or a solid to a vapor, as in heating water to steam.

vapor pressure. The pressure at any given temperature of a vapor in equilibrium with its liquid or solid form.

wet milling. A process similar to dry milling except that the various components of grain are separated in water.

whey. The watery part of milk separated from the curd in the process of making cheese; it is produced commercially in large quantities and can be used as fertilizer, animal feed, or feedstock in the production of ethanol.

whole stillage The undried "bottoms" from the beer well comprised of non-fermentable solids, distiller's solubles, and the mashing water.

wine gallon. A United States gallon of liquid measure equivalent to the volume of 231 cubic inches.

wood alcohol. See **methanol.**

working capital. Capital which is used for initial training, inventory, and labor.

wort. The liquid remaining from a brewing mash preparation following the filtration of fermentable beer.

yeast. Single-cell microorganisms (fungi) that produce alcohol and CO_2 under anaerobic conditions and acetic acid and CO_2 under aerobic conditions; the microorganism that is capable of changing sugar to alcohol by fermentation.

zymosis. See **fermentation.**

ACRONYMS

AAFCO	Association of American Feed Control Officials
AOF	Office of Alcohol Fuels, DOE
AQCR	Air Quality Control Regions
BATF	Bureau of Alcohol, Tobacco, and Firearms
BOD	Biochemical Oxygen Demand
CFR	Code of Federal Regulations
DCF-IROR	Discounted Cash Flow-Interest Rate of Return
DDG	Distiller's Dried Grains
DDGS	Distiller's Dried Grains with Solubles
DDS	Distiller's Dried Solubles
DOE	Department of Energy
DPRA	Development Planning Research Associates, Manhattan, Kansas
EDA	Economic Development Administration
EDR	Economic Development Representative
EPA	Environmental Protection Agency
FDA	Food and Drug Administration
FmHA	Farmers Home Administration
GRAS	Generally Recognized as Safe
HUD	Housing and Urban Development
INEL	Idaho National Engineering Laboratory
LPG	Liquid Petroleum Gas
NAAQS	National Ambient Air Quality Standards
NAFI	National Alcohol Fuels Institute
NEPA	National Environmental Policy Act
NPDES	National Pollutant Discharge Elimination System
NSPS	New Source Performance Standards
OSHA	Occupational Safety and Health Administration
POTW	Publicly Owned Treatment Works
PSD	Prevention of Significant Deterioration
PSR	Pressure Stripper-Rectifier
RCRA	Resource Conservation and Recovery Act
SBA	Small Business Administration
SBIC	Small Business Investment Company
SEIDB	Solar Energy Information Data Bank
SERI	Solar Energy Research Institute
UCRS	Uniform Contracts Reporting System
UDAG	Urban Development Action Grant
USDA	U.S. Department of Agriculture

Appendix E

Technical Reference Data

Conversion Table
Properties of Gasoline, Methanol, and Ethanol
Water, Protein, and Carbohydrate Content of Selected Farm Products
Comparison of Raw Materials for Ethanol Production
Corn Area, Yield, and Production, 1977-1978
Grain Sorghum Area, Yield, and Production, 1977-1978
Grain Sorghum Production Costs, 1978
Grain Sorghum Silage and Forage Use, 1976-1978
Barley Production Costs, 1978
Red Winter Wheat and All Wheat Production Costs, 1978
Sugar Beet Area, Yield, and Production
Components of Molasses
Production of Selected Fruit Crops
Sugar Beet Production and Value, 1977
Utilization of DDG, DDS, and DDGS
Components of DDG, DDS, and DDGS

Table E-1. Conversion Table

To convert from	To	Multiply By
barrels (oil)	gallons	42
°C	°F	1.8 x (°C + 32)
centimeters	inches	0.394
cubic feet	bushels	0.804
cubic feet	cubic meters	0.028
cubic feet	cubic yards	0.037
cubic meters	bushels	28.377
cubic meters	cubic feet	35.314
cubic meters	cubic yards	1.308
cubic meters	gallons	264.17
cubic meters	liters	1000
cubic meters/kilogram	cubic feet/pound	16.02
cubic meters/second	million gallons/day	22.83
grams/liter	parts per million	1000
grams/liter	pounds/cubic foot	0.062
grams/liter	pounds/1000 gallons	8.35
hectares	acres	2.471
kilograms	pounds	2.205
kilograms/cubic meter	pounds/cubic foot	0.062
kilojoules	Btu	0.948
kilojoules/cubic meter	Btu/cubic foot	33.5
kilojoules/kilogram	Btu/pound	0.447
kilojoules/liter	Btu/gallon	3.589
kilopascals	atmospheres	9.869×10^3
kilopascals	pounds/square inch	0.145
kilowatthours	Btu	3.413
kilowatthours	joules	3.6×10^6
liters	gallons	0.264
meters	feet	3.281
metric tons	pounds	2205
metric tons	tons	1.102
micrometers	inches	0.000039
millimeters	inches	0.039
pounds/square inch	atmospheres	0.068
proof	percent alcohol by volume	0.5

Table E-2. Properties of Gasoline, Methanol, and Ethanol

CHEMICAL PROPERTIES	GASOLINE	METHANOL	ETHANOL
Formula	C_4–C_{12}	CH_3OH	C_2H_5OH
Molecular Weight	Varies	32.04	46.10
% Carbon (by Weight)	85–88	38.70	52.10
% Hydrogen (by Weight)	12–15	9.70	13.10
% Oxygen (by Weight)	Indefinite	51.60	34.70
C/H Ratio	5.6–7.4	3.00	4.00
Stoichiometric Air-to-Fuel Ratio	14.2–15.1	6.45	9.00
PHYSICAL PROPERTIES	**GASOLINE**	**METHANOL**	**ETHANOL**
Specific Gravity	0.70–0.78	0.79	0.794
Liquid Density			
lb/ft^3	43.6 Approx.	49.30	49.30
lb/gal	5.8–6.5	6.59	6.59
Vapor Pressure			
psi at 100°F (Reid)	7–15		2.50
psi at 77°F	0.3 Approx.		0.85
Boiling Point (°F)	80–440	149	173
Freezing Point (°F)	-70 Approx.	-208	-173
Solubility in Water (ppm)	240	Infinite	Infinite
Solubility of Water in Compound (ppm)	88	Infinite	Infinite
Viscosity at 68°F (Centipoise)	0.288		1.17
THERMAL PROPERTIES	**GASOLINE**	**METHANOL**	**ETHANOL**
Lower Heating Value			
Btu/lb	18,900 (Avg.)	9,066	11,500
Btu/gal	115,400 (Avg.)		73,560
Higher Heating Value			
Btu/lb at 68°F	20,260	10,258	12,800
Btu/gal	124,800		84,400
Heat of Vaporization			
Btu/lb	150	506	396
Btu/gal	900	3,340	3,378
Octane Ratings			
Research	91–105	106	106–108
Pump (RON + MON)/2	86–90	92	98–100
Flammability Limits (% by Volume in Air)	1.4–7.6		3.3–19.0
Specific Heat (Btu/lb – °F)	0.48		0.60
Autoignition Temperature (°F)	430–500		685
Flash Point	-50		55
Coefficient of Thermal Expansion at 60°F and 1 atm	0.0006		0.00112

Source: *Fuel From Farms – A Guide to Small-Scale Ethanol Production,* SERI, Denver, 1980.

Table E-3. Water, Protein, and Carbohydrate Content of Selected Farm Products

Crop	%Water	%Protein	% Carbo-hydrate
Apples, raw	84.4	0.2	14.5
Apricots, raw	85.3	1.0	12.8
Artichokes, French	85.5	2.9	10.6
Artichokes, Jerusalem	79.8	2.3	16.7
Asparagus, raw	91.7	2.5	5.0
Beans, lima, dry	10.3	20.4	64.0
Beans, white	10.9	22.3	61.3
Beans, red	10.4	22.5	61.9
Beans, pinto	8.3	22.9	63.7
Beets, red	87.3	1.6	9.9
Beet greens	90.9	2.2	4.6
Blackberries	84.5	1.2	12.9
Blueberries	83.2	0.7	15.3
Boysenberries	86.8	1.2	11.4
Broccoli	89.1	3.6	5.9
Brussels sprouts	85.2	4.9	8.3
Buckwheat	11.0	11.7	72.9
Cabbage	92.4	1.3	5.4
Carrots	8.2	1.1	9.7
Cauliflower	91.0	2.7	5.2
Celery	94.1	0.9	3.9
Cherries, sour	83.7	1.2	14.3
Cherries, sweet	80.4	1.3	17.4
Collards	85.3	4.8	7.5
Corn, field	13.8	8.9	72.2
Corn, sweet	72.7	3.5	22.1
Cowpeas	10.5	22.8	61.7
Cowpeas, undried	66.8	9.0	21.8
Crabapples	81.1	0.4	17.8
Cranberries	87.9	0.4	10.8
Cucumbers	95.1	0.9	3.4
Dandelion greens	85.6	2.7	9.2
Dates	22.5	2.2	72.9
Dock, sheep sorrel	90.9	2.1	5.6
Figs	77.5	1.2	20.3
Garlic cloves	61.3	6.2	30.8
Grapefruit pulp	88.4	0.5	10.6
Grapes, American	81.6	1.3	15.7
Lamb's-quarters	84.3	4.2	7.3
Lemons, whole	87.4	1.2	10.7
Lentils	11.1	24.7	60.1
Milk, cow	87.4	3.5	4.9
Milk, goat	87.5	3.2	4.6
Millet	11.8	9.9	72.9
Muskmelons	91.2	0.7	7.5
Mustard greens	89.5	3.0	5.6
Okra	88.9	2.4	7.6
Onions, dry	89.1	1.5	8.7
Oranges	86.0	1.0	12.2
Parsnips	79.1	1.7	17.5
Peaches	89.1	0.6	9.7
Peanuts	5.6	26.0	18.6
Pears	83.2	0.7	15.3
Peas, edible pod	83.3	3.4	12.0
Peas, split	9.3	1.0	62.7
Peppers, hot chili	74.3	3.7	18.1
Peppers, sweet	93.4	1.2	4.8
Persimmons	78.6	0.7	19.7
Plums, Damson	81.1	0.5	17.8
Poke shoots	91.6	2.6	3.1
Popcorn	9.8	11.9	72.1
Potatoes, raw	79.8	2.1	17.1
Pumpkin	91.6	1.0	6.5
Quinces	83.8	0.4	15.3
Radishes	94.5	1.0	3.6
Raspberries	84.2	1.2	13.6
Rhubarb	94.8	0.6	3.7
Rice, brown	12.0	7.5	77.4
Rice, white	12.0	6.7	80.4
Rutabagas	87.0	1.1	11.0
Rye	11.0	12.1	73.4
Salsify	77.6	2.9	18.0
Soybeans, dry	10.0	34.1	33.5
Spinach	90.7	3.2	4.3
Squash, summer	94.0	1.1	4.2
Squash, winter	85.1	1.4	12.4
Strawberries	89.9	0.7	8.4
Sweet potatoes	70.6	1.7	26.3
Tomatoes	93.5	1.1	4.7
Turnips	91.5	1.0	6.6
Turnip greens	90.3	3.0	5.0
Watermelon	92.6	0.5	6.4
Wheat, HRS	13.0	14.0	69.1
Wheat, HRW	12.5	12.3	71.7
Wheat, SRW	14.0	10.2	72.1
Wheat, white	11.5	9.4	75.4
Wheat, durum	13.0	12.7	70.1
Whey	93.1	0.9	5.1
Yams	73.5	2.1	23.2

Source: *Fuel From Farms — A Guide to Small-Scale Ethanol Production,* SERI, Denver, 1980.

Table E-4. Comparison of Raw Materials for Ethanol Production

Raw material	Gallons of ethanol	Pounds of protein yield	Percent protein dry
Corn	2.6/bu	18/bu	29-30
Wheat	2.6/bu	20.7/bu	36
Grain sorghum	2.6/bu	16.8/bu	29-30
Average starch grains	2.5/bu	17.5/bu	27.5
Potatoes (75 percent moist) 12-14	1.4/cwt	14.8/cwt	10
Sugar beets	20.3/ton	264/ton	20
Sugarcane	17/ton	–	–
Sugar beet molasses (50 percent sugar)	0.35/gal	–	–
Sugarcane molasses (55 percent sugar)	0.4/gal	–	–

Source: *Fuel from Farms – A Guide to Small-Scale Ethanol Production,* SERI, Denver, 1980.

Table E-5. Corn Area, Yield, and Production, 1977–1978

STATE	AREA PLANTED FOR ALL PURPOSES		CORN FOR GRAIN					
			AREA HARVESTED		AVERAGE YIELD PER HARVESTED ACRE		PRODUCTION	
	1977	1978	1977	1978	1977	1978	1977	1978[1]
	THOUSAND ACRES	THOUSAND ACRES	THOUSAND ACRES	THOUSAND ACRES	BUSHELS	BUSHELS	THOUSAND BUSHELS	THOUSAND BUSHELS
AL	840	640	375	544	29.0	50.0	10,875	27,200
AZ	65	70	50	50	100.0	115.0	5,000	5,750
AR	55	40	43	30	53.0	58.0	2,279	1,740
CA	430	420	247	281	116.0	126.0	28,652	35,406
CO	960	1,000	695	720	116.0	110.0	80,620	79,200
CT	54	53	–	–	–	–	–	–
DE	203	187	185	175	56.0	96.0	10,360	16,800
FL	623	430	299	370	35.0	52.0	10,465	19,240
GA	2,240	1,700	1,000	1,500	24.0	50.0	24,000	75,000
ID	120	123	28	39	86.0	87.0	2,408	3,393
IL	11,350	11,000	11,080	10,730	105.0	111.0	1,163,400	1,191,030
IN	6,400	6,100	6,210	5,900	102.0	108.0	633,420	637,200
IA	13,800	13,300	12,700	12,500	86.0	117.0	1,092,200	1,462,500
KS	2,030	1,820	1,680	1,500	96.0	102.0	161,280	153,000
KY	1,650	1,570	1,470	1,410	90.0	85.0	132,300	119,850
LA	86	65	65	47	52.0	59.0	3,380	2,773
ME	51	50	–	–	–	–	–	–
MD	730	690	600	590	72.0	97.0	43,200	57,230
MA	42	43	–	–	–	–	–	–
MI	2,800	2,670	2,320	2,250	85.0	81.0	197,200	182,250
MN	6,900	7,000	6,000	6,190	100.0	104.0	600,000	643,760
MS	250	215	160	135	36.0	56.0	5,760	7,560
MO	2,900	2,400	2,650	2,200	76.0	87.0	201,400	191,400
MT	90	88	11	5	68.0	72.0	748	360
NE	7,150	7,100	6,550	6,550	99.0	113.0	648,450	740,150
NV	3	–	–	–	–	–	–	–
NH[2]	26	27	–	–	–	–	–	–
NJ	149	135	95	95	70.0	91.0	6,650	8,645
NM	135	90	114	72	90.0	105.0	10,260	7,560
NY	1,375	1,300	640	600	80.0	90.0	51,200	47,400
NC	2,000	1,760	1,740	1,600	51.0	76.0	88,740	121,600
ND	620	600	237	253	73.0	79.0	17,301	19,987
OH	3,900	3,850	3,620	3,610	105.0	105.0	380,100	379,050
OK	140	120	95	73	82.0	65.0	7,790	4,745
OR	45	45	12	13	95.0	95.0	1,140	1,235
PA	1,615	1,615	1,160	1,190	92.0	95.0	106,720	113,050
RI	4	4	–	–	–	–	–	–
SC	825	640	690	550	36.0	55.0	24,840	30,250
SD	3,000	3,250	2,150	2,560	59.0	67.0	126,850	171,520
TN	900	820	730	660	65.0	66.0	47,450	43,560
TX	1,800	1,600	1,650	1,440	98.0	100.0	161,700	144,000
UT	80	92	13	16	89.0	90.0	1,157	1,440
VT	110	112	–	–	–	–	–	–
VA	855	825	560	615	55.0	82.0	30,800	50,430
WA	128	130	64	65	119.0	121.0	7,616	7,865
WV	100	93	54	58	74.0	77.0	3,996	4,466
WI	3,850	3,750	2,800	2,750	104.0	98.0	291,200	269,500
WY	89	87	30	34	85.0	81.0	2,550	2,754
U.S.	83,568	79,719	70,872	69,970	90.7	101.2	6,425,457	7,081,849

[1] Preliminary.

[2] Estimates discontinued after 1977 crop.

Source: *USDA Agricultural Statistics 1979.*

Table E-6. Grain Sorghum Area, Yield, and Production, 1977–1978

STATE	AREA PLANTED FOR ALL PURPOSES		SORGHUM FOR GRAIN				PRODUCTION	
			AREA HARVESTED		AVERAGE YIELD PER HARVESTED ACRE			
	1977	1978	1977	1978	1977	1978	1977	1978*
	THOUSAND ACRES	THOUSAND ACRES	THOUSAND ACRES	THOUSAND ACRES	BUSHELS	BUSHELS	THOUSAND ACRES	THOUSAND ACRES
AL	75	65	27	34	27.0	37.0	729	1,258
AZ	100	80	90	73	80.0	78.0	7,200	5,694
AR	285	230	252	200	52.0	60.0	13,104	12,000
CA	150	210	132	185	73.0	71.0	9,636	13,135
CO	460	490	263	280	31.0	81.0	8,153	8,860
GA	75	85	24	43	28.0	29.0	672	1,247
IL	80	80	64	68	64.0	68.0	4,096	4,624
IN	23	25	15	15	78.0	65.0	1,170	975
IA	45	36	32	24	74.0	75.0	2,368	1,800
KS	4,850	4,700	4,050	4,020	60.0	52.0	243,000	209,040
KY	50	37	32	23	57.0	62.0	1,824	1,426
LA	35	30	20	17	33.0	34.0	660	576
MS	60	65	24	21	32.0	38.0	768	798
MO	1,050	930	930	850	73.0	80.0	67,890	68,000
NE	2,300	2,000	2,070	1,830	71.0	75.0	146,970	137,250
NM	297	336	245	267	48.0	46.0	11,760	12,282
NC	110	125	72	86	37.0	52.0	2,664	4,472
OK	765	700	565	485	38.0	36.0	21,470	17,460
SC	23	26	12	15	16.0	32.0	192	480
SD	29	29	343	340	49.0	50.0	16,807	17,000
TN	40	45	20	24	51.0	51.0	1,020	1,224
TX	5,600	5,700	4,800	4,650	48.0	49.0	230,400	227,850
VA	24	25	10	11	43.0	47.0	430	517
U.S.	16,526	16,049	14,092	13,561	56.3	55.1	792,983	747,788

*Preliminary.

Source: *USDA Agricultural Statistics 1979.*

Table E-7. Grain Sorghum Production Costs, 1978

COST ITEM	CENTRAL PLAINS	SOUTHERN PLAINS	SOUTHWEST	UNITED STATES
COSTS PER ACRE				
Variable Items				
Seed	$ 3.56	$ 3.46	$ 7.57	$ 3.60
Fertilizer	12.14	11.89	29.70	12.37
Lime	0.02	(1)	(1)	0.01
Chemicals[2]	3.83	1.85	2.00	2.95
Custom Operations[3]	4.73	4.82	4.96	4.77
All Labor	10.42	14.07	39.04	12.53
Fuel and Lubrication	7.37	12.51	26.80	9.94
Repairs	6.70	10.75	12.55	8.54
Drying	2.17			1.20
Purchased Irrigation Water			23.28	0.45
Interest	1.29	1.61	5.32	1.50
Total Cost, Variable Items	$52.23	$60.96	$151.22	$57.86
Machinery Ownership				
Replacement	15.82	25.06	25.96	19.95
Interest	6.96	10.13	11.41	8.40
Taxes and Insurance	2.12	3.00	3.39	2.52
Total Cost, Machinery Ownership	$24.90	$38.19	$40.76	$30.87
General Farm Overhead	6.63	6.52	11.56	6.67
Management[4]	8.38	10.57	20.35	9.54
TOTAL COSTS PER ACRE, EXCLUDING LAND	$92.14	$116.24	$223.89	$104.94
Land Allocation, Composite With:				
Current Value[5]	42.00	28.56	82.01	37.04
Average Acquisition Value[6]	27.58	17.74	56.42	23.94
COSTS PER BUSHEL				
Variable	$ 0.91	$ 1.34	$ 2.16	$ 1.10
Machinery Ownership	0.43	0.84	0.58	0.59
Farm Overhead	0.12	0.14	0.16	0.13
Management	0.15	0.23	0.29	0.18
Total Cost, Excluding Land	$ 1.61	$ 2.55	$ 3.19	$ 2.00
Land Allocation, Composite With:				
Current Value	0.73	0.63	1.17	0.71
Average Acquisition Value	0.48	0.39	0.80	0.46
TOTAL PER BUSHEL COSTS OF PRODUCTION TO A RENTER				
Cost to Share Renter[7]	2.36	3.48	4.54	2.83
Cost to Cash Renter[8]	2.05	3.03	3.71	2.63
Weighted Renter Cost[9]	2.33	3.42	4.20	2.81
YIELD PER ACRE (BUSHELS)	57.2	45.5	70.2	52.5
PERCENT OF U.S. PRODUCTION	58.9	36.0	2.5	97.4

[1]Not applicable.

[2]Includes herbicides, insecticides, and rodenticides not otherwise included under custom operations.

[3]Includes custom application of crop chemicals, the cost of chemicals in some cases, and custom harvesting and hauling.

[4]Based on 10 percent of above costs.

[5]Based on prevailing tenure arrangements in 1974, reflecting actual combinations of cash rent, net share rent, and owner-operator land allocations, land values, land tax rates, and cash rates updated to current year.

[6]Same as footnote 5, except average value of cropland during the last 35 years is used for owner-operated land instead of current land value.

[7]Share-renter portion of cost divided by share-renter portion of crop.

[8]Cash-renter costs including cash rent divided by total yield.

[9]Weighted average of share renter and cash renter based on prevailing tenure arrangements in 1974.

Source: *USDA Agricultural Statistics 1979.*

Table E-8. Grain Sorghum Silage and Forage Use, 1976–1978

STATE	SILAGE									FORAGE AREA HARVESTED		
	AREA HARVESTED			AVERAGE YIELD PER ACRE			PRODUCTION					
	1976	1977	1978*	1976	1977	1978*	1976	1977	1978*	1976	1977	1978*
	1,000 ACRES	1,000 ACRES	1,000 ACRES	TONS	TONS	TONS	1,000 ACRES	1,000 ACRES	1,000 ACRES	1,000 ACRES	1,000 ACRES	1,000 ACRES
AL	18	15	18	11.5	10.5	10.0	207	158	180	11	18	10
AZ	5	6	5	16.5	17.0	17.0	83	102	85	2	2	1
AK	13	17	12	9.5	10.5	10.5	124	179	126	10	11	13
CA	11	7	12	17.0	17.0	17.0	187	123	204	10	10	8
CO	21	21	19	11.0	7.0	11.0	231	147	209	165	146	131
GA	30	18	29	12.0	7.0	12.5	360	126	363	8	11	8
IL	10	11	7	11.5	12.5	11.5	115	138	81	7	5	5
IN	8	6	7	12.5	12.0	13.0	100	72	91	—	—	—
IA	13	11	9	11.0	13.0	14.0	143	143	126	2	1	2
KS	290	300	230	8.6	12.5	10.0	2,494	3,750	2,300	340	430	355
KY	13	11	8	12.5	12.5	12.0	163	138	96	5	5	4
LA	11	9	9	10.0	11.0	10.0	110	99	90	5	5	2
MS	20	27	35	11.5	12.5	13.0	230	338	455	9	4	8
MO	30	44	33	9.5	11.0	11.5	285	484	380	35	56	39
NE	80	90	70	7.5	11.5	13.0	600	1,035	910	60	70	50
NM	5	8	4	11.0	12.0	13.0	55	96	52	41	37	47
NC	25	22	24	11.5	8.0	13.5	288	176	324	7	12	13
OK	41	34	45	8.0	11.0	9.0	328	374	405	139	121	115
SC	12	11	11	9.0	6.5	9.5	108	72	105	2	3	2
SD	45	56	55	3.0	8.0	7.5	135	448	413	115	74	55
TN	12	8	11	11.0	10.0	10.5	132	80	116	8	10	8
TX	50	100	50	12.0	8.0	10.5	600	800	515	900	600	700
VA	9	10	11	10.0	9.0	11.0	90	90	121	2	2	2
U.S.	772	842	714	9.3	10.9	10.9	7,168	9,168	7,747	1,883	1,633	1,578

*Preliminary.

Source: *USDA Agricultural Statistics 1979.*

Table E-9. Barley Production Costs, 1978

COST ITEM	NORTHEAST	NORTHERN PLAINS	SOUTHERN PLAINS	SOUTHWEST	NORTHWEST	UNITED STATES
COSTS PER ACRE						
Variable Items						
Seed	$ 6.23	$ 4.05	$ 4.57	$ 7.32	$ 5.43	$ 4.75
Fertilizer	14.16	5.60	7.07	8.52	10.73	7.01
Lime	2.74	–	–	–	–	0.04
Chemicals[1]	1.78	1.26	2.75	1.50	2.35	1.54
Custom Operations[2]	3.24	2.60	2.52	4.38	3.26	2.95
All Labor	9.76	7.90	13.24	21.38	12.42	10.62
Fuel and Lubrication	4.95	4.83	12.68	5.87	4.12	5.19
Repairs	5.23	5.89	7.60	7.23	5.80	6.11
Purchased Irrigation Water	–	0.97	–	25.35	4.57	4.63
Miscellaneous	–	0.11	–	–	–	0.07
Interest	2.29	0.71	1.57	3.53	1.14	1.20
Total Cost, Variable Items	$50.38	$33.92	$52.00	$ 85.08	$49.82	$44.11
Machinery Ownership						
Replacement	14.38	14.20	21.61	16.63	15.38	15.03
Interest	6.02	6.37	9.04	7.05	7.07	6.68
Taxes and Insurance	1.81	1.93	2.79	2.12	2.13	2.02
Total Cost, Machinery Ownership	$22.21	$22.50	$33.44	$ 25.80	$24.58	$23.73
General Farm Overhead	8.35	5.55	5.60	11.15	7.05	6.56
Management[3]	8.09	6.20	9.10	12.20	8.15	7.44
TOTAL COST PER ACRE, EXCLUDING LAND	$89.03	$68.17	$100.14	$134.13	$89.60	$81.84
Land Allocation, Composite With:						
Current Value[4]	109.79	34.31	40.23	71.92	60.55	44.82
Average Acquisition Value[5]	40.56	18.46	25.12	46.61	30.52	24.66
COSTS PER BUSHEL						
Variable	$ 1.16	$ 0.78	$ 1.09	$ 2.02	$ 0.88	$ 0.97
Machinery Ownership	0.51	0.51	0.70	0.62	0.43	0.52
Farm Overhead	0.19	0.13	0.12	0.26	0.12	0.14
Management	0.19	0.14	0.19	0.29	0.14	0.16
Total, Excluding Land	$ 2.05	$ 1.56	$ 2.10	$ 3.19	$ 1.57	$ 1.79
Land Allocation, Composite With:						
Current Value	2.52	0.79	0.85	1.71	1.06	0.98
Average Acquisition Value	0.93	0.42	0.53	1.11	0.53	0.54
TOTAL PER BUSHEL COSTS OF PRODUCTION TO A RENTER						
Cost to Share Renter[6]	3.59	2.31	2.95	4.89	2.20	2.72
Cost to Cash Renter[7]	2.74	2.17	3.23	3.84	2.11	2.46
Weighted Renter Cost[8]	2.83	2.26	2.98	4.63	2.19	2.64
YIELD PER ACRE (BUSHELS)	43.5	43.7	47.6	42.1	57.1	45.8
PERCENT OF U.S. PRODUCTION	1.3	56.5	4.0	10.8	18.5	91.1

[1] Includes herbicides, insecticides, and rodenticides not otherwise included under custom operations.

[2] Includes custom application of crop chemicals, the cost of chemicals in some cases, and custom harvesting and hauling.

[3] Based on 10 percent of above costs.

[4] Based on prevailing tenure arrangements in 1974, reflecting actual combinations of cash rent, net share rent, and owner-operator land allocations, land values, land tax rates, and cash rents updated to current year.

[5] Same as footnote 4, except average value of cropland during the last 35 years is used for owner-operated land instead of current land value.

[6] Share-renter portion of cost divided by share-renter portion of crop.

[7] Cash-renter costs including cash rent divided by total yield.

[8] Weighted average of share renter and cash renter based on prevailing tenure arrangements in 1974.

Source: *USDA Agricultural Statistics 1979.*

Table E-10. Red Winter Wheat and All Wheat Production Costs, 1978

COST ITEM	CENTRAL PLAINS	SOUTHERN PLAINS	NORTHERN PLAINS	SOUTHWEST	UNITED STATES	ALL WHEAT TOTAL
COSTS PER ACRE						
Variable Items						
Seed	$ 2.26	$ 2.79	$ 2.75	$ 9.19	$ 2.64	$ 3.89
Fertilizer	4.27	5.80	6.18	18.38	5.28	7.81
Lime	NA	NA	NA	NA	NA	0.12
Chemicals[1]	0.61	1.21	1.05	4.74	0.94	1.16
Custom Operations[2]	2.36	3.03	2.66	7.49	2.72	2.92
All Labor	9.14	9.01	7.19	21.60	9.15	8.63
Fuel and Lubrication	5.37	5.80	4.25	6.24	5.40	5.19
Repairs	5.81	5.65	5.91	7.63	5.81	5.89
Purchased Irrigation Water	NA	NA	0.73	33.16	0.80	0.55
Interest	1.28	1.38	1.36	4.66	1.40	1.48
Total Cost, Variable Items	$31.10	$34.67	$32.08	$113.09	$34.14	$37.64
Machinery Ownership						
Replacement	13.52	13.64	14.51	17.69	13.76	14.15
Interest	5.90	5.59	6.71	7.63	5.93	6.17
Taxes and Insurance	1.81	1.68	2.05	2.31	1.80	1.87
Total Cost, Machinery Ownership	$21.23	$20.91	$23.27	$ 27.63	$21.49	$22.19
General Farm Overhead	6.63	5.17	6.16	13.44	6.36	6.55
Management[3]	5.90	6.08	6.15	15.42	6.20	6.64
TOTAL COST PER ACRE, EXCLUDING LAND	$64.86	$66.83	$67.66	$169.58	$68.19	$73.02
Land Allocation, Composite With:						
Current Value[4]	38.72	29.02	44.36	86.28	37.27	43.68
Average Acquisition Value[5]	23.96	15.96	21.71	62.39	21.96	25.07
COSTS PER BUSHEL						
Variable	$ 1.13	$ 1.66	$ 1.00	$ 1.84	$ 1.28	$ 1.27
Machinery Ownership	0.77	1.00	0.73	0.45	0.80	0.75
Farm Overhead	0.24	0.25	0.19	0.22	0.24	0.22
Management	0.21	0.29	0.19	0.25	0.23	0.22
Total, Excluding Land	$ 2.35	$ 3.20	$ 2.11	$ 2.76	$ 2.55	$ 2.46
Land Allocation, Composite With:						
Current Value	1.40	1.39	1.39	1.41	1.40	1.47
Average Acquisition Value	0.87	0.76	0.68	1.02	0.82	0.84
Value of Pasture	0.06	0.35	NA	NA	0.12	0.06
TOTAL PER BUSHEL COSTS OF PRODUCTION TO A RENTER						
Cost to Share Renter[6]	3.48	4.65	3.09	3.66	3.73	3.62
Cost to Cash Renter[7]	3.44	3.73	3.13	3.72	3.57	3.61
Weighted Renter Cost[8]	3.48	4.44	3.10	3.69	3.71	3.62
YIELD PER ACRE (BUSHELS)	27.6	20.9	32.0	61.3	26.7	29.7
PERCENT OF U.S. PRODUCTION	54.7	24.6	13.3	4.9	97.5	96.9

[1] Includes herbicides, insecticides, and rodenticides not otherwise included under custom operations.

[2] Includes custom application of crop chemicals, the cost of chemicals in some cases, and custom harvesting and hauling.

[3] Based on 10 percent of above costs.

[4] Based on prevailing tenure arrangements in 1974, reflecting actual combinations of cash rent, net share rent, and owner-operator land allocations, land values, land tax rates, and cash rents updated to current year.

[5] Same as footnote 4, except average value of cropland during the last 35 years is used for owner-operated land instead of current land value.

[6] Share-renter portion of cost divided by share-renter portion of crop.

[7] Cash renter costs including cash rent divided by total yield.

[8] Weighted average of share renter and cash renter based on prevailing tenure arrangements in 1974.

NA – Not applicable.

Source: *USDA Agricultural Statistics 1979.*

Table E-11. Sugar Beet Area, Yield, and Production

STATE	AREA PLANTED		AREA HARVESTED		AVERAGE YIELD PER HARVESTED ACRE		PRODUCTION	
	1977	1978[1]	1977	1978[1]	1977	1978[1]	1977	1978[1]
	THOUSAND ACRES	THOUSAND ACRES	THOUSAND ACRES	THOUSAND ACRES	TONS	TONS	THOUSAND ACRES	THOUSAND ACRES
AZ[2]	12.9	15.7	12.8	15.0	22.3	20.5	285	308
CA[2]	227.0	207.0	217.0	195.0	26.1	24.5	5,664	4,778
CO	77.0	89.0	72.0	84.0	19.5	18.3	1,404	1,538
ID	115.4	136.3	107.4	134.1	19.5	20.3	2,094	2,722
KS	26.0	18.0	24.0	26.0	16.7	17.0	401	442
MI	92.3	93.0	85.5	91.0	21.0	19.3	1,796	1,756
MN	264.0	265.0	260.0	263.0	18.2	18.9	4,732	4,971
MT	46.4	45.4	45.0	44.0	19.9	19.8	896	885
NB	75.0	79.0	67.7	67.7	20.0	18.0	1,354	1,368
NM	1.3	2.1	1.2	1.2	19.2	20.6	23	37
ND	157.8	156.2	155.2	155.2	17.8	19.7	2,769	3,056
OH	24.9	24.5	23.3	23.3	20.3	16.9	457	394
OR	8.9	9.2	8.9	8.9	25.1	24.0	206	214
TX	19.9	28.0	23.5	23.5	17.3	17.6	309	414
UT	10.4	14.9	14.7	14.7	17.7	17.0	173	250
WA	63.9	69.2	68.5	68.5	24.3	26.5	1,495	1,815
WY	49.5	49.5	48.8	48.8	19.6	18.9	949	922
U.S.	1,272.6	1,302.0	1,235.5	1,263.9	20.6	20.3	25,007	25,870

[1]Preliminary.
[2]Relates to year of harvest.

Source: *USDA Agricultural Statistics 1979.*

Table E-12. Components of Molasses

COMPONENTS	COMPOSITION %	
	RANGE	AVERAGE
Water	17 – 25	20
Sucrose	30 – 40	35
Dextrose (Glucose)	4 – 9	7
Levulose (Fructose)	5 – 12	9
Other Reducing Substrates	1 – 5	4
Ash	7 – 15	12
Nitrogenous Compounds	2 – 6	4.5
Non-Nitrogenous Acids	2 – 8	5
Waxes, Sterols, Phospholipids	0 – 1	0.4

Source: Paturau, J.M., *By-products of the Cane Sugar Industry.* Amsterdam, The Netherlands: Elsevier Publishing; 1969.

Table E-13. Production of Selected Fruit Crops

YEAR	APPLES (COMMERCIAL CROP)	PEACHES	PEARS	GRAPES
	THOUSAND TONS	THOUSAND TONS	THOUSAND TONS	THOUSAND TONS
1964	3,160	1,726	728	3,478
1965	3,070	1,737	500	4,351
1966	2,881	1,695	748	3,734
1967	2,718	1,344	464	3,069
1968	2,735	1,818	624	3,549
1969	3,410	1,842	727	3,899
1970	3,199	1,498	549	3,103
1971	3,187	1,441	749	3,994
1972	2,939	1,186	612	2,579
1973	3,133	1,295	730	4,198
1974	3,290	1,459	742	4,199
1975	3,765	1,419	749	4,366
1976	3,237	1,510	841	4,398
1977	3,336	1,492	787	4,298
1978	3,817	1,351	727	4,567

Source: *USDA Agricultural Statistics 1979.*

Table E-14. Sugar Beet Production and Value, 1977

STATE	PRODUCTION	SEASON AVERAGE PRICE PER TON RECEIVED BY FARMERS	VALUE OF PRODUCTION
	THOUSAND TONS	DOLLARS	THOUSAND DOLLARS
AZ	285	$24.40	$ 6,954
CA	5,664	26.40	149,530
CO	1,404	26.30	36,925
ID	2,094	25.50	53,397
KS	401	21.90	8,792
MI	1,796	21.10	36,100
MN	4,732	20.60	97,479
MT	896	29.10	26,074
NB	1,365	27.00	36,558
NM	23	25.00	575
ND	2,769	21.40	59,257
OH	457	20.20	9,231
OR	206	23.00	4,738
TX	309	23.40	7,231
UT	173	26.50	4,619
WA	1,495	26.50	39,618
WY	949	28.80	27,331
U.S.	25,048	$24.20	$604,409

*Relates to year of harvest; includes some acreage planted previous fall.

Source: *USDA Agricultural Statistics 1979.*

Table E-15. Utilization of DDG, DDS, and DDGS

RATION	DISTILLERS DRIED GRAINS	DISTILLERS DRIED SOLUBLES	DISTILLERS DRIED GRAINS AND SOLUBLES
Cattle			
DE, kcal/kg	3,408	3,608	3,570
ME, kcal/kg	2,794	2,959	2,927
TDN, %	83	80	82
NEmilk	2,150	2,210	2,210
NEm	2,050	2,030	2,035
NEp	1,347	1,335	1,335
Poultry			
ME, kcal/kg	1,631	2,750	2,620
Swine			
DE, kcal/kg	2,030	3,305	3,085
ME, kcal/kg	1,835	2,985	2,790
TDN, %	46	75	70

Source: Distillers Feed Research Council

Table E-16. Constituents of DDG, DDS, and DDGS

CONSTITUENT	DISTILLERS DRIED GRAINS	DISTILLERS DRIED SOLUBLES	DISTILLERS DRIED GRAINS AND SOLUBLES
Moisture	7.50	4.50	9.00
Protein	27.00	28.50	27.00
Fat	7.60	9.00	8.00
Fiber	12.80	4.00	8.50
Ash	2.00	7.00	4.50
Amino Acids			
Lysine	0.60	0.95	(Available 0.60)
Methionine	0.50	0.50	(Available 0.60)
Cystine	0.20	0.40	0.40
Histidine	0.60	0.63	0.60
Arginine	1.10	1.15	1.00
Aspartic Acid	1.68	1.90	1.70
Threonine	0.90	0.98	0.95
Serine	1.00	1.25	1.00
Glutamic Acid	4.00	6.00	4.20
Proline	2.60	2.90	2.80
Glycine	1.00	1.20	1.00
Alanine	2.00	1.75	1.90
Valine	1.30	1.39	1.30
Isoleucine	1.00	1.25	1.00
Leucine	3.00	2.60	2.70
Tyrosine	0.80	0.95	0.80
Phenylalanine	1.20	1.30	1.20
Tryptophan	0.20	0.30	0.20
Fatty Acids			
Linoleic			
Fat	47.20	49.10	48.50
Ingredient	3.60	4.40	3.90
Linolenic			
Fat	5.20	5.10	5.00
Ingredient	0.38	0.46	0.40

Source: Distillers Feed Research Council

Index

A

acid
  acetic, 186, 258-259
  amino, 219
  lactic, 258-259
ACR Process Corp., 251
adsorption/absorption, 271-272
Agricultural Products Industrial Utilization Committee, 11
Agrol, 9
air quality, 395
air quality control regions (AQCR), 389
Alabama, 154, 217
Alaska, 3
alcohol
  anhydrous, 19, 245-246
  biomass-derived, 6
  low-proof, 419-420
  pure, 4
  shipping of, 365
  U.S. goal for production, 18
  weight of, 443-444
alcohol-fuel
  production, 513-523
    U.S. goals for, 2, 3
  program, legislation for, 524-529
alcohol fuels
  history, 9-12
  technical reference data, 556-569
  U.S. strategy for use, 24
alcohol industry, business plan, 456-477
alcohol plants
  economic analysis, 484-485, 490-491
  feasibility analysis, 478-495
  management plan, 485, 491
  market analysis, 480-481
  production factors, 481-483
  regulatory requirements, 486, 492
  site selection, 483-484
  technical and engineering design, 481, 489-490
alcohol production
  funding, 46
  impacts on agriculture, 175-182
  socioeconomics, 20
  supply and demand, 175-178
Alcoline, 9
aldehydes, 246, 361
aluminum, 34
American Automobile Association (AAA), 31-32
American Feed Control Officials, 201, 219
American Maize, 188
ammonia, 36, 88
ammonium sulfate, 41, 365
Amoco Oil Company, 46, 187
amylopectin, 85
amylose, 85
analysis, sensitivity, 232-234
annual operating variables
  large community DDGS, 243
  large on-farm ethanol plant, 237
  small community DDGS, 241
  small community, wet stillage plant, 239
antiknock index, 1
Apple Agri-Sales Tri-Star farm-scale still, 276-279
Archer-Daniels-Midland (ADM) plant, 245-246
Arkansas/Gulf process, 254-256, 262-268
Ashland Oil, 188
aspiration, 239
Assistant Secretary for Rural Development (USDA), 20
Atlantic Richfield Oil Company (ARCO), 46, 188

Australia, 501-502
azeotrope, 90, 104

B

bagasse, 43, 83, 143, 152, 172, 497
barley, 4, 147, 171, 181, 274
Bartlesville Energy Research Center, 415
Beckman Construction Company farm-scale ethanol plant, 279-282
beef, prices of, 175
beer, 89, 102, 193
beer slurry, 256
benzene, 105, 245, 384-385
bibliography, 533-538
biochemical oxygen demand (BOD), 369-370, 381, 387
bioconversion, 258, 381-383
biomass, 2, 3
biomass alcohol
  growth, 19
  industry, 19
biomass conversion, 39-40
biomass energy, development of, 28
Biomass Energy Development Plans, 18
biomass feedstocks, 50
boilers
  biomass, 112, 116
  coal-fired, 356-357, 381-383
  emissions of, 110
Brazil, 4, 5, 37, 185, 417, 420, 496-498, 501, 502-508
Brazilian National Alcohol Program, 4
brewer's yeast slurry, 205
Btu, content in motor fuels, 28
Brown's Alcohol Motor Fuel Cookbook, 442
Bureau of Alcohol, Tobacco, and Firearms (BATF), 51, 57, 120, 132, 421, 458, 492, 521-522, 524
butanol, 184

C

Cameroon, 508
Canada, 37, 508
capital cost
  large community DDGS, 244
  large on-farm ethanol plant, 238
  small community DDGS, 242
  small community, wet stillage plant, 240
  small on-farm ethanol plant, 236
capital recovery factors, ethanol plants, 227
carbohydrates, 13
  conversion of, 13, 15, 99-101
carbon dioxide, 154-155, 220, 245, 302, 381, 384
carburetor modifications, 33, 441-455
Caribbean Islands, 496
Carter, President, 3, 11, 18, 163
cascading, 308
cassava, 497, 503-504, 510
cellobiose, 257
cellulose, 13, 83, 85, 104, 254-267
  production processes, 38
cellulose hydrolysis processes, 252, 254-267
cellulose pretreatment processes, 252-254
Central America, 496
cereal grains, 9, 13
cetane number (CN), 422
cheese whey plant, 245
Chemapec process, 249-250
chemicals, exposure to, 387-388
Chevrolet, 451
Chevron, 188
Chicago, 210-211, 219
China, 10, 38
Cities Service Company, 188
Clean Air Act, 31, 33, 389
closed-loop feedback system, 412, 415
coal, 39, 40
Colorado, 154
Columbia University, 269
commercial-scale facilities, 6, 42
  defined, 4

compression ratio, 418-419
condensation, 90-91
condensed distiller's solubles, 202
Coordinating Lubricants Research, 420-421
copper, 34
coproducts
  animal consumption and, 158-159
  animal feed, 201-220
  ethanol, 201-220
  human food, 205-206
corn, 4, 36-37, 42, 51, 111, 112, 118, 122, 123, 142, 147, 152-154, 156, 160, 163, 249, 251, 274-276, 283, 284, 351, 385, 497
  demand for, 163-164, 179
  nutrient composition, 212
  prices, 168-169, 171, 174-175, 177, 179, 201-211, 223
  processing of, 147-148, 151
  production of, 437
Cornbelt region, 160
corn starch, 8, 271
corn yield, 167
Corporate Average Fuel Economy (CAFE), mileage standards, 8
Costa Rica, 508
costs, environmental control, 388-389
cotton, 38
CPC International, 188
crops
  agricultural, 1
  fruit, 144
  rotation of, 37
  starch, 145-149, 151
  sugar, 143-145
Crude Oil Windfall Profit Tax Act of 1980 (P.L. 96-223), 20, 463
Curtis, Senator Carl, 11
Czechoslovakia, 9

D

DDG (*See* distiller's dried grains)
DDGS (*See* distiller's dried grains with solubles)
DDGS plants
  large community, 132-135, 243-244
  small community, 130-132, 241-242
dehydration, 270-272, 347-348, 384
  centralized, 230, 231
dehydration plant, central topping cycle, 135-138
delignification, 260
denaturing, 158-159
Department of Energy (DOE), U.S., 4, 11, 18, 20, 21, 24-27, 28-29, 33, 34, 38, 39, 51, 53, 57, 178, 179, 190, 273, 289, 309, 346, 349, 393, 524
  funding, 27
  responsibility for biomass energy, 21-22
Department of Housing and Urban Development (HUD), 53, 524
destoring, 83
dewatering, 317
dextrins, 86, 145
Diamond Shamrock, 188
diesel engines, ethanol in, 431-432
diesel-ethanol mixtures, 425-427
diesel fuel, use on farms, 439
diffusion, 112
Director of Economics, Policy Analysis and Budget (USDA), 21
distillation, 13, 90-93, 103-104, 261, 270, 278, 281, 284, 287-288, 300-302, 313-314, 357-361, 402-403
  controls for, 108
  extractive, 270
  hazards of, 384
  vacuum, 249
distillation process, 15-16
distiller's dried grains (DDG), 35-37, 41, 202, 206-211, 214, 221, 251, 265, 278, 281, 284, 288, 349, 351, 365-366, 372, 373, 383, 385, 392
  prices, 209-211

distiller's dried grains with solubles (DDGS), 35-37, 116, 156, 177, 202, 206-209, 212-213, 214, 223-224
distiller's wash, 249
DOE (*See* Department of Energy, U.S.)
Dole, Senator Robert, 11

E

Economic Development Administration (EDA), 53, 57, 524
ecosystems, ethanol and, 395-396
EG&G, 273, 289, 304, 346
  small-scale ethanol plant, 306-317
Egypt, 500, 508-509
electrodialysis, 269
energy balance, 39-46
Energy Information Administration (EIA), 23
Energy Research and Development Administration (ERDA), 415
Energy Security Act (P.L. 96-294), 18-29, 42, 46, 47, 179, 180-182
Energy Tax Act of 1976 (P.L. 95-618), 20
engines
  cold starting, 417, 442
  diesel, 239
  spark ignition (SI), 408-421, 425, 429-432
environmental assessments, 392-393
environmental control costs, 388-389
environmental control systems, hazards of, 386
Environmental Protection Agency (EPA), 31, 33, 34, 51, 389, 392
enzyme hydrolysis, 39, 252, 254
enzymes
  liquefying, 13
  saccharifying, 13
EPA (*See* Environmental Protection Agency)
ERDA (*See* Energy Research and Development Administration)
erosion, 381, 393
ethanol
  anhydrous, 1, 4, 32, 185-186, 510
  as a fuel, 183-187, 433-441
  azeotropic, 425
  blending of, 187
  carbureted, 427-429
  conversion rate from feedstocks, 152-154
  coproducts of, 201-220
  distributors of, 187-188
  drying, 104, 109
  fuel-grade, 47, 273
  government policies, 192-193
  handling of, 187
  hazards of, 394-396
  hydrated, 185
  industrial applications of, 186
  market for, 183-201, 221
  prices of, 192-193, 256, 258
  producers of, 187-188
  production in other countries, 496-512
  properties of, 183-184
  R&D, 29-30
  storage, 187
  straight, 237-239
  200-proof, 430, 439, 441
  unblended, 416-424
  uses of, 118, 184, 432-435
  volumetric heat content, 406, 410, 416
ethanol coproducts, market for, 183-221
ethanol feedstock, 13
ethanol-gasoline blends, 185
ethanol plants
  commercial-scale, 349-379
  criteria for, 110-116
  decisionmaking phase, 54-63
  development of, 54-63
  decision and planning worksheets, 63-82
  design criteria, 93-110, 273-379
  development cost, 57, 60-62

economics of, 42, 50-53, 61, 62
emergency operations, 320, 327-330, 334-336, 348
equipment for, 292-294, 343-346
farm-scale, 274-289
feasibility, 62-63
fuel-grade, 37, 47
hazards, 97
implementation phase, 55-56
investors and, 56-57, 60-63
large on-farm, 124-127, 237-238
maintenance of, 304-306, 320, 328, 336-337, 348-349
operating cost comparisons, 115
safety considerations, 327, 348, 396-397, 401-405
shutdown, 302, 320, 326, 333
small on-farm, 121-124, 236
small-scale, 289
start-up, 320-321, 324-325, 329-332
taxes on, 120-123
ethanol production, 12-18, 50-82, 83-138, 307-308
codes and standards, 400
costs, 223, 224, 226, 229-230, 235
economics of, 222-224, 253, 263, 265
environmental and safety factors, 380-405
feasibility, 55
feedstocks for, 4
hazards, 18
marketing aspects, 51-52, 60
technical aspects, 51, 60-61
ethyl alcohol plants
costs, 374-375
economics of, 376
ethylene, 186, 192
Europe, 10, 38, 424
exports
corn, 160, 207-209
soybean, 207-208
wheat, 160
extraction, solvent, 271
extruder, 262-264

F

Facet Carburetor Corp., 452
farm machinery, diesel, 406-409
farmers
energy costs and, 274
energy supply and, 274
Farmers Home Administration (FmHA), 26-27, 53, 524
feasibility study, 55
federal gasohol plan, 18-30
Federal Water Pollution Control Act of 1972 (Clean Water Act), 389, 390-391
feedstock (*See also* specific feedstocks)
availability of, 160-168
cellulose, 151-152
characteristics of, 140-141
climate and, 166-167
cost of, 168-175, 275, 376, 377, 379
economics of, 169
fundamentals of, 139-182
government incentives and, 172
government policies and, 167
handling and storage, 98-99
hazards of storage, 383
historic trends, 167-168
investment costs, 114
market demand for, 167
preparation of, 83-85
process factors, 113
procurement of, 169-171
production of, 173, 393-394
regional potential for, 162
selection of, 139
supply and disappearance, 161
transportation costs and, 173
types of, 140, 143-152
U.S. locations, 221
wastes, 154
fermentation, 13, 86-90, 101-103, 246-247, 249, 254, 257, 259-260, 269, 300, 312, 361
batch, 101, 106
continuous, 101

coproducts from, 154-155
defined, 83, 86
extractive, 270-271
hazards of, 384
fermentation alcohol
costs, 5
legislation for, 3
objectives, 40
production, 6
transportation, 6
U.S. production of, 4
fermentation ethanol, 1, 139
capacity, 2, 3
government subsidy, 8
plants, 8
production of, 50-82
success of, 8
fermentation, temperature and, 89, 102-103
fermentation process, 16
Fiat, 509
fire protection, 403-404
Fish, John Robert, 454
flue gas desulfurization (FGD), 386
flue gas scrubber, 370-372
flush, car tank, 34
FmHA (*See* Farmers Home Administration)
fodder beets, 4, 99, 144
Food and Drug Administration (FDA), 524
food versus fuel (grains), 34-38
Ford, Henry, 4
Ford diesel, 426
France, 9, 497
fuel-air equivalence ratio, 413
fuels, heating values of, 406-407
properties of, 407
fuel use, agricultural, 440
fumigation, alcohol, 428-429
funding, 24-26
fungal amylase production, 355-356
fusel oil, 41, 154-155, 245-246, 248, 251, 361-362

G

gasohol
advantages of, 46-47
availability, 195
benefits of, 196, 200-201
compared to unleaded gasoline, 1
defined, 1
driving problems and, 32
fleet tests and, 31-33
issues, 30-47
legislation, 11
marketing, 46
public response to, 1, 3, 50, 194
rational program for, 36
spark ignition engines and, 31-33
stations in U.S., 7
uses of, 406-455
vehicle performance and, 31-34
weather problems and, 32, 196-197
Gasohol Acceptance Study, 193-201
gasoline
cost, 6
demand for, 6, 8, 160
unleaded, 1, 4, 192
use on farms, 438
weight of, 443-444
gasoline blends, effects of leaning, 414-416
gasoline-ethanol
distillation curve, 412
effects on carburetor, 411
efficiency, 411®412
engine modifications for, 415-416
heat content, 410
heat of combustion, 411
phase separation, 410
volatility, 411
gelatinization, 86, 311
Gene Schroder farm-scale ethanol plant, 273, 285-289
General Motors, 185
General Motors diesels, 424
Georgia Pacific Corp., 245
Georgia Tech, 253

Germany, 9, 10
glossary and acronyms, 539-555
glucoamylase, 86, 119, 317
glucose, 13, 140, 252
Gosset, Larry, 279, 281
grains
cost of, 172-173
drying of, 241
export of, 36-38
policy issues, 38
Great Plains, 160
Gregor, Dr. Harry, 269
grinding, 85
Guatemala, 509
Gulf process, 42

H

health and safety, worker, 387-388
Hitler, Adolf, 10
homofermentation, 258
HUD (*See* Department of Housing and Urban Development)
Hungary, 9
hydrocarbons, 360-361, 385, 406, 441, 451
hydrolysis, 39, 85-86, 106-107, 112, 145, 151, 257-260
acid, 39, 99-100, 252, 265
continuous acid, 262-264
enzymatic, 99-100
hazards of, 384

I

Idaho, 4, 221, 273, 289, 309, 346
ignition delay, 421
Illinois, 31, 32, 41, 42, 173, 224, 245, 427, 437
imports, petroleum, 186-187
India, 37, 509
Indiana, 11, 173, 188, 276
Indianapolis 500, 10
Indonesia, 510
industry, beverage, 160
infections, microbial, 89-90
insurance loss, 400
intermediate-scale plants, funding, 26, 27
international development, biomass-derived fuels and, 496-511
International Harvester, 424
Iotech Process, 252
Iowa, 4, 12, 31, 32, 194, 245
Iowa Development Commission, 193-201
iron, 34
Ireland, 509
issues, policy, 38
Italy, 9, 509-510

J

Jamaica, 510
Japan, 510
Jerusalem artichokes, 99, 101, 144
John Deere, 426

K

Kansas, 9, 213, 253, 254, 263, 265, 383, 437
Katzen plant design, 42
Katzen process, 349-350, 381, 389
Kentucky, 204
Kentucky Agricultural Experiment Station, 204

L

Ladisch, Dr., 105
Lake States, 160, 163
land, use of, 394
large-scale plants, funding, 27
leaching, 83
lead, 34
lean air/fuel mixture, 33
legislation (*See also* specific legislation)
federal, 20, 23, 24
state, 20, 23-24

lignin, 252-253, 260
lignocellulose technology, 38-39, 42, 508
facility requirements, 39
liquefaction, 85-86, 145, 299
liquefied petroleum gas (LPG), 235, 241, 406-408, 427, 437-441
liquid fuels, on-farm use, 433, 436-441
Little, Arthur D., 271
loan guarantee programs, 24-29, 42, 53, 393, 523-524

M

M&W Gear Company, 427, 429
Madison process, 264
Maine, 11
MAN diesels, 424-425
markets
animal feed, 207-208
ethanol, 221
Maryland, 7
Massachusetts Institute of Technology (MIT), 258-260, 268
membrane technology, 268-269
Michigan, 282-283
Milbrew, Inc., 245-246
Miles Laboratory, 280
milling
dry, 149
wet, 16, 150
milling and cooking, hazards of, 383
milo, 274, 279, 282, 286
Minnesota, 437
molasses, 111, 112, 144, 153, 496-498, 508-509
Montana, 464
Mother Earth News, 442
municipal solid waste, 42, 95, 110, 256, 262
mutation, 259

N

Natick Laboratories, 253-254, 257-258
National Ambient Air Quality Standards, 389
National Energy Act (NEA), 11
National Energy Plan (NEP), 11
National Fire Protection Association (NFPA), 405
National Historic Preservation Act, 492
National Pollutant Discharge Elimination System (NPDES), 390-391
National Research Council, 213
Nebraska, 11, 31, 32, 408, 415, 417
Nebraska Tractor Test Laboratory, 408
net energy gain, 40, 44
net fuel gain, 40, 42-45
net petroleum gain, 40, 44
net premium fuel gain, 40
New England, 163
New Guinea, 510
New Source Performance Standards (NSPS), 389-390
New York University, 262-264
New York University Process, 262-264
New Zealand, 510
nitrogen oxide (NO_x), 305
noise, hazards of, 383

O

Occupational Safety and Health Administration (OSHA), 53, 495
octane, gasoline-ethanol, 410-411
octane enhancers, 31, 235-236
octane rating, 187, 406-407
Office of Alcohol Fuels (DOE), 21
Office of Technology Assessment (U.S. Congress), 34
Ohio, 154, 173, 224
oil
embargo of 1973, 11
imported, 3, 40
U.S. consumption, 3
U.S. reserves, 3
100-proof ethanol, production cost, 232
160-proof ethanol, production cost, 230-232

OPEC (*See* Organization for Petroleum Exporting Countries)
Organization for Petroleum Exporting Countries (OPEC), 11
Otto cycle engine, 9

P

Paraguay, 510
Pasteur efficiency, 36
PEDCo, 111
Pennsylvania, University of, 271
petroleum, 4
petroleum, crude, 39
pH factor, 13, 85-86, 88, 99, 103, 106, 107, 317, 371, 404
Philippines, 510, 511
Phillips Petroleum Company, 46, 188
Plymouth, 419
pot still, operating variables, 233
pot still plants, 118-121, 233
potable alcohol, 8
potatoes, 4, 13, 83, 85, 99, 111, 112, 146-147, 151, 163-164, 171, 509-510
  as a coproduct, 219
premium fuel, savings, 40
pressure-cascading technique, 247
pressure stripper-rectifier (PSR), 359-360, 361
prevention of significant deterioration (PSD), 389
PROALCOOL program, 504-508
process control, 106-110
processing, batch, 99-100, 109
propane, 406-407
Publicker Industries, 188
publicly owned treatment works (POTW), 387-389, 390
pulp mill, sulfite, 246
pumps and drives, controls for, 109
Purdue/Tsao Process, 264-266
Purdue University, 105

Q

quicklime, 271

R

Radian Corp., 384
Randy Butters farm-scale ethanol plant, 282-285
Raphael Katzen Associates International, Inc., 247-248, 273, 349, 361, 373, 379
Rask, Norman, 499
reflux, 91, 104
regulations, fuel-alcohol plants and, 524-525
Reid vapor pressure, 411
"The Report of the Alcohol Fuels Policy Review," 12
research, biological, 266, 268
residues, agricultural, 95, 110
Resource Conservation and Recovery Act (RCRA) of 1976, 389, 392
rice, 147, 160
rubber, synthetic, 246
rye, 147, 171

S

Saccharification, 36, 86, 254, 261, 300, 311-312, 317, 322-323, 353-354
SBA (*See* Small Business Administration)
Securities and Exchange Commission (SEC), 462
Senegal, 511
simultaneous saccharification and fermentation (SSF) reactor, 254, 256
Small Business Administration, 53, 57, 524
small-scale facilities, defined, 4
small-scale plants, 116-138
  business plan, 464-477
  funding, 25-27
Solar Energy Research Institute (SERI), 273, 289, 464
solids
  separation of, 105
  transport of, 105

solubles, distiller's 105
sorghum, 4, 37, 42, 99, 101, 112, 144, 153, 171, 176, 223, 504-508
South Africa, 511
South America, 496
soybeans, 37, 168-169
  nutrient composition, 212
Standard Oil Company, 187
starch, 83, 95, 104, 145
  conversion, 85-86
stillage, 105-106, 155
  drying costs of, 202
  evaporation and drying, 362-363, 365
  potential for, 215, 217
  prices, 209, 212
  problems of using, 218-219
  processing, 105-106
  production rates, 158
  recovery and use, 16-17
  thin, 17, 105-106, 155, 362
  use as a feed ingredient, 203-205
  wet, 118, 127-130
stillage solids, 16-17
straw, 152
sugar, 38
  concentration in mash, 88-89
sugar beets, 4, 15, 83, 99, 101, 112, 113, 164, 497
  as a coproduct, 219
sugarcane, 4, 9, 13, 15, 83, 99, 101, 112, 139, 143, 145, 160, 164, 171, 173, 497, 503, 508, 510
  as a coproduct, 219-220
sugars, simple, 83
sulfur dioxide (SO_2), 385
Sweden, 9, 511
synfuel, 6

T

tax credit, 234, 379, 523
tax incentives, 11, 12, 23, 47, 52, 193, 197-198
Texaco, 6, 46, 187
Texas, 4, 223, 279, 281
Textile Research Institute, 272
Thailand, 510
thermocompression, 251
Toyota, 415
tractors, diesel, 275
tubers, 151

U

Union of Soviet Socialist Republics (U.S.S.R), 38
U.S. Army Natick R&D Command, 253-254, 257-258, 268
United States Department of Agriculture (USDA), 4, 9, 11, 18, 20, 21, 24-27, 28-29, 38, 39, 51, 57, 116, 124, 164, 173, 178-180, 190, 273, 390, 394, 408, 433, 437, 524
United States Department of Agriculture (USDA), funding, 26, 27
United States Department of Agriculture (USDA), responsibility for biomass energy, 20, 22
United States Department of Commerce (DOC), 11, 53, 524
United States Department of Treasury (DOT), 524
United States National Alcohol Fuels Commission (NAFC), 273, 274
University of California at Berkeley, 269
University of Pennsylvania/GE Reentry Systems Division Process, 260-262, 268
USDA (*See* United States Department of Agriculture)

V

vacuum techniques, 269-270
vaporization
  ethanol, 442
  heat of, 406, 408, 416-417

venturi tube, 444
Virginia, 11, 152
vitamin A, 213
vitamin B, 36, 88, 213, 219
vitamin D, 213
Vogelbusch Division-Bohler Brothers of America, Inc., 249
Volkswagen, 185
Volvo, 511
Vulcan-Cincinnati, Inc., 111, 248-249

W

Walker, Hiram, 9
Washington, 245
waste
  bakery, 279
  citrus, 163, 171
wastewater, 369-370, 387, 388, 391
water resources, 381, 386, 394
weir, 92
West Germany, 511-512
wet stillage plant, small community, 127-130, 239-240
wheat, 4, 37, 160, 171, 176, 274
Windfall Profit Tax Act, 42, 172
Wisconsin, 245
wood, 13
World Bank, 508
World War II, 4, 10, 246

X

xylose, 259

Y

yeast, 36-37, 83, 85, 87-88, 140, 146, 154-155, 256, 261, 276
  strains, 87-88
  types of, 87-88

Z

zinc, 34